Praxis Manned Spaceflight Log

Tim Furniss and David J. Shayler with
Michael D. Shayler

Praxis Manned Spaceflight Log 1961–2006

Published in association with
Praxis Publishing
Chichester, UK

Tim Furniss
Spaceflight Correspondent
Flight International
Bideford
Devon
UK

David J. Shayler
Astronautical Historian
Astro Info Service
Halesowen
West Midlands
UK

Michael D. Shayler
Editor and Designer
Astro Info Service
Birmingham
UK

SPRINGER–PRAXIS BOOKS IN SPACE EXPLORATION
SUBJECT *ADVISORY EDITOR*: John Mason B.Sc., M.Sc., Ph.D.

ISBN 10: 0-387-34175-7 Springer Berlin Heidelberg New York
ISBN 13: 978-0-387-34175-0 Springer Berlin Heidelberg New York

Springer is part of Springer-Science + Business Media (springer.com)

Library of Congress Control Number: 2006937359

Apart from any fair dealing for the purposes of research or private study, or criticism or review, as permitted under the Copyright, Designs and Patents Act 1988, this publication may only be reproduced, stored or transmitted, in any form or by any means, with the prior permission in writing of the publishers, or in the case of reprographic reproduction in accordance with the terms of licences issued by the Copyright Licensing Agency. Enquiries concerning reproduction outside those terms should be sent to the publishers.

© Praxis Publishing Ltd, Chichester, UK, 2007
Printed in Germany

The use of general descriptive names, registered names, trademarks, etc. in this publication does not imply, even in the absence of a specific statement, that such names are exempt from the relevant protective laws and regulations and therefore free for general use.

Cover design: Jim Wilkie
Project Copy Editor: Mike Shayler
Typesetting: Originator Publishing Services, Gt Yarmouth, Norfolk, UK

Printed on acid-free paper

Contents

Authors' Preface	xiii
Acknowledgements	xv
Foreword	xvii
List of illustrations	xxiii
Prologue	xxxiii

1 Reaching the Heavens ... 1
  Access and method ... 1
   The atmosphere ... 2
  Space flight methods ... 2
   Rocket planes ... 2
   Sub-orbital flights ... 3
   Flight paths ... 4
   Launch sites ... 5
   Landing methods ... 6
   Emergency escape ... 8
  Launch systems ... 10
   Astro-flights ... 10
   Sub-orbital flight ... 11
   American orbital launchers ... 12
   Soviet/Russian orbital launchers ... 16
   Soviets on the Moon ... 18
   Chinese orbital launch vehicle ... 18

2 The Quest for Space ... 19
  X-15 – A Rocketplane to space ... 19
   X-15 – A winged marvel ... 20

	X-15 flights over 50 miles	20
	The sub-orbital Mercury flights	22
	Mercury Redstone 3	23
	Mercury Redstone 4	25
	Apollo Block I	27
	Apollo 1	28
	Recovery from Apollo 1	30
	Soyuz Pad Abort – A fast ride and a stiff drink	31
	Soyuz T10-1	31
	The X-Prize – The dawn of private spaceflight	33
	The X-Prize challenge	33
	A record in two stages with a bonus	34
	White Knight and Spaceship One	36
3	**The Orbital Programmes**	37
	Into Space	37
	Vostok and Voskhod	37
	Mercury	39
	To the Moon	40
	Gemini	40
	Apollo	41
	Soviet lunar plans	43
	Long-duration spaceflight	44
	Soyuz	44
	Salyut	46
	Skylab	48
	Mir	50
	For all mankind	50
	Space Shuttle	53
	Military manned space flight	53
	International manned space flight	53
	Shenzhou	55
4	**The First Decade: 1961–1970**	57
	Vostok 1	58
	Vostok 2	60
	Mercury Atlas 6	62
	Mercury Atlas 7	65
	Vostok 3 and 4	67
	Mercury Atlas 8	70
	Mercury Atlas 9	72
	Vostok 5 and 6	75
	Voskhod 1	78
	Voskhod 2	81
	Gemini 3	84

Gemini 4 87
Gemini 5 90
Gemini 7 and 6A 93
Gemini 8 96
Gemini 9 98
Gemini 10 101
Gemini 11 104
Gemini 12 107
Soyuz 1 110
Apollo 7 113
Soyuz 3 115
Apollo 8 117
Soyuz 4 and 5 120
Apollo 9 123
Apollo 10 126
Apollo 11 129
Soyuz 6, 7 and 8 132
Apollo 12 136
Apollo 13 139
Soyuz 9 142

5 The Second Decade: 1971–1980 145
Apollo 14 146
Soyuz 10 149
Soyuz 11 152
Apollo 15 155
Apollo 16 158
Apollo 17 161
Skylab 2 163
Skylab 3 166
Soyuz 12 169
Skylab 4 171
Soyuz 13 174
Soyuz 14 176
Soyuz 15 178
Soyuz 16 181
Soyuz 17 183
Soyuz 18-1 185
Soyuz 18 187
Soyuz 19 and Apollo 18 (ASTP) 190
Soyuz 21 194
Soyuz 22 196
Soyuz 23 198
Soyuz 24 201
Soyuz 25 204

viii **Contents**

Soyuz 26	206
Soyuz 27	209
Soyuz 28	211
Soyuz 29	213
Soyuz 30	216
Soyuz 31	218
Soyuz 32	220
Soyuz 33	223
Soyuz 35	226
Soyuz 36	229
Soyuz T2	231
Soyuz 37	233
Soyuz 38	235
Soyuz T3	237

6 The Third Decade: 1981–1990 ... 239

Soyuz T4	240
Soyuz 39	243
STS-1	245
Soyuz 40	248
STS-2	250
STS-3	252
Soyuz T5	254
Soyuz T6	257
STS-4	259
Soyuz T7	262
STS-5	264
STS-6	267
Soyuz T8	270
STS-7	272
Soyuz T9	275
STS-8	278
STS-9	281
STS 41-B	284
Soyuz T10	287
Soyuz T11	290
STS 41-C	292
Soyuz T12	296
STS 41-D	299
STS 41-G	302
STS 51-A	305
STS 51-C	308
STS 51-D	310
STS 51-B	313
Soyuz T13	316

STS 51-G 319
STS 51-F 321
STS 51-I 324
Soyuz T14 327
STS 51-J 330
STS 61-A 332
STS 61-B 335
STS 61-C 338
STS 51-L 341
Soyuz T15 345
Soyuz TM2 348
Soyuz TM3 351
Soyuz TM4 353
Soyuz TM5 356
Soyuz TM6 358
STS-26 361
Soyuz TM7 364
STS-27 367
STS-29 369
STS-30 371
STS-28 374
Soyuz TM8 377
STS-34 380
STS-33 383
STS-32 385
Soyuz TM9 388
STS-36 391
STS-31 394
Soyuz TM10 397
STS-41 400
STS-38 403
STS-35 406
Soyuz TM11 409

7 The Fourth Decade: 1991–2000 413
STS-37 414
STS-39 417
Soyuz TM12 420
STS-40 423
STS-43 426
STS-48 429
Soyuz TM13 432
STS-44 435
STS-42 438
Soyuz TM14 441

STS-45	444
STS-49	447
STS-50	451
Soyuz TM15	454
STS-46	457
STS-47	460
STS-52	463
STS-53	467
STS-54	470
Soyuz TM16	473
STS-56	476
STS-55	479
STS-57	482
Soyuz TM17	485
STS-51	488
STS-58	491
STS-61	494
Soyuz TM18	498
STS-60	501
STS-62	504
STS-59	507
Soyuz TM19	510
STS-65	513
STS-64	516
STS-68	519
Soyuz TM20	522
STS-66	526
STS-63	529
STS-67	533
Soyuz TM21	536
STS-71	539
Mir EO-19	542
STS-70	544
Soyuz TM22	547
STS-69	550
STS-73	554
STS-74	557
STS-72	560
Soyuz TM23	563
STS-75	566
STS-76	569
STS-77	572
STS-78	575
Soyuz TM24	578
STS-79	581

STS-80	584
STS-81	587
Soyuz TM25	590
STS-82	593
STS-83	596
STS-84	599
STS-94	602
Soyuz TM26	605
STS-85	608
STS-86	611
STS-87	614
STS-89	617
Soyuz TM27	620
STS-90	623
STS-91	626
Soyuz TM28	629
STS-95	632
STS-88	635
Soyuz TM29	638
STS-96	641
STS-93	644
STS-103	647
STS-99	650
Soyuz TM30	653
STS-101	656
STS-106	659
STS-92	662
Soyuz TM31	665
STS-97	668

8 The Fifth Decade: 2001–2006 671

STS-98	672
STS-102	675
ISS EO-2	678
STS-100	681
Soyuz TM32	684
STS-104	687
STS-105	690
ISS EO-3	693
Soyuz TM33	696
STS-108	699
ISS EO-4	702
STS-109	704
STS-110	708
Soyuz TM34	711

STS-111	714
ISS EO-5	717
STS-112	720
Soyuz TMA1	723
STS-113	726
ISS EO-6	729
STS-107	732
Soyuz TMA2	736
Shenzhou 5	739
Soyuz TMA3	742
Soyuz TMA4	745
Soyuz TMA5	748
Soyuz TMA6	751
STS-114	754
Soyuz TMA7	757
Shenzhou 6	760
Soyuz TMA8	763
STS-121	766
STS-115	769
Soyuz TMA9	772

9 The Next Steps 775
   The immediate future 777
   Summary 778

Appendix A The Log Book 1961–2006 781

Appendix B Cumulative Space Flight and EVA Experience 795

Appendix C Future Flight Manifest 2006–2011 (as at 1 October 2006) . . . 813

Appendix D A Selected Timeline 819

Bibliography 827

Authors' Preface

One of the most frustrating and time consuming chores to do with collating data on each manned space flight is in finding original source material that is consistent. Questions are constantly being raised that require a definitive answer, or at least standard application, if you want to make sense of it all. To give you some examples: "Where does 'space' begin?" "What distinguishes a high-altitude research pilot from a space explorer or a 'tourist'?" "Are the recent 'X-Plane' flights really sub-orbital space-flights?" "In multi-person crews, which one enters 'space' first?" "Upon landing, does a Shuttle mission end when the wheels touch the runway, or when they come to a stop?" "Does an EVA start from when the space walker puts a suit on, or when they step out of the airlock?" All of these questions find different answers even in official data and this can make a space author's job that much harder.

What *is* clear is that when a spacecraft enters orbit, it is assigned a specific orbital object catalogue number. Therefore, one can follow these orbital flights in chronological order, even if the details are open to interpretation. To most crews, "the mission" is one of the most important objectives for their flight and their future careers, and they are assessed by their performance and achievements on "the mission" and its specific objectives or tasks. Usually, records, milestones and ceremonies are not as important to the flight crew as they are to watchers on the ground.

This book, therefore, is not (nor intends to try to be), a definitive record of all manned space flight aspects. Indeed, it is doubtful that such a tome could actually be written, and certainly not in the tight confines of 900 pages. What we have tried to do instead is to present is a single, handy, quick reference source of who did what on which mission, and when they accomplished it, in the 45 years between 1961 and 2006. For more detailed information, other books in this Springer-Praxis series can be referred to, as can those cited in the bibliography of this or other books in the series.

The objective of this book was to keep things simple, so we have therefore focused mostly on *orbital* missions (or in a few cases, those which were intended for orbital flight and had left the pad, but never made it into space). The other "sub-orbital"-type missions are listed in context, but are detailed in the opening sections.

By way of introduction, an overview of the methods used to reach space or fly particular types of mission is presented. This is followed by a look at those missions which essentially bridged the gap between aeronautical flight and space flight. Finally, the programmes that have actually been conducted are overviewed, before each orbital space flight is addressed, starting with Yuri Gagarin aboard Vostok 1 in April 1961 and ending with the launch of the 14th resident crew to ISS in September 2006, a span of 45 years. We have also started recording the missions leading towards the 50th anniversary of Gagarin's flight with the currently manifested missions of 2006-2011, reminding us all that the log is an ever-expanding account of the human exploration of space. As one mission ends, another is being prepared for flight.

In the detail of the main log entries, we have focused on the highlights and achievements for each mission, as this book was always intended to supplement the more in-depth volumes in the Praxis series, as well as other works. It is also intended as a useful starting volume for those who are just becoming interested in human space flight activities and who have not had the opportunity to collect the information from past missions or completed programmes. We also hope that this work will help to generate other, more detailed works on past and current programmes, and in time on those programmes that are even now being planned and will write the future pages of space history – and further entries in the *Praxis Log of Manned Space Flight*.

Tim Furniss
Dave Shayler
Mike Shayler
September 2006

Acknowledgements

Assembling a book of this nature would be impossible without a network of fellow space sleuths and journalists. In particular, the assistance over many years of the following friends and colleagues is much appreciated:

- Australia: Colin Burgess
- Europe: Brian Harvey (Ireland), Bart Hendrickx (Belgium) and Bert Vis (The Netherlands)
- UK: Phil Clark, Rex Hall, David Harland, Gordon Hooper, Neville Kidger, Andy Salmon
- USA: Michael Cassutt, John Charles, James Oberg and Asif Siddiqi.

The authors also wish to express their appreciation for the on-going help and support of the various Public Affairs departments of NASA, ESA, and the Russian Space Agency.

The assistance of the Novosti Press Agency and US Information Service was of great help in detailing the pioneering years of human space flight.

Various national and international news organisations were also often consulted, including the publications, *Flight International*, *Aviation Week and Space Technology*, and *Soviet Weekly*.

The staff of the British Interplanetary Society (with its publication *Spaceflight*) have continued to support our research for many years. We have fond memories of Ken Gatland, past President of the BIS and space flight author, who was an inspiration to many with his documentation of various missions and space activities.

We must express our thanks to Colonel Al Worden (CMP Apollo 15) for his generous foreword.

We also appreciate the help and support of our families during the time it took to compile and prepare this book from its original idea to the finished format.

Acknowledgements

Last but not least, we appreciate the support and understanding of Clive Horwood, Publisher of Praxis, with a project that took a lot out of all of us. Thanks to the staff of Springer-Verlag in both London and New York for post production support; to Neil Shuttlewood and staff at Originator Books for their typesetting skills; to Jim Wilkie for his continued skills in preparing the cover for the project, and to the book printer for the final result.

Foreword

I was born during the great American Depression, in 1932, at a time when our telephone had a hand crank to call the operator and there were six other families on the line, the bathroom was outdoors, there was no running water and our drinking water came from a hand-dug well in the front yard. Money was tight, but there was a lot of work and fun on my grandparent's farm. We lived nicely and I can still hear the rain beating on the tin roof at night.

As I grew up, my parents bought a small farm in Jackson, Michigan, and I, along with my five brothers and sisters, lived there during my teen years. One of the most memorable days during those times was the crash of a small airplane behind our house. I was awed by the laid back spirit of the pilots, and thought that would be a great thing to do sometime. The problem was that I did not see myself living my life on a farm. So, when I graduated from high school I searched for the right college to attend, considering that my parents did not have the money to send me to a good one. I ended up going to the United States Military Academy at West Point, graduating there in 1955. Since there was no Air Force Academy at the time, West Point needed to send a third of each class to the Air Force. I elected to go to the Air Force because I thought promotions would be quicker. I found out that was not going to be true, but in the meantime I discovered that I had a real talent for flying, something with which I had very little prior experience.

I never considered a career in the space program, because the possibility of getting in the program was so remote. I flew fighter aircraft for a number of years, and became my squadron's armament officer because I spent a lot of time in the hanger learning the maintenance business for high performance fighters. While there I rebuilt the armament shop into a very modern work place to motivate the technicians and increase the quality of work. It was a successful effort and the squadron became the role model for others. In fact, I was asked to go to headquarters to help other squadrons do the same thing. Instead, I asked for and received an assignment back to college to learn about guided missiles. While in college I was the operations officer for all the Air Force pilots, and that fact helped me to get into the Test Pilot School at Farnborough, England. I was transferred to the RAF for a year while at the school, and I returned to the United States to teach at the USAF Research Pilots School at Edwards AFB in California. I still did not believe I had a chance to become an astronaut, but I wanted to be the best test pilot possible. However, NASA had a selection program, and I applied in late 1965. Because of my academic and flight background, I was lucky enough to be selected in April of 1966.

I found out very quickly that one does not become an astronaut by being selected. You have to make a space flight to really and truly be an astronaut, and there was a long training period to finish before assignment to a flight. After that period, which included all the spacecraft operations and special geology training, I was assigned to the support crew of Apollo 9. My job was to check out the spacecraft at the factory, and to complete the build up and check of the hatch that would be used between the Command Module and the Lunar Module. Subsequently, I was assigned to the Apollo 12 back-up crew as Command Module Pilot (CMP), and then to the Apollo 15 prime crew as CMP. Apollo 15 has been proclaimed the most scientific flight of the Apollo program. We trained hard for the extensive science we would accomplish on the flight, and the results were to confirm our efforts were worthwhile.

During the course of our flight training and preparations it was quite clear to us that a vast amount of data was being accumulated. However, we were focused on the flight and what we had to do to make it successful. Once the flight was under way, we concentrated on the science and experiments we were assigned, and how we would keep on the time line so we would not miss anything. At the same time, Mission Control recorded and maintained the down link data for scientific and post-flight analysis. We kept minimal written data on board because of the crush of schedule and the attempt to get all the data we could from both observations and science equipment.

After the flight, all the data was reduced at Johnson Space Center in the form of written reports and Prime Investigator research papers. This process took many months, and in some cases years before any comprehensive knowledge became clear. Because of this process, our knowledge of the Moon has been enhanced tremendously.

Our business was not record keeping, but completing the mission in a successful fashion. Others were responsible for the data and records of our flight. Today, the records are the most important historical evidence of the flight of Apollo 15.

There have been many flights to near space, almost space, and long distance space. They all require a very high level of competence and extraordinary engineering. The X-15, for example, was a magnificent machine, and it opened the way to space. Yuri Gagarin and Al Shepard started the human space initiative, and since then well over two hundred flights have been launched. Each is unique in its own way, with different mission objectives and goals. Humans are curious about what is over the horizon, and they have been exploring for thousands of years to find new continents, new routes to markets, better places to live and work or to find new riches to take back home. Space is also part of our exploration dream, and has been since Jules Verne opened our minds to the possibility of space flight. He even had his lunar crew of three men launch from a site near Cape Kennedy, go to the Moon and return and land in the ocean.

Maybe fact follows science fiction, but here we are today launching crews from Cape Kennedy, and we will soon be sending them back to the Moon.

My journey to space is pretty typical of the American Astronauts. We all had flight and academic experience, but none of us understood what it would take to go into space until we were actually involved in the program. It turned out that hard work was the key, and that training was non-stop before any flight. We also had to maintain a certain degree of calm and fatalism. I remember thinking, the night before launch, that as I talked to my family it just might be the last conversation I would have with them. But the rewards were worth the risk and we did our jobs gladly and freely.

To really understand how all this came about, this book is essential reading. Starting with Yuri Gagarin and following on through the years, this book will educate you on the fast progression of the space programs of several countries. Understanding where we have been will help you understand where we are going. Enjoy!

Colonel Alfred M. Worden USAF Ret.
NASA Group 5 (1966) Pilot Astronaut
Command Module Pilot Apollo 15, 1971

Offician portrait of Al Worden for Apollo 15

To Fallen Heroes
The crews of Apollo 1, Soyuz 1, Soyuz 11, Challenger and Columbia
And all the other space explorers who are gone, but never forgotten.

Every journey begins with the first small step. Each small step into space contributes to a larger leap to colonise the cosmos. Each mission's achievements contribute to the success of the next entry in the world's manned space flight log book. What started as national rivalry has evolved into international cooperation where each successive space crew can genuinely claim they "came in peace for all mankind."

Illustrations

FOREWORD
Portrait of Al Worden xix

DEDICATION
The first small step xxi

REACHING THE HEAVENS
Mike Adams with the X-15 3
The options for reaching the Moon 4
Launch pad at Cape Canaveral 5
Soviet/Russian launch pad at Baikonur 6
Apollo entry configuration 7
The Shuttle Landing Facility 8
An X-15 launch 10
Mercury launch vehicles 11
US space capsules 13
The Shuttle stack 14
Launch of the Space Shuttle 15
A Soyuz launch 17

THE QUEST FOR SPACE
Launch of Mercury-Redstone 3 24
Grissom in the Mercury pressure garment 26
Apollo 1 after the fire 29
The Soyuz T10 abort crew 32
White Knight and Spaceship One 34
Spaceship One's first glide flight 35
View of Earth from Spaceship One 36

THE ORBITAL PROGRAMMES

Vostok and Voskhod cosmonauts	38
The Original Seven Mercury astronauts	39
The Apollo Service Module	42
The Soyuz craft	45
Salyut 1	47
Salyut 7	47
The Skylab crews	49
Mir in 1998	51
The seven US Mir resident astronauts	52
ASTP artwork	54

THE FIRST DECADE

Yuri Gagarin	59
Titov's medical check	61
The launch of Friendship 7	63
Scott Carpenter	66
Nikolayev and Popovich during their flight	68
Wally Schirra	71
Launch of Mercury Atlas 9	73
Tereshkova and Korolyov	76
Ceremonies for the first Voskhod crew	79
The first spacewalk	82
Young and Grissom	85
Ed White on EVA	88
Conrad and Cooper after recovery	91
Gemini 6 in orbit	94
Lovell and Borman	95
Gemini 8's docking target	97
The "angry alligator"	99
Successful docking during Gemini 10	102
Earth from Gemini 11	105
Cernan jokes with the Gemini 12 crew	108
Komarov in training	111
The Apollo 7 crew receives a phone call	114
Georgi Beregovoy	116
Earth rise	118
Soyuz 5 EVA cosmonauts	121
Scott's stand-up EVA on Apollo 9	124
Stafford and Young on Apollo 10	127
Deploying the flag on Apollo 11	130
The Soyuz 6 crew	133
Group shot of the "troika" crews	134
Conrad with Surveyor 3	137
Apollo 13 crew at recovery	140
The Soyuz 9 crew	143

THE SECOND DECADE

Al Shepard on the Moon	147
The Soyuz 10 crew during training	150
The ill-fated Soyuz 11 crew	153
Apollo 15 landing area	156
Duke working near the LRV	159
Splashdown of Apollo 17	162
Skylab in orbit	164
The Apollo spacecraft docked to Skylab	167
The Soyuz 12 crew during training	170
The Skylab 4 crew after recovery	172
Inside the cramped Soyuz 13 module	175
Soyuz 14 crew in Sokol suits	177
The Soyuz 15 prime crew	179
Filipchenko in the OM during Soyuz 16	182
Gubarev and Grechko at TsPK	184
Lazarev and Makarov before their flight	186
The second Salyut 4 crew	188
Apollo 18 launches	191
The Russian ASTP crew	192
Soyuz 21 cosmonauts at Baikonur	195
Aksenov at work during Soyuz 22	197
The only Russian crew to splash down	199
The final military Soyuz crew	202
The Soyuz 25 crew studying for their mission	205
Romanenko and Grechko	207
Medical tests aboard Salyut 6	210
Remek at work in Salyut 6	212
The first Salyut EVA	214
Hermaszewski and colleagues aboard Salyut 6	217
Soyuz 29 and 31 crews pose for the camera	219
Lyakhov and Ryumin during training	221
The Bulgarian Interkosmos crew	224
Ryumin is helped out of the Soyuz	227
Moving equipment between Soyuz craft	230
The end of a successful test flight	232
The Vietnamese cosmonaut at work on Salyut	234
The Cuban Interkosmos mission	236
Safe recovery of the Soyuz T3 crew	238

THE THIRD DECADE

The end of Salyut 6 operations	241
The Soyuz 39 crew in training	244
STS-1 coming in to land	246
Prunariu wears the Chibis garment	249
Columbia returns to orbit	251
STS-3 lands in New Mexico	253

Berezovoy studies star charts.. 255
Chrétien and Ivanchenkov on Salyut 7.. 258
The STS-4 crew greeted by President Reagan.................................... 260
Berezovoy and Savitskaya in Salyut 7.. 263
Satellite deployment during STS-5... 265
The first Shuttle EVA... 268
The Soyuz T8 crew in training... 271
The STS-7 crew.. 273
The Soyuz T9 crew relax after recovery.. 276
Dr. Bill's clinic... 279
The Spacelab 1 module aboard Columbia... 282
McCandless flies the MMU.. 285
A busy EVA programme for the T10 crew... 288
Gymnastics aboard Salyut 7.. 291
Testing the MMU in the payload bay.. 293
A full crew complement aboard Salyut 7.. 297
The launch pad abort of STS 41-D.. 300
The STS 41-G crew leaving for the launch pad.................................. 303
Joe Allen retrieves Palapa.. 306
USAF astronauts Onizuka and Shriver... 309
STS 51-D crew with fly swatters... 311
The STS 51-B Gold and Silver shifts... 314
Savinykh and Dzhanibekov wearing thermals..................................... 317
The multi-national STS 51-G crew.. 320
The STS 51-F crew having fun.. 322
Van Hoften launches the Leasat.. 325
The T13 and T14 crews aboard Salyut 7... 328
One of the few images released for STS 51-J................................... 331
Payload Specialist Reinhard Furrer.. 333
Constructing the EASE/ACCESS hardware... 336
Bill Nelson exercises aboard Columbia... 339
The Teacher in Space candidates... 342
EVA training for the Soyuz T15 crew... 346
The Soyuz TM2 crew review training plans...................................... 349
Soyuz TM3 prime and back-up crews... 352
The Soyuz TM4 crew.. 354
The crew for the second Bulgarian space flight................................ 357
Portrait of the Soyuz TM6 crew.. 359
Return to flight for the Shuttle.. 362
Chrétien and his Soyuz TM7 colleagues... 365
At work during STS-27... 368
View from the flight deck... 370
Magellan deployment... 372
STS-28 crew in "starburst" pose... 375
Viktorenko and Serebrov... 378
The Galileo probe is deployed... 381
"Maggot on board"... 384
LDEF is finally retrieved... 386
Solovyov and Balandin review EVA equipment.................................... 389

Filming out of the windows	392
The Hubble Space Telescope prior to deployment	395
The crew of Soyuz TM10	398
Ulysses begins its mission	401
Thumbs up from the STS-38 crew	404
Ships that pass in the night	407
The crew of Soyuz TM11	410

THE FOURTH DECADE

The Compton Observatory held by the RMS	415
Some of the payload of STS-39	418
The Soyuz TM12 crew included the first Briton in space	421
Vestibular experiments aboard STS-40	424
TDRS-E being deployed	427
Constructing experiments aboard STS-48	430
The Soyuz TM13 crew	433
Observations through the windows during STS-44	436
A "starburst" pose from the crew inside Spacelab	439
Portrait of the TM14 crew	442
A busy flight deck for STS-45	445
EVA astronauts hold the Intelsat VI satellite	448
USML-1 being fitted into Columbia	452
The international Soyuz TM15 crew	455
The EURECA satellite is lifted out of the payload bay	458
Prime and alternative crew members for Spacelab J	461
The Space Vision System experiment	464
Discovery lands at the end of the last DoD mission	468
Harbaugh carries Runco in the payload bay	471
Dockings at Mir	474
Assembly of payloads for STS-56	477
Walter at work in Spacelab D-2	480
Low and Wisoff on EVA during STS-57	483
The Soyuz TM17 crew	486
The ACTS satellite with TOS upper stage	489
Fettman takes a spin on the rotating chair	492
Servicing the Hubble Space Telescope	495
Record-breaker Polyakov with the Soyuz TM18 crew	499
Krikalev operates SAREX during STS-60	502
USMP-2 and OAST-2 experiments in the payload bay	505
MS Jones monitors several cameras	508
The 16th main Mir crew	511
Mukai floats into the IML-2 Spacelab module	514
Meade tests the SAFER system	517
Space Radar Laboratory-2 aboard Endeavour	520
The crew of Soyuz TM20	523
Tanner at work on the mid-deck	527
Foale and Harris on EVA during STS-63	530

Illustrations

Deployment of the Astro-2 payload	534
Thagard in his sleep restraint on Mir	537
A Russian–American handshake in space	540
The 19th resident Mir crew	543
Henricks operates the HERCULES-B geolocation system	545
The Mir EO-20 crew in the new Docking Module	548
The Wake Shield Facility on its second flight	551
Thornton and Bowersox working at the Drop Physics Module	555
Atlantis docked to Mir	558
Retrieval of the Japanese SFU	561
Lucid and "the two Yuris"	564
The broken tether of the TSS	567
Clifford on EVA during STS-76	570
The Inflatable Antenna completing its inflation	573
Multiple experiments underway in Spacelab	576
The crew of Soyuz TM24	579
Carl Walz among the stowage bags	582
Jones works at the aft flight deck station	585
Unpacking logistics to transfer to Mir	588
Tsibliyev and Linenger on EVA at Mir	591
The fifth and final EVA of STS-82	594
Science during a shortened mission	597
Linenger briefs Foale aboard Mir	600
A planning session aboard STS-94	603
The Soyuz TM26 crew	606
Tryggvason at work in the mid-deck	609
Mir, showing the damage to Spektr	612
Scott releases the AEROCam camera	615
Sharipov signs the Mir roster	618
The Soyuz TM27 crew	621
Preparation of Neurolab, the final Spacelab Long Module	624
The end of US residency aboard Mir	627
The Soyuz TM28 crew	630
John Glenn, the oldest man in space	633
The embryonic ISS is created	636
The first Slovakian in space	639
Moving supplies aboard Zarya	642
Eileen Collins, the first female Shuttle commander	645
Another service for the Hubble telescope	648
Shuttle Radar Topography mission hardware	651
The final Mir resident crew	654
Williams uses the handrails during EVA	657
The ISS configuration after STS-106	660
The new additions to ISS after STS-92	663
Fresh fruit for the first ISS residents	666
ISS with new solar arrays	669

THE FIFTH DECADE

The Destiny laboratory is lifted out of the Shuttle's payload bay	673
A crowded ISS during the first crew exchange.	676
The second ISS resident crew	679
Canadian contributions to ISS.	682
The first space "tourist" comes home.	685
The Quest airlock about to be installed	688
Another crew exchange inside Destiny	691
Formal picture of the ISS-3 crew.	694
The Soyuz TM33 crew at work on ISS.	697
Another change of shift on ISS	700
The ISS-4 crew.	703
Grunsfeld and Linnehan finish the latest Hubble service programme	705
Smith at work on the S0 Truss	709
The Soyuz Taxi-3 crew	712
Endeavour docked to the ISS	715
ISS tool kit	718
ISS configuration October 2002.	721
Soyuz TMA1 crew join the ISS-5 crew.	724
EVA astronauts at work on the P1 Truss.	727
The ISS-6 crew in Sokol suits	730
The ill-fated STS-107 crew	733
The first ISS caretaker crew	737
The first Chinese national in space.	740
The Soyuz TMA3 crew about to launch.	743
The TMA4/ISS-9 crew	746
A review of equipment for the ISS-10 crew.	749
The ISS-11 crew takes over	752
Robinson works on the tiles of STS-114.	755
Olsen and the ISS-7 crew prior to launch	758
The crew of Shenzhou 6	761
Brazil's first astronaut, aboard ISS.	764
Farewells between the crews of STS-121 and ISS-13	767
ISS is expanded once again following STS-115	770
Space flight participant Ansari.	773

Other Works by the Authors

Other manned space exploration books by Tim Furniss

A source book of Rockets, Spacecraft and Spacemen (1972) ISBN 0-7063-1494-8
The Story of the Space Shuttle (1984) ISBN 0-340-35280-9
Guinness Spaceflight: the Records (1985) ISBN 0-85112-451-8
Space Shuttle Log (1986) ISBN 0-7106-0360-6
Manned Spaceflight Log (1986) ISBN 0-7106-0402-5
One Small Step (1989) ISBN 0-854290586-0

Other space exploration books by David J. Shayler

Challenger Fact File (1987), ISBN 0-86101-272-0
Apollo 11 Moonlanding (1989), ISBN 0-7110-1844-8
Exploring Space (1994), ISBN 0-600-58199-3
All About Space (1999), ISBN 0-7497-4005-X

With Harry Siepmann

NASA Space Shuttle (1987), ISBN 0-7110-1681

Other books by David J. Shayler in this series

Disasters and Accidents in Manned Spaceflight (2000), ISBN 1-85233-225-5
Skylab: America's Space Station (2001), ISBN 1-85233-407-X
Gemini: Steps to the Moon (2001), ISBN 1-85233-405-3
Apollo: The Lost and Forgotten Missions (2002), ISBN 1-85233-575-0
Walking in Space (2004), ISBN 1-85233-710-9

Other Works by the Authors

With Rex Hall

The Rocket Men (2001), ISBN 1-85233-391-X
Soyuz: A Universal Spacecraft (2003), ISBN 1-85233-657-9

With Rex Hall and Bert Vis

Russia's Cosmonauts (2005), ISBN 0-38721-894-7

With Ian Moule

Women in Space: Following Valentina (2005), ISBN 1-85233-744-3

With Andy Salmon and Mike Shayler

Marswalk: First Steps on a New Planet (2005), ISBN 1-85233-792-3

With Colin Burgess

NASA's Scientist Astronauts (2006), ISBN 0-387-21897-1

Prologue

PRAXIS LOG OF MANNED SPACEFLIGHT – A USER'S GUIDE

Each log entry was compiled to the same basic layout. The missions are given their official designation but are not numbered chronologically. With variations in defining exactly what constitutes a space flight, and with the increasing tendency for international crews to launch and/or land on separate missions, we have found it far simpler to list the missions in launch sequence and to describe their achievements, than to say superficially which world mission or national mission it was.

The International Designation is the official orbital identification number issued by the International Committee on Space Research (COSPAR). COSPAR gives all satellites and fragments an international designation, based on the year of the launch and the number of *successful orbital launches* in that calendar year (1 Jan–31 Dec). For example, Apollo 11 received the designation 1969-59A, indicating that it was the 59th orbital launch during the year 1969. The letter code at the end of the designation refers to the type of vehicle launched. Normally, the letter "A" is given to the *main* instrumented spacecraft; "B" to the rocket; and "C", "D," "E" and so on assigned to fragments or ejections. Letters "I" and "O" are not used. If there are more than 24 pieces (such as debris from an explosion), the sequence after "Z" becomes "AA, AB and so on up to "AZ", and then "BA", "BB", etc. For this volume, we have listed only the "A" designations. These items are tracked by the North American Aerospace Defense command (NORAD) which supplies orbital data elements (via NASA) on all traceable satellites – very useful in the identification of potential space debris impacts. In the years 1957–1962, a different system was used, with designations utilising the symbols of the 24 letters of the Greek alphabet. For the years 1961–1962 in this volume, we iterate these Greek letters in full for clarity.

The launch date, launch site and landing date and site are given as local time; we have not tried to convert to GMT or UT. We have omitted local times for clarity wherever possible, although for some of the more historic missions in the days before data was accessible at the click of a mouse button, we have kept some of this data in as a useful reference point. The launch vehicle details have been included where known. It is likely that further data will come to light in future years that will enable us to give a more complete picture of such information.

Durations are given from official sources (NASA or Soviet/Russian) and for Shuttle missions, this is from lift-off to wheel stop at the end of its runway landing. Callsigns (when used) and mission objectives are also presented for information.

Crew details are for the *PRIME*, or flight, crew only and are presented in the order commander; pilot; then specialists in numerical sequence. Each crew entry lists their full name, age at time of launch, military affiliation or civilian, position on this crew, the number of times they have flown into space, and their previous missions for quick cross-reference. All crew members are either American (astronauts) or Soviet (cosmonauts) unless their nationality is noted.

The flight log records key mission events and, where necessary, pre- and post-flight operations. When an X-15, sub-orbital or X-prize flight occurred, it is mentioned briefly for continuity in the main text. The details of such missions are included in the opening sections.

When a crew is launched on one mission and returns on another, their whole flight is reported under their *launch* mission and only briefly mentioned under their *landing* mission. Therefore, when a space station crew is launched with a core crew of two with a third passenger, the passenger's activities are recorded along with that of the core space station crew in the same "mission log." This process evolved during the Mir programme, in which guest cosmonauts would fly with an expedition crew who remained on the station, while the guest returned home after about a week in the older spacecraft and with the previous core crew.

On ISS, there have been several occurrences of a complete ISS core crew being launched as "passengers" on a Shuttle mission, and landing "as passengers" on a separate Shuttle mission. Here, we have covered the launch of the Shuttle mission separately, followed by the resident crew's activities as second entry and the landing mission as a third.

Milestones are significant events, achievements and celebrations relating to that crew or mission's flight into space.

We have not provided references as there are just so many to collate all this data from. The most referred to sources are listed in the bibliography and further details of sources of information can be obtained from the authors if so desired.

Following these guidelines, the Quest for Space section covers those missions that did not reach orbital flight but are part of the story of human space exploration: the 13 launches between 1962 and 1968 of the X-15 that exceeded the then-designated 50 mile (80 km) limit; the two Mercury Redstone sub-orbital missions in 1961; the Apollo 1 pad fire that claimed the lives of three American astronauts on 27 January 1967 just two weeks prior to their planned mission; the Soyuz T10-1 pad abort which occurred just seconds prior to the planned lift-off; and the recent X-Prize flights of Spaceship 1 in 2004.

The launch abort of the Soyuz 18-1 mission in April 1975 is included in the log entries, as is the loss of Challenger during the STS 51-L mission in January 1986. Both of these missions had launched and were "missions in progress" when they encountered their specific difficulties. Had they continued in their planned trajectory, both would have reached orbit.

Wherever possible, we have followed the metric system of weights and distances.

The Appendices review orbital space flight between 1961 and 2006; the cumulative time that astronauts and cosmonauts have spent in space in the order of most experienced; and a brief timeline of historic and key missions in the exploration of space.

Call signs: In the early days of manned space flight, there was no requirement to identify one spacecraft from another because there was never more than one in orbit at a time. Mercury astronauts, however, following the tradition of pilots naming their aircraft, assigned names to their Mercury capsules, adding the number 7 to signify the seven original Mercury astronauts. Thus, the Mercury missions were also known as Friendship 7, Sigma 7, Aurora 7 etc. Had Deke Slayton flown, he would have used the call sign Delta 7

The Gemini spacecraft used the spacecraft's number as a call sign (though for a while the Gemini 4 astronauts tried to assign the name "American Eagle" to the flight and it was also known as "Little Eva" – for the EVA or spacewalk). The early Apollo missions also did not require a call sign but by now, distinctive mission emblems were being worn by the crews (from Gemini 5). These have become a traditional part of any manned space flight and are descriptive and colourful. The names of the crew are usually displayed on the emblem, though not always. Programme emblems, activity emblems (such as the EVA badge), payload and support teams emblems and (from 1978) Astronaut Group selection emblems have evolved from these. Russian cosmonauts and Chinese yuhangyuans have displayed similar types of emblems.

From Apollo 9 and the first manned flight test of the Lunar Module, it was necessary to be able to clearly identify both the Command and Lunar modules during radio conversations as both would be flying separately at some stage during the mission, with members of the crew aboard each module. Thus, the Command Module became "Gumdrop" and the Lunar Module "Spider." This practice continued throughout Apollo up to Apollo 17. For Skylab and the American Apollo spacecraft used during the ASTP flight, the crews used the call signs "Skylab" or "Apollo". When the Americans began to fly the Space Shuttle in 1981, the call sign became the name of the individual orbiter that was being used, as each has its own moniker.

For the Soviet and Russian missions, each pilot cosmonaut chose their own call sign. When in command of a mission, they adopted that call sign for the flight, with other crewmembers appending "2" or "3" to it to identify themselves individually during the mission. When engineer cosmonauts began to fly as mission commanders in 1978, they too were assigned personal call signs, and resident Soviet/Russian space station crews were also known by the call sign of the commander. For ISS missions, it appears that cosmonaut Soyuz TMA commander call signs are used for contact over Russian ground stations and during flights of the Soyuz spacecraft independent of the ISS. It is unclear if Chinese Shenzhou missions or yuhangyuans have adapted a call sign.

1

Reaching the Heavens

Space – the so-called "final frontier" or the "new ocean" – has been beckoning humans since ancient times. Our ancestors looked to the heavens and saw their Gods as bright sparking pin-picks of light in the night sky. By day, the position of the Sun was often used for ceremonies and worship rituals as it moved across the sky. The changing face of the Moon, the occasional blocking of the sunlight and the movements of some of those small speckles of light all frightened and intrigued our forebears for thousands of years.

In the first four hundred years or so of the second millennium AD, our understanding of the heavens, planets, stars, moons and "space" began to grow, and while our interpretations were often wrong, or were subjugated to the religious beliefs of the day, developments in scientific instrumentation, medical advances, engineering capabilities, industrial processes and human curiosity began to focus on what lay beyond the confines of our own world. Fanciful stories were conceived and published, weird and wonderful machines proposed, and myths and monsters imagined.

In the twentieth century, the development of flight and research into rocketry gradually opened up the possibility of exploring the void of space. Advances in miniaturisation, computation, medicine, pressurised chambers and life support systems to explore the upper atmosphere and deepest oceans were the stepping stones to get there. But the driving force behind the final stage would be the military, at least at first. The desire for supremacy over a rival superpower was the catalyst behind our first tentative steps into the "new ocean". Eventually, what had started as a race became a team event, a combination of what both sides had learned – sometimes at painful and tragic cost – into a new era of understanding and exploration.

ACCESS AND METHOD

Ask most people "How do you get into space?" and they would reply "by rocket", not realising the fact that they are already "in space" on planet Earth, travelling in orbit around the Sun. We are all "astronauts", it's just that most of us haven't left the planet yet. For a lucky few though, leaving the planet has afforded them some of the most spectacular sights and experiences yet known to mankind. But there is more to it than simply "flying into space". There are various ways of doing this, depending on your mission and the type of spacecraft you have.

The atmosphere

Exactly where the atmosphere ends and space begins is a subject that has long been debated. Our atmosphere consists of roughly seventy-eight per cent nitrogen, twenty per cent oxygen, one per cent argon and trace amounts of other gases. It is not, however, uniform all the way up and has significant variations in temperature and pressure with increasing altitude. This defines the layers of the atmosphere. Our atmosphere can be divided into five regions of increasing altitude: the troposphere (0–16 km), the stratosphere (16–50 km), the mesosphere (50–80 km), the thermosphere (80–640 km) and the exosphere (640–10,000 km). Humans can survive with varying degrees of ease without assistance in the lower-most region, but require pressurised aircraft compartments or balloons up in the stratosphere. Above that is the realm of "almost space". The air here is much too thin to support an air-breathing engine, yet is sufficient to cause atmospheric drag on vehicles travelling through it. Above this, in the thermosphere, is where most of the spacecraft and satellites orbit the Earth, and the method used by most vehicles to travel in this region is by rocket thrust in the vacuum conditions.

SPACE FLIGHT METHODS

As the quest for space began, two methods of getting there were investigated, both requiring rockets for power. One was to develop winged craft which would access space from a carrier aircraft in the upper atmosphere and would perform a guided entry and landing for the return. The other was to use blunt-ended capsules on top of former military missiles shot through the atmosphere on ballistic trajectories, relying on the increasing density of the atmosphere to slow the return sufficiently for parachutes to finish the landing.

Rocket planes

The first powered steps towards space were made by a series of American rocket-propelled aircraft. In October 1947, the first supersonic flight was made by the X-1, piloted by Chuck Yeager. In 1963, an X-15 rocket plane piloted by Joe Walker reached an altitude of 106 km (66 miles). The majority of rocket planes were indeed released from carrier aircraft at high altitude before igniting their onboard rocket engines for a quick climb to the fringes of space and then a gliding landing on a runway. One of the X-15s flew at over Mach 6 in 1967. Together with several strangely shaped, blunt-bodied, wingless vehicles known as lifting bodies (which evaluated the

Pilot Mike Adams with the X-15

technological possibilities of a vehicle that could survive heat of re-entry, fly at subsonic speeds and still make a controlled horizontal landing), such programmes would lay the groundwork for what eventually became the Space Shuttle. The X-Prize winners Spaceship One drew upon the same legacy.

Sub-orbital flights

The first rockets to fly into space flew simple "up and down" ballistic flights to high altitude, but did not reach the velocity or height to enter Earth orbit. Many were used for science missions in the upper atmosphere (starting in the 1940s) and often carried animals. The payloads were usually, but not always, recovered. These trajectories gave engineers the opportunity to put their equipment though the stresses of launch, re-entry and recovery without committing it to orbital flight. The first two Americans in space in 1961 flew this type of trajectory, although the Russians abandoned their sub-orbital programme in order to place the first person in orbit ahead of the Americans.

4 Reaching the Heavens

Flight paths

In order to save fuel and weight, the rotation of the Earth is used to give any launch a speed boost. Therefore, most launch sites are situated on or as close as possible to the equator to take maximum advantage of this and launch their spacecraft into a west-to-east trajectory. An east-to-west launch is possible, but it would be very expensive in terms of launch weight and fuel, and therefore cost. A rocket reaching a speed of approximately 29,000 kph (18,000 mph) can overcome the effects of Earth's gravity pulling it back to the ground and enter an arc around the planet, known as an orbit. Strictly speaking of course, an orbiting spacecraft is still falling towards the Earth – it's just travelling fast enough to keep missing it. There are many types of orbits, and all crewed flights to date have entered relatively low altitude Earth orbit, even if they have later further boosted their speed in order to head out into deeper space. An orbit has a high and low point, known respectively as the apogee and perigee, and an orbit with a very high apogee is known as a highly elliptical orbit.

Satellites are launched into different inclinations, or angles to the equator. A polar orbit flies approximately over the north and south poles, effectively at almost a 90-degree inclination. Most orbits feature inclinations that are much lower than this. Geostationary, or geosynchronous, orbiting satellites orbit around the equator at approximately 35,000 km (22,000 miles) and travel at the same speed as the Earth rotates, giving the perception of being "stationary" in the sky as seen by an observer. Such orbits are mainly used by communications and weather satellites. All crewed vehicles have so far been placed in relatively low-altitude and low-inclination Earth orbits, although there was a proposal to fly an Apollo mission into a highly elliptical Earth orbit and plans for polar-orbiting Shuttle missions were abandoned in 1986.

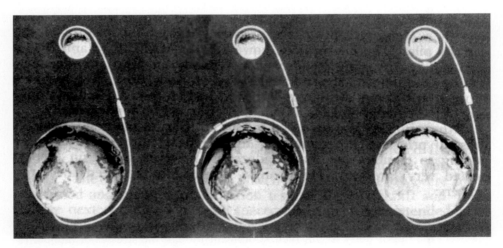

An illustration of the options that Apollo could use to reach the Moon: (left) Direct Ascent; (centre) Earth Orbital Rendezvous, or EOR; (right) Lunar Orbital Rendezvous, or LOR – the method eventually chosen.

The Apollo lunar flights (1968–1972) used low-Earth parking orbits before heading out towards the Moon, using an additional engine burn of the upper stage to begin their trans-lunar trajectory. Once at the Moon, the spacecraft could either fly once around the Moon and back to the Earth (called a circumlunar trajectory – this was used for Soviet unmanned Zond missions which were precursors of planned but unflown manned missions), or orbit the Moon or land on it. The original flight plans for US manned lunar landings included several options: A direct ascent to the Moon surface using a powerful booster, joining two craft together in Earth orbit before flying out to the Moon, or a lunar orbit rendezvous (which was the method chosen for Apollo) with the Command Service Module (CSM) and two-stage Lunar Module (LM) flying together on one booster. After the Moon landing expedition was over, the ascent stage of the LM rendezvoused with the CSM in lunar orbit, the crew transferred between vehicles and the CSM headed home for parachute landing in the ocean. The new plans for the return to the Moon resemble aspects of the Earth Orbital Rendezvous method that was considered for Apollo.

Launch sites

Manned space flights have thus far been launched from Pad 1 – "The Gagarin Pad" – at Site 5 or Pad 31 at Site 6 at the Baikonur Cosmodrome in Kazakhstan (Russian launches); Pads 5, 14, 19 and 34 at Cape Canaveral and Pads A and B of Launch Complex 39 at the Kennedy Space Center in Florida (American launches); and most recently at Jiuquan in China (Chinese launches).

The USAF test pilot base at Edwards Air Force Base (AFB) in California was the home of the X-15 rocket plane, which made "astroflights", while Mojave Airport,

One of several pad facilities at Cape Canaveral

6   Reaching the Heavens

Pad 1 at the Baikonur Cosmodrome in Kazakhstan

California was the home base for the Spaceship One "space tourism" test flights. Mojave may also be the home base for the Virgin Galactic Spaceship Two tourist vehicle. Vandenberg AFB, California was proposed as the Space Shuttle's west coast base for military (polar orbit) missions, but the first mission was cancelled after the Challenger accident in 1986 and the Shuttle never flew from Vandenberg.

Landing methods

All US manned space mission landings up to the Space Shuttle programme were "splashdowns" in either the Atlantic or Pacific Ocean. The crew vehicles descended under parachutes after re-entry into the atmosphere, and a recovery crew would

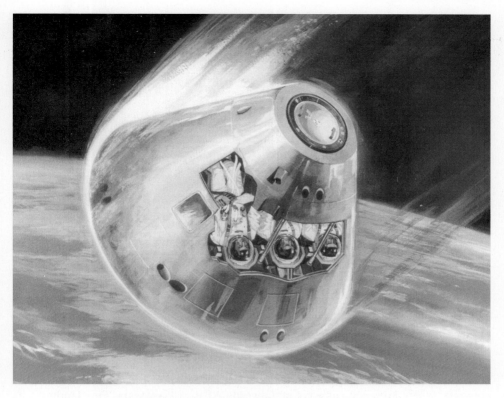

Diagram of an Apollo re-entry configuration

attach a flotation device to the capsule once it had hit the ocean. This was not always the case with some Mercury and Gemini missions, which landed outside the planned recovery area. This method was easier than developing terrain landing capabilities, although Gemini and Apollo spacecraft development teams did investigate this option. Splashdowns still carried the risk of drowning the crew or the possibility of a parachute failure, and there was also the added expense of drafting a flotilla of military ships for each mission recovery.

Soviet or Russian missions have always been targeted for landing in home territory, normally Kazakhstan. Vostok pilots ejected from their capsules and landed by separate parachute as developing a soft landing system to protect the occupant was deemed too complicated and time-consuming in the race to beat the Americans to orbit. Voskhod crews landed in their craft under parachutes, aided by retro-rockets in the parachute support system. A more effective system was incorporated in the base of the Soyuz, with four solid propellant rocket engines fired 1.5 metres off the ground to effect what is known as a "dust-down" or a "soft-landing". While this is usually reliable, it has occasionally proven otherwise. This method is also used by the Chinese Shenzhou, which is based on the Russian Soyuz design.

The Shuttle Landing Facility (SLF) at the Kennedy Space Center

The Space Shuttle orbiter lands on a runway and has used three sites – mainly the concrete Shuttle Landing Facility at the Kennedy Space Center in Florida and the concrete or dry lake bed runways at Edwards AFB in California. One landing was made at the Northrop Strip (gypsum) at White Sands in New Mexico. There is also a network of contingency landing sites around the globe, all with long runways for use as emergency return options for the Shuttle.

Emergency escape

As the early spacecraft were developed from military missiles, they had a very nasty tendency to explode on or shortly after lift-off. Getting these vehicles safe enough to carry a crew – man-rating – was a priority, but to be on the safe side, additional escape methods were developed. All systems are tested many times to ensure their correct operation in the event of an emergency and though no crew would wish to experience launch escape systems in action, it is comforting knowing that there is an option to get out of a bad situation if necessary, assuming you have time to do anything about it, of course.

The Soviet Vostok one-man craft employed an ejector seat, which would be used in the event of a launch failure or operationally for pre-landing ejection and parachute descent. In contrast, the Voskhod was the most dangerous of all spacecraft. Adapted from the basic Vostok one-man vehicle, it could carry two cosmonauts with spacesuits, or three without, but it didn't have any room for ejection seats.

The US Mercury and Apollo spacecraft, and the Russian Soyuz spacecraft, utilised a rocket launch escape system during the final stages of the countdown and in the early stages of ascent. The Chinese Shenzhou has a similar rocket escape system to the Russian Soyuz. The Russians have experienced two launch aborts during the forty years of Soyuz operations. During the abort in 1975, the escape tower had been ejected and the crew had to rely on the ballistic abort and return separation sequence, enduring up to 20-G loads for a few seconds. Then in 1983, a Soyuz launch pad abort system was activated when a launch vehicle caught fire on the pad. It exploded seconds after the crew were boosted to safety by the escape tower.

The American Gemini spacecraft adopted ejector seats, which could be used prior to launch or during ascent if required. The first four Shuttle orbital test missions also had ejection seats fitted, but these were deactivated after the fourth and final test mission and removed after the fifth flight.

The Space Shuttle also has a series of abort modes available during the ascent phase, although, when asked whether they wanted to fly a test mission of these aborts, the astronauts declined, indicating that they would test them when they needed them.

In the event of a main engine failure prior to ignition of the twin Solid Rocket Boosters (SRBs), there are options to abort the launch and this has occurred on five occasions in the history of the Shuttle programme. The three official abort modes are Return to Launch Site (RTLS), Trans-Atlantic Abort (TAL) and Abort to Orbit (ATO).

RTLS occurs early in the flight if one or more engines are lost. This is the most risky, and many astronauts don't believe it is survivable, as the stack has to turn around and fly back to the Kennedy Space Center. This would be a severe strain on the structural integrity of the vehicle. A TAL abort is a preferred option, with three prime locations in Europe, while an ATO is the only one to have actually occurred thus far. This was in 1985 during mission STS 51-F, when one engine shut down early during launch. The mission continued normally. It is obvious to all who fly the Shuttle, however, that a failure of an SRB (which occurred on the STS 51-L launch in January 1986) is non-survivable. There are other "contingency" aborts which are not "official" but are nonetheless trained for. The orbiter could shed its External Tank (ET) and conceivably land at Bermuda or other east coast sites in the USA, but this, too, would provide a real challenge for the flight crew – and quite a spectacle for the locals.

As the Shuttle launch pads were adapted from the older Apollo Saturn V launch pad, some of the facilities are still available to the crew on the pad in the event of an emergency situation. Should an emergency occur before launch, such as fire after a launch pad abort, the crew would evacuate the Shuttle and use "baskets" on slide wires to descend to the ground, then go through tunnels into an underground bunker. There is also the option of using an armoured personnel carrier to vacate the pad as quickly as possible. Any countdown demonstration test during training on the pad ends with this emergency drill, although it does not include the actual slide wire ride. During the Apollo era, and due to the height of the tower, the crew could have used a slide tube system leading to an underground protected bunker, if they had time to get there!

For resident crews on the International Space Station (ISS), and previously on the Soviet Salyut and Mir stations, an emergency return vehicle has always been available – the Soyuz. There is always one attached to the station and when crews become larger as more of the station is built, there will be two attached. NASA has been planning and designing its own Crew Rescue Vehicle but it is unlikely that this will ever be built. The Soyuz is perfectly adequate for the job. It is probable, however, that if NASA build a Crew Exploration Vehicle for the Vision for Space Exploration programme, one or two of these will be docked to the ISS.

LAUNCH SYSTEMS

In the history of manned space flight, there have been numerous designs for systems to carry people into space. Many have reached the point of almost making a manned space launch, but have been cancelled prior to the event. Between 1961 and 2006, there have been just two "rocket planes" (X-15 and Spaceship One) that have touched space, while only eight launch systems (seven rockets and the Shuttle) have actually achieved manned space launcher status.

Astro-flights

Throughout the space age, there has been a worldwide uncertainty as to precisely where the atmosphere ends and space begins. Some say 50 miles (80.45 km), others 62 miles (100 km), and there are those who claim it doesn't happen until you are in orbit. However, the X-15 rocket plane reached altitudes of between 50.70 and 66.75 miles

An X-15 is launched from beneath a B-52 bomber

(81.59 and 107.42 km) on thirteen "astro-flights" by eight pilots between July 1962 and August 1968. In the early 1960s, the USAF decided that a military pilot making a flight over 50 miles (80.45 km) would be eligible for the rating of Air Force Astronaut Pilot and awarded Astronaut Wings to those who achieved it. The five US Air Force pilots were awarded Astronaut Wings at the time, but the three civilian pilots had to wait until 2006 to receive theirs. The award should also therefore be given to Mike Melvill and Brian Binnie, who flew the space tourist prototype vehicle Spaceship One in 2004.

The X-15 flights used the B-52 aircraft to "air-launch" the rocket research plane by dropping it from beneath the wing, usually at about 45,000 ft (13,716 m), where it began its descent to the ground either as a glide flight or by igniting its engines and completing its mission. Spaceship One was carried to 13,716 m and 14,356 m by the White Knight launch aircraft for its two record-breaking missions.

Sub-orbital flight

The first US astronauts, Alan Shepard and Gus Grissom, flew sub-orbital Mercury test flights in 1961 aboard Redstone rockets, reaching over 160 km (99 miles) altitude. They were recognised as astronauts. When a Soviet Soyuz R7 booster failed in 1975,

Mercury launch vehicles

causing the abort that led to Soyuz 18-1 making a sub-orbital flight to about 145 km (90 miles) altitude, the flight was credited as a "space flight" to the two cosmonauts.

The Redstone was the USA's first intermediate-range ballistic missile and was powered by one Rocketdyne A-7 engine, with a thrust of 35,380 kg (78,013 lb) burning liquid oxygen, ethyl alcohol and water. The Mercury–Redstone was 25.29 m (82.97 ft) high. Two such vehicles were used in the Mercury programme for manned flights, while a third was cancelled with the desire to press on to the first orbital manned flight using Atlas.

American orbital launchers

The Atlas ICBM was used to launch four manned Mercury missions in 1962–3, while the Titan II ICBM launched ten Gemini crews between 1965–6. A modified Titan II would have been used to fly the manned DynaSoar military space plane in the mid-1960s, but this was cancelled in 1963 and replaced by the Gemini-based military Manned Orbital Laboratory. This was due to be launched on a Titan IIIM starting in 1966 but was also cancelled (in 1969), with some of its astronauts transferring to NASA.

The Atlas D intercontinental ballistic missile (ICBM) had a thrust of 166,470 kg (367,066 lb) from two Rocketdyne LR89 engines – which were burnt out and separated at about $T + 2$ min 14 sec – and an LR105 central sustainer engine. These were powered by liquid oxygen and kerosene. The Atlas was stabilised at lift-off by two powerful vernier engines. The Mercury–Atlas combination was 29 m (95 ft) high. The Atlas booster for the fifth manned mission was the first of a new model to be used for Mercury and was static test-fired on the pad because the US Air Force was concerned about turbo-pump failures that had occurred on some military ICBM launches. Atlas 113D would also ascend on ignition, rather than remaining on the pad for the previously prescribed two-second hold down period.

The Gemini Launch Vehicle (GLV) was a modified ICBM. Its twin first-stage LR-87 engines burned nitrogen tetroxide and hydrazine hypergolic propellants which ignited spontaneously on contact. The first-stage engines had a thrust of 195,046 kg (430,076 lb). The second stage, with a smaller LR-91 engine, had a thrust of 45,359 kg (100,017 lb). First-stage cut-off came at $T + 2$ min 30 sec, with a "fire in the hole" second-stage ignition following immediately. Orbit was achieved in 5 min 30 sec after launch. The launch vehicle was 3.04 m (10 ft) in diameter and with Gemini on top, was 33.22 m (109 ft) tall.

The Saturn family of launch vehicles was developed for civilian space launches by a team led by Werner von Braun. The series built upon the successes and proven hardware of its early variants (Saturn 1 and 1B) before the huge Saturn V was used to send American astronauts to the Moon. Other variants were proposed but none were funded or built. Following a series of unmanned launches, the Saturn 1 manned missions were cancelled as unnecessary. After unmanned test flights, the Saturn 1B

Comparison of the US Mercury, Gemini and Apollo space capsules, and their launchers

launched one Apollo crew on a test flight in 1968, three Skylab space station crews in 1973–4 and the US part of the Apollo–Soyuz Test Program in 1975. Just two unmanned test flights were flown before the Saturn V carried a crew aloft for the first time. Apollo 8 and 10–17 launched their crews to the Moon in 1968–72, while Apollo 9 launched to Earth orbit as planned.

The Saturn 1B launch vehicle, with the launch escape system on top of the Command Module, was 74.37 m (244 ft) tall. The launch escape system comprised a 10 m (33 ft) high tower with a 66,675 kg (147,018 lb) thrust solid propellant motor, which could be used on the pad or during the first 100 seconds of launch. When ejected, it pulled away a conical blast shield from the Command Module, exposing the latter's five windows. The first stage of the Saturn 1B comprised eight H1 engines, developing a thrust of 743,899 kg (1,640,297 lb) and burning the RP1 and liquid oxygen propellants for the first 150 seconds. The second stage was the S-IVB cryogenic liquid oxygen/liquid hydrogen stage that would also form the third stage of the Saturn V booster. The S-IVB was powered by the J2, 102,059 kg (225,040 lb) thrust engine with a burn time of 450 seconds. This engine could be restarted. The Saturn S-IVB also included the all-important Instrument Unit, the vehicle's guidance and performance system. The

14 Reaching the Heavens

Saturn 1B was also the first manned launch vehicle used that was not a converted ballistic missile.

The Saturn V launch vehicle was 110.64 m (363 ft) high from the base of the F1 engines to the tip of the launch escape system tower. The first stage, called the S-1C, had five F1 engines developing a thrust of 3,442,801 kg (7,591,376 lb) and burning liquid oxygen and RP1 propellants at a rate of 15 tonnes a second. The second stage, the S-11, also had five engines, called J2, with a thrust of 498,956 kg (1,100,198 lb). The third stage was the S-IVB from the Saturn 1B launch vehicle. The whole vehicle weighed 2,903,020 kg (6,401,159 lb) at lift-off. The third stage was used for the Trans-Lunar Injection (TLI) burn to take the spacecraft out of Earth orbit and towards the Moon. The last Saturn V launch was a two-stage variant that carried the unmanned Saturn Workshop (Skylab) – itself a modified former S-IVB stage – in 1973. The earlier and larger Nova launch vehicle was abandoned in favour of the Saturn class of vehicles, which would be developed much quicker.

The Space Shuttle flew its first mission in 1981 and will be retired in 2010, although the programme may be extended if there are any further delays to the completion of the International Space Station. The "Space Shuttle" is a combination of boosters, fuel tank and orbital vehicle, often termed "the stack". The orbiter is the manned portion of the vehicle and there have been six orbiters built: OV-101 (Enterprise) was used for

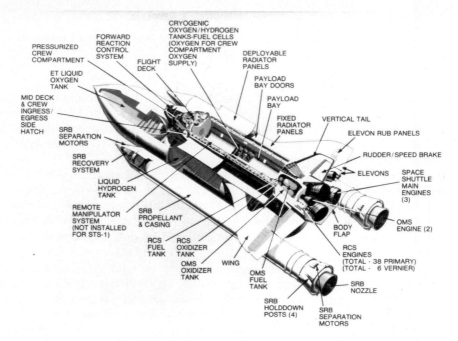

The US Shuttle system, known as the stack

With the Solid Rocket Boosters lit, the Shuttle is committed to launch

atmospheric and ground tests; OV-102 (Columbia 1981–2003) was the first to launch to space and was lost in the STS-107 re-entry accident; OV-099 (Challenger 1983–1986) was a former structural test article and was lost in the STS 51-L launch accident; OV-103 (Discovery) is the oldest remaining vehicle and has been in service since 1984; OV-104 (Atlantis) has been used since 1985; and OV-105 (Endeavour) was built as a replacement for Challenger and was introduced in 1992.

The Space Shuttle orbiter Enterprise, which was to have been refurbished for space flight later, made five approach and landing atmospheric glide flights over Edwards Air Force Base in 1977, being air-launched from the back of a Boeing 747. Three flights were piloted by NASA astronauts Fred Haise and Gordon Fullerton and the other two by Joe Engle and Richard Truly. The longest glide flight, ALT 3, lasted 5 min 34 sec. Columbia was the first space flight-worthy orbiter and weighed 99,454 kg (219,296 lb) at orbital insertion. The orbiter measured 37.24 m (122 ft) long, with a wingspan of 23.79 m (78 ft). Three liquid hydrogen/liquid oxygen main engines, with a maximum thrust rating of 100 per cent, 170,098 kg (375,066 lb) each at sea level, took the orbiter into an initial orbit, which was then augmented by firing the Orbital Manoeuvring System (OMS) engines. These were powered by nitrogen tetroxide and UDMH, which also powered the reaction control thrusters. Power was provided by a liquid oxygen/liquid hydrogen fuel cell system. A Thermal Protection System (TPS) heat shield tile system, comprising over 35,000 tiles, covered the orbiter to protect it from re-entry temperatures of between 370 and 1,260°C. A 756,453 kg (1,667,979 lb), 47 m (154 ft) long, 3.7 m (12 ft) wide External Tank (ET), painted white on the first two missions, held the SSME liquid oxygen and liquid hydrogen. Attached to it were two 45.46 m (149 ft) long, 3.7 m (12 ft) wide Solid Rocket Boosters (SRB), weighing 586,502 kg (1,293,237 lb). Tail to nose, the Shuttle stack measured 56.144 m (184 ft) tall. The orbiter, an unpowered glider, had conventional flaps, rudder and ailerons for control in the atmosphere. Columbia's STS-1 landing speed was 344 kph (214 mph). Improvements were made to the Shuttle fleet all the time. For example, Challenger was equipped with SRBs with an uprated thrust of 1,469,200 kg (3,239,586 lb) on its first mission, as well as main engines which were throttled up to 104 per cent for the first time. It also carried a new lighter weight external tank. Much of the areas covered with Low Temperature Surface Insulation tiles on Columbia were covered by lighter blankets of Advanced Flexible Reusable Surface Insulation on Challenger.

Soviet/Russian orbital launchers

The Russian ICBM R7-based Vostok booster with an upper stage was used to launch Yuri Gagarin on the first manned space flight on 12 April 1961. Five more missions followed until an "up-rated" Voskhod craft was launched using the Voskhod rocket, another derivative of the R7. A series of Soyuz boosters, also based on the R7 first stage, have been used to launch crewed missions since 1967 and the launcher will continue for many more years to send Soyuz International Space Station crew ferry and emergency return vehicles aloft.

The basic R7 launch vehicle had its origins in 1947, with the idea of grouping (clustering) rocket stages together instead of stacking them on top of each other (as with Saturn V).The final configuration from studies became Izdelie 8K71 (Product 8K71). Originally developed as an ICBM, it featured a central sustainer engine (Blok A) to which four strap-on boosters were attached (Blok B, V, G and D) forming stage 1. The central engine was the RD-108, providing 75 tons of thrust at lift off and burning for 304 seconds, while the four RD-107 boosters delivered 83 tons of thrust and burned for 122 seconds, separating at burn-out to allow the central sustainer to continue the

Launch systems 17

A Soyuz launch

ascent. Both engines used LOX/kerosene and remained the core propulsion vehicle for all Soviet and Russian manned launches and a significant number of unmanned ones. Depending on the payload, additional upper stages were added to propel the payload into orbit. The central core of the R7 stands 26 m (85 ft) with a diameter of 2.95 m (9.5 ft) tapering to 2.15 m (7 ft) to accommodate the four strap-on stages. Each of the boosters was 19 m (62 ft) long.

Manned variants

8K72K (Vostok 1961–3). This version featured an upper stage (Blok E) with a single RO-7 engine, burning LOX/kerosene with a thrust of 5.6 tons and a 430-second burn time. Used to launch the six manned Vostok missions, its design was not revealed to the west until it appeared at the 1967 Paris Air Show.

11A57 (Voskhod 1964–5). This was an improved variant of the Vostok launcher, with an upper stage powered by the RD-108 engine with a vacuum thrust of 30.4 tons and 240-second burn time. This was used to launch the two manned Voskhod spacecraft.

11A511 (Soyuz 1967–76). This version was developed from 1963 to specifically launch the Soyuz spacecraft. The upper stage (Blok I) had an RD-110 engine with a 30.4 ton thrust and 246-second burn time. These vehicles launched the early Soyuz missions, starting with Soyuz 1 in 1967. Its final use was for the Soyuz 23 spacecraft in 1976. With the Soyuz payload and launch shroud, this vehicle measured about 49.3 m (162 ft) in height.

11A511U (Soyuz-U). An upgraded variant of the standard Soyuz booster, this was first used for the launch of Soyuz 16 in 1974 and was in service for over 27 years. It was used for launching the Soyuz, Soyuz T and Soyuz TM variants, as well as the Progress and Progress M re-supply vessels. It was also used to deliver the Pirs facility to ISS in September 2001. This vehicle used improved engines, ground and support facilities, increasing the payload mass and orbital delivery altitude.

11AIIU2 (Soyuz U2). Further improvements to payload delivery mass led to this variant of launcher being used for the first time on a Soyuz launch with Soyuz T12 in July 1984. It was last used on Soyuz TM22, after which the production of Sintin (synthetic kerosene) for improved first-stage launch performance ceased in 1996.

Soyuz FG. Upgrades to the engines resulted in the RD-108A central core engine, which developed a vacuum thrust of 101,931 kg and had a 286-second burn time. The RD-107A engines provided a thrust of 104.1 tons and a 120-second burn time. Both engines burned LOX/kerosene. This variant was first used for manned launches on Soyuz TMA1 in October 2002 and is the variant currently in use.

Soviets on the Moon

The abandoned Soyuz manned lunar programme would have featured circumlunar missions under the L1 programme on the Proton launch vehicle (later flown unmanned by Zond spacecraft), and manned lunar landing (L3) missions launched on the massive N1. Though the N1 was launched unmanned four times between 1969 and 1972, each failed just seconds into flight, effectively putting the final nail in the coffin of the Soviet manned lunar programme that had been beaten by the success of Apollo.

Chinese orbital launch vehicle

The Long March 2F booster will be the workhorse of the Chinese manned spacecraft programme, carrying further Shenzhou craft into orbit. Unmanned launches commenced in November 1999, with the first manned flight made in 2003 and the second in 2005.

This vehicle is an adaptation of the Long March 2E, which was upgraded for manned flight. In 2002, it received the official name of *Shenjian* (Magic Arrow). The height of the vehicle (with shroud and launch tower) is 58.34 m and it features a central core first stage of four YF-20B engines (300-ton thrust) and four strap-on boosters each with the YF-20 engine (300-ton thrust). The second stage features a single YF-22 engine with 93.5-ton thrust.

2

The Quest for Space

The real quest for space has existed for centuries, ever since man first noticed the stars and began to wonder about them, gradually fostering the desire to visit them. Centuries of Earth-based observations of the cosmos evolved into the science of astronomy. The desire for human "flight" was often intertwined in the early years with a passion for the written word and a vivid imagination, creating numerous stories of fantasy and adventure in the heavens. With the development of the balloon, the chance to actually ascend into the atmosphere gave scientists real experience and data about the difficulties of high-altitude flight, as well as the realisation that there was a limit to our atmosphere. In the closing years of the nineteenth century and opening decades of the twentieth, the development of life support systems, pressurised compartments, diving apparatus, and of course the aircraft, were the important steps along the way to the series of high-altitude, high-speed aircraft and stratospheric balloons that set and surpassed record after record from the 1920s through to the 1950s. The knowledge gained, and the sacrifices made, were the final link in the chain that led to Gagarin's pioneering journey into orbit.

X-15 – A ROCKETPLANE TO SPACE

During the 1940s and 1950s, proposals for rocket-propelled aircraft designed to exceed the speed of sound were instigated in several countries. The most successful were the American X-series of aircraft, which included the X-1 that broke the sound barrier (Mach 1) in October 1947 and the X-15 hypersonic research aircraft. The X-20 was a USAF proposal for a one-man orbital space plane which would have been rocket-launched, but would have landed on a runway utilising the technique of dynamic soaring, hence its nickname – the DynaSoar. The concept never flew. Similar designs were developed in the Soviet Union and both countries eventually used the technology to develop a manned reusable space shuttle, though the Soviet craft (Buran) only flew one short unmanned mission in 1988.

X-15 – A winged marvel

The X-15 programme began in 1954 and three aircraft were constructed. The programme was designed to provide data on aerodynamics, structural and control problems, and the physiological aspects of high-speed, high-altitude flight. Once these primary objectives had been met and exceeded, the programme began to gather data from various experiments and materials carried by the X-15 to the fringes of Earth's atmosphere, many of which helped in the development of techniques and material for future programmes such as Apollo and the Space Shuttle. To assist in meeting this new objective the second aircraft was modified (after a serious accident in 1962), improving its performance to attain speeds of Mach 8 and support the development of supersonic combustion ramjet engines (scramjets).

Measuring 50 feet (15.24 metres) long and with a short stubby wingspan of 22 feet (6.70 metres), the X-15 featured a wedge-shaped vertical tail. The vehicle weighed about 14,000 lbs (6,350.4 kg) unfuelled and 34,000 lbs (15,422.40 kg) with a full fuel load. The high speeds that the X-15 would attain posed a problem for protecting the structural skin of the aircraft. This was solved with a heat-resistant skin made from Inconel-X nickel–steel alloy over a titanium and stainless steel structure. Control during atmospheric flight was by conventional aerodynamic control surfaces, but "in space", eight hydrogen-peroxide thrusters on the nose controlled the pitch and yaw, with a further four on the wings used for roll control. Difficulties with the X-15's primary propulsion system (Thiokol XLR99) resulted in the decision to use two Reaction Motors XLR11 rocket engines for the initial powered test flights. Each XLR11 engine produced 2,000 lb (907.2 kg) of thrust. The manually-controlled XLR99 rocket engine had a thrust of about 60,000 lbs (27,216 kg).

Air-launched from under the wing of a B-52 at about 45,000 ft (13,700 metres) at an air speed of 500 mph (804.5 kph), the rocket motor was ignited by the pilot after dropping from the B-52 wing and burned for about 80 seconds. There were two flight profiles. The one for high altitude included a steep rate of climb while the other was used for level attitude high-speed objectives. Each free-flight lasted for about 10–11 minutes, mostly unpowered, and at the peak altitude of the astro-flights, only two or three minutes of "weightlessness" was experienced. However, the pilots did get a clear view of the curvature of the Earth's surface and the blackness of space above the thin layer of the atmosphere. The X-15's landing gear consisted of rear skids and a nose wheel, and landings were usually at about 200 mph (321.8 kph).

X-15 flights over 50 miles

The X-15, whose programme operated between June 1959 and October 1968, was a rocket-powered aircraft built by North American Aviation. The programme was operated as a joint NASA/USAF/USN venture, for aeronautical research at speeds in excess of Mach 6 and altitudes up to and beyond 50 miles. The fastest recorded speed was eventually 4,520 mph (Mach 6.7) and the highest altitude achieved was 354,200 feet (66.8 miles).

Table 2.1. X-15 flights over 50 miles altitude

Date	Free-flight	Pilot	Aircraft	Altitude (miles)
1962 Jul 17	62	White	3	59.16 (95.18 km)

First FAI-certified world altitude record; 3 aborts preceded the attempt. The rocket engine fired for one second longer than planned resulting in a speed 248 mph (399 kph) faster than planned. At peak altitude White could see a panorama that stretched from San Francisco in California down to Mexico.

1963 Jan 17	77	Walker	3	51 (82.05 km)

Walker's flight was to study the handling of the X-15 without its ventral fin at extreme altitudes and to conduct an infra-red experiment.

1963 Jun 27	87	Rushworth	3	55 (88.49 km)

This flight was aimed at providing the pilot with experience of high-altitude handling and phenomena.

1963 Jul 19	90	Walker	3	65.3 (105.06 km)

On this flight, Walker was to study the expansion of the airframe during re-entry with the ventral fin removed. He also deployed and towed a nitrogen-filled balloon and conducted horizon-scanning, photometric, infra-red and ultraviolet observations, all in ten minutes.

1963 Aug 22	91	Walker	3	66.75 (107.40 km)

Walker's third "astro-flight" reached the highest altitude attained by an X-15 in 199 free-flights. He also attained a speed of Mach 5.58 (3,794 mph or 6,104.5 kph).

1965 Jun 29	138	Engle	3	53.14 (85.50 km)

Engle's first "astro-flight" included a horizon-scanning experiment.

1965 Aug 10	143	Engle	3	51.7 (83.18 km)

Engle's second "astro-flight" occurred just 8 months prior to his selection as a NASA astronaut. He was the first astronaut selected who already held "astronaut–pilot wings".

1965 Sep 29	150	McKay	3	56 (90.10 km)

After surviving a crash of the X-15 #2 aircraft that almost killed him in 1962, McKay finally completed an "astro-flight" which investigated boundary-layer noise and structural loads on the horizontal tail, as well as horizon-scanning experiments.

1965 Oct 14	153	Engle	1	50.17 (80.72 km)

In his third "astro-flight", Engle completed a programme that included taking measurements of atmospheric pressure and further experiments in the scanning of Earth's horizon.

1966 Nov 1	174	Dana	3	58 (93.32 km)

This flight included the objectives of collecting micrometeorites, and tests of a dual-channel radiometer and a tip-pod accelerometer. Precise measurements of the attitude and density of the atmosphere were also taken.

1967 Oct 17	190	Knight	3	53.4 (85.92 km)

Further collection of micrometeorites, the recording of wing-tip pod deflection during re-entry, observations of the ultraviolet plume of the XLR99 rocket exhaust, and studies of the solar spectrum above 200,000 ft (60,960 metres) were all objectives assigned to this mission.

1967 Nov 15	191	Adams	3	50.4 (81.09 km)

The scientific objectives of this 12th "astro-flight" of the programme, included a UV study of the rocket exhaust plume, observations of the solar spectrum and the bow shockwave of the wing-tip pod. Nose-gear loads were to be observed, micrometeoroids collected and an ablative material tested for use on the Saturn 5 booster. Adams was killed on this flight and was awarded his USAF Astronaut Wings posthumously.

1968 Aug 21	197	Knight	1	50.7 (81.57 km)

The final X-15 "astro-flight" was just two missions prior to the end of the programme. The planned 200th flight was cancelled. In 199 free-flight missions, the three X-15s had logged 30 hours 13 minutes 49.4 seconds in flight and had flown 41,763.8 miles (67,197.95 km). Pilot experience at Mach 4 was almost 6 hours, with a further 90 minutes at Mach 5 and 78 seconds at Mach 6.

Fifteen pilots were selected to fly the X-15, although there was no formal selection process. They were all qualified test pilots prior to assignment to the programme. Eventually, only twelve flew X-15 missions, of which there were 199 completed by the three X-15 vehicles. In addition, several captive flights were executed, where the X-15 was not released from under the wing of the B-52 launch aircraft.

Although not considered as a spacecraft, the X-15 did operate in a region of the upper atmosphere whose conditions were only fractionally different from those encountered by a vehicle in Earth orbit. In the early 1960s, the USAF had declared that flights above 50 miles (80.45 km) would be classified as a space flight. They would award USAF Astronaut Wings to honour those USAF pilots that attained this altitude. In contrast, the Federation Aeronautique International (FAI), the international aeronautical record-keeping body, decided that flights over 100 kilometres (or 62 miles) would be classified as space flights.

Of the 199 X-15 flights, thirteen surpassed the 50-mile altitude barrier, and these have been designated astro-flights, rather than space flights. Eight of the X-15 pilots (Walker, White, Rushworth, Engle, McKay, Dana, Knight and Adams) flew these thirteen missions. Of these, only five (White, Rushworth, Engle, Knight and Adams) were USAF pilots who received the USAF wings. The remaining three (Walker, McKay and Dana) were civilians and did not qualify for the USAF title. However, Walker completed two X-15 flights in excess of the FAI qualification altitude.

THE SUB-ORBITAL MERCURY FLIGHTS

The "stepping stones to space" process of developing the Space Shuttle from the rocket planes and lifting bodies had its precedent in the ballistic capsule programmes. Unmanned variants were tried out in pad aborts, sub-orbital and orbital missions, before committing them to manned missions. In Russia, dogs were used in the Vostok (and Voskhod) programme as test subjects prior to sending cosmonauts into space, while in the United States, primates performed a similar function for the Mercury programme. These "animal space explorers" paved the way for humans to venture into space by qualifying launch, orbital and entry systems and flight profiles. Though several biological payloads and research subjects have been flown on other missions, never again have animal tests preceded human flights in order to qualify new spacecraft variants.

One way to qualify human flight systems prior to committing to a more challenging orbital mission was by means of a "sub-orbital" flight – essentially a simple boosted ascent and almost immediate re-entry and landing. This profile was first proposed in both the Russian and American programmes, but the Russians abandoned the idea in favour of securing orbital flight prior to the Americans. The Americans went through with their sub-orbital test, but reduced the number of flights from seven to three, and finally two when they proved so successful. Though the two astronauts (Alan Shepard and Gus Grissom) surpassed 185 km (well over the FAI or USAF criteria), they did not enter orbit, both making only 15-minute "space flights" with about 5 minutes in weightlessness. All the record books accredit these two flights as "official space flights" as do we, but we have categorised them in this chapter as part of the "quest for space".

MERCURY REDSTONE 3

Int. Designation	None – sub-orbital flight
Launched	5 May 1961
Launch Site	Pad 5, Cape Canaveral, Florida
Landed	5 May 1961
Landing Site	Atlantic Ocean
Launch Vehicle	Redstone No. 7; capsule no. 7
Duration	15 min 28 sec
Callsign	Freedom 7
Objective	First sub-orbital test of Mercury spacecraft with a human occupant; first US manned ballistic space flight

Flight Crew

SHEPARD, Alan Bartlett Jr., 38, USN, pilot

Flight Log

The first of what were originally to be seven manned sub-orbital Mercury flights, then reduced to three, could have taken place in March 1961, before Gagarin, had the programme not hit technical problems. Al Shepard decided to name his spacecraft Freedom. Adding the number seven to the name became too irresistible, as the capsule and rocket were both serial number seven and there were seven astronauts. This established a precedent for later manned flights. Shepard simulated the flight inside Freedom on Cape Canaveral's Pad 5 three times before the first launch attempt on 2 May was thwarted by bad weather.

On 5 May, the astronaut was up at 01:10 hours and inside Freedom 7 at 05:20 hours. Compelled to urinate in his spacesuit because of the unforeseen 2 hr 34 min launch holds, the laconic Shepard finally got airborne at 09:34 hours, uttering the first of 78 statements, practised so many times in the simulator, announcing lift-off. His heartbeat was monitored at 126 beats per min. The period of maximum dynamic pressure reached at $T + 58$ sec buffeted the vehicle and caused some concern. The launch escape system tower separated at $T + 2$ min 32 sec, as Shepard was experiencing a maximum 6.3 G force. The Redstone shut down at $T + 142$ sec and Shepard arced even higher over the Atlantic Ocean, at a maximum speed of 8,262 kph (5,134 mph), reaching a maximum altitude of 185.6 km (115.3 miles).

During his 4 min 45 sec period of weightlessness, Shepard fired his thrusters to orientate the spacecraft in yaw, pitch and roll movements for a period of 40 sec. He only saw the Earth as black and white out of his periscope and not the porthole, and then he moved the craft to a nose down angle of 34° before firing the retros, although they were not needed during this sub-orbital flight. The descent was uneventful, the 0.5 G light coming on at 60,960 m (200,000 ft) and with Shepard enduring 11 G deceleration. The

Mercury Redstone 3 is launched on a sub-orbital trajectory from Cape Canaveral, with America's first astronaut Alan B. Shepard aboard

drogue chute deployed at 6,400 m (21,000 ft) and the main chute at 3,048 m (10,000 ft). Freedom hit the sea at a speed of 10.7 m/sec (35.1 ft/sec), 475.2 km (295 miles) downrange from the Cape at $T + 15$ min 28 sec, the shortest manned space flight in history. Shepard removed the hatch and was hauled aboard a helicopter from the recovery ship *Lake Champlain*.

Milestones

2nd manned space flight
1st US manned space flight
1st to make orientation manoeuvres
1st flight to splashdown in the sea
1st flight to end with the crew aboard

MERCURY REDSTONE 4

Int. Designation	None – sub-orbital flight
Launched	21 July 1961
Launch Site	Pad 5, Cape Canaveral, Florida
Landed	21 July 1961
Landing Site	Atlantic Ocean
Launch Vehicle	Redstone No. 8; capsule no. 11
Duration	15 min 37 sec
Callsign	Liberty Bell 7
Objective	Second sub-orbital test of Mercury spacecraft with a human occupant, further system qualification towards manned orbital missions

Flight Crew

GRISSOM, Virgil Ivan "Gus", 35, USAF, pilot

Flight Log

Grissom was aboard Mercury spacecraft Liberty Bell 7 on 19 July, when the launch attempt was scrubbed. The next attempt on 21 July was delayed for 2 hours 15 min before the Redstone ignited at 07:20 hours. Grissom admitted to being "a little scared" and his pulse rate rose to 162 beats per minute during the ascent. The Redstone cut off at 2 min 22 sec and Liberty Bell separated at a speed of 8,317 kph (5,168 mph), reaching an altitude of 189.6 km (117.8 miles). Grissom oriented the

Gus Grissom wearing the Mercury pressure garment, showing the neck dam and inflatable jacket used during water survival training and, for Grissom, during the recovery of Liberty Bell 7

spacecraft, and the urge to look out of the window was strong during his short space flight.

As the retro-rockets fired, Grissom's heart rate shot up to 171. After an uneventful re-entry, the main chute was deployed a little higher than planned and Grissom noticed a small tear in it. Landing came at $T + 15$ min 37 sec, 484.8 km (301 miles) downrange of the Cape, and the recovery ship USS *Randolph* was on hand for the pick-up. While Grissom prepared to evacuate the capsule as planned for a helicopter recovery, he armed explosive bolts on the hatch. Suddenly the hatch blew off, and water rushed into the slightly listing capsule. Grissom got out quickly but he had forgotten to seal his oxygen hose connector and water seeped into his spacesuit.

Floundering and close to drowning, the distraught astronaut watched as a helicopter concentrated efforts on recovering the capsule. Eventually Grissom grabbed a lowered line and was dragged underwater for about 10 m (32.8 ft), before being hauled aboard like a drowned rat. Liberty Bell had by this time taken in about 900 kg (1,985 lb) of seawater, and as a helicopter tried to raise it a warning light came on in the pilot's cockpit and he let the capsule go, only to discover later that the warning light was in error and the capsule, now 552 m (1811 ft) under the Atlantic Ocean, could have been recovered. Grissom claimed that he never touched the plunger that would explosively separate the hatch and was just "lying there when it blew". The cause of the failure was never officially established. After years of discussions outside of NASA into the possibility of recovering the capsule, the Liberty Bell 7 capsule was finally located on the sea bed of the Atlantic at 2,850 fathoms in May 1999. On 20 July 1999, exactly 30 years after Apollo 11 landed on the Moon, Liberty Bell 7 was winched aboard the prime recovery vessel. The capsule was returned to Cape Canaveral the next day – the 38th anniversary of its launch.

Milestones

3rd manned space flight
2nd US manned space flight
1st space flight to end with the loss of the spacecraft

APOLLO BLOCK I

Early planning for Apollo included a series of manned missions designed to evaluate the systems and procedures of the Apollo parent craft (the Command and Service Module, or CSM) in Earth orbit, prior to committing it to lunar distance flights or flights with the Lunar Module. These capsules were termed Block 1 and did not feature the docking and transfer tunnel system utilised on the lunar missions. Subsequent Block II CSMs were designed to fly in conjunction with the LM in Earth orbit or deep space, or to support the lunar landing flights. More advanced missions that fell under the Apollo Applications Program banner would use a proposed (but unflown) Block III series of CSMs. Some of the amendments proposed to support extended-duration lunar missions were actually incorporated into the "J" series of scientific Apollo missions flown in 1971–1972 using upgraded Block II CSMs and LMs. Block III CSMs were also planned to support flights to orbital workshops (later Skylab), but none were fabricated. There was also a Block I mission known as Apollo 2, but this was cancelled in 1966 when it became apparent that it was too much of a duplication of the Apollo 1 mission, given the desire to press on with qualifying the Block II series of CSMs, the Lunar Module and the Saturn V for manned flights.

APOLLO 1

Int. Designation	None – fatal pad fire accident prior to planned launch
Launched	Planned 21 February 1967
Launch Site	Pad 34, Kennedy Space Center, Florida
Landed	Planned 7 March 1967
Landing Site	Pacific Ocean
Launch Vehicle	Saturn 1B
Duration	Planned 13 days 18 hours 50 minutes
Callsign	Apollo 1
Objective	First manned qualification test of Apollo (Block I) CSM in Earth orbit for up to 14 days; test firings of the Service Propulsion System; evaluation of systems and procedures by crew and vehicle

Flight Crew

GRISSOM, Virgil Ivan "Gus", 40, USAF, commander, 3rd mission
Previous missions: Mercury Redstone 4 (1961); Gemini 3 (1965)
WHITE II, Edward Higgins, 36, USAF, senior pilot, 2nd mission
Previous mission: Gemini 4 (1965)
CHAFFEE, Roger Bruce, 31, USN

Flight Log

Such was the frenetic success of the US manned space programme in 1966 that the buoyant NASA even suggested that the first Apollo spacecraft make a rendezvous and crew transfer flight with Gemini 12. Apollo 2 could then make a lunar looping flight, some thought. Beneath this over-confidence lay shocking flaws in the Apollo spacecraft systems. Apollo 1 was to have flown a 10-day shakedown flight in the autumn of 1966. Problems with checking out the spacecraft – even returning it to the Rockwell factory – delayed the flight to December, then to 21 February 1967. In early January, the three-man crew posed happily in front of their Saturn 1B with Apollo on top of it at Pad 34. But they and several engineers were concerned. Apollo 1 was a tetchy spacecraft. On Friday 27 January, Gus Grissom, Ed White and Roger Chaffee waited patiently to begin a thorough countdown demonstration test, which at last began at 13:00 hours.

On entering the capsule, however, the crew noticed a strange smell. A delay of 1 hr 42 min ensued while this was investigated before the crew was entombed in the capsule behind a tightly bolted hatch. The air in the cabin was purged with pure oxygen, which pressurised the capsule at 16.7 psi. The countdown proceeded with a few niggling problems until it reached $T - 10$ min at 18:20 hours. Communications between the spacecraft and the control centre became so bad that the tetchy Grissom wondered

The charred inside of Apollo 204 CSM-012 following the pad fire which claimed the lives of Apollo 1 crew Grissom, White and Chaffee

aloud how on Earth they were going to get to the Moon if they couldn't even talk between buildings. Eleven minutes later, at 18:31 hours, the exposed wiring in a cable beneath Grissom's seat short-circuited.

In the pure oxygen atmosphere, the flash ignited nylon netting and Velcro and within seconds, sheets of flames covered the left-hand side of the craft. Grissom shouted "Fire!" as the flames spread across the ceiling. Chaffee shouted "Fire in the spacecraft!" The pressure inside the capsule rose to bursting point, the release of pressure caused a surge of flame right across the spacecraft, towards the collapsed wall of the spacecraft's right-hand side. "We've got a bad fire ... we're burning up!" White meanwhile was desperately trying to unlock the complicated three-tier hatch which, even in the best of circumstances, could only be opened in 90 seconds. The crew was doomed. Carbon monoxide and other poison from the inferno choked them to death. Grissom died on the floor with his feet still in the couch. Chaffee died in his right-hand seat and White had been welded to his centre seat by the heat. The irony was that at the end of the countdown test, the crew were to have practised an emergency egress.

A manned lunar landing before the end of Kennedy's decade seemed doubtful as an investigation was mounted. It revealed shocking deficiencies in the design, workmanship and quality control. Apollo was redesigned – no flammable materials, no pure oxygen and a hatch that could be opened at the pull of a lever. But for the fire and the imperfections it exposed, it is doubtful that Apollo 11 would have achieved the manned landing on the Moon within Kennedy's deadline. Some have stated that the fire was not the only delay to the programme, as there were on-going difficulties with the Saturn V and Lunar Module that probably would have delayed the first lunar landing into 1969, even if the fire had not occurred.

RECOVERY FROM APOLLO 1

Following the loss of the Apollo 1 crew and a review of future operations, a sequence of seven steps was proposed in September 1967 which could be assigned to an Apollo mission, or series of missions, to form a distinctive, step-by-step advance towards the landing missions. Though amended and expanded as the programme developed, these seven steps were as follows (with the flown missions of each step given in brackets):

A missions – Unmanned development missions of the Saturn V and CSM Block II (Apollo 4 and 6)
B missions – Unmanned development missions of the Lunar Module (Apollo 5)
C missions – Manned CSM and Saturn 1B missions in low-Earth orbit for up to 11 days (Apollo 7)
D missions – Manned Saturn V or Saturn 1B missions flying the CSM and LM in low-Earth orbit for up to 11 days (Apollo 9)
E missions – Manned CSM and LM operations using the Saturn V in high-apogee Earth orbits for up to 11 days (this flight plan was amended in August 1968 to fly a manned CSM around the Moon for ten orbits – Apollo 8)
F missions – Deep space evaluation of manned CSM/LM combinations, including lunar orbital missions utilising the Saturn V, of up to 10 days (Apollo 10)
G missions – The initial lunar landing attempts. Eight-day missions with one three-hour surface EVA, a small surface experiment package deployment and limited sample collection (Apollo 11).

This sequence was later expanded to include:

H missions – Ten-day lunar landing missions, with two surface EVAs on foot, deployment of a more sophisticated Apollo Lunar Surface Experiment Package; extended geological sampling traverses, future landing site surveys from orbit (Apollo 12, 13 and 14)
I missions – Lunar orbit survey missions (none flown or manifested)
J missions – Extended-duration "super-science" lunar landing missions. Three periods of surface EVA, additional surface modes of transport (Lunar Roving Vehicles); expanded surface and orbital scientific packages; geological survey and collection from landing sites of high scientific interest (Apollo 15, 16 and 17).

SOYUZ PAD ABORT – A FAST RIDE AND A STIFF DRINK

Following Apollo 1, all manned space flights which attempted orbital flight succeeded in doing so up to 1975 and the aborted launch of Soyuz 18A (this mission exceeded 192 km and since it is recorded as a "space mission", it is detailed in the main log section). However, space flight is not without risk and two Soviet missions ended in tragedy and one Apollo lunar mission almost claimed the lives of the crew. In April 1967, the maiden manned flight of Soyuz ended in the tragic loss of its sole cosmonaut, Vladimir Komarov, after a difficult flight. In June 1971, the three Soyuz 11 cosmonauts died during the entry and landing phase after completing a successful 23-day mission to Salyut 1 – the first space station. In April 1970, an onboard explosion in the Service Module of Apollo 13 *en route* to the Moon caused the landing to be aborted as the mission became a fight to keep the crew alive and return them to Earth. The struggles and ingenuity of the crew, in conjunction with the sterling efforts of controllers and contractors and the support of their families, has become part of space folklore. These three missions can also be found in the main log. There was, however, another mission that was supposed to carry its crew into space in 1983, but it failed to even get off the ground – at least, not in the way the crew had expected to.

SOYUZ T10-1

Int. Designation	None – pad abort seconds prior to launch
Launched	Planned 27 September 1983
Launch Site	Pad 1, Baikonur Cosmodrome, Kazakhstan
Landed	Planned for late December 1983
Landing Site	Planned for Kazakhstan
Launch Vehicle	R7 (Soyuz U); spacecraft serial number 16L
Duration	Planned for about 90 days (abort duration was 5 min 13 sec)
Callsign	Okean (Ocean)
Objective	Replace T9 crew on Salyut 7 to complete their original Soyuz T8 three-month resident programme with two EVAs to install additional solar array panels

Flight Crew

TITOV, Vladimir Grigoryevich, 36, Soviet Air Force, commander, 2nd mission
Previous mission: Soyuz T8 (1983)
STREKALOV, Gennady Mikhailovich, 43, civilian, flight engineer, 3rd mission
Previous missions: Soyuz T3 (1980); Soyuz T8 (1983)

Flight Log

Fresh from their Soyuz T8 abortive docking attempt with Salyut the previous April, Vladimir Titov and Gennady Strekalov were paired for a possible long-duration mission later in the year, swapping with Soyuz T9's Lyakhov and Aleksandrov. When Salyut 7 began to malfunction, the Soyuz T10 mission started to take on a repair status, with Titov and Strekalov trained to perform a spacewalk to place new solar panels on the outside of the station. At 01:36 hours local time at Baikonur, the countdown of the Soyuz rocket reached $T - 80$ seconds and all seemed to be normal. A valve within the propellant valve failed to close and the base of the booster caught fire. Gradually a large fire rose up the side of the booster and a massive explosion was imminent.

The Soyuz T abort system was damaged by the fire, however, and it was another ten seconds before the ground control team recognised that there was a serious problem. A back-up abort procedure was put into action by two ground controllers

The Soyuz T10 abort cosmonauts Titov and Strekalov hoping to continue the programme they started on their T8 mission, five months before the pad abort

in separate rooms. Titov and Strekalov must have known that there was a problem because the booster was by now pitching over 35° and engulfed in flames. It could not be seen by the controllers. With barely a second remaining before the cosmonauts disappeared into the conflagration, the launch escape system at last sprang into life.

The Descent and Orbital Modules of Soyuz T10 were electronically severed from the instrument section inside the payload shroud and the twelve solid propellant rockets at the top of the escape tower ignited, producing a thrust of 80,000 kg (176,400 lb). Titov and Strekalov were airborne, as the booster exploded. Pulling 18 G, the cosmonauts were powerless at first, then five seconds later, four aerodynamic panels were folded out to stabilise the strange projectile. Twenty-five smaller propellant rockets fired to maintain stabilisation.

Instantly, the Descent Module containing Titov and Strekalov was severed and literally fell out of the payload shroud at an altitude of about 1,050 m (3,444 ft). The back-up parachute was deployed, because its swifter opening time was well-suited for the low-altitude opening, and the re-entry heatshield deployed to expose the soft-landing rockets. Soyuz T10 hit the ground about 3.2 km (2 miles) away, as the pad was a sea of flames, burning for over 20 hours. Titov and Strekalov were administered a stiff glass of vodka but did not need hospital treatment. The experience did not put them off space flight as each flew again, probably believing the old adage that lightning never strikes twice. That lightning had struck once took time to filter through to the west, for details were not released for some time.

Milestones

1st use of crew launch escape system

THE X-PRIZE – THE DAWN OF PRIVATE SPACEFLIGHT

Test pilot and private astronaut Brian Binnie created history on 4 October 2004, the 47th anniversary of the launch of Sputnik 1. He was the pilot of Spaceship One on the second of two flights to exceed an altitude of 320,000 feet twice within 14 days, thereby claiming the $10 million Ansari X-Prize. In doing so, Spaceship One became the first private manned spacecraft to fly above 100 km altitude and briefly enter space. The flights of Spaceship One were part of the Tier One private manned space programme operated by Scaled Composites.

The X-Prize challenge

The X-Prize was created in 1996 with the challenge of placing the same privately funded manned spacecraft in space on a sub-orbital trajectory twice within two weeks. The X-Prize was modelled on the Orteig Prize, which was won by Charles Lindbergh in 1927 by flying solo non-stop across the Atlantic Ocean.

34 The Quest for Space

First captive flight of White Knight and Spaceship One. © 2004 Mojave Aerospace Ventures LLC, photographed by Scaled Composites. Spaceship One is a Paul G. Allen Project. Used with permission.

The idea of privately funded manned space flights instead of government sponsored ones was attractive to commercial and entrepreneurial organisations across the globe and has its routes in classic science fiction stories of the 1940s and 1950s, and in attempts to get a private citizen into space as early as the 1980s. The original X-Prize was proposed in order to advance the goal of manned space flight by private flights, using private contributions, entrance fees from teams and user fees from a proposed X-Prize credit card. By May 1998 fourteen teams had entered the competition. There were ten American entrants, three from England and one from Argentina.

The programme and its success also boosted support for the idea of "space tourism" and, coupled with the flights of "space flight participants", or "space tourists" on Soyuz flights, has seen a growth in interest in the idea of private citizens making flights into space in recent years. In 2004, the X-Prize was renamed the Ansari X-Prize when Anousheh Ansari and her brother-in-law contributed a "significant donation" to the foundation on 5 May 2004. Ansari of course became the fourth "space tourist" on a Soyuz mission to ISS in September 2006, and is the last entry in this edition of the log. A future edition is highly likely to include many more orbital space flights by private citizens in the coming years.

A record in two stages with a bonus

The initial X-Prize flight occurred on 29 September 2004 when Mike Melvill piloted Spaceship One for 24 minutes and reached an altitude of 337,500 ft (102.87 km). The

Spaceship One on its first glide flight. © 2004 Mojave Aerospace Ventures LLC, photographed by Scaled Composites. Spaceship One is a Paul G. Allen Project. Used with permission.

second 24-minute flight, by Binnie, reached 367,442 ft (111.99 km) above the surface of the Earth and, in addition to securing the X-Prize, also broke the altitude record set by X-15 NASA test pilot Joseph Walker on 22 August 1963. An initial flight, and first attempt at the X-Prize, had taken place on 21 June 2004, when Mike Melvill flew a 24-minute flight to 328,491 ft (100.12 km) to the fringes of space. However, a flight control malfunction during the flight meant that the second attempt, to win the X-Prize, would have to be delayed beyond the required two-week turnaround. The Spaceship One programme was completed with the attainment of the X-Prize and the team is now working with Virgin Galactic on Spaceship Two designs to extend the objectives and opportunities further than the original prize.

Spaceship One "astro-flights"

Date	Launch	Powered flight	Pilot	Altitude (km)
2004 Jun 21	60th	15th	Melvill	100.12
2004 Sep 29	65th	16th	Melvill	102.87
2004 Oct 4	66th	17th	Binnie	111.99

A view of the curvature of the Earth taken from one of the Spaceship One X-Prize flights. © 2004 Mojave Aerospace Ventures LLC, photographed by Scaled Composites. Spaceship One is a Paul G. Allen Project. Used with permission.

White Knight and Spaceship One

The launch aircraft for the Spaceship One vehicle was a manned twin-turbojet research aircraft designed for high-altitude flights. It carried Spaceship One to an altitude of about 50,000 ft (15.24 km) although its ceiling was 53,000 ft (16.15 km) with a payload capacity of 8,000 lbs (3,628 kg). With a crew capacity of three, it could increase its 82 ft (25 m) wing to 93 ft (28.3 m) for increased climb capability.

Spaceship One featured a three-seat (although for the X-Prize flights, two were ballasted) 60 inch (152 cm) diameter shirt sleeve environment cabin, with space-qualified ECS and dual-pane windows. The unique bullet-shape-with-wings configuration gave aircraft-like qualities for the boost phase, glide and landing. Its "care-free" configuration allowed a hands-off re-entry and reduced aerodynamic and thermal loads by converting to a pneumatic-actuated "feather" configuration, offering a stable, high-drag shape for entry. The motor powering the system was a "new non-toxic liquid nitrous oxide/rubber-fuelled hybrid propulsion system" specifically developed for Spaceship One.

3

The Orbital Programmes

There have been countless proposals and plans for programmes to support the manned exploration of space. Some never left the drawing boards, while others got as far as having hardware produced, only to be cancelled for a variety of reasons prior to the first manned flight. The following are the *main* manned programmes that have been conducted since 1961. For more in-depth information about these programmes, see the Bibliography.

INTO SPACE

The Cold War-inspired space race launched man into space sooner than was perhaps planned, and with rapidly developed hardware. America developed the bell-shaped Mercury capsule and the Soviet Union came up with a "space ball", all to achieve the goal of "Man in Space Soonest", or "MISS", as the Americans called it. The Soviet Union won this particular race, with their one-man Vostok capsule, shaped like a ball. It had an ejection seat to allow emergency escape and for the cosmonaut to eject prior to landing. It was one way of saving development time in order to get their "Man in Space Soonest".

Vostok and Voskhod

Vostok 1 weighed 4,726 kg (10,419 lb) and comprised a spherical flight module and an instrument section, shaped like a double cone, containing batteries and a retro-rocket. The total length of the spacecraft was 4.4 m (14.4 ft), with a maximum diameter of 2.43 m (7.97 ft). The habitable module was 2.3 m (7.55 ft) in diameter and weighed 2,240 kg (4,938 lb). Vostok was designed to support life for ten days in an orbit low enough to guarantee a re-entry due to natural decay in that period, in case of retro-rocket failure. The instrument section, weighing 2,270 kg (5,004 lb) and measuring

The first cosmonauts of Vostok and Voskhod – l to r Komarov, Feoktistov, Gagarin, Leonov, Titov, Bykovsky, Tereshkova, Popovich, Belayev, Yegorov, Nikolayev (AIS collection)

2.25 m (7.38 ft) long, utilised the TDU 1 retro-engine, powered by nitrous oxide and an amine-based fuel, with a thrust of 1.6 tonnes and a burn time of 45 seconds. The flight module landed at 10 m/sec (32.8 ft/sec) – enough to injure the passenger seriously – so the pilot ejected at a height of 7 km (4 miles) and parachuted separately to the ground, landing at a speed of 5 m/sec (16.4 ft/sec). The Vostok capsule had been tested three times successfully in six attempts on Korabl–Sputnik 1–5 (known in the West as Sputnik 4, 5, 6, 9 and 10), four of which carried canine passengers. Another canine-carrying mission failed to reach orbit. Other missions were planned but were cancelled in favour of upgrading the spacecraft into what became Voskhod as an interim measure to compete with the US Gemini series of missions.

Voskhod was essentially a Vostok spacecraft without crew ejector seats and with a back-up retro-rocket pack on top. It also had a landing retro-rocket system. The spacecraft weighed 5,320 kg (11,730 lb), comprising the 2,900 kg (1,802 lb) flight Descent Module, a 2,280 kg (5,027 lb) Instrument Module and a 145 kg (320 lb) back-up retro-pack. It was 5 m (16 ft) high, with a maximum diameter of 2.43 m (8 ft). The back-up retro-rocket – needed because Voskhod's orbit, higher than Vostok on the more powerful SL-4 booster, would not naturally decay in ten days in the event of retro-fire failure – had a thrust of 12 tonnes and comprised 87 kg (192 lb) of solid propellant. It resembled an inverted cup on top of the Descent Module. To enable a 0.2 m/sec (0.65 ft/sec) landing, compared with the Vostok landing speed of 8 to 10 m/sec (26–32 ft/sec), a landing retro-rocket was added, deployed with the parachute so that it fired downwards in front of the Descent Module. Voskhod flew one unmanned test flight, Cosmos 47, six days before the first manned flight. Voskhod 2 weighed 5,683 kg (12,531 lb). The main difference compared with Voskhod 1 was the flexible airlock, which was approximately 2.13 m (7 ft) long and 0.91 m (3 ft) in diameter. This was jettisoned after the EVA. The payload fairing of the SL-4 launch vehicle was modified to include a blister-like covering for the stowed airlock which protruded slightly from the flight capsule. Again, an unmanned craft (Cosmos 57) flew three weeks before Voskhod 2. There was also a 22-day canine flight flown as Cosmos 110 in early 1966, and a series of other manned Voskhod flights were planned, but they were cancelled in a desire to move on to the more advanced Soyuz programme.

Mercury

The US Mercury capsule was even smaller, with the pilot "putting it on", rather than getting into it, as some astronauts described the entry. It would splashdown at sea under a single parachute.

The Mercury capsule was a bell-shaped spacecraft, 2.87 m (9.41 ft) high, with a maximum diameter across the heat shield base of 1.85 m (6.07 ft). At lift-off, with a launch escape system tower on its top, Mercury weighed about 1,905 kg (4,200 lb). The attitude of the heavily instrumented but severely cramped capsule would be changed by the release of short bursts of hydrogen peroxide gas from 18 thrusters located on the

The Original Seven Mercury astronauts pose by a Mercury capsule and an Apollo CM at the Manned Spacecraft Center in Houston, Texas, during *Look* magazine's coverage of the Collier Trophy Award in June 1963. L to r are Cooper, Schirra (partially hidden), Shepard, Grissom, Glenn, Slayton and Carpenter

craft. These movements could be controlled by the Automatic Stability and Control System (ASCS, which acted as the craft's "autopilot" from the ground), through the Rate Stabilisation Control System (RSCS), or manually by the astronaut using a hand controller connected to a fly-by-wire system. At the back of the capsule, over an ablative heat shield, was a retro-pack containing three solid propellant rockets. These were held to the heat shield by three metal "straps" which were deployed along with the heat shield during re-entry. Mercury descended to a sea landing, or "splashdown" under one main parachute. Just before splashdown, the heat shield was dropped 1.21 m (4 ft), pulling out a rubberised fibreglass landing bag to reduce shock. Mercury had been tested unmanned 17 times previously on various rockets, only seven times successfully. Freedom 7 was Mercury capsule 7, the only first-production run, man-rated capsule and the only one to fly manned with one circular porthole, rather than a larger rectangular window, one of several new features requested by the astronauts but not included in time for MR3. Mercury capsule No. 11 (for MR4 the second sub-orbital mission) was in fact the first operational capsule designed for orbital flights, and included a rectangular window and an explosive side hatch, recommended by the astronauts for safety purposes.

Both Vostok and Mercury made six single-person flights between 1961 and 1963. Mercury was succeeded by a larger two-person craft called Gemini, while Vostok basically became Voskhod, into which a three-man crew was crammed for the maiden flight in 1964. A second flight included the first spacewalk, performed in 1965. These two flights in a sense diverted the Soviet Union from its lunar goal, because they were flown for short-term prestige.

TO THE MOON

The space race ultimately turned into the Moon race after President Kennedy's challenge in May 1961 for the USA to land a man on the Moon before the end of the decade. Human space exploration wasn't going to involve step-by-step advances, but a crash programme.

Gemini

To cover the steps that still needed to be learned a new programme, Gemini, was devised. Now, the Americans could develop the technologies and experience required to go the Moon, including spacewalks, rendezvous and docking and long-duration flights. Ten crewed flights were launched between 1965 and 1966.

The distinctive black and white Gemini spacecraft consisted of two components; the re-entry module, of similar configuration but larger than Mercury, with a pressurised cabin, re-entry control, and rendezvous and recovery sections; and the adapter module, with retro-rockets and equipment. Gemini 3 – which did not carry rendezvous systems – weighed 7,111 kg (15,680 lb) and measured 5.58 m (18 ft) long with a base diameter of 2.28 m (7.5 ft). The re-entry module was 3.35 m (11 ft) long and 2.28 m (7.5 ft) at its heat shield base.

Gemini 3's systems included a 100 per cent oxygen environmental control system, electrical batteries – fuel cells would be fitted for the first time on Gemini 5 – sixteen liquid-fuelled orbital attitude and manoeuvring system thrusters, and four solid propellant 1,133 kg (2,498 lb) thrust retro re-entry control system rockets. Gemini was also equipped with ejection seats and did not have a launch escape system. There was a drogue and one main parachute, and the landing sequence ended with Gemini moving from vertical to 30° horizontal position for splashdown.

Apollo

The three-man Apollo would be launched to the Moon using a Saturn V megabooster, with two of the crew landing on the surface of the Moon. The trip utilised the Lunar Orbit Rendezvous (LOR) method, in which an Apollo Command and Service Module (CSM) mother ship and a Lunar Module (LM) would fly to the Moon together. Initially, Apollo Command and Service Modules, termed Block I, were to be tested in Earth orbit for up to 14 days, before sending improved spacecraft (Block II) into deep space or around the Moon on solo test flights prior to man-rating the LM in a series of flight tests. Following the initial lunar landing, a series of exploration missions (up to nine) were planned to explore the local vicinity, using flying and roving vehicles to assist the astronauts who would spend four or five days on the Moon and about a week in orbit. A follow-on programme called Apollo Applications (Block III) would see expansion of the lunar landing programme to missions of up to two weeks on the surface, supported by a small S-IVB space station and leading to the creation of a lunar research base in the 1980s.

The loss of the Apollo 1 crew in 1967 cancelled all Block I Apollo missions. The first manned flight (Apollo 7) was a CSM-only mission, as was Apollo 8, the first manned lunar orbiting mission. Apollo 9 tested the CSM/LM combination in Earth orbit, and Apollo 10 repeated the feat in lunar orbit. This led to the Apollo 11 lunar landing in July 1969. By then, ten landings were manifested and others planned. However, a variety of social, political and hardware issues, notwithstanding the Apollo 13 aborted lunar mission that almost cost the lives of the crew, terminated the Apollo lunar landing programme with the sixth landing – Apollo 17.

Apollo 7, weighing 14,694 kg (32,400 lb) in orbit, comprised two of the three Apollo Moon landing flight modules, namely the Command and Service Modules. The Command Module was 3.48 m (11.4 ft) high and 3.91 m (12.8 ft) in diameter. It weighed 5,556 kg (12,251 lb) and comprised heat shields, 12 reaction control system thrusters, a triple main parachute landing system, computers, waste management, hot water and food. The atmosphere of the 5.95 m^2 (64 ft^2) cabin was gradually changed from a 60/40 oxygen/nitrogen mix at 15 psi to a 5 psi pure oxygen atmosphere after liftoff. The Service Module, including the UDMH–nitric oxide, 9,752 kg (21,503 lb) thrust Service Propulsion System with its 2.8 m (9 ft) long nozzle, was 7.49 m (24.5 ft) long and the same diameter as the Command Module. It contained the fuel tanks for the SPS, fuel cells for the electrical and water generation system, pumps, radiators and a series of RCS thrusters. Apollo 7 Command Module 101 was a lighter weight CM/SM combination than on later flights. Apollo 8, CM103, weighed 28,901 kg (63,727 lb)

The Apollo Service Module and Service Propulsion System engine

and included, for the first time, the high-gain steerable S-band antenna of four 78 cm (31 in) diameter parabolic dishes mounted on a folding boom at the aft end of the Service Module.

The combined Apollo 9 CM104, LM3 modules weighed 36,559 kg (80,613 lb) in orbit. The two-stage aluminium–aluminium alloy LM, which was not designed to

Table 3.1. Apollo Lunar Landing mission details

Apollo mission	Landing site	Landing date	Surface EVAs	Total EVA time (H:M)	Surface stay time (H:M)	
11	Sea of Tranquillity	1969 Jul 20	1	2:23	21:36	
12	Ocean of Storms	1969 Nov 19	2	7:45	31:31	
13	[Fra Mauro planned	1970 Apr – mission aborted two days out from Earth]				
14	Fra Mauro	1971 Feb 05	2	9:22	33:30	
15	Hadley-Apennine	1971 Jul 30	4	18:34	66:54	
16	Descartes	1972 Apr 20	3	20:14	71:02	
17	Taurus-Littrow	1972 Dec 11	3	22:03	74:59	

Cancelled original Apollo lunar landing mission plans as of late 1969

18	Copernicus?	1972 Feb/Mar	3	21:00 approx	70:00 approx	Cancelled Sep 1970
19	Hadley Rille?	1972 Jul/Aug	3	21:00 approx	70:00 approx	Cancelled Sep 1970
20	Tycho Crater?	1972 Nov/Dec	3	21:00 approx	70:00 approx	Cancelled Jan 1970
Total:	10 missions, 7 flown, 6 landings		15	80:21	299:32	

withstand re-entry, was 6.98 m (23 ft) tall. The descent stage was an octagonal structure, 3.22 m (10.5 ft) high, with a maximum diameter of 4.29 m (14 ft). With four landing legs deployed, the LM measured 9.44 m (31 ft) tall. The landing struts were made of crushable aluminium honeycomb and each 0.9 m (3 ft) diameter footpad had surface sensing probes to signal descent engine shutdown. One of the legs had a ladder, extending from the porch of the ascent stage crew hatch. The descent engine was surrounded by equipment bays, holding fuel and gas tanks, navigation and guidance systems, and science equipment used on lunar missions. The descent stage was covered in Mylar–aluminium alloy for thermal and meteoroid protection. The ascent stage was 3.75 m (12 ft) high, with a maximum diameter of 4.44 m (14.5 ft). It contained a 7.16 m^3 (253 ft^3) pressurised crew compartment, measuring 2.33 m (7.5 ft) by 1.06 m (3.5 ft). The ascent stage also included an ascent engine, docking port and tunnel, guidance and navigation, and life support systems. The total weight of the first manned Lunar Module was 4,450 kg (9,812 lb). Apollo 11 (CM107, LM5) weighed 43,869.6 kg (96,732 lb). Apollo 15 (CM112, LM8) weighed 46,785.35 kg (103,162 lb). This was a new J-series mission which incorporated enhanced Command, Service and Lunar Modules.

Soviet lunar plans

The Soviet Union had an even more ambitious plan for a manned landing on the Moon. An N1 mega-booster would launch two cosmonauts to the Moon aboard a combined Soyuz orbiter–lander vehicle. After entering orbit, one cosmonaut would spacewalk from the orbiter to the attached lunar lander inside the upper stage and enter the lander. He would then separate the lander and descend to the surface, spending a few minutes on the ground planting a flag and collecting some rocks before heading back for a rendezvous with the mother ship, where he would spacewalk back to the cabin. The orbiter would then fly home for a Soyuz-type landing. The N1 programme was a disaster, with catastrophic launch failures, and the Moon landing plan was simply over-ambitious. A series of unmanned Zond missions tested various elements of the circumlunar manned programme with varying success between 1968 and 1970, but the whole thing was left without a purpose after the Americans reached the Moon in 1969.

LONG-DURATION SPACEFLIGHT

With the loss of the race to the Moon, the Soviets reported that they had actually never intended to go there anyway. Their plan was to develop a long-duration orbital station. It was years before the truth came out and the details of their abandoned lunar programme became known. However, their statement was partially correct, as a military-based space platform called Almaz had been in development for years, supported by other Soyuz-type military variants. Almaz would not be the first station launched, however. To hasten the launch of the first Soviet space station, elements of Soyuz were added to a civilian variant called DOS and amalgamated into the world's first space station – Salyut. This was launched two years before the Americans launched Skylab, which was itself fabricated from left over Apollo lunar hardware.

Soyuz

Critical to sustaining long-duration space flight is the supply of sufficient logistics and the rotation of the crews. For this, the Soviets called upon their orbital lunar spacecraft Soyuz, adapting it to fly as a space station ferry craft (in manned and unmanned versions) and to serve as a crew rescue craft while docked to the station. The Soyuz vehicle was one of the most successful programmes in space history. Although the first manned mission in 1967 was a failure and resulted in the first casualty of space flight, a series of variants – Soyuz, Soyuz T, TM and TMA – have carried many crews to the Salyut and Mir national space stations and continue to do so to the current International Space Station. The programme will soon be entering its 40th year. After recovering from the loss of Soyuz 1 and the death of its cosmonaut, the Soviets evolved a series of missions to develop the rendezvous and docking technique they had intended to use on the way to the Moon, now amended for the space station programme. In addition, a short series of solo Soyuz flights flew space station equipment, conducting a series of test and supplementary flights to the often troubled Salyut series of stations.

The "original" Soyuz spacecraft was designed as a Vostok successor in about 1962. It weighed 6,450 kg (14,222 lb) and was 8.85 m (29 ft) long from the base of its instrument section to the tip of its docking probe. The 2.3 m (7.5 ft) long, 2.3 m (7.5 ft) diameter instrument section, called the Equipment Module (EM), included a UDMH–nitric oxide prime and back-up propulsion system, for orbital manoeuvres and retro-fire. The prime engine had a burn time of 500 seconds and a thrust of 417 kg (919 lb). The instrument section included two 3.6 m (12 ft) by 1.9 m (6 ft) solar panels.

The Soyuz, workhorse of the Russian space programme, is photographed on approach to a space station

The flight and Descent Module (DM) was shaped like an inverted cup and measured 2.2 m (7 ft) long and 2.3 m (7.5 ft) in diameter. It included up to three seats and systems such as hydrogen peroxide ACS thrusters, a beacon, sun and infra-red sensors, and rendezvous radar beacons. It was equipped with one drogue and one main parachute (plus a reserve), which opened at about 8,500 m (28,000 ft) altitude, and, beneath a jettisonable heat shield, a soft-landing retro-rocket to reduce speed to 0.3 m/sec (1 ft/sec) at 1 m (3 ft) altitude. Attached to the flight module was an Orbital Module (OM), a spherical capsule containing extra housekeeping and science equipment and which acted as an airlock for EVAs. This was 2.65 m (8.69 ft) long and 2.25 m (7.3 ft) in diameter. The OM was discarded after retro-fire. It also included a 1.2 m (4 ft) long docking probe at its tip. The Soyuz 12 spacecraft was basically the same as the earlier Soyuz craft, except the crew wore spacesuits (following the loss of the Soyuz 11 crew who hadn't). The craft was equipped with only batteries for power, and no solar panels, as it was intended as a space station ferry with only a two-day independent flight capability.

Soyuz T – for Transport – was introduced in 1979 and weighed about 6,850 kg (15,104 lb). It was a redesigned Soyuz ferry vehicle, reconfigured to take a crew of three and with two solar panels which allowed independent flight for four rather than two days. Also included were new computers, controls and telemetry systems. The major change to the Soyuz was its fully integrated fuel system, with attitude control thrusters using the same fuel source as the main propulsion unit. The thrust of the main engine was reduced to 315 kg (695 lb) but there were now 26 ACS thrusters aboard. The main reason for this was that some previous docking failures could have been overcome had the cosmonauts been able to transfer fuel from the ACS system to the main spacecraft engine. Soyuz T2 was preceded by three unmanned tests under the Cosmos label (869, 1001 and 1074) and one Soyuz (T1) in 1979.

Soyuz TM – Transport Modification – was introduced in 1986 and weighed about 7,100 kg (15,653 lb). This was an uprated and heavier Soyuz T spacecraft, incorporating new primary and back-up parachutes, improved power systems and retro-rockets, and the capability to carry 200 kg (441 lb) more payload and return to Earth with 50 kg (110 lb). Soyuz TM was also equipped with a rendezvous and docking system compatible with the Mir Kurs system. Soyuz TM1 was an unmanned test flight to Mir in 1986.

Soyuz TMA – Transport Modification Anthropometric – was introduced in 2002 and was more of a systems and internal upgrade than a structural one, measuring and weighing about the same as the TM. The requirement for a new version of Soyuz was in part due to larger (American) crew members being assigned to Soyuz missions. New seating support structures and modifications to the descent landing engines meant a slightly greater landing mass was possible, allowing regular three-person crews to be flown. In addition, the controls and displays now featured more computer displays and smaller electronics systems. There were no unmanned TMA precursor flights.

An unmanned variant called Progress was introduced in 1978 and has also been upgraded (Progress M, M1). This has been used to re-supply Soviet space stations with fuel, logistics and orbital re-boost capability and is still an integral element in the ISS programme.

Salyut

This highly successful programme began with the first unmanned launch in April 1971 and ended with the de-orbiting of Salyut 7 in February 1991. Two of the stations (Salyut 3 and Salyut 5) were military bases called Almaz. The differences between the "military" (Almaz, or Diamond) and the "civilian" (DOS, Russian for Permanent Orbital Station) stations were in their orbital parameters, durations, the composition of crew members (military officers and engineers for Almaz against civilian flight engineers and guest cosmonauts for Salyut), the openness of reporting of crew activities, and the research programmes. There were setbacks – the first crew to Salyut 1 could not enter the station, the second lost their lives in a re-entry accident, and there were five Soyuz missions which failed to achieve a docking with the station. In addition, there was a launch abort and a pad abort which cancelled Salyut missions. An unmanned Salyut was lost in a launch failure in July 1972, and in April 1973, Salyut (Almaz) 2 was lost shortly after entering orbit. This was followed the next month by the loss of Cosmos 557, which failed even before it received a Salyut identification. Salyut 4, 5, 6 and 7, however, pushed the boundaries of space endurance and demonstrated the wide range of experiments that could be conducted. The missions also demonstrated how much maintenance and trouble-shooting could be conducted. Salyut 6 and 7 also housed a series of visitors from Interkosmos and other countries, flying one-week visiting missions under the command of a veteran Soviet cosmonaut.

Long-duration spaceflight 47

An artists impression of Salyut 1 in orbit, with a Soyuz on docking approach

Salyut 7 in orbit

Table 3.2. Salyut and Almaz stations

Station	EO crew	Variant	Launched	Re-entry
Salyut 1	1	DOS-1	1971 Apr 19	1971 Oct 11
Salyut	–	DOS-2	1972 Jul 29	Failed to reach orbit
Salyut 2	–	Almaz-1	1973 Apr 3	1973 May 28
Cosmos 557	–	DOS-3	1972 May 11	1973 May 22
Salyut 3	1	Almaz-2	1974 Jun 25	1975 Jan 24
Salyut 4	2	DOS-4	1973 Dec 26	1977 Feb 2
Salyut 5	2	Almaz-3	1976 Jun 22	1977 Aug 8
Salyut 6	6	DOS 5-1	1977 Sep 29	1982 Jul 29
Salyut 7	6	DOS 5-2	1982 Apr 19	1991 Feb 7
Almaz	–	Almaz-4	Cancelled	

Featuring dual-docking ports, the later Salyut stations allowed crews to dock a pair of Soyuz craft to the same station, or dock a Progress unmanned re-supply craft to re-stock the resident crew or re-supply the consumables on board, thus prolonging the operational life of the station. The US Skylab programme had no such facilities.

The hybrid Salyut was a cylindrical structure, featuring two habitable compartments (transfer and work), with an internal docking port at the front and a modified Soyuz propulsion system at the rear. Power was supplied by two pairs of Soyuz solar panels. The overall length of the station was 15.8 m, with a maximum diameter of 4.15 m and a 90 m^3 habitable volume. The average mass at launch was 18,900 kg.

The Salyut 3 and 5 (Almaz) stations were different in design. Still cylindrical, the Soyuz docked with a rear port instead of a forward port. There was an airlock chamber for EVAs (although none were ever conducted from either Almaz), a work compartment and a living compartment. Almaz was 14.55 m long, with a maximum diameter of 4.15 m and a similar habitable volume to that of Salyut. The station had larger arrays, however, and incorporated a detachable data capsule that could be ejected at the end of the military-orientated mission.

Salyut 4 was very similar to Salyut 1 but featured three steerable solar arrays, with a larger surface area and more capacity to produce electricity for the increased number of science experiments.

Salyut 6 and 7 resembled the earlier Salyuts but featured docking ports at the front and back, giving the capacity to re-supply and refuel the station using Progress freighters. Two Soyuz vessels could dock at the same time and the first EVAs for Soviet cosmonauts since 1969 were conducted from these stations. These record-breaking vehicles laid the ground work for Mir, testing hardware, techniques and systems for the larger and more capable station that would follow.

Skylab

America's follow-on programme to Apollo was launched in 1973 and housed three crews of three astronauts on long-duration missions of 28, 59 and 84 days. This would

The Skylab crews. Top: Skylab 2; left: Skylab 3; and right: Skylab 4

be the first step towards gaining long-duration space flight experience for the Americans, one that would not be followed up until Shuttle–Mir operations over 20 years later. Skylab 1 was fully fitted laboratory and crew quarters in an S-IVB stage that was launched on a two-stage Saturn V. The cylindrical workshop included two large solar arrays, an airlock for EVAs, a multiple docking adapter with two Apollo CSM docking ports (one for a two-man CSM rescue craft to bring home a stranded three-person Skylab crew if required) and the Apollo (Solar) Telescope Mount, which was a converted Apollo Lunar Module descent stage with four extendable solar arrays. Inside the workshop, the former hydrogen tank was divided into crew quarters and a large experiment area. The working volume of the station was 367.9 m^3. The Skylab Apollo Command and Service Module weighed 13,782 kg (30,389 lb). It was similar to the Apollo CSMs, but was outfitted for extended-duration missions. Modifications included an additional 680 kg (1,499 lb) propellant tank for the RCS system, and three 500-ampere batteries.

Skylab suffered damage during launch and was almost lost before a crew could be launched to it. However, sterling efforts by ground crews and the astronauts restored the station to operational use and it became one of the success stories of the space programme, though it is often forgotten in the shadow of Apollo. Skylab was designed to research the potential for a wide range of experimentation in medicine and industrial applications, such as the manufacture of electronic components and pharmaceuticals, as well as astronomy and solar physics observations and Earth remote sensing. Other missions were planned, including a rendezvous with an early Shuttle mission, but the station could not remain in orbit and re-entered in July 1979. Plans for Skylab B and Skylab C were supported but never fully funded.

Mir

The core module of Mir was launched in February 1986 and the fully assembled multi-module station was de-orbited fifteen years later in March 2001. In between, it housed 28 main crews and a host of visiting crews, including 16 international guest cosmonauts and 7 NASA astronauts on long-duration missions. A total of 41 Russian cosmonauts also visited, lived and worked on the station, as did 37 Shuttle astronauts during 9 docking missions. The station eventually acted as the link between Russia and the USA. The spiralling cost of the original US-led Space Station Freedom programme, and the collapse of the Soviet Union, led to Russia becoming part of the International Space Station programme which, despite national calls for extending the life of the station or launching a second Mir, eventually took over from the historic Mir station in 2000 as the main focus for Russian space efforts.

Mir's base block was 13.13 m long with a maximum diameter of 4.15 m. It featured a forward docking node, with five other ports allowing transport craft to dock at the forward or aft ports of the module. Four separate science modules were later added. Kvant was docked permanently at the rear port and had its own aft facility to allow continued docking at the rear of the complex by Soyuz and Progress craft. The other modules were located around the forward docking node. Extensive EVA work from the node, and then from Kvant 2 enhanced and supported Mir's research programme and capabilities.

FOR ALL MANKIND

Between 1961 and 1977, all manned space flight orbital programmes had been completed by either American (NASA) astronauts (Mercury–Gemini–Apollo–Skylab–ASTP) or Russian cosmonauts (Vostok–Voskhod–Soyuz–Salyut). American plans to fly the Space Shuttle, with reduced stress on launch and landing and with non-pilot crew positions, offered the chance to fly non-professional crew members – scientists or engineers who could operate specialist equipment on the orbital stages of the flight, either in a dedicated research laboratory (Spacelab), on the mid-deck, or on pallets in the payload bay from the aft flight deck. These opportunities were offered to foreign space agencies as well, including those of Australia, Canada, Europe and Japan.

Mir in 1998 showing the four add-on modules, a docked Progress and Kvant at rear, and a Soyuz TM spacecraft in centre. Clockwise from left: Kvant-2 showing unused MMU outside on support frame; Priroda; Spektr (showing damaged arrays); and Kristall with attached Shuttle Docking Module

Table 3.3. Mir and its modules

Module	Length	Max dia.	Hab. volume	Mass (kg)	Launched	Docked
Core	13.13 m	4.15 m	90.0 m^3	20,400	1986 Feb 20	N/A
Kvant 1	5.8 m	4.15 m	40.0 m^3	11,000	1987 Mar 30	1987 Apr 11
Kvant 2	13.73 m	4.35 m	61.3 m^3	19,565	1989 Nov 26	1989 Dec 6
Kristall	13.73 m	4.35 m	60.8 m^3	19,640	1990 May 31	1990 Jun 10
Spektr	14.4 m	4.36 m	62.0 m^3	19,340	1995 May 23	1995 Jun 1
Priroda	12.0 m	4.35 m	66.0 m^3	19,700	1996 Apr 23	1996 Apr 26

The four latter modules were initially docked to the front port before being relocated to the required side docking port using a small robot arm.

The seven NASA astronauts who lived and worked on Mir between 1995 and 1998: l to r – Thagard, Thomas, Blaha, Lucid, Linenger, Foale, Wolf

The Soviet Union, wishing to steal some of the thunder from flying the first mission and payload specialists on the Shuttle, initiated a programme in 1976 to fly "guest cosmonauts" under the Interkosmos programme, starting in 1978. Interkosmos was a cooperative space science programme of the Eastern bloc countries, which was expanded to include flights of minimally-trained "cosmonauts" from these member countries alongside flight-experienced Soviet cosmonauts on short "visiting missions" to space stations. These "international" missions were gradually expanded to include other space-partner nations (France, India) and later with a more commercial eye following the break-up of the Soviet Union. The US Shuttle flight opportunities also expanded to include representative crew members from countries who booked a majority payload, representatives from US governmental departments, or military officers supporting DoD payloads. There were also plans to fly a teacher and a journalist on the Shuttle, in what was hoped to be the beginning of more "routine" access to space for anyone suitably qualified and trained from across the planet.

With the International Space Station, such opportunities have again been promoted "for the benefit of all mankind", although the programme still remains to realise its status as a truly international space complex.

Table 3.4. Shuttle orbiters summary 1977–2006

OV	Name	In service	Missions	Status
101	Enterprise	1977	ALT/Ground	Retired in 1985
102	Columbia	1981–2003	28	Lost during STS-107
099	Challenger	1983–1986	10	Lost during STS 51-L
103	Discovery	1984–	32	Active
104	Atlantis	1985–	27	Active
105	Endeavour	1992–	19	Active

Space Shuttle

The Space Shuttle was to have been the workhorse of the US space programme, flying over fifty times a year, with 26 launches from two pads at KSC and 26 launches from Complex 6 at Vandenberg AFB in California. It would provide a platform for astronomical research, Earth observation, materials processing, medicine and other applications, leading to a US space station. But it soon became obvious that the Shuttle was not going to be able to meet this objective and its actual launch rates were much less. Additional volume was made available for the crew by flying the Spacelab science laboratory (introduced in 1983), and a mid-deck augmentation module called SpaceHab (from 1993), initially offering commercial locker space and, for space station missions, additional logistics storage facilities. The Shuttle programme has been a great success, especially in terms of space repair and the assembly of the International Space Station, but many missions were under-utilised. Two missions ended with the loss of the vehicle and the deaths of 14 crew. The Shuttle will be retired in 2010, or earlier if there is another accident. Conversely, its career may have to be extended if the final assembly of the International Space Station falls behind schedule.

Military manned space flight

The USA planned the military DynaSoar space plane in the early 1960s and the Manned Orbital Laboratory later in the decade. Neither materialised. However, there have been ten classified military Shuttle missions, including the deployment of reconnaissance satellites. The Soviet Union flew two military Almaz space station missions, aboard Salyut 2 and 3 in 1973–74. The Russian Buran space shuttle was slated to conduct a series of military missions, but the programme collapsed after just one unmanned launch.

International manned space flight

In 1972, after several years of negotiations, the first joint mission between the Soviet Union and the USA was agreed, and was launched in 1975 as the Apollo–Soyuz Test Project, or ASTP. In order for the Apollo and Soyuz crews to join together in orbit, a docking module was developed to allow a physical link between the two spacecraft.

Artwork depicting the highlights of the historic ASTP international space mission. Dual launches, a docking in space and the five crew members. l to r Slayton, Brand, Stafford, Kubasov, Leonov with the programme logo in the centre

The docking module was an airlock 3 m (10 ft) long and over 1 m (3 ft) in diameter. It had an Apollo LM-type (drogue) docking port at one end to take the Apollo docking probe, and at the other end was the androgynous docking system. This consisted of an extendable guide ring with three petal-like plates on its circumference, each plate

Table 3.5. ISS major elements dimensions as of 2006

Element	Launched	Vehicle	Length	Max dia.	Hab. volume	Mass (kg)
Zarya	1998 Nov 20	Proton	12.5 m	4.1 m	71.5 m^3	11,600
Unity	1998 Dec 4	STS-88	5.4 m	4.5 m	70 m^3	11,612
Zvezda	2000 Jul 12	Proton	13.10 m	4.15 m	89.0 m^3	19,051
Z1 Truss	2000 Oct 11	STS-92	5.79 m	4.6 m	n/a	8,100
P6 Truss	2000 Nov 30	STS-97	13.7 m	4.6 m	n/a	7,900
Destiny	2001 Feb 7	STS-98	8.5 m	4.26 m	10.75 m^3	14,514
Quest	2001 Jul 12	STS-104	6.0 m	4.0 m	34 m^3	6,064
Pirs	2001 Sep 14	Progress	2.55 m	2.20 m	13 m^3	2,876
S0 Truss	2002 Apr 8	STS-110	13.4 m	4.6 m	n/a	12,247
S1 Truss	2002 Oct 7	STS-112	13.7 m	4.6 m	n/a	14,061
P1 Truss	2002 Nov 23	STS-113	13.7 m	4.6 m	n/a	14,061
P3/4 Truss	2006 Sep 21	STS-115	13.7 m	4.6 m	n/a	15,876

having a capture latch inside it. Once the latches of both craft were engaged, the active vehicle retracted the guide ring, pulling the craft together. The docking module was jettisoned by the Apollo crew prior to retro-fire.

The Soviet Union also flew a series of Interkosmos missions with cosmonauts from several Soviet bloc countries, starting in 1978. This led to a series of international commercial missions and cooperative missions with European and US astronauts, and the creation of the opportunity to sell seats to rather affluent "tourists".

NASA's Space Shuttle Spacelab programme has featured several international missions using a series of modules in the payload bay and these have particularly assisted the crews to gain the experience needed for the International Space Station. Spacelab 1 consisted of a 15,088 kg long module, single pallet and a 1 m (3.3 ft) diameter, 5.8 m (19 ft) long tunnel from the modular Spacelab hardware kit. Over the years, there would be sixteen module flights of Spacelab, six pallet-only missions and a further eleven Shuttle missions featuring Spacelab pallets that supported a variety of payloads, many of them international in nature.

The International Space Station programme evolved from a 1984 US presidential initiative to create a large space station. A cooperative team of 16 nations worked to launch separate elements of the station and construct it in orbit over several years, re-supplied by American Shuttle and Russian Soyuz and Progress vehicles. The Zarya module was a descendant of the add-on Mir modules, whilst Zvezda was similar to the Mir base block.

Shenzhou

This Chinese manned programme aims to establish a manned "space station" in orbit within a few years. Two Shenzhou orbital modules will be linked together to establish a precursor space laboratory, before a more permanent station is launched.

After years of speculation and four unmanned test flights, China achieved what it had planned for decades – the capability to put Chinese citizens in space independently.

Forty years after the Soviet Union and America started the human space flight programme, China became only the third nation to claim manned space flight capability. Such capability was envisaged in the mid-1960s, and in the 1970s, a programme called Dawn (Shuguang) was planned. Even though astronaut trainees were selected in April 1971 (Project 714), the project progressed very little and was terminated in 1972. It remained a guarded secret for over thirty years. A serious space biomedical programme continued in China, as did a series of flights to develop the technology to recover satellite capsules from orbit, and as the launch infrastructure improved, the idea for conducting a manned space programme resurfaced. Rumours persisted throughout the 1980s, even to the point of releasing photos of Air Force system testers undergoing simulations and pressure suit checks. In 1984, the US offered China a flight for a payload specialist on the Shuttle, and even went so far as to select a group of candidates who toured JSC just days before the loss of Challenger in January 1986. After that tragedy, the whole payload specialist programme was suspended indefinitely. An offer to fly a Chinese cosmonaut to Mir was not taken up.

However, in 1992, a new project, called Project 921, was created to support a national Chinese manned programme. Cooperative agreements with the Russians, including visits to the Russian space centres and the cosmonaut training centre, helped the Chinese to advance their own plans and programmes. In 1998, a team of trainees were selected to train for the new flight programme, which began with a series of four unmanned missions, the first of which was placed in orbit in 1999. The first manned orbital flight occurred in October 2003. The Chinese programme is a measured and deliberate one, building on the success of the previous mission and learning from the failures as well as from US and Russian experiences.

Their Shenzhou vehicle can carry up to three crew members. It resembles the Russian Soyuz TM, but differs in several respects. The Chinese vessel is larger, being 2.8 m in diameter and 8.8 m long (Soyuz TM is 7.5 m long). It is also heavier at 7.8 tons (Soyuz TM is 7.2 tons). Shenzhou also has solar panels attached to the Orbital Module in addition to the Instrument Module, with the reported capacity of up to three times the power of the Soyuz arrays. The OM itself is two tons heavier and has the capability of independent orbital flight with its own manoeuvring engines. The Descent Module is also slightly larger than the Soyuz, at 2.5 m diameter by 2 m long (Soyuz is 2.17 m × 1.9 m). Finally, the Instrument Module is about 70 cm longer than the Russian equivalent.

4
The First Decade: 1961–1970

VOSTOK 1

Int. Designation	1961 mu 1
Launched	12 April 1961
Launch Site	Pad 1 (later known as the Gagarinskiy Start or Gagarin Pad), Site 5, Baikonur Cosmodrome, Kazakhstan
Landed	12 April 1961
Landing Site	30 km southwest of Engels, near Smelovako, Saratov
Launch Vehicle	R7 (8K72K serial #E103-16); spacecraft serial number (11F63/3KA) #3
Duration	108 min
Callsign	Kedr (Cedar)
Objective	First manned test of Vostok spacecraft and launch vehicle systems, procedures and profiles

Flight Crew

GAGARIN, Yuri Alekseyevich, 27, Soviet Air Force, pilot

Flight Log

12 April 1961 dawned a fine spring day at the Baikonur Cosmodrome. Yuri Gagarin and his back-up, Gherman Titov, awoke in their tiny log cabin alongside a second cottage where the home of their mentor, the space designer Sergei Korolyov had found it hard to sleep. The two were suited up in the MIK building nearby, and by 08:50 Baikonur time were at Pad 1, exiting from a bus in which they had travelled with many fellow cosmonauts. Traditional hugs and kisses followed at the foot of the rocket, before Gagarin left Titov grounded. Gagarin was inside his capsule (known as the "sharik" – sphere) by 09:10 hours.

Instruments indicated that the hatch had failed to seal properly, delaying the launch by seven minutes. At 11:07 hours, the engines of the Vostok booster sprang to life, building up to full thrust, while from the nearby blockhouse, Korolyov bid Gagarin farewell. The booster clamps were released and the cosmonaut was airborne. *"Poyekhali" ("Off we go")* Gagarin exclaimed as the rocket lifted off. The rocket's ascent was excellently filmed from the launch pad and its shimmering shadow moving across the flat steppe land of Kazakhstan is one memorable shot from the now frequently shown film, which was strictly off-limits in the West until 1967.

By 11:21 hours (09:21 hours Moscow time, 06:21 hours GMT) Gagarin was in orbit, a mere passenger on an automatic flight, in which even the retro-fire would be triggered from the ground. Manual control was possible but not planned for the flight. Indeed, the manual controls were locked on Vostok (East) 1 (the Russians do not officially number the first spacecraft of a new type, although Soyuz TMA1 in 2002 was designated as such) but could have been released in an emergency, provided Gagarin was radioed a three-digit code to give him control. Gagarin sent messages of goodwill

Yuri Gagarin, shortly after his pioneering flight into space on 12 April 1961 in Vostok, the first entry in the manned space flight log book

and the unsurprising news that the Earth was "blue and the sky very, very dark." Vostok, in its 65.07° inclination orbit (the highest inclination of a manned spacecraft to date), with a maximum altitude of 302 km (188 miles), sailed over the Pacific Ocean, South America and the Atlantic.

Over Africa, the retro-engine fired, initiating the re-entry, with the time on the clock at $T + 1$ hour 18 minutes into the flight, Moscow Radio having announced the launch just minutes earlier. The flight module initially did not separate cleanly, which could have resulted in the loss of Gagarin at the moment of his triumph. Fortunately the descent module did finally separate and made a 10-minute re-entry, ending at an altitude of 7 km (4 miles) with drogue parachute deployment. Gagarin ejected and reportedly landed close to bemused farm workers and a cow, 26 km (16 miles) southwest of Engels, near Smelovako, Saratov, at $T + 1$ hour 48 minutes, for the last 20 minutes of which Gagarin had not been aboard Vostok 1. The flight times of the Vostok cosmonauts and their capsules were released by Energiya in 2000. This revealed that the shariks landed about 10 minutes *after* the cosmonauts had parachuted to the ground. The time for Gagarin's capsule was not revealed but this would imply that Gagarin's "flight" landed at 1 hour 38 minutes – 98 minutes after launch. The fact that Gagarin had landed separately was not realised at the time in the West and not confirmed officially until 1978. Gagarin thus became the first of only six people so far – the Vostok pilots – to end a space flight without their spacecraft. Vostok 1 is also the shortest Soviet one-crew flight and the shortest manned orbital flight.

Milestones

1st manned space flight
1st Vostok manned flight

The next two flights after Gagarin were the two American sub-orbital flights by Alan Shepard on 5 May 1961, and Gus Grissom on 21 July 1961. Details of these flights can be found in the Quest for Space chapter.

VOSTOK 2

Int. Designation	1961 tau 1
Launched	6 August 1961
Launch Site	Pad 1, Site 5, Baikonur Cosmodrome, Kazakhstan
Landed	7 August 1961
Landing Site	724 km southeast of Moscow, near to the village of Krasny Kut, close to where Gagarin had landed
Launch Vehicle	R7 (8K72K serial #E103-17); spacecraft serial number (11F63/3KA) #4
Duration	1 day 1 hr 18 min
Callsign	Oryel (Eagle)
Objective	Seventeen-orbit manned mission – 24 hours

Flight Crew

TITOV, Gherman Stepanovich, 25, Soviet Air Force, pilot

Flight Log

Titov was in his flight cabin at 09:30 hours and waited for the planned lift-off at 11:00 hours Baikonur time. Vostok 2 was inserted into a 64.9° inclination orbit, with an apogee of 232 km (144 miles). Soon afterwards, Titov began to feel sick, as weightlessness impaired the otolithic functions of his inner ear. His nausea became quite uncomfortable and meant that several experiments planned for the 24-hour mission could not be operated. The cosmonaut did, however, manage to sleep and found it quite disconcerting to wake with his arms outstretched, almost touching the controls. He later operated those same controls to perform manual changes to the spacecraft's orientation, using the attitude control system thrusters. Despite all this, however, he enjoyed the view through a porthole which magnified the Earth.

Vostok 2's descent module also did not separate cleanly from the retro section but the connections were finally severed to allow a safe entry. Titov became the first cosmonaut officially to land separately from his spacecraft, as Gagarin's planned exit had remained a secret to ensure that the pioneering Vostok flight could enter the international record books. The relatively enormous leap from Gagarin's flight to a 24-hour flight for Vostok 2 was dictated by the need to land in the prime recovery zone, which was overflown only every 16–17 orbits, or 24 hours.

Vostok 2 61

Titov undergoes his pre-flight medical check prior to Vostok 2

Milestones

4th manned space flight
2nd Soviet manned space flight
2nd Vostok manned flight
1st manned space flight to last more than one day
Aged 25 years 10 months 27 days, Titov became the youngest person to fly in space, a record he held for over 45 years

MERCURY ATLAS 6

Int. Designation	1962 gamma 1
Launched	20 February 1962
Launch Site	Pad 14, Cape Canaveral, Florida
Landed	20 February 1962
Landing Site	Northwest of San Juan, Puerto Rico, Atlantic Ocean
Launch Vehicle	Atlas 109D; spacecraft serial number SC-13
Duration	4 hrs 55 min 23 sec
Callsign	Friendship 7
Objective	Three-orbit mission

Flight Crew

GLENN, John Herschel Jr., 40, USMC, pilot

Flight Log

This was one of the most heroic space missions in history. For a start, the Atlas booster was not very reliable, and blew up on the first unmanned Mercury orbital flight attempt in March 1961. The next two unmanned orbital missions, which actually reached orbit, were only partially successful. Mercury Atlas 6 was originally another unmanned test before the decision was made to man it with Glenn.

The mission was originally scheduled for December 1961, before being rescheduled first for 13 January 1962, then 16 January. On 27 January, Glenn lay in the capsule for almost six hours before the launch attempt was scrubbed. The launch was rescheduled for 1 February, then 15 February, and finally 20 February when, after a further 3 hours 44 min in the capsule and six launch holds, Glenn was at last committed to launch. The Atlas thundered away on the pad, reached full thrust and, at 09:27 hours local time, finally became airborne. Glenn, his heart beating at a reasonable 110 beats per minute, announced the start of the mission in the now customary style, confirming that the spacecraft clock had started. Observers watched nervously as the Atlas 109D reached the point at which MA3 had exploded in March 1961.

The vehicle went through Max Q – the point of maximum dynamic pressure on the vehicle – at $T + 100$ seconds and the two outboard engines cut off at $T + 2$ minutes 14 seconds. Glenn was subjected to 7.7 G acceleration during the five minute ascent, which ended at orbital velocity of 28,233 kph (17,544 mph). Orbital inclination was 32.5° and maximum altitude was 265 km (165 miles). The capsule, Friendship 7, turned around and Glenn saw his Atlas tumbling about 30 m (98 ft) away. The view took his breath away as he looked back at the Cape, travelling backwards towards Africa on the first of his three planned orbits. Because his orbital status gave him "go for at least seven orbits", according to flight controllers, this has sometimes been misinterpreted to mean that this was the plan.

The launch of Mercury Atlas 6 and the mission of Friendship 7

Glenn experienced problems with the automatic orientation system and continually had to manually correct a yaw motion. He also saw strange "fireflies" on the outside of the spacecraft, the source of which could not be explained at the time. The mission was proceeding tolerably well, until mission controllers received a signal with the disastrous news that the heat shield on Friendship 7 might be loose. If this was so, then Glenn would be killed during re-entry.

After the three retros had fired as Glenn was completing his third orbit, mid-way between Hawaii and Los Angeles (giving him the impression that he was heading back to the former rather than towards the latter), he was recommended to keep the retro-pack attached through the re-entry, although he was not told why. His heart rate peaked at 132. The change to the flight plan resulted in a more spectacular re-entry than envisaged, as first the straps holding the retro-pack and then the retro-pack itself were burnt away during the 1,650°C peak re-entry temperatures, at a speed of about 24,000 kph (14,912 mph) and an altitude of 40 km (25 miles).

After a nerve-tingling wait during which communications were cut off by incandescent gases surrounding the craft, Glenn's hale and hearty voice was at last heard. The heat shield had not been loose after all. The main chute came out at 3,291 km (10,800 ft) altitude and Friendship 7 descended into the Atlantic Ocean, 9.6 km (6 miles) from the recovery ship, the USS *Noa*, northwest of San Juan, Puerto Rico, but 64 km (40 miles) away from the prime ship, USS *Randolph*. The capsule with the astronaut inside was picked up, and Glenn injured his hand slightly when he blew the hatch. The flight time of 4 hours 55 minutes 23 seconds made this the shortest US manned orbital flight.

Milestones

5th manned space flight
3rd US manned space flight
3rd Mercury manned flight
1st US orbital manned space flight

MERCURY ATLAS 7

Int. Designation	1962 tau 1
Launched	24 May 1962
Launch Site	Pad 14, Cape Canaveral, Florida
Landed	24 May 1962
Landing Site	Northeast of Puerto Rico, Atlantic Ocean
Launch Vehicle	Atlas 107D; spacecraft serial number SC-18
Duration	4 hrs 56 min 5 sec
Callsign	Aurora 7
Objective	Three-orbit mission

Flight Crew

CARPENTER, Malcolm Scott, 37, USN, pilot

Flight Log

Mercury Atlas 7, which was originally to have been the first manned orbital flight, was assigned to astronaut Deke Slayton (who would have used the call sign *Delta 7*), with Wally Schirra as his back-up. On 16 March 1962, it was announced that Slayton would be dropped from flight status because of a "heart flutter", but his replacement was not Schirra but Glenn's back-up, Scott Carpenter. This was mainly because Carpenter had more simulator training experience and MA7 was to be a three-orbit affair, like Glenn's.

The mission – the first to be flown by a replacement astronaut – was also to be the first science flight, with experiments to study visibility from space, the behaviour of liquid in weightlessness, to take hand-held photos of weather patterns and the airglow phenomena, and to deploy a five-colour, 74 cm (29 in) diameter balloon extending on a tether about 30 m (98 ft) from the capsule, which Carpenter named Aurora 7 after his hometown street. In retrospect, the astronaut was given too much to do on the flight and as a result, fell behind the schedule, became a bit flustered and was impatient and careless. He was rather unfairly blamed by some fellow astronauts.

The flight began at 07:45 hours at Cape Canaveral after being delayed three times, each for only 15 minutes, because ground fog would make launch monitoring and photography difficult. Carpenter reached 32.5° orbit, with an apogee of 267 km (166 miles) and a maximum speed of 28,242 kph (17,550 mph) and got to work. His spacesuit overheated, he had to make several manual orientation manoeuvres to correct the automatic control system, and before retro-fire he had not even read his pre-retro-fire checklist. While struggling to do three things at once, he inadvertently hit the roof of Aurora 7, discovering the source of Glenn's "fireflies" as ice particles on the outside. He switched to manual control in preparation for the orientation manoeuvre and forgot to switch off the automatic system, wasting precious fuel.

Scott Carpenter, pilot of Mercury Atlas 7

The spacecraft was at the wrong yaw and pitch angle at retro-fire, which came three seconds late. Carpenter lost his orientation propellant during the re-entry. He gave a continuous, somewhat excitable commentary into a tape recorder during the re-entry and landed 400 km (249 miles) off target, northeast of Puerto Rico at $T + 4$ hours 56 minutes 5 seconds. He was spotted in a life raft 38 minutes after splashdown, but another two and a half hours elapsed before a helicopter from the USS *Pierce* picked him up, giving rise to public worries that he had been lost.

Milestones

6th manned space flight
4th US manned space flight
4th Mercury manned flight
1st flight by a replacement crew

On 17 July 1962, the first X-15 astro-flight to 95.9 km (pilot Robert White, 38, USAF) was achieved using the number 3 aircraft.

VOSTOK 3 AND 4

Int. Designation	1962 alpha upsilon 1 (Vostok 3), alpha nu 1 (Vostok 4)
Launched	11 (Vostok 3) and 12 (Vostok 4) August 1962
Launch Site	Pad 1, Site 5, Baikonur Cosmodrome, Kazakhstan (both vehicles)
Landed	15 August 1962
Landing Site	South of Karaganda, Kazakhstan (Vostok 3), Vostok 4 landed a further 190 km away
Launch Vehicle	R7 (8K72K); spacecraft serial number (11F63/3KA) #5 (Vostok 3); and #6 (Vostok 4)
Duration	3 days 22 hrs 22 min (Vostok 3); 2 days 22 hrs 57 min (Vostok 4)
Callsign	Sokol (Falcon) – Vostok 3; Berkut (Golden Eagle) – Vostok 4
Objective	Simultaneous extended-duration flight of two spacecraft

Flight Crew

NIKOLAYEV, Andrian Grigoryevich, 32, Soviet Air Force, pilot Vostok 3
POPOVICH, Pavel Romanovich, 31, Soviet Air Force, pilot Vostok 4

Flight Log

The dual flight of Vostok 3 and 4 resulted from a desire to demonstrate the ability to control two separate spacecraft in orbit at the same time (crucial to Soviet plans for multi-spacecraft exploration of the Moon and the creation of space stations) and to monitor the condition of two cosmonauts simultaneously during and after relatively long duration flights. This was not seen as the prime objective publicly, however, which was proved by the spectacular and ill-informed coverage of the missions in the western media in expectation of a space docking by the two spacecraft, and which only served to perpetuate the myth of a Soviet lead in space technology.

Vostok 3, with pilot Andrian Nikolayev, was launched at 13:30 hrs Baikonur time on 11 August and was soon in a 64.93° orbit, with an apogee of 227 km (141 miles). The mission was described as a long-duration one by Soviet officials, who sprang a shock in the west at 13:02 hrs the following day by launching Vostok 4 crewed by Pavel Popovich, as Vostok 3 flew overhead. As Vostok 4 entered orbit, it passed to within 6.5 km (4 miles) of Vostok 3. The relatively close encounter was brief, and with no manoeuvring ability it was impossible to achieve a rendezvous in space. The western media, however, lapped it all up. The dual mission of "Nik and Pop", as the cosmonauts were dubbed, was described as a rendezvous in space and the mission as a huge leap forward by the Soviets towards a manned landing on the Moon in a matter of years.

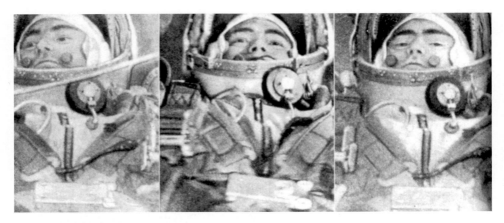

Nikolayev (Vostok 3, top) and Popovich (Vostok 4, bottom) shown inside their respective spacecraft during their historic "group flight", demonstrating the wonders of microgravity.

In their individual orbits – Vostok 4's apogee was 234 km (146 miles), with a 64.98° inclination – Nikolayev and Popovich monitored their health and were allowed to undo their straps to float about freely in the rather spacious cockpit. This was not merely a luxury, but an experiment to see whether the unrestrained movement would bring about inner ear disturbance and cause nausea, which in the case of Nikolayev and Popovich it did not. They ate proper packaged food, such as cutlets, pies and fruit, and Nikolayev was the first cosmonaut to be featured on national TV programmes from his cockpit. The official objectives of the two missions were to maintain radio contact with Earth; carry out regular psychological, physiological and vestibular tests; orientate the spacecraft using attitude control thrusters; make observations using binoculars and the naked eye; float free during the fourth and each second orbit for a period of between 50 to 60 minutes at a time; regulate cabin atmosphere; conduct biological experiments; take food four times a day; and record in a log book and tape recorder their observations and progress of the flight plan.

The missions were eagerly used by Premier Nikita Khrushchev for propaganda purposes, hammering home the Soviet lead over the USA. By the end of the Vostok 3 mission, after 64 orbits, Vostok 4 had drifted 2,720 km (1,690 miles) away. Nikolayev landed south of the town of Karaganda at $T + 3$ days 22 hours 22 minutes on 15 August, and the same day, Popovich landed 190 km (118 miles) away at $T + 2$ days 22 hours 57 minutes. Neither had succumbed to space sickness and this led to the conclusion that the affliction was experienced by only some space travellers and not all who made long journeys. Even longer Vostok missions were then planned.

Milestones

7th and 8th manned space flights
3rd and 4th Soviet manned space flights
3rd and 4th Vostok manned flights
1st joint manned space flight
1st in-flight public TV

MERCURY ATLAS 8

Int. Designation	1962 beta delta 1
Launched	3 October 1962
Launch Site	Pad 14, Cape Canaveral, Florida
Landed	3 October 1962
Landing Site	756.35 km northeast of Midway Island, Pacific Ocean
Launch Vehicle	Atlas 113D; spacecraft serial number SC-16
Duration	9 hrs 13 min 11 sec
Callsign	Sigma 7
Objective	Six-orbit mission

Flight Crew

SCHIRRA, Walter Marty Jr., 39, USN, pilot

Flight Log

Carpenter's science-packed flight plan was being changed until almost launch day, so it was not surprising that he had trouble in orbit. One of his main critics was MA8 pilot Wally Schirra, who was extremely pleased with the smooth running of his own six-orbit mission, which he had made a modest engineering test flight with the minimum of experiments and manoeuvring. Indeed, his flight plan had been cast in stone on 8 August. What was surprising was the conservatism of mission planners, who decided to increase orbital flight experience by just 50 per cent, with a first-time landing in the Pacific Ocean, just short of a full six orbits.

MA8 could have been the first aborted launch, for just ten seconds after leaving the pad at 07:15 hrs local time, Atlas 113D developed an alarming clockwise roll rate just 20 per cent short of an abort. The launch was the first to be shown on British television on the same day, three hours later, thanks to the Telstar communications satellite. It was also watched from the Cape by nine new NASA astronauts, who had been selected the previous month. Schirra had been in his capsule, with the engineering name Sigma 7, since 04:14 hrs, and reached the highest Mercury orbit of 282 km (175 miles) and a speed of 28,256 kph (17,558 mph).

There had been concern that Schirra would be affected by the radiation belt created the previous July by the horrendous US upper-atmosphere nuclear test, Project Dominic. An overheating spacesuit almost forced an early return after just one orbit, but fortunately spacesuits were Schirra's speciality. He deployed the MA7-type tethered multicoloured balloon, this time with success, and spent at least one orbit letting the spacecraft drift as it pleased. The retros were fired at $T + 8$ hours 56 minutes 22 seconds and Sigma 7 splashed down in the Pacific Ocean 756.35 km (470 miles) northeast of Midway Island, and just 7.2 km (4 miles) from its target, close to the recovery ship, USS *Kearsage*. Schirra's mission lasted 9 hours 13 minutes

Mercury Atlas 8 pilot Wally Schirra

11 seconds. He stayed with the ship until winched aboard the *Kearsage*. Media interest in the flight was minimal.

Milestones

9th manned space flight
5th US manned space flight
5th Mercury manned flight

MERCURY ATLAS 9

Int. Designation	1963-015A
Launched	15 May 1963
Launch Site	Pad 14, Cape Canaveral, Florida
Landed	16 May 1963
Landing Site	128 km southeast of Midway Island, Pacific Ocean
Launch Vehicle	Atlas 130D; spacecraft serial number SC-20
Duration	1 day 10 hrs 19 min 49 sec
Callsign	Faith 7
Objective	First US 24-hour space flight

Flight Crew

COOPER, Leroy Gordon Jr., 36, USAF, pilot

Flight Log

Such was the increased confidence in the Mercury spacecraft and manned space flight, that NASA not only planned a flight three times as long as Schirra's, but also increased the duration again in November 1962 to a full 22 orbits. The man in the hot seat, Gordon Cooper, was named the same month, with a May 1963 launch date as the target. Cooper, who affirmed his faith in God, his country and the Mercury team by naming his spacecraft Faith 7, had a packed flight plan, with emphasis on photography. He called the mission, the "flying camera". The camera was fixed on to the tripod on 22 April and was ready to go on 14 May.

Unfortunately, the gantry tower refused to budge because water had seeped into its diesel fuel pump and when the gantry was moved away two hours later, radar data from the Bermuda tracking station was insufficient and the launch was scrubbed. Not so on 15 May, when the relaxed Cooper awoke from a catnap in Faith 7 in time to be launched at 08:04 hrs local time. He reached his 32.5° orbit with an apogee of 267 km (166 miles) and a peak velocity of 28,238 kph (17,547 mph) five minutes later. Cooper remained extremely unruffled and calm throughout the flight, which featured the first in-flight television from a US spacecraft, although the pictures were disappointing.

Cooper's photography from Faith 7, however, was a revelation, confirming to observers his own reports of being able to see the wakes of ships and smoke from a log cabin in the Himalayas with the naked eye. Cooper deployed a small flashing beacon from Faith 7, the first deployment in history, as well as a tethered balloon like Schirra's. The flight went swimmingly, with Cooper becoming the first American to sleep in space, but during the nineteenth orbit, the astronaut noticed the one-G light coming on, which apparently detected the onset of gravity.

Tracing the cause, the astronaut discovered that the attitude and stabilisation control system a/c converter had failed. The astronaut would have to perform an

Gordon Cooper heads for orbit aboard Mercury Atlas 9

entirely manual re-entry, which he did perfectly, splashing down just 7 km (4 miles) from the USS *Kearsage*, 128 km (80 miles) southeast of Midway Island in the Pacific Ocean, at $T + 1$ day 10 hours 19 minutes 49 seconds, the longest launch-to-landing solo US manned space flight in history. A planned three-day mission (Mercury–Atlas 10/Freedom II, flown by Alan Shepard) was mooted but scrapped, and the Mercury programme ended officially on 12 June 1963

Milestones

10th manned space flight
6th US manned space flight
6th and final Mercury manned flight
1st satellite deployment from manned spacecraft

VOSTOK 5 AND 6

Int. Designation	1963-020A (Vostok 5); 1963-023A (Vostok 6)
Launched	14 (Vostok 5) and 16 (Vostok 6) June 1963
Launch Site	Pad 1, Site 5, Baikonur Cosmodrome, Kazakhstan (both launches)
Landed	19 June 1963
Landing Site	619.4 km northeast of Karaganda, Kazakhstan (Vostok 6); 539 km northwest of Karaganda, Kazakhstan (Vostok 5)
Launch Vehicle	R7 (8K72K); spacecraft serial number (11F63/3KA) #7 (Vostok 5); and #8 (Vostok 6)
Duration	4 days 23 hrs 6 min (Vostok 5); 2 days 22 hrs 50 min (Vostok 6)
Callsign	Yastreb (Hawk) – Vostok 5; Chaika (Seagull) – Vostok 6
Objective	Second group flight; five-day solo flight and first female space flight

Flight Crew

BYKOVSKY, Valeri Fyodorovich, 28, Soviet Air Force, pilot Vostok 5
TERESHKOVA, Valentina Vladimirovna, 26, Soviet Air Force, pilot Vostok 6

Flight Log

The much-rumoured launch of Vostok 5 was delayed by bad weather on 13 June but the following day, at 17:00 hrs local time at Baikonur, launch pad 1 reverberated to the sound of another SL-3 ignition as cosmonaut Valeri Bykovsky began what was planned as a long-duration mission. At 4 days 23 hours 6 minutes, it actually became (and still remains) the longest manned solo space flight in history. Vostok 5 entered a 65° inclination orbit with an apogee of 209 km (130 miles) as rumours persisted that another Vostok would be launched the following day. It was to be a Vostok that would overshadow Bykovsky's feat.

Vostok 6 carried the first woman (and tenth human) to venture into space. Valentina Tereshkova was launched at 14:30 hrs Baikonur time. Reflecting the frenetic space activity of the 1960s, in between the Vostok 5 and 6 launches, the USA had performed six satellite launches, all from California. Vostok 6 entered a 65° inclination orbit with a peak altitude of 218 km (135 miles) and almost immediately came to within 5 km (3 miles) of Vostok 5 for a brief encounter, which according to the western press went much further, with such headlines as "Valya chases her space date".

As Tereshkova was not a pilot, it was perhaps inevitable that she reportedly had difficulties in adapting to weightlessness, but the rumours of her being so ill that she pleaded to come home seem far-fetched, as it appears that the flight, originally

Tereshkova and Korolyov in discussion prior to her historic space flight

planned to be a 24-hour affair, was in fact extended. The launch of a woman into space was undoubtedly a major propaganda coup for Premier Khrushchev, who may have ordered such a mission, a theory supported by the fact that the next woman to fly into space was not launched until 1982.

Tereshkova was the first of the space pair to land, 624 km (388 miles) northeast of Karaganda at mission elapsed time of 2 days 22 hours 50 minutes. As the landing was a nominal one, she is thus the only woman to end a space flight outside her spacecraft, as well as the only one to make a solo female space flight. Bykovsky was the third Vostok pilot to experience a partial separation of the descent module but the separation occurred prior to the worst part of the re-entry profile and he returned to Earth about 540 km (336 miles) northwest of Karaganda. Plans were set in motion for a Vostok 7 mission lasting a week, by "non-cosmonaut" doctor Boris Yegorov, in the summer of 1964 but, like the US Mercury programme, the Soviet Vostok project ended after six flights. However, the next series of spacecraft (Soyuz, or "Union") would not be ready for some time and so in order to appear ahead in the space race with the Americans, the Vostok was converted into what seemed to outsiders to be a radically new and improved spacecraft – Voskhod.

Milestones

11th and 12th manned space flights
5th and 6th Soviet manned space flights

> 5th and 6th Vostok manned flights
> 1st space flight with female crew (Vostok 6)
> 1st joint male–female space flight
> Bykovsky has held the solo space flight record for over 43 years

On 27 June 1963, Robert Rushworth, 39, of the USAF flew X-15-3 on the third astro-flight, to 88 km. Less than a month later, on 19 July 1963, Joe Walker, 42, flew the same vehicle on the fourth astro-flight, this time to 105 km. Finally this year, on 22 August 1963, Walker flew X-15-3 to 107 km in the fifth astro-flight, the highest altitude any X-15 would attain.

VOSKHOD 1

Int. Designation   1964-065A
Launched   12 October 1964
Launch Site   Pad 1, Site 5, Baikonur Cosmodrome, Kazakhstan
Landed   13 October 1964
Landing Site   3,105 km northeast of Kustanai
Launch Vehicle   R7 (11A57); spacecraft serial number (11F63/3KV) #3
Duration   1 day 17 min 3 sec
Callsign   Rubin (Ruby)
Objective   Multi-crew flight

Flight Crew

KOMAROV, Vladimir Mikhailovich, 37, Soviet Air Force, commander
FEOKTISTOV, Konstantin Petrovich, 38, civilian, flight engineer
YEGOROV, Boris Borisovich, 27, civilian, doctor

Flight Log

Voskhod ("Sunrise") 1 provided the classic illustration of how the secret Soviet space programme completely misled the west. After the Vostok missions, Sergei Korolyov, the then anonymous space designer, considered improvements to the basic spacecraft to allow longer missions by more than one passenger. These studies led to the design of a new spacecraft, Soyuz, which would perform Earth orbital and lunar looping missions and support a possible lunar landing programme. Delays to Soyuz meant that there would be a hiatus in the manned space programme, to which Premier Khrushchev reacted in customary fashion, demanding a multi-crewed space flight before the United States launched its two-man Gemini spacecraft in early 1965.

As Soyuz could not be accelerated, Korolyov responded with a version of the uprated Vostok. But in order to launch three men rather than two, as the Americans were planning, practically all the "stuffing" had to be taken out of Vostok and the crew would have to fly without spacesuits and without any means of emergency escape. Voskhod would, however, carry a back up retro-rocket. Despite the imperfections of Voskhod, seven cosmonauts seemed happy to be assigned to train for the most risky manned space flight in history. The three to be chosen were a commander, Vladimir Komarov, a scientist, Konstantin Feoktistov – who, it turned out, was the man who helped design Vostok to fly with three passengers – and a doctor, Boris Yegorov.

They arrived at the launch pad wearing cotton overalls and leather flying helmets, about to board the first SL-4 booster to fly a manned crew a few days after the one and only "test flight" of the "new" Voskhod, as Cosmos 47. Launch came at 12:30 hrs Baikonur time and soon after the spacecraft had reached its 65°, 409 km (254 miles)

The first "space crew" walk down the red carpet to report on the success of their mission to welcoming officials in Moscow. L to r Feoktistov, Komarov, Yegorov

maximum altitude orbit, the western media went wild, reporting that Russia had launched a "mammoth" new spaceship in which the scientist and doctor would perform experiments while the commander controlled the mission. In truth, the conditions were so cramped inside the Voskhod that it must have been hard to eat and go to the toilet, let alone perform experiments, although Yegorov apparently performed some basic medical checks.

Khrushchev had the propaganda success he wanted, but as he was congratulating the crew by telephone, the receiver was taken from his hands. The Brezhnev–Kosygin takeover had begun and it was they who greeted the fortunate cosmonauts after they had landed safely. The crew remained in the spherical capsule as small retro-rockets fired just before touchdown to cushion the impact, some 310 km (193 miles) northeast of Kustanai. The mission lasted just 1 day 17 minutes 3 seconds, the shortest three-crew flight in history. The three-man crew apparently requested an extension but were refused by Korolyov, who quoted Shakespeare: "There are more things in heaven and Earth, Horatio ..." At the time, no one knew the name of the chief designer, who had one more spacecraft to design before he succumbed to ill health in January 1966.

Milestones

13th manned space flight
7th Soviet manned space flight
7th Vostok manned flight
1st Voskhod manned variant flight
1st three-person crew
1st flight by crew without spacesuits
1st flight with no launch escape or ejection system
1st Soviet space flight to end with crew inside spacecraft
1st space flight with non-pilot, civilian crew

VOSKHOD 2

Int. Designation	1965-022A
Launched	18 March 1965
Launch Site	Baikonur Cosmodrome, Kazakhstan
Landed	19 March 1965
Landing Site	180 km northeast of Perm, Siberia
Launch Vehicle	R7 (11A57); spacecraft serial number (11F63/3KD) #4
Duration	1 day 2 hrs 2 min 17 sec
Callsign	Almaz (Diamond)
Objective	First Extra-Vehicular Activity (EVA – spacewalk) demonstration

Flight Crew

BELYAYEV, Pavel Ivanovich, 39, Soviet Air Force, commander
LEONOV, Alexei Arkhipovich, 30, Soviet Air Force, pilot

Flight Log

The second of a planned series of multi-crew Voskhod missions got underway at 12:00 Baikonur time, entering a 65° inclination orbit with the highest manned apogee to date of 495 km (308 miles). Instead of three crewmen without spacesuits, there were two, this time suitably attired. This Voskhod had been reconfigured to carry a telescopic airlock leading from the crew cabin. The space where the third seat had been was left free to give one of the crewmen, Alexei Leonov, the room in which to don an emergency oxygen backpack and connect an umbilical air and communications tether to his spacesuit, before crawling through the airlock after the spacecraft had been depressurised.

The first walk into open space began at the start of the second orbit of Voskhod, as Leonov emerged from the airlock at the end of his 5 m (16 ft) tether, watched by two television and film cameras attached to the end of the airlock and on top of the back-up retro-rocket. Leonov cavorted in space doing cartwheels, not for show but because he was essentially out of control as his umbilical snaked around. His official free spacewalk lasted 12 minutes 9 seconds, but he was in the vacuum environment for about 20 minutes, since he couldn't get back into the airlock. His spacesuit had ballooned more than anticipated and he had to squeeze himself back into the airlock quite forcibly before closing the hatch and re-pressurising the spacecraft.

Unfortunately, he forgot to retrieve the film camera which would have shown clear photos of the EVA rather than the blurred and fuzzy television reproductions. Nonetheless, Leonov's exploits had a dramatic effect on the watching world, capturing more headlines than Gagarin himself, and this mission was one of the highlights of the "Space Age" of the 1960s. The rest of the flight went quietly until they

The first "walk" in space

attempted retro-fire at the end of the seventeenth orbit. The prime retro-rocket in the instrument module of Voskhod failed to fire because of a sensor attitude control malfunction.

The cosmonauts made one more orbit as, not without a certain amount of drama, preparations were made to fire the back-up retro-pack on the next pass. The instrument section was apparently jettisoned (again not cleanly) before this, and the Voskhod spherical flight cabin was manually orientated by Belyayev, who punched the retro-pack arming device. The re-entry was quite dramatic and the capsule naturally missed the main recovery area by 960 km (597 miles), landing in the thick, snow-covered forest near the city of Perm. A damaged telemetry antenna made it impossible for the rescue teams to locate the craft, so the cosmonauts put their emergency landing training into effect, lighting a fire and waiting for rescue. However, ravenous wolves compelled their return to the capsule and it was two and a half hours before a helicopter spotted the capsule, thanks to the parachutes which were splayed out across the tree tops. Ground vehicles rescued the crew after they had spent a night in the forest.

Observers in the west, expecting a landing announcement to be made at the end of the seventeenth orbit, suspected that something was wrong and were only told of the touchdown when the crew had been located, four hours later. The drama of the landing events was only fully revealed a year later, rather perversely, after the emergency US landing of Gemini 8. Flight time was 1 day 2 hours 2 minutes 17 seconds. This proved to be the last Voskhod manned mission. There had originally been plans for a series of at least seven manned Voskhod flights. Voskhod 3 was to have been a two-man 15–20-day extended scientific mission, and then Voskhod 4 would have flown a 15-day biomedical mission with a cosmonaut doctor in the crew. Voskhod 5 would be an all female crew with an EVA, Voskhod 6 was a 14-day EVA mission featuring the use of a small manoeuvring unit, and Voskhod 7 would attempt tether dynamics with the spent upper stage before flying a 10–15 day mission. There was also a plan to include a professional Soviet journalist on board a Voskhod but all these flights were cancelled.

Milestones

14th manned space flight
8th Soviet manned space flight
8th Vostok manned flight
2nd Voskhod manned variant flight
1st manned space flight with two crew
1st space flight with EVA operations
1st extended mission

GEMINI 3

Int. Designation	1965-024A
Launched	23 March 1965
Launch Site	Pad 19, Cape Kennedy, Florida
Landed	23 March 1965
Landing Site	Western Atlantic Ocean
Launch Vehicle	Titan II Gemini Launch Vehicle-3 (GLV-3); spacecraft serial number 3
Duration	4 hrs 52 min 51 sec
Callsign	Molly Brown/Gemini Three
Objective	Three-orbit manned test flight; test of orbital manoeuvring system

Flight Crew

GRISSOM, Virgil Ivan "Gus", 39, USAF, commander, 2nd mission
Previous mission: Mercury–Redstone 4 (1961)
YOUNG, John Watts, 35, USN, pilot

Flight Log

Project Gemini was born as the logical follow-on to the Mercury programme, but its *raison d'être* was changed by President Kennedy's pledge to land a man on the Moon before 1970. Gemini was to act as the testing ground for all the manoeuvres and operations to be performed on an Apollo mission, but in Earth orbit – orbital manoeuvring, rendezvous, docking, extended flights and spacewalking. The task of Gemini 3 was straightforward: with the aid of the first computer on a manned spacecraft, Gemini 3 would change its orbit.

The crew was chosen at the time of the Gemini 1 unmanned test flight and was well into training by the time unmanned Gemini 2 had become the first to be recovered. Command pilot Gus Grissom and his pilot John Young, the Taciturn Twins as they were dubbed, were overshadowed by the exploits of Voskhod 2 five days earlier, particularly as Gemini 3 was to make only a modest three orbits. The crew were lying in their ejector seats inside Gemini 3 at 07:30 hrs, waiting for a 09:00 hrs launch. At $T - 35$ minutes, the Titan II first-stage oxidiser line sprang a leak and a handy wrench was required, delaying the launch by 24 minutes. The hypergolic engines of the Titan gave out a high-pitched whine and sprang to life and the mission lifted off.

As the second stage ignited while still attached to the first stage, its exhaust spewing out of the lattice framework between the two, the rocket was surrounded by a bright aurora, disconcerting the pilot. After reaching the 32.5°, 224 km (139 miles) peak apogee orbit, the crew was immediately assigned to a series of science

Young (left) and Grissom aboard Gemini 3

experiments, including a sea urchin cell growth experiment, which failed because Grissom was rather heavy-handed with it.

At last, the frustrated astronauts had some real space flying to do as, at $T + 1$ hour 30 minutes, Grissom performed the first orbital manoeuvring system burn, for 75 seconds. Two more burns followed, with the last placing Gemini 3's perigee at 72 km (45 miles), low enough to ensure re-entry even if the retros failed to fire, which they didn't. Grissom tried to use Gemini's lift capability to reduce a predicted landing miss but the capability of the spacecraft was less than anticipated and resulted in a 111 km (69 miles) miss. As Gemini assumed splashdown position, it literally yanked from vertical to an almost horizontal position, pitching the crew forward and smashing Grissom's faceplate against the instrument panel.

The flight ended near Grand Turk Island in the Atlantic Ocean at $T + 4$ hours 52 minutes 51 seconds, the shortest US two-crew mission. The landing miss meant a long wait in heaving seas but Grissom – remembering Liberty Bell – elected to stay on board with the hatch well and truly closed. Grissom lost his pre-launch breakfast and both doffed their spacesuits in the heat. They later walked rather ignominiously along the deck of the carrier *Intrepid*, to which they had been helicoptered, in sporting underwear beneath bathrobes. After the Liberty Bell 7 incident, Grissom named his next space craft "Molly Brown" after the hit Broadway show "The Unsinkable Molly

Brown". NASA was not happy about this and asked him to change the name, but when he indicated that his second choice was "Titanic" they relented. "Molly Brown" became the last named American spacecraft until Apollo 9 in March 1969.

Gemini 3 became known as the "corned beef sandwich flight", when afterwards it was revealed that Young had been reprimanded for carrying food aboard and offering it to Grissom who, on taking a hefty bite, spread crumbs around the cabin. The prank was, not surprisingly, hatched by the back-up command pilot, Wally Schirra, who put the sandwich into Young's spacesuit, but the joke got out of hand and became the subject of a Congressional inquiry. The cost of the sandwich, which Schirra had bought in Cocoa Beach, escalated, and it became known as the "$30 million sandwich".

Milestones

15th manned space flight
7th US manned space flight
1st Gemini manned flight
1st manned space flight to perform orbital manoeuvres
1st US two-man crew mission
1st flight by crewman on second mission

GEMINI 4

Int. Designation	1965-043A
Launched	3 June 1965
Launch Site	Pad 19, Cape Kennedy, Florida
Landed	7 June 1965
Landing Site	Atlantic Ocean
Launch Vehicle	Titan II GLV #4; spacecraft serial number 4
Duration	4 days 1 hr 56 min 12 sec
Callsign	Gemini Four
Objective	Four-day extended-duration mission; first US EVA excursion

Flight Crew

McDIVITT, James Alton, 35, USAF, command pilot
WHITE II, Edward Higgins, 34, USAF, pilot

Flight Log

When the flight plan for Gemini 4 was initially worked out, station-keeping with the Titan second stage and spacewalking were not on the agenda. Indeed, doctors were doubtful that the mission should last four days and recommended a two-day mission. The astronauts supported an EVA, but initially this was only a stand-up EVA in the hatch. After Leonov's exploits, they got what they wanted, but with just nine days to spare – for confirmation of the planned spacewalk was only made on 25 May 1965. Station-keeping with the second stage of the booster was the idea of Gus Grissom and Gordon Cooper, who had light-heartedly suggested such a manoeuvre during space-to-ground communications during Gemini 3.

A misbehaving gantry tower got stuck and spoiled the launch day slightly, delaying the ascent of the rookie astronauts James McDivitt and Edward White by 1 hour 16 minutes. The launch was shown live on television in Britain and the rest of Europe via the Early Bird communications satellite, at 11:15 hrs Cape time and 10:15 hrs Houston time, where the new Manned Space Flight Center and flight control room was situated, ready to take command of its first mission. Gemini 4 entered a 32° inclination orbit with a peak apogee of 296 km (184 miles). McDivitt's station-keeping with the second stage of the Titan was not altogether a success, with 42 per cent of the Orbital Attitude Manoeuvring System (OAMS) propellant being consumed. The experiment was called off and the EVA delayed for an extra orbit.

On orbit No. 3, Edward White exited Gemini 4 for a 21-minute adventure that featured some of the finest space photography, courtesy of McDivitt. White's 7.62 m (25 ft) long tether provided oxygen and he had a ventilator control module on his chest to provide nine minutes worth of emergency oxygen, if required. The excited and

Ed White takes a stroll during Gemini 4

enthusiastic White controlled his movements using an oxygen-powered hand-held manoeuvring unit, and had to be ordered back into the capsule because night was approaching. The hatch was closed 36 minutes after it had been opened, but only after some strenuous pulling by the two crewmen.

The rest of the flight, lasting a US record 62 orbits, passed quietly as the crew performed 11 scientific experiments and took a fine photo of Cape Kennedy from the cramped confines of the spacecraft. The onboard computer failed towards the end of the flight and McDivitt performed a two-phase manual re-entry, first lowering the orbit to 76 by 158 km (47 by 98 miles) before firing the retros to initiate an 8-G re-entry. Splashdown at $T + 4$ days 1 hour 56 minutes 12 seconds was 81 km (50 miles) off target, about 625 km (388 miles) east of Cape Kennedy. The jubilant crew, having almost caught up with the Russians, were recovered by a helicopter from USS *Wasp*.

Milestones

16th manned space flight
8th US manned space flight
2nd Gemini manned flight
1st US and second flight with EVA operations
1st US manned launch seen live in Europe

On 29 June 1965, USAF pilot Joseph Engle, 32, flew the sixth X-15 astro-flight in the number 3 aircraft to an altitude of 85 km. Six weeks later, on 10 August 1965, he was again at the controls of X-15-3 on the seventh astro-flight, this time to 83 km.

GEMINI 5

Int. Designation	1965-068A
Launched	21 August 1965
Launch Site	Pad 19, Cape Kennedy, Florida
Landed	29 August 1965
Landing Site	Western Atlantic
Launch Vehicle	Titan II GLV No 5; spacecraft serial number 5
Duration	7 days 22 hrs 55 min 14 sec
Callsign	Gemini Five
Objective	Eight-day extended-duration mission

Flight Crew

COOPER, Leroy Gordon, 38, USAF, command pilot, 2nd mission
Previous mission: Mercury–Atlas 9 (1963)
CONRAD, Charles "Pete" Jr., 35, USN, pilot

Flight Log

Gemini 5 was America's bid to exceed the Soviet five-day space endurance record. Indeed, such was their determination that the first official astronauts' flight badge featured an image of a covered wagon of the "Old West" whose slogan was "California or bust". The Gemini 5 crew emblem carried the motto "Eight Days or Bust". When this was proposed, the crew were told to cover the slogan, in case they should "bust" before the eight days were reached. Cooper's connection with the Mercury programme was perpetuated when, after a countdown rehearsal on Pad 19, the crew had to be rescued by the "cherry picker" crane used at the Mercury–Redstone Pad 5, after the main gantry failed to erect itself.

The launch was delayed on 19 August by threatening storms and was recycled by 48 hours. At 09:00 hrs local time, Gemini 5 thundered into the skies right on time, entering a record US altitude of 303 km (188 miles) in its 32.6° inclination orbit. A 5 m (16 ft) segment of the Titan first stage was recovered in the Atlantic, marking another US space first. The major plan for Gemini 5 was to eject a 34.4 kg (76 lb) radar evaluation pod from the rear adapter section and for the astronauts to back away 84 km (52 miles), then rendezvous with it. These plans were almost immediately thwarted when the fuel cell oxygen pressure decreased from 800 psi to 120 psi. Spacecraft power had to be conserved drastically and plans were made to bring the crew home after just three orbits.

The pressure finally dropped to 60 psi but mission planners decided to keep the crew aloft for a lazy, boring drifting flight. This seemed interminable to the crew, who in their months of training had covered almost every topic imaginable and didn't therefore talk to each other – or the ground – much. Surprisingly, mission control

Conrad (left) and Cooper smile broadly upon their successful recovery after 8 days in space

planned a five-orbit-change "phantom rendezvous" as a practice, which took them even higher to 349 km (217 miles) in the 32.6° orbit. The crew were able to perform 17 science experiments, one of which was to evaluate their ability to see things on the ground, and although they did not see a special "chessboard" target, they did see a rocket launched from Vandenberg AFB.

They also saw the wake of their prime recovery ship USS *Lake Champlain* on which they would later beam proudly after a flight of 7 days 22 hours 55 minutes 14 seconds, shortened by one orbit because of fears of a hurricane in the splashdown zone. Gemini 5 missed its target by 170 km (106 miles), but it did beat the Soviet endurance record. More importantly, the Americans had flown a mission lasting as long as it would take to fly to the Moon and back. The crew reportedly ripped off the patch covering their emblem slogan, having surpassed their objective.

Milestones

17th manned space flight
9th US manned space flight

> 3rd Gemini manned flight
> 1st US on-time lift-off
> 1st flight to be curtailed
> 1st manned spacecraft to be powered by fuel cells
> 1st flight to feature a personal crew emblem

On 28 September 1965, NASA civilian test pilot John McKay, 42, flew the X-15 number 3 aircraft on the eighth astro-flight, to 90 km. The next astro-flight occurred on 14 October 1965, when USAF pilot Joe Engle, 33, flew the X-15 number 1 aircraft on its first such flight. The programme's ninth astro-flight attained an altitude of almost 81 km.

GEMINI 7 AND 6A

Int. Designation	1965-100A (Gemini 7); 1965-104A (Gemini 6A)
Launched	4 and 15 December 1965
Launch Site	Pad 19, Cape Kennedy, Florida
Landed	16 December 1965
Landing Site	Both spacecraft splashed down in the western Atlantic
Launch Vehicle	Titan II GLV No. 7 (Gemini 7) and GLV No. 6 (Gemini 6A); spacecraft serial number 7 (Gemini 7) and 6 (Gemini 6A)
Duration	13 days 18 hrs 35 min 1 sec (Gemini 7); 1 day 1 hr 51 min 54 sec (Gemini 6A)
Callsign	Gemini Seven; Gemini Six
Objective	Fourteen-day extended-duration mission (Gemini 7); first space rendezvous (Gemini 6A with Gemini 7)

Flight Crew

BORMAN, Frank, 37, USAF, command pilot Gemini 7
LOVELL, James Arthur Jr., 37, USN, pilot Gemini 7
SCHIRRA, Walter Marty Jr., 42, USN, command pilot Gemini 6A, 2nd mission
Previous mission: Mercury–Atlas 8 (1962)
STAFFORD, Thomas Patten Jr., 35, USAF, pilot Gemini 6A

Flight Log

NASA continued its pre-Apollo rehearsals with plans for Gemini 6 to perform the first docking in space and then for Gemini 7 to keep two men "in the can" for 14 days. The first objective was to be met on 25 October 1965 when an Atlas Agena was to place the Agena second stage (housing a docking port and a rendezvous radar antenna) in orbit as a target for Gemini 6, which would be launched 90 minutes later. With astronauts Wally Schirra and Tom Stafford already in Gemini 6 at Pad 19, the Atlas thundered away from Pad 14, but the Agena exploded and the frustrated astronauts were grounded. NASA hatched a plan to overcome the setback. They would launch Gemini 7 first, on Gemini 6's original Titan, then launch Gemini 6 to rendezvous with Gemini 7. The plan was announced by President Johnson himself, the space supporter who had persuaded President Kennedy to shoot for the Moon.

So first Gemini 7 – with crewmen Frank Borman and James Lovell looking like aliens in their lightweight, 8 kg (18 lb) spacesuits, with strange hoods rather than helmets – took off at 14:30 hrs on Saturday 4 December, sharing US television screens with a football match. The astronauts entered a 28.9° inclination orbit with a maximum altitude of 327 km (203 miles) and sat it out in the tight confines, waiting

Gemini 6 photographed from Gemini 7

for Gemini 6 to be launched on 12 December. Lovell was allowed to take off his spacesuit, while Borman had to endure the flight with electrodes fixed to his head and suffer the indignity of bursting his urine bag after filling it, rather than before. Contrary to the media coverage, pioneering space flight was an endurance, not a picnic.

The Gemini 6 astronauts had a new experience to endure on 12 December, when at 09:54 hrs local time their Titan II ignited, only to shut down 1.2 seconds later when a dust cap left in a gas generator caused imperfect combustion. This was spotted by the malfunction detection system. Although the spacecraft clock had started, Schirra knew instinctively that he had not lifted off. He elected not to pull the ejection lever, which would have subjected him and Stafford to a 20-G ride, killing the rendezvous mission and probably crippling them. Stafford had been to the launch pad twice and had not lifted off. However, at 08:37 hrs on 15 December, he finally did so, and the space chase was on. Gemini 6 entered an 8.9° orbit which would reach a maximum apogee of 311 km (193 miles).

Lovell (left) and Borman look tired but happy after their 14-day marathon flight

Seven very carefully planned and controlled manoeuvres brought Gemini 6 to within 15 cm (6 in) of Gemini 7. Officially, the rendezvous had been achieved at 14:33 hrs Cape time. It was the greatest feat in manned space flight so far, and the media coverage epitomised the excitement of the 1960s space race. Five hours 18 minutes and a lot of good natured bantering (and a seasonal "Jingle Bells" from Schirra and Stafford) later, Gemini 6 backed away and made a landing at $T + 1$ day 1 hour 51 minutes 54 seconds, just 11.2 km (7 miles) from USS *Wasp*. Borman and Lovell continued their endurance flight and the operation of 18 science experiments, finally landing 10.4 km (6 miles) from the USS *Wasp* at $T + 13$ days 18 hours 35 minutes 1 second. This is the longest US two-crew space flight. They were light-headed and stooping as they walked across the carrier deck, but had proved beyond a doubt that man had a place in space.

Milestones

18th and 19th manned space flights
10th and 11th US manned space flights
4th and 5th Gemini manned flights
1st space rendezvous
1st flight cancellation (Gemini 6)
1st launch pad abort (Gemini 6A)
1st four-man joint mission

GEMINI 8

Int. Designation	1966-020A
Launched	16 March 1966
Launch Site	Pad 19, Cape Kennedy, Florida
Landed	16 March 1966
Landing Site	Western Pacific Ocean
Launch Vehicle	Titan II GLV No. 8; spacecraft serial number 8
Duration	10 hrs 41 min 26 sec
Callsign	Gemini Eight
Objective	Docking with Agena unmanned target vehicle

Flight Crew

ARMSTRONG, Neil Alden, 35, civilian, command pilot
SCOTT, David Randolph, 34, USAF, pilot

Flight Log

The first space docking was on the agenda for Gemini 8, and Scott was to make a two-hour spacewalk "around the world" at the end of a 28 m (92 ft) tether and attached to a 42 kg (93 lb) Extravehicular Support Package. The crew had been inside Gemini 8 for 14 minutes when the Agena target thundered away from Pad 14. Their own launch came at 10:41 hrs local time, although the Titan II seemed a bit sluggish to start with. Perfect orbit was achieved, with a 28.9° inclination and an apogee–perigee of 292–160 km (181–99 miles). The Agena was 1,963 km (1,220 miles) away and the space chase began. It ended with a "real smoothie" of a docking, as Armstrong described it, at $T + 6$ hours 32 minutes and at a speed of about 8 cm (3 in) per second.

The matter-of-fact docking complete, the first US space emergency then began in a rather insidious manner. First, the two spacecraft rolled 30° out of position and the crew thought that the Agena, which was causing some concern on the ground anyway, was at fault. They disengaged its control system and brought the two craft under control using Gemini's thrusters. Suddenly, a faster roll developed and the crew decided to separate from the Agena barely 27 minutes after docking, backing away as they did with a short burst of the thrusters. Then things got pretty violent. Gemini went into a 70 rpm roll and yaw combined, and the crew came close to their physiological limits. Thruster 8 had short-circuited and was firing intermittently, the crew discovered later. There was only one thing to do, which was to cut off the OAMS thrusters and fire the re-entry control system.

Mission rules dictated an emergency return to Earth and Gemini 8 splashed down about 800 km (497 miles) east of Okinawa at $T + 10$ hours 41 minutes 26 seconds, glad to have made water and not a remote jungle. After an uncomfortable three-hour wait, the crew was met by the USS *Leonard F. Mason* and they climbed aboard from the

Gemini 8 approaches the Agena docking target

rolling sea up a Jacob's ladder. Both astronauts would have another ladder later in their careers, this time to step down, as Gemini 8 was the only flight whose crew members both subsequently walked on the Moon.

Milestones

20th manned space flight
12th US manned space flight
6th Gemini manned flight
1st space docking
1st emergency return to Earth

GEMINI 9

Int. Designation	1966-047A
Launched	3 June 1966
Launch Site	Pad 19, Cape Kennedy, Florida
Landed	6 June 1966
Landing Site	Western Atlantic Ocean
Launch Vehicle	Titan II GLV No. 9; spacecraft serial number 9
Duration	3 days 20 min 50 sec
Callsign	Gemini Nine
Objective	Rendezvous and docking mission; EVA activities

Flight Crew

STAFFORD, Thomas Patten Jr., 35, USAF, command pilot, 2nd mission
Previous mission: Gemini 6 (1965)
CERNAN, Eugene Andrew, 32, USN, pilot

Flight Log

Gemini 9 certainly seemed to be a jinxed 13th US manned space flight, even before it got airborne. The prime Gemini 9 crew were killed in an air crash in St Louis on 28 February 1966, when their T-38 aircraft hit the roof of the building housing the Gemini 9 spacecraft, before bouncing off and crashing into a car park. Command pilot Elliott See and pilot Charles Bassett were replaced by the back-up crew, Tom Stafford and Eugene Cernan, who had landed at St Louis in a second aircraft shortly afterwards.

On 17 May, the Agena 9 target rocket for the three-day rendezvous, docking and spacewalking mission flew into the Atlantic Ocean after a second-stage malfunction, and the mission was scrubbed. After the Agena 6 failure in October 1965, NASA began to develop an alternative target without its own engine, called the Augmented Target Docking Adapter, ATDA, should the Agena fail again. On 1 June, an Atlas booster carried ATDA into orbit, while Stafford and Cernan waited for their blast-off within the tight, 40-second window. Computer problems grounded them at $T - 2$ minutes. Stafford, the Gemini 6 pilot, had now been to the pad five times and had lifted off only once.

However, he and Cernan were at last airborne at 08:39 hrs on 3 June, heading for what scientists were expecting to be a rather unusual sight, as signals from the ATDA indicated that its payload shrouds had not separated fully. They were right. Stafford gave Gemini 9 its trademark, describing the ATDA as an "angry alligator". They couldn't dock but, as planned, backed off and conducted two more rendezvous. During the manoeuvres in the 28.9° inclination orbit, the astronauts reached a peak

The ATDA docking target was dubbed the "angry alligator" by the crew of Gemini 9

altitude of 311 km (195 miles). Cernan's planned spacewalk was delayed to 5 June, because the astronauts felt exhausted.

The Gemini pilot was equipped with an enlarged EVA Life Support System (ELSS) on his chest, with its own heat exchanger to cool ventilator air and to provide 30 minutes of emergency oxygen. Cernan's tether was 7.62 m (25 ft) long. He had planned to don the US Air Force manned manoeuvring unit, still attached to a tether, which was housed in the adapter section of the spacecraft. The AMU weighed 76 kg (168 lb) and was powered by nitrogen peroxide thrusters. Cernan's legs were protected from the AMU's exhaust by dark grey leggings of eleven layers of aluminised film. He donned the device but never flew it because Stafford called him back after a record 2 hours 8 minutes outside. Cernan found the spacewalk utterly exhausting, as controlling his movement in weightlessness was almost impossible. Compounding the problem were the snaking umbilicals of his spacesuit, the AMU, poor communications when Cernan switched to the AMU circuit, plus the inability of his spacesuit's environmental control system to handle his body heat. Cernan ended up with a fogged faceplate and couldn't see out.

Gemini 9 did, however, perform a party piece at the end of the 3 day 20 minute 50 second mission, splashing down just 1.44 km (1 mile) from the recovery ship

USS *Wasp*, a target miss of just 704 m (2,300 ft) some 552 km (343 miles) east of Bermuda.

> **Milestones**
>
> 21st manned space flight
> 13th US manned space flight
> 7th Gemini manned flight
> 1st US manned space flight by back-up crew
> 2nd US and 3rd flight with EVA operations
> Closest splashdown of a Gemini to a recovery vessel (0.38 nautical miles)

GEMINI 10

Int. Designation	1966-066A
Launched	18 July 1966
Launch Site	Pad 19, Cape Kennedy, Florida
Landed	21 July 1966
Landing Site	Western Atlantic Ocean
Launch Vehicle	Titan II GLV No. 10; spacecraft serial number 10
Duration	2 days 22 hrs 46 min 39 sec
Callsign	Gemini Ten
Objective	Rendezvous and docking mission; 2nd rendezvous objective; high-apogee profile; EVA activities

Flight Crew

YOUNG, John Watts, 36, USN, command pilot, 2nd mission
Previous mission: Gemini 3 (1965)
COLLINS, Michael, 36, USAF, pilot

Flight Log

The 299th Atlas (and the 100th NASA Atlas) vehicle took off from Pad 15 on 18 July, taking with it the Agena 10 target stage, which duly entered its programmed orbit. Astronauts John Young and Michael Collins were launched 100 minutes later at 17:20 hrs, and within 5 hours 52 minutes were docked with Agena 10. Young had used rather too much fuel, however, and practice dockings were cancelled. The Taciturn Two, as the astronauts were described, used the Agena 10 engine to boost them into a record 763 km (474 miles) apogee in the 28.9° orbit, increasing the speed of the docked combination by 129 m/sec during the Agena's 80-second burn. The boost captured the attention of the crew more than the scenery did, because with the huge stage before their windows, the view was limited. The crew rested for nine hours, and on the next mission "day" relit the Agena to reduce the orbital height in preparation for a rendezvous with the second target – the dead Agena 8 stage used during the Gemini 8 mission the previous March.

Before the rendezvous, Collins performed a stand-up EVA, standing on his seat and poking his head and shoulders out of the spacecraft hatch, mainly to set up an astronomical camera and to retrieve cosmic dust particle collectors from the outside of the spacecraft, two of the 16 science experiments being flown on the mission. During the 49-minute exercise, both astronauts were badly affected by leaking lithium hydroxide from the spacecraft's environmental control system. Their eyes streaming and throats burning, they cut short the EVA. The Agena 10 was undocked after 38 hours 47 minutes attached to Gemini 10, while the latter, using its OAMS thrusters, made an optical rendezvous with the Agena 8 without the use of radar.

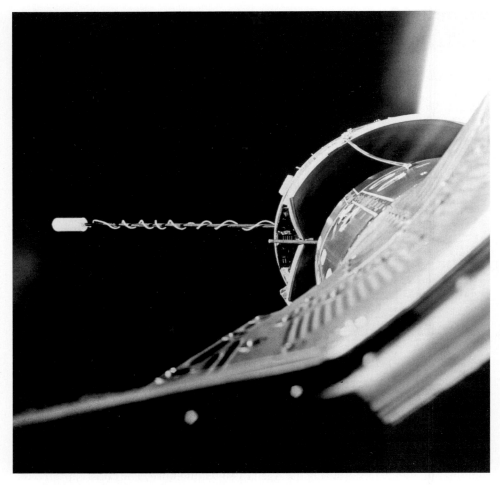

Gemini 10 docking with its Agena target, as seen from Mike Collins' window

On 20 July, Young piloted Gemini to within 3 m (10 ft) of the rocket stage and Collins opened the hatch for his full EVA, during which he recovered a cosmic dust particle collector from its side. He lost a camera in the process as he wrestled with a nitrogen gas-powered hand-held manoeuvring unit and the annoying tendency to float up and away from where he meant to be. A fault on another camera meant that no pictures were produced of Collins next to the Agena. The planned one-and-a-half-hour spacewalk was called off after 39 minutes because of concern over usage of the dwindling station-keeping thruster fuel. Getting the 15 m (49 ft) tether and the astronaut back into the spacecraft took a lot of effort and having done so, just before retro-fire, the crew opened the hatch for three minutes to dump some rubbish into orbit, namely the chest pack and tether.

Gemini 10 splashed down 846 km (526 miles) east of Cape Kennedy, 5.4 km (3 miles) from the recovery ship, USS *Guadalcanal* at $T + 2$ days 22 hours 46 minutes 39 seconds.

Milestones

22nd manned space flight
14th US manned space flight
8th Gemini manned flight
1st re-boost into high orbit
1st mission to rendezvous with two separate spacecraft
1st astronaut to make bodily contact with another spacecraft
1st US mission to launch on the day originally scheduled
3rd US and 4th flight with EVA operations

GEMINI 11

Int. Designation	1966-081A
Launched	12 September 1966
Launch Site	Pad 19, Cape Kennedy, Florida
Landed	15 September 1966
Landing Site	Western Atlantic Ocean
Launch Vehicle	Titan II GLV No. 11; spacecraft serial number 11
Duration	2 days 23 hrs 17 min 8 sec
Callsign	Gemini Eleven
Objective	Rendezvous and docking mission; high-apogee orbit and tether dynamics exercises; EVA activities

Flight Crew

CONRAD, Charles "Pete" Jr., 36, USN, command pilot, 2nd mission
Previous mission: Gemini 5 (1965)
GORDON, Richard Francis Jr., 37, USN, pilot

Flight Log

Gemini 11's task was to simulate a take-off by an Apollo lunar lander, and rendezvous and docking with an Apollo Command Module in lunar orbit, by taking off and docking with the Agena 11 target within 94 minutes, or one orbit. This meant that after the successful ascent of Agena 11 on the delayed launch day of 12 September, Gemini 11 would have just two seconds in its launch window. Its Titan II booster met its target within half a second, spewing into life with that characteristic high-pitched whine as the hypergolic propellants ignited spontaneously on contact.

Once in 28.8° inclination orbit, Conrad and Gordon switched on their rendezvous radar and computer and the space chase began. Rendezvous was achieved within 85 minutes and the ebullient Conrad jockeyed Gemini's nose inside the Agena target. For the first time in the programme, the crew performed docking practice, with Gordon being the first Gemini pilot to dock. The Agena was given a quick test burn in preparation for a longer burn later in the mission which was designed to take the crew to a record altitude.

On 13 September, after a rest, Gordon opened the hatch to begin an EVA, which everyone hoped would at last dispel doubts about man's ability to work in space. The burly Gemini pilot floated towards the Agena docking collar and sat astride it to attach a tether. He used a 10 m (33 ft) tether and a hand-held manoeuvring unit. Riding like a cowboy, as Conrad described him, Gordon had great difficulty remaining in one place and not floating upwards. The effort proved too much for the spacesuit's environmental control system and Gordon became blinded with sweat. Conrad called him back after just 38 minutes. Gemini had just one more flight to prove

Gemini 11 flies over India and Sri Lanka

that effective EVA was possible. The crew overcame their disappointment with the big Agena burn, lasting 26 seconds, over the Canary Islands.

En route to a high point of 1,372 km (853 miles) over Western Australia, the highest Earth orbit manned apogee, Gordon took an epic photo of India and Sri Lanka from a height of about 800 km (497 miles). The Agena was used to reduce the orbital height, in preparation for a stand-up EVA by Gordon lasting 128 minutes, during which he conducted several of the mission's science experiments and also had a cat nap.

Finally, Gemini 11 undocked from Agena 11, which was left dangling at the end of a 30 m (98 ft) tether. Conrad fired the thrusters to put the combination into a spin and thus created artificial gravity. After separating from the Agena and backing away, the crew performed further rendezvous exercises and a two-minute "equipment jettison" out of an opened hatch, before sitting back for a fully computer-controlled retro-fire, re-entry and landing. The eventful mission ended 1,120 km (696 miles) east of Miami, within 3 km (2 miles) of the recovery ship, USS *Guam*, at $T + 2$ days 23 hours 17 minutes 8 seconds.

Milestones

23rd manned space flight
15th US manned space flight
9th Gemini manned flight
1st manned mission to create artificial gravity in orbit
1st computer-controlled retro-fire re-entry and landing
4th US and 5th flight with EVA operations

On 1 November 1966, NASA civilian test pilot William Dana, 35, flew X-15-3 to 93 km during the programme's tenth astro-flight.

GEMINI 12

Int. Designation	1966-104A
Launched	11 November 1966
Launch Site	Pad 19, Cape Kennedy, Florida
Landed	16 November 1966
Landing Site	Western Atlantic Ocean
Launch Vehicle	Titan II GLV No. 12; spacecraft serial number 12
Duration	3 days 22 hrs 34 min 31 sec
Callsign	Gemini Twelve
Objective	Rendezvous and docking mission; expanded EVA activities

Flight Crew

LOVELL, James Arthur Jr., 38, USN, command pilot, 2nd mission
Previous mission: Gemini 7 (1965)
ALDRIN, Edwin Eugene "Buzz" Jr., 36, USAF, pilot

Flight Log

Apart from the now customary Agena rendezvous and docking, the most important task of Gemini 12 was to overcome the perennial spacewalking problem. For this, astronaut Buzz Aldrin was to have the additional assistance of straps, harnesses, Velcro patches and even "golden slippers" spacesuit boot holders, to help keep him in place, so he could do really effective work without overworking his environmental control system. The crew walked to the pad with "The" and "End" cards on their backs and were told when they reached Gemini 12 that theirs was "the last chance ... no reruns ... show will close after this performance". The show began with the Agena 12 launch from Pad 14 and followed that with Gemini 12's lift-off at 15:26 hrs local time on 11 November. The flight achieved a 28.9° inclination orbit, which would at one point of the flight reach an apogee of 301 km (187 miles).

Fortunately, pilot Aldrin had written a thesis on manned orbital rendezvous before he was selected as an astronaut, so when the spacecraft's rendezvous radar failed, his slide rule and sextant came out. To the admiration of ground controllers, rendezvous was achieved at $T + 3$ hours 46 minutes with a docking 28 minutes later. The astronauts had planned to use Agena's engine for a modest re-boost to an altitude of 640 km (398 miles), but controllers were concerned about a potentially dangerous malfunction and instead ordered the crew to use the Gemini engines to perform "rendezvous" with a solar eclipse over South America, which was an added bonus to the mission. Aldrin then performed a 2 hour 39 minute stand-up EVA (SUEVA), taking photographs casually leaning over the spacecraft like a tourist.

The following day, 12 November, mission planners watched with some trepidation as Aldrin began the most important EVA so far. It went swimmingly and Aldrin

Astronaut Eugene Cernan (left) jokes with the Gemini 12 crew as they prepare to board the capsule

did everything he planned during the successful 2 hour 8 minute spacewalk. One of his tasks had been to attach a tether to the Agena 12 target, which later allowed the astronauts to perform an artificial gravity test. On the third day, Aldrin opened the hatch again and made a 51-minute SUEVA, conducting some of Gemini's 14 science experiments. Thruster problems continued to be a concern throughout the mission but did not prevent a safe re-entry and landing at $T + 3$ days 22 hours 34 minutes 31 seconds, just 4.16 km (3 miles) from USS *Wasp*. The Gemini programme had been concluded with ten manned missions, clocking up 80 man-days in space. What was remarkable was that throughout the programme, the Soviets did not launch one cosmonaut into space. America was on its way – to the Moon.

Milestones

24th manned space flight
16th US manned space flight
10th Gemini manned flight
1st manned mission to witness solar eclipse in space
5th US and 6th flight with EVA operations

The next US mission after the end of the Gemini programme should have been the mission of Apollo 1 in February 1967. The tragic events on 27 January 1967 which prevented this mission are covered in detail under the chapter Quest for Space (Chapter 2).

SOYUZ 1

Int. Designation	1967-037A
Launched	23 April 1967
Launch Site	Pad 1, Site 5, Baikonur Cosmodrome, Kazakhstan
Landed	24 April 1967
Landing Site	65 km east of Orsk (fatal crash landing)
Launch Vehicle	R7 (11A511); spacecraft serial number (7K-OK) #4
Duration	1 day 2 hrs 47 min 52 sec
Callsign	Rubin (Ruby)
Objective	Manned test flight of Soyuz 7K-OK variant; intended active craft for docking with passive "Soyuz 2" (cancelled)

Flight Crew

KOMAROV, Vladimir Mikhailovich, 40, Soviet Air Force, pilot, 2nd mission
Previous mission: Voskhod 1 (1964)

Flight Log

Three months after the shocking Apollo 1 fire, what ostensibly began as a Soviet triumph, the flight of Soyuz 1, also ended in tragedy. The cause of the first fatal space mission was, like Apollo 1, over-confidence and bad workmanship. In fact, it could be called sheer foolhardiness. Four unmanned Soyuz test flights under the guise of the all-embracing Cosmos programme had failed. When he arrived at the Baikonur Cosmodrome, cosmonaut Vladimir Komarov must have been aware that he was laying his life on the line. But for what? It appears that Soyuz 1 was to attempt a propaganda coup to overshadow the US space programme still in mourning over Apollo 1. It was to fly into orbit and await the arrival of Soyuz 2, which would not only dock with it but would transfer two crewmen externally by EVA.

The space spectacular began with the launch of Soyuz 1, at 05:35 hrs local time. Komarov entered a 201–244 km (125–152 miles) orbit, with a unique manned inclination of 51°, and hit trouble. One solar panel did not deploy, and without necessary power many systems were degraded. Soyuz was the first manned spacecraft to carry solar panels. On another launch pad about 32 km (20 miles) away, another Soyuz booster was ready to launch Soyuz 2, carrying Valeriy Bykovsky, Yevgeniy Khrunov and Aleksey Yeliseyev. The latter two would be the EVA transfer crewmen. The launch was at first scrapped, then dramatically, plans were set in motion for an extraordinary rescue mission, during which the two Soyuz 2 EVA crewmen would pull out Soyuz 1's stuck solar panel.

The Soyuz 2 trio went to bed to rest before the following day's rescue bid. Meanwhile, attempts were made to terminate the Soyuz 1 mission. Komarov apparently tried to fire the retro-rocket on the sixteenth and seventeenth orbits, but

Komarov in training for Soyuz 1

probably had difficulty orienting the spacecraft. He succeeded on the eighteenth orbit and during the southbound re-entry towards an emergency landing zone, the spacecraft may have been out of full control, so much so that when the main landing parachute was deployed, it tangled.

Soyuz plummeted to the ground as Komarov awaited his fate. The capsule smashed into the ground, at $T + 1$ day 2 hours 47 minutes 52 seconds, at 08:22 hrs on 24 April, causing a large crater and catching fire. Komarov had no ejection seat and had made the ultimate sacrifice. When the Soyuz 2 crew awoke, they were told the news. At the time, all the world knew was that Soyuz 1 was on a solo mission and its parachute had tangled. The full facts have never emerged and the planned Soyuz 2 mission was never officially confirmed by the Soviets until 1989, despite heaps of evidence that included photos of the two crews together.

Milestones

25th manned space flight
9th Soviet manned space flight
1st Soyuz manned flight
1st fatal space mission

During the recovery period in manned space flight following the twin tragedies of Apollo 1 and Soyuz 1, the final astro-flights of the X-15 programme took place – and suffered a tragedy of their own. On 17 October 1967, William "Pete" Knight, 38, of the USAF, flew X-15 aircraft number 3 on the eleventh astro-flight to an altitude of almost 86 km. The following month, on 15 November 1967, USAF pilot Michael Adams, 37, was killed in the crash of X-15 aircraft number 3 after attaining 81 km during the twelfth astro-flight. The final such flight occurred on 21 August 1968, with NASA civilian test pilot William Dana, 37, at the controls. The thirteenth astro-flight of the programme saw X-15 aircraft number 1 reach a peak altitude of 81 km.

APOLLO 7

Int. Designation	1968-089A
Launched	11 October 1968
Launch Site	Pad 34, Cape Kennedy, Florida
Landed	22 October 1968
Landing Site	Western Atlantic Ocean southeast of Bermuda
Launch Vehicle	Saturn 1B/AS-205; spacecraft CSM-101
Duration	10 days 20 hrs 9 min 3 sec
Callsign	Apollo Seven
Objective	Earth orbital demonstration of Block II CSM performance including multiple Service Propulsion System (SPS) burns; crew, spacecraft and mission support facility performance; CSM rendezvous capability

Flight Crew

SCHIRRA, Walter Marty Jr., 45, USN, commander, 3rd mission
Previous missions: Mercury–Atlas 8 (1962); Gemini 6 (1965)
EISELE, Donn Fulton, 38, USAF, command module pilot
CUNNINGHAM, Walter, 36, civilian, lunar module pilot

Flight Log

Like a phoenix rising from the ashes, 623 days after the Apollo 1/204 spacecraft disaster, Apollo 7/205 lifted off from Pad 34 at Cape Kennedy, three minutes late, at 11:03 hrs on 11 October 1968, on a Saturn 1B. Redesigned, tested and re-tested, Apollo 7 was to conduct a thorough shakedown flight of the revised system before the USA could be confident again about going forward to the Moon. The 10 days 20 hours 9 minutes 3 seconds-long mission (which ended with the Command Module upside down but righted by buoyancy balloons, close to the recovery ship USS *Essex*), was brilliant, termed 101 per cent successful by NASA chiefs. Indeed so successful was it that the media, having nothing much to write about, zeroed in on the mood of the crew, thus misrepresenting the true nature of the flight.

True, the zealous commander Wally Schirra, the first to make three space flights, and his senior pilot Donn Eisele caught colds – probably during a hunting trip a few days before blast-off, thus introducing strict quarantine conditions for future crews. Schirra was distinctly very irritable at times during the mission, refusing to turn on in-flight television, making burns and re-entering with his crew not wearing their helmets. Schirra, Eisele and the healthy pilot Walt Cunningham, who did not catch a cold, had entered a 31.64° inclination orbit and separated from the S-IVB second stage, turned around and simulated the extraction of a Lunar Module (which was not flown) that would take place on Saturn V-boosted Moon flights, in what was called the

Eisele, Schirra and Cunningham receive a phone call from President Johnson

transposition and docking manoeuvre. An added bonus of this 20-minute long station-keeping manoeuvre was superb photography showing the S-IVB flying right over the Cape and the nearby Kennedy Space Center.

The extremely busy flight plan – which to the chagrin of Schirra, several controllers tried to make flexible at rather short notice, leading the commander to announce that he would become the onboard flight director for the rest of the mission – included eight ignitions of the service propulsion system engine, so vital to lunar orbital insertion and trans-Earth injection burns. The longest of these burns lasted 66 seconds and enabled Apollo 7 briefly to reach an altitude of 430 km (267 miles). In-flight television shows were very well received on the ground and featured much light-hearted banter between the mission control team and the crew on orbit.

Milestones

26th manned space flight
17th US manned space flight
1st Apollo CSM manned flight
1st US three-crew space flight
1st space flight by a crewman on third mission

SOYUZ 3

Int. Designation	1968-094A
Launched	26 October 1968
Launch Site	Pad 31, Site 6, Baikonur Cosmodrome, Kazakhstan
Landed	30 October 1968
Landing Site	Near to the city of Karaganda
Launch Vehicle	R7 (11A511); spacecraft series number 7K-OK #10
Duration	3 days 22 hrs 50 min 45 sec
Callsign	Argon (Argon)
Objective	Manned qualification of Soyuz spacecraft; intended docking with unmanned Soyuz 2 (cancelled)

Flight Crew

BEREGOVOY, Georgy Timofeyevich, 47, Soviet Air Force, pilot

Flight Log

The remarkable statistic regarding the Soyuz 1–2 debacle was that, had it been successful, the first Soviet space docking would have been achieved on a manned mission, against all previous Soviet traditions. The Soviets brought things back to normal with the unmanned, automatic docking flights of Cosmos 186–188 and 212–213 in late 1967 and the spring of 1968. It was assumed, naturally, that a manned docking was to follow. First Soyuz 2 was launched (from Pad 1) – on 25 October – secretly and unmanned. Then the following day, the oldest man in space to date, Georgy Beregovoy, boarded Soyuz 3, which was launched at 13:34 hrs local time from the other Soyuz pad (31), the first time this was used for a manned launch, and injected into a 51.6° inclination orbit. By the time it arrived, recorded pictures of his ascent appeared on Soviet television, together with the delayed announcement of the launch of Soyuz 2.

The manned docking seemed to be on, but it was not to be. Beregovoy's Soyuz merely made an automatic approach to within 167 m (548 ft). It was revealed in 1989 that the test pilot cosmonaut had been trying to dock with Soyuz 2 while flying Soyuz 3 upside down! He had to be "rescued" by ground control from his precarious predicament and further attempts to dock were called off. A further rendezvous was conducted before Soyuz 2 returned to Earth on 28 October. Beregovoy spent the rest of the mission making observations and showing television viewers around his spaceship, which even featured little curtains on the window of the Orbital Module. It was no coincidence that Apollo 7 had just returned to Earth having featured the "Wally, Donn and Walt" television shows that had earned them accolades from the US TV industry. Beregovoy, who had reached a maximum altitude of 252 km (157 miles) during the mission, the twenty-fifth manned orbital space flight, fired his retros for

Former Soviet test pilot and Soyuz 3 cosmonaut Georgi Beregovoy

145 seconds on 30 October and landed safely near Karaganda, after a flight of 3 days 22 hours 50 minutes 45 seconds.

Milestones

27th manned space flight
10th Soviet manned space flight
2nd Soyuz manned space flight
1st manned launch from Pad 31
1st Soviet launch to be shown on network television

APOLLO 8

Int. Designation	1968-118A
Launched	21 December 1968
Launch Site	Pad 39A, Cape Kennedy, Florida
Landed	28 December 1968
Landing Site	Pacific Ocean
Launch Vehicle	Saturn V AS-503; spacecraft designations: CSM-103; Lunar Module Test Article: LTA-B
Duration	6 days 3 hrs 0 min 42 sec
Callsign	Apollo 8
Objective	Manned qualification of Saturn V; demonstration of crew, spacecraft and ground support team over lunar mission distance to include lunar orbital operations.

Flight Crew

BORMAN, Frank, 40, USAF, commander, 2nd mission
Previous mission: Gemini 7 (1965)
LOVELL, James Arthur, 40, USN, command module pilot, 3rd mission
Previous missions: Gemini 7 (1965); Gemini 12 (1966)
ANDERS, William Alison, 35, USAF, lunar module pilot

Flight Log

The epic Apollo 8 mission began life in 1966 as what would have been Apollo 3, a flight test of the Lunar Module in a highly elliptical Earth orbit, launched on the first manned Saturn V vehicle. This was to follow Apollo 2, which would have been the first low-Earth orbit test of the Lunar Module, following two launches of the Saturn 1B with the Apollo CM and LM respectively, and a rendezvous and docking in Earth orbit. After the Apollo 1 fire, Apollo 2 became Apollo–Saturn 503, to conduct an all-up, Saturn V-boosted test of the modules in Earth orbit; and Apollo 3, with its original objectives, became Apollo-Saturn 504. AS-503, which was to become known as Apollo 8, was due to take place in late 1968, soon after Apollo 7 returned to Earth, but it became clear that the Lunar Module for the mission would not be ready in time.

At the same time, the Soviets had completed an unmanned, recoverable, lunar looping mission with Zond 5, basically an unmanned Soyuz craft without the Orbital Module, launched on a Proton vehicle. A hyped-up NASA clearly expected a manned Zond lunar looping flight soon and offered the commander of Apollo 8 the chance to make a lunar orbital mission, without his Lunar Module. The commander, James McDivitt, refused and decided to wait for his module. NASA therefore brought the Apollo–Saturn 504 mission forward as Apollo 8, and its crew was prepared for the lunar mission, which was given even more impetus when Zond 7 made another lunar

Earth rise

loop, this time making a landing inside the Soviet Union after a relatively benign 7-G re-entry.

As Apollo 8 was readied for history, there were rumours that Zond 8 was on the pad, with a lone cosmonaut atop a Proton preparing to beat the USA to the Moon. Despite the rumours, it is unlikely that the flight was ready in time. So, with much ado, Frank Borman, Michael Collins and William Anders made ready, only for Collins to be replaced by James Lovell when it was discovered that he needed back surgery.

Apollo 8 captured the imagination of the world. Watched worldwide live on television, the Saturn V thundered off the pad at 07:51 hrs local time on 21 December. Apollo 8 was in 32.6° inclination, 182–188 km (113–117 miles) Earth orbit in 11 minutes 25 seconds, still attached to the S-IVB third stage, which reignited at $T + 2$ hours 50 minutes. The 5 minute 17 second burn propelled Apollo 8 to a speed of 38,761 kph (24,086 mph) – escape velocity. Man was leaving the Earth for the first time. Gradually the Earth reduced in size and the crew beamed back live TV to incredulous viewers, caught up in the emotion and history of the epic mission. A major milestone occurred as Apollo's speed had dwindled to 3,556 kph (2,215 mph) some 62,240 km (38,676 miles) from the Earth, as lunar gravity took control and the speed increased. Two mid-course manoeuvres later, on Christmas Eve at $T + 68$ hours 58 minutes 45 seconds, the Earth lost contact with the Apollo crew as it disappeared behind the Moon, the first people to be cut off from their home planet.

The SPS engine ignited for 242 seconds, reducing the speed of the spacecraft from 8,000 kph (4,971 mph) to 5,953 kph (3,699 mph) and lunar orbit was achieved, only confirmed to the rest of the world when Apollo 8 reappeared on the other side of the Moon. The initial orbit of 112–308 km (70–191 miles) was circularised at 112 km (70 miles) by a second SPS burn on the third orbit. The highlight of the twenty lunar orbits was the crew's message of goodwill on Christmas Day, including the reading of passages from Genesis from the Bible. Twenty hours 11 minutes in lunar orbit ended with an SPS burn of 203.7 seconds, again on the unseen far side of the Moon.

Thankfully, Apollo 8 emerged triumphantly from the other side and began the journey home, which was to end with a dramatic, double-skip re-entry, at a speed of 39,513 kph (24,553 mph), during which the Command Module gained 9,144 m (30,000 ft) altitude, losing momentum before descending into the atmosphere for the final plunge. The mission ended with a splashdown at 8° north 165° west in the Pacific Ocean at mission elapsed time of 6 days 3 hours 0 minutes 42 seconds. The three men, who were hailed as the Columbuses of the Space Age, were aboard the USS *Yorktown* 88 minutes later.

Milestones

28th manned space flight
18th US manned space flight
2nd Apollo CSM manned mission
1st manned space flight to achieve escape velocity, approx 22,500 mph (36,202 kph)
1st manned flight to the Moon
1st manned flight to orbit the Moon
1st manned spacecraft to enter atmosphere from lunar distance, at 24,681 mph (39,713 kph)

SOYUZ 4 AND 5

Int. Designation	1969-004A (Soyuz 4)/1969-005A (Soyuz 5)
Launched	14 (Soyuz 4) and 15 (Soyuz 5) January 1969
Launch Site	Pad 31, Site 6 (Soyuz 4), Pad 1, Site 5 (Soyuz 5), Baikonur Cosmodrome, Kazakhstan
Landed	17 (Soyuz 4) and 18 (Soyuz 5) January 1969
Landing Site	Soyuz 4 – 40 km (25 miles) northwest of Karaganda; Soyuz 5 – 200 km (124 miles) southwest of Kustanai
Launch Vehicle	R7 (11A511); spacecraft serial numbers (7K-0K) #12 (Soyuz 4) and #13 (Soyuz 5)
Duration	2 days 23 hrs 20 min 47 sec (Soyuz 4); 3 days 54 min 15 sec (Soyuz 5)
Callsign	Amur (Amur – Soyuz 4); Baikal (Baikal – Soyuz 5)
Objective	Docking of two manned Soyuz spacecraft and the EVA transfer of two crew members from Soyuz 5 to Soyuz 4

Flight Crew

SHATALOV, Vladimir Aleksandrovich, 42, Soviet Air Force, pilot Soyuz 4
VOLYNOV, Boris Valentinovich, 34, Soviet Air Force, commander Soyuz 5
YELISEYEV, Aleksey Stanislovich, 34, civilian, flight engineer Soyuz 5
KHRUNOV, Yevgeny Vasilyevich, 35, civilian, research engineer Soyuz 5

Flight Log

Cosmonaut Vladimir Shatalov was launched alone aboard Soyuz 4 at about 12:29 hrs local time. Within ten minutes he was in his initial 51.7° inclination orbit, from which he would eventually manoeuvre to a new orbit with a maximum altitude of 222 km (138 miles). The next day, Soyuz 5 entered its initial 51.6° orbit after a launch from the freezing Baikonur at about 12:05 hrs local time. It carried Boris Volynov and the two cosmonauts who should have flown Soyuz 2 in 1967, Aleksey Yeliseyev and Yevgeny Khrunov. On 16 January, the two craft docked. Soyuz 4 was the active spacecraft both during the automatic approach to a distance of 100 m (328 ft) and for the manual, Shatalov-controlled soft dock, followed by a hard dock minutes later.

The whole event had been seen on television via a camera on Soyuz 4, and was accompanied by ribald comments from the crew, much to the chagrin of ground control. The Soviets claimed that they had achieved an "experimental space station", but at 12,926 kg (28,502 lb), the combined weight of the two spacecraft was lighter than a single Apollo. The first docking between two manned spacecraft was followed by an even more eventful space transfer, which was made externally because the docking mechanism prevented an internal transfer and there were no internal hatches.

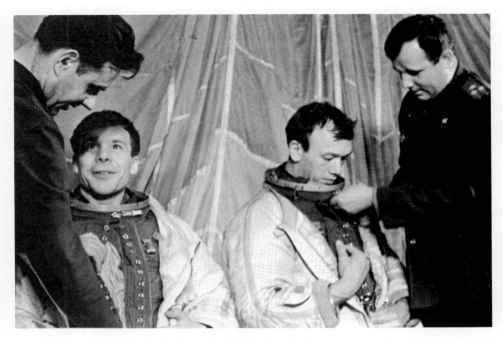

Soyuz 5 cosmonauts Khrunov (left) and Yeliseyev wearing EVA suits. Except that this is a pre-Soyuz 1 image, with Komarov on the far left and Gagarin far right. The Soyuz 5 EVA was originally to have been attempted in a docking between Soyuz 1 and 2, but was cancelled when Komarov's fatally flawed spacecraft developed problems.

First, Yeliseyev and Khrunov floated into the Soyuz 5 Orbital Module, donned spacesuits and depressurised the module, leaving Volynov alone in the flight cabin. Wearing upgraded spacesuits to that worn by Leonov with a lifeline tether and a small breathing air pack strapped to the legs, Khrunov opened the Orbital Module hatch and floated towards the depressurised Orbital Module of Soyuz 4, followed closely by Yeliseyev and both watched by a television camera (which unfortunately returned very poor pictures of the historic event).

The first EVA to involve two spacewalkers ended after 37 minutes. After sealing the outer hatch and re-pressurising the OM, the hatch to the Descent Module was opened and Shatalov welcomed his new crew, receiving some post and reports on the launch of Soyuz 4. The event that should have occurred between Soyuz 1 and 2 two years earlier had been achieved, ahead of the USA which was, coincidentally, planning a similar EVA exercise for Apollo 9 in March 1969.

The combined spacecraft undocked on 16 January after 4 hours 33 minutes 49 seconds together. The crews conducted separate experiments in geography, geology, navigation, medicine and radio communications before Soyuz 4 came home on 17 January, carrying two crewmen who had been launched in another craft. They came down 40 km (25 miles) northwest of Karaganda, in bitter temperatures of $-35°$, with a flight time of 2 days 23 hours 20 minutes 47 seconds. Yeliseyev and Khrunov, the hitchhikers, had clocked up a space time of just 1 day 23 hours 45 minutes. The deserted Volynov, who reached a maximum altitude of 237 km (147 miles) in Soyuz 5, came home on 18 January, 200 km (124 miles) southwest of Kustanai, with a flight time of 3 days 0 hours 54 minutes 15 seconds.

The re-entry of Soyuz 5 almost ended in disaster as the Propulsion Module failed to separate cleanly from the Descent Module. This caused the spacecraft to begin its entry the wrong way round, with the sealed inner hatch facing forward instead of the heat shield. Volynov, who was not wearing a pressure suit, heard the separation charges fire but also saw the PM still attached out of the side window. Smelling the burning rubber of the hatch gasket, Volynov thought he would not survive re-entry. As the G forces increased, the PM suddenly separated by atmospheric friction, causing the DM to swing around to the correct orientation for re-entry. Volynov realised he would indeed survive re-entry after all but then found, like Komarov on Soyuz 1, that his main parachute had tangled. This time it untangled, but the landing was so hard, despite the soft-landing rockets working, that he broke several teeth in his upper jaw. He had landed 600 km from the intended landing site due to the difficulties in separating the components. He got out of the capsule shaken, but able to walk to a nearby peasant hut to await the rescue team. It took some time for him to fully recover from the ordeal.

Milestones

29th and 30th manned space flights
11th and 12th Soviet manned space flights
3rd and 4th Soyuz manned space flight
1st docking between two manned spacecraft
1st crew transfer
1st landing by crew launched in another spacecraft
1st "spacecraft" with four crew
2nd Soviet and 7th flight with EVA operations

APOLLO 9

Int. Designation	1969-018A
Launched	3 March 1969
Launch Site	Pad 39A, Kennedy Space Center, Florida
Landed	13 March 1969
Landing Site	Western Atlantic Ocean
Launch Vehicle	Saturn V AS-504; spacecraft designations: CSM-104, LM-3
Duration	10 days 1 hr 0 min 54 sec
Callsign	CSM – Gumdrop; LM – Spider
Objective	Demonstration of crew, spacecraft, and mission support facilities during a manned Saturn V mission in Earth orbit with a CSM and LM; demonstration of LM crew and vehicle performance in Earth orbit

Flight Crew

MCDIVITT, James Alton, 39, USAF, commander, 2nd mission
Previous mission: Gemini 4 (1965)
SCOTT, David Randolph, 36, USAF, command module pilot, 2nd mission
Previous mission: Gemini 8 (1966)
SCHWEICKART, Russell Louis "Rusty", 33, lunar module pilot

Flight Log

If Apollo 11 was going to make the first manned landing on the Moon, Apollo 9 would have to be a spectacular success. And so it was. The main objective of the mission was to test-fly the Lunar Module in Earth orbit. Bad colds delayed the launch of the all-up Apollo stack from 28 February to 3 March, at 11:00 hrs local time. Once in orbit, command module pilot Dave Scott separated from the S-IVB stage and performed the first transposition and docking manoeuvre to extract the LM, which had been nicknamed Spider because of its arachnid-like appearance. The Command Module was called Gumdrop after the appearance of the CM when it was covered in blue wrappings as it was transported across the US. The individual names were chosen because of the need to identify the communications sources during the joint flight, a procedure that continued to the end of the Apollo lunar programme in 1972.

Interestingly, the S-IVB stage was restarted twice for the injection into solar orbit, but with slightly less speed than planned. Had the burn been for a manned trans-lunar injection, a Moon-landing mission could have been aborted. Meanwhile, in Earth orbit, Jim McDivitt commanded the SPS engine to fire four times, changing the altitude parameters of the 32.6° inclination orbit and testing the structural dynamics of the joint spacecraft. The maximum altitude achieved during the mission was 200 km (124 miles). On the third day, dressed in full space gear, McDivitt and Rusty

Dave Scott performs a stand-up EVA during Apollo 9

Schweickart entered Spider for the first checkout, while it was still attached to Gumdrop. This included a 367-second firing of the descent engine, which for the final 59 seconds was manually throttled by McDivitt, the first such manoeuvre in space history.

The SPS engine was fired again to fine-tune the orbit for the joint Spider–Gumdrop rendezvous and docking mission, but space sickness hit Schweickart, cancelling his EVA wearing the fully independent Apollo Portable Life Support System (PLSS) spacesuit, during which he planned an external transfer from the porch of the LM to the Command Module. However, he did recover enough to perform a 37-minute EVA standing on the porch on 7 March. The EVA resulted in some classic photographs. On 8 March came the big test. Spider was separated from

Gumdrop and fired its descent engine twice, ending up 19.2 km (12 miles) higher. Then, for the first time, the LM ascent engine was fired, after separation of the descent stage, placing it 120 km (75 miles) behind and 16 km (10 miles) below Gumdrop, to simulate lunar ascent and the rendezvous and docking manoeuvre.

Six hours later, Spider and Gumdrop were together, but not before McDivitt's eye-straining final docking, which resulted in the recommendation that on future flights this should be performed by the CMP. The ascent stage of Spider was separated as its engine fired again, to place it in a high-Earth orbit as the crew in Gumdrop wound down the mission with detailed Earth observations and photography. Re-entry was delayed one orbit because of fears of high seas in the splashdown area, giving Apollo 9 another first – the first extended US manned space flight. After a 3.6 m/sec (12 ft/sec) burn of the SPS, reducing the speed by 353 kph (219 mph), enough to induce re-entry, Apollo 9 splashed down safely at 23.25° north 68° west at $T + 10$ days 1 hour 0 minutes 54 seconds, some 5 km (3 miles) from USS *Guadalcanal*. Only one more test remained before the Moon landing.

Milestones

31st manned space flight
19th US manned space flight
3rd manned Apollo CSM flight
1st manned Apollo LM flight
1st manned flight in spacecraft unable to return to Earth
1st manual engine throttling
6th US and 8th flight with EVA operations

APOLLO 10

Int. Designation	1969-043A
Launched	18 May 1969
Launch Site	Pad 39B, Kennedy Space Center, Florida
Landed	26 May 1969
Landing Site	Pacific Ocean
Launch Vehicle	Saturn V AS-505; spacecraft designations: CSM-106; LM-4
Duration	8 days 0 hrs 3 min 23 sec
Callsign	CSM – Charlie Brown; LM – Snoopy
Objective	Demonstration of crew, spacecraft, mission support facilities during a manned Saturn V mission to lunar orbit with a CSM and LM; demonstration of LM crew and vehicle performance in the cis-lunar, and lunar (orbital) environment

Flight Crew

STAFFORD, Thomas Patten Jr., 38, USAF, commander, 3rd mission
Previous missions: Gemini 6 (1965); Gemini 9 (1966)
YOUNG, John Watts Jr., 38, USN, command module pilot, 3rd mission
Previous missions: Gemini 3 (1965); Gemini 10 (1966)
CERNAN, Eugene Andrew, 34, USN, lunar module pilot, 2nd mission
Previous mission: Gemini 9 (1966)

Flight Log

The riskiest space flight yet, Apollo 10 was to simulate a Moon landing in the final test before Apollo 11. Had development of the Lunar Module not been delayed, it is quite possible that Apollo 10 would have made the first real landing, making its commander Tom Stafford and LMP Eugene Cernan the first men on the Moon. Apollo 10 left new launch pad 39B at the Kennedy Space Center at 12:49 hrs local time on 18 May 1969 and placed the S-IVB and Apollo stack in a 32.5°, 184 km (114 miles) apogee Earth-parking orbit. Then the orbital speed was increased from 7,800 m/sec to 11,171 m/sec (25,593 ft/sec to 36,651 ft/sec) by the S-IVB's engine.

Soon after, Apollo 10 became Charlie Brown and Snoopy (named after the popular *Peanuts* cartoon characters created by Charles L. Schultz). For the mission, Charlie Brown exchanged his WWI flying ace goggles and scarf for a space helmet, while Snoopy the beagle was a symbol of quality performance. As the LM was extracted from the spent stage, it was seen live on the first colour television show from space. Happy TV shows were beamed from the light-hearted crew *en route* to the Moon, which needed only one SPS mid-course manoeuvre, rather than the planned

Stafford (left) and Young in the Apollo 10 Command Module

four, such was the accuracy of the flight profile. At about $T + 76$ hours, Apollo 10 reached lunar orbit, which was circularised at 110 km (68 miles). Some 14 hours later, the risky, untried part of the mission began.

Snoopy undocked and flew in station-keeping mode for a while before firing its descent engine for a brief 27.4 seconds, simulating a lunar landing and taking Stafford and Cernan to within 15.52 km (10 miles) of the lunar surface. Amid high excitement, the crew described the scene of boulders bigger than houses and a magnificent Earthrise, as Snoopy flew over the Sea of Tranquillity – Apollo 11's target – testing the all-important LM radar. The descent engine was fired again, twice, before staging. Because a switch had been left in the wrong position in Snoopy's cockpit, the staging, achieved at the second attempt, placed the ascent stage in an uncontrollable gyration, which at least led the LMP Cernan volubly to consider his fate.

Control regained, Stafford fired the ascent engine for 15 seconds, to simulate the rise from the lunar surface to rendezvous with Charlie Brown – and the lonely John Young. Careful RCS thruster firings gently nudged the LM towards the CM and at $T + 106$ hours, docking was achieved. After 31 lunar orbits, in 61 hours 31 minutes, Apollo 10 leapt from the Moon, and three days later flew into Earth's atmosphere at a record manned speed of 39,897 kph (24,792 mph), landing at $T + 8$ days 0 hours 3 minutes 23 seconds, at 165° west 5°south, some 6.4 km (4 miles) from the USS *Princetown*.

Milestones

32nd manned space flight
20th US manned space flight
4th Apollo manned flight
4th Apollo CSM manned flight
2nd Apollo LM manned flight
1st flight by experienced multi-crew
1st flight by two manned craft in lunar orbit
1st crewman to fly solo in lunar orbit (Young)
2nd manned flight to and orbit of the Moon
Fastest Apollo re-entry speed from lunar distance – 39,897 kph

APOLLO 11

Int. Designation	1969-059A
Launched	16 July 1969
Launch Site	Pad 39A, Kennedy Space Center, Florida
Landed	24 July 1969
Landing Site	Pacific Ocean
Launch Vehicle	Saturn V AS-506; spacecraft designations: CSM-107; LM-5
Duration	8 days 3 hrs 18 min 35 sec
Callsign	CSM – Columbia; LM – Eagle
Objective	The primary objective of the Apollo programme: a manned lunar landing and a safe return to Earth

Flight Crew

ARMSTRONG, Neil Alden, 38, civilian, commander, 2nd mission
Previous mission: Gemini 8 (1966)
COLLINS, Michael, 38, USAF, command module pilot, 2nd mission
Previous mission: Gemini 10 (1966)
ALDRIN, Edwin Eugene "Buzz", 39, USAF, lunar module pilot, 2nd mission
Previous mission: Gemini 12 (1966)

Flight Log

Commander Neil Armstrong reckoned that the chances of total success for the first attempt to land on the Moon on Apollo 11 were 50:50. Six hundred million people all over the world watched on television as Apollo 11 began its journey at 09:32 hrs local time from the Kennedy Space Center. Eleven minutes 46 seconds later, Armstrong and his crew of Mike Collins and Buzz Aldrin were over the first hurdle – 184 km (114 miles) above the Earth in a 32.7° inclination orbit. The mission was starting quietly and it continued this way, with the rather sombre crew keeping comments to a minimum. The trans-lunar injection burn, lasting 5 minutes 47 seconds, was a success, as was the transposition and docking manoeuvre by Collins.

Some TV broadcasts were made and the avidly-followed mission continued with Armstrong and Aldrin, in their spacesuits, checking out the LM, which had been named Eagle. Apollo 11 achieved lunar orbit with a 347-second SPS burn at $T + 75$ hours 50 minutes. The orbit was circularised by a second SPS "tweak" at 110 km (68 miles). The climax approached as Eagle separated from the Command Module Columbia at $T + 100$ hours 12 minutes. One hour 20 minutes later, with Eagle on the far side of the Moon, the descent engine fired for 30 seconds to begin descent orbit insertion (DOI). Fifty-seven minutes later, both Eagle and Columbia emerged from

Apollo 11 astronauts Aldrin and Armstrong deploy the Stars and Stripes at Tranquillity Base

the far side, with Eagle now approaching its 14.56 km (9 miles) low point or perilune, when the powered descent initiation burn (PDI) was to begin.

The 756.3-second long burn seemed interminable to the waiting world, but to the crew it went so fast that neither could recall very much about it, other than the computer alarms that nearly aborted the landing. An overloaded computer was protesting, but ground controller Stephen Bales reported that all was well. The mission was given a go to land, but with seconds left Armstrong could see that the autopilot was taking Eagle into a boulder-strewn crater. He took partial control and amid clouds of dust, landed with between 15 and 20 seconds of fuel left, at $T + 102$ hours 45 minutes. The time in the UK was 21:18 hrs on 20 July.

The landing site, named Tranquillity Base by Armstrong, was about 6.4 km (4 miles) downrange of the planned touchdown point, at $0°41'15''$ north $23°0'26''$ east. Having reached the Moon, Armstrong could start to think about what his first words would be when he stepped upon it. Although in the Gemini programme the pilot went for the walks, on Apollo, because of the design of the Lunar Module hatch which opened towards the LMP thus trapping him, the commander would go out first, which was regarded as a logical thing to do anyway given the prestigious nature of the event. Armstrong's right boot touched the lunar dust at $T + 109$ hours 42 minutes, 03:56 hrs British time on 21 July. As he stepped onto the lunar surface, Armstrong said: "That's one small step for a man, one giant leap for mankind." Armstrong was joined by Aldrin, who took the TV camera and placed it on a tripod some distance away so that both astronauts could be seen, looking like ghosts on the black and white TV.

The flag was raised and the short moonwalk was interrupted by a gushing President Nixon. The astronauts rushed to finish the deployment of EASEP experiments but Armstrong did find time to take some classic photographs of Aldrin, the first person to make two spacewalks. Although he had the camera briefly, to take pictures of his boot print and a panorama of the surface, Aldrin did not take a formal still of the first man on the Moon, although he did happen to feature in one of the panoramas with his back to camera and standing in the shadow of the Lunar Module. The moonwalk lasted 2 hours 31 minutes 40 seconds, during which Armstrong was on the surface for 2 hours 14 minutes and Aldrin for 1 hour 33 minutes.

After 21 hours 36 minutes on the Moon, the critical ascent engine burn began, firing for 435 seconds to place Eagle's ascent stage in orbit for its rendezvous with Collins. The SPS engine fired for 2 minutes 29 seconds and after 59 hours 30 minutes in lunar orbit, Columbia was *en route* for its landing at 169°west 13°north, coming down near the USS *Hornet* at $T + 8$ days 3 hours 18 minutes 35 seconds. The epic mission was over, rather ironically overshadowed by the antics of the late President Kennedy's younger brother Edward, who was involved in a fatal traffic accident at Chappaquidick.

Milestones

33rd manned space flight
21st US manned space flight
5th Apollo manned space flight
5th Apollo CSM manned flight
3rd Apollo LM manned flight
1st manned landing on the Moon
1st walk on the Moon
3rd manned flight to and orbit of the Moon
7th US and 9th flight with EVA operations

SOYUZ 6, 7 AND 8

Int. Designation	1969-085A (Soyuz 6), 086A (Soyuz 7), 087A (Soyuz 8)
Launched	11 (Soyuz 6), 12 (Soyuz 7) and 13 (Soyuz 8) October 1969
Launch Site	Pad 1, Site 5 (Soyuz 7); Pad 31, Site 6 (Soyuz 6, Soyuz 8), Baikonur Cosmodrome, Kazakhstan
Landed	16 (Soyuz 6), 17 (Soyuz 7) and 18 (Soyuz 8) October 1969
Landing Site	Soyuz 6 – 179.2 km (111 miles) northwest, Soyuz 7 – 153.6 km (95 miles) northwest and Soyuz 8 – 144 km (89 miles) north of Karaganda
Launch Vehicle	R7 (11A511) for all three launches; spacecraft serial numbers (7K-0K) #14 (Soyuz 4); #15 (Soyuz 5); #16 (Soyuz 8)
Duration	4 days 22 hrs 42 min 47 sec (Soyuz 6); 4 days 22 hrs 40 min 23 sec (Soyuz 7); 4 days 22 hrs 50 min 49 sec (Soyuz 8)
Callsign	Soyuz 6 – Antey (Antaeus); Soyuz 7 – Buran (Snowstorm); Soyuz 8 – Granit (Granite)
Objective	Soyuz "troika" group flight; rendezvous and docking between Soyuz 7 and 8; space welding experiments on Soyuz 6

Flight Crew

SHONIN, Georgy Stepanovich, 34, Soviet Air Force, commander Soyuz 6
KUBASOV, Valery Nikoleyevich, 34, civilian, flight engineer Soyuz 6
FILIPCHENKO, Anatoly Vasilyevich, 41, Soviet Air Force, commander, Soyuz 7
VOLKOV, Vladislav Nikoleyevich, 33, civilian, flight engineer Soyuz 7
GORBATKO, Viktor Vasilyevich, 34, Soviet Air Force, research engineer, Soyuz 7
SHATALOV, Vladimir Aleksandrovich, 42, Soviet Air Force, commander Soyuz 8 and group commander, 2nd mission
Previous mission: Soyuz 4 (1969)
YELISEYEV, Aleksey Stanislovich, 35, civilian, flight engineer Soyuz 8, 2nd mission
Previous mission: Soyuz 5 (1969)

Flight Log

Soyuz 6 was to have been a solo mission but was flown together with Soyuz 7 and 8 which were to perform a Soyuz 4/5-type rendezvous, docking and transfer mission. Soyuz 6 – without a docking probe – set off first at 16:10 hrs local time on 11 October. It carried two cosmonauts, Shonin and Kubasov, and entered a 51.7° inclination

The Soyuz 6 crew of Kubasov (left) and Shonin

orbit, which would, after four manoeuvres, reach a maximum altitude of 242 km (150 miles). Their objectives were the usual Soviet ones of "testing, checking, perfecting and conducting" plus a unique experiment called Vulcan, in which automatic welding would be attempted inside the unpressurised Orbital Module. On the 77th orbit of Soyuz 6, three processes were attempted: electron beam, fusible electrode and compressed arc welding, under the control of Kubasov. The samples were returned to Earth. In 1990, some 21 years later, it was revealed that the low-pressure compressed arc had inadvertently almost burned a hole right through the inner compartment flooring and damaged the hull of the Orbital Module. The crew were at first unaware of this as they were sealed in the DM during the welding operation, but found the damage when they entered the OM towards the end of their mission.

When Soyuz 7 was launched at 15:45 hrs local time from Baikonur the day after, most observers felt that a docking was likely since, at the time, it was not known that

The crews of Soyuz 6–8 pose for a "group shot". Back row from left: Gorbatko, Filipchenko and Volkov (Soyuz 7). Front row from left: Kubasov and Shonin (Soyuz 6), Shatalov and Yeliseyev (Soyuz 8)

Soyuz 6 could not do so. Indeed, one of Soyuz 7's stated objectives was "manoeuvring and navigation tests" with Soyuz 6. But Filipchenko, Gorbatko and Volkov were supposed to dock not with Soyuz 6 but with Soyuz 8, which was duly launched at 15:19 hrs local time on 13 October, with Shatalov and Yeliseyev, the first Soviet space-experienced crew.

Problems with the Igla rendezvous system were experienced, and a manual attempt at docking was not successful. The nearest the two craft came to one another was 487 m (1,600 ft), observed for 4 hours 24 minutes by Soyuz 6 from about 1.6 km (1 mile) away. Maximum altitudes achieved by Soyuz 7 and 8 were 244 and 235 km (152 and 146 miles) respectively during their missions which, with Soyuz 6, entailed detailed Earth and celestial observations under the group command of Shatalov.

The "mystery missions", which in total involved 31 orbital change manoeuvres, ended on 17, 18 and 19 October, 179.2 km (111 miles) northwest, 153.6 km (95 miles) northwest and 144 km (89 miles) north of Karaganda respectively.

Milestones

34th, 35th and 36th manned space flights
13th, 14th and 15th Soviet manned space flights
1st three-manned-spacecraft mission
1st time with seven people in space at once
Shortest turnaround between missions – ten months, for Shatalov and Yeliseyev

APOLLO 12

Int. Designation	1969-099A
Launched	14 November 1969
Launch Site	Pad 39A, Kennedy Space Center, Florida
Landed	24 November 1969
Landing Site	Pacific Ocean
Launch Vehicle	Saturn V AS-507; spacecraft designations: CSM-108; LM-6
Duration	10 days 4 hrs 36 min 25 sec
Callsign	CSM – Yankee Clipper; LM – Intrepid
Objective	Second manned lunar landing mission (H-1)

Flight Crew

CONRAD, Charles "Pete" Jr., 39, USN, commander, 3rd mission
Previous missions: Gemini 5 (1965); Gemini 11 (1966)
GORDON, Richard Francis Jr., 40, USN, command module pilot, 2nd mission
Previous mission: Gemini 11 (1966)
BEAN, Alan LaVern, 37, USN, lunar module pilot

Flight Log

Flying to the Moon a second time wasn't any easier, but it seemed that way after the euphoria of Apollo 11. Indeed, Apollo 12 had two particular hazards, one deliberate and one unpredictable but none the less avoidable. The deliberate hazard was to be the hybrid trajectory to the Moon, which did not guarantee Apollo 12 a "free return" by lunar loop if there was a major systems failure *en route*. The second hazard could have been avoided had NASA not decided to launch the mighty Saturn V in heavy rain and dark storm clouds, seemingly to please the space budget-cutter, President Richard Nixon, who had come to the KSC to watch.

About 36 seconds after 11:22 hrs local time, with the Saturn already out of view, Pad 39A was hit by lightning. So was Apollo 12. Commander Conrad saw the multicoloured control panel displaying systems shorts and said that it seemed that "everything in the world had dropped out." LMP Bean restored systems as the second and third stages proceeded effortlessly into 199 km (124 miles) 32° orbit. All the electrical circuits were checked and the go for the Moon was given. The S-IVB burned for 5 minutes 45 seconds and the transposition and docking manoeuvre was successful, but the S-IVB was placed into an unusual and highly elliptical orbit of the Earth, rather than into solar orbit, due to a malfunction.

The TV shows were jocular and informative. Conrad and Bean checked out the Lunar Module, and one mid-course correction was made to place Apollo out of the free return and on course for a lunar orbit with desirable lighting conditions at the landing point. Apollo 12's SPS lit up on the lunar far side and placed the spacecraft

Pete Conrad examines the Surveyor 3 spacecraft. The Apollo 12 Lunar Module can be seen on the horizon.

into a 110 by 312 km (68 × 194 miles) orbit, which was adjusted two orbits later to an eventual 110 km (68 miles). At $T + 107$ hours 54 minutes, the Lunar Module became Intrepid and the Command Module Yankee Clipper, illustrating that this was an all-Navy crew. DOI began at $T + 109$ hours 23 minutes with a 29-second firing placing Intrepid at a perilune of 14.4 km (9 miles) for the PDI. Before this, there was hectic activity between the ground and the crew to update the LM's navigation programme, which continued two minutes into the burn that began at $T + 110$ hours 20 minutes.

The high-spirited crew came into their Ocean of Storms landing site, close to the unmanned Surveyor 3 spacecraft which had landed there in 1967. Conrad landed Intrepid about 856 m (2,808 ft) northwest of Surveyor at $T + 110$ hours 32 minutes at 3°11′51″ south 23°23′7.5″ west. CMP Gordon spotted both Intrepid and Surveyor from orbit in Yankee Clipper. The first moonwalk began at $T + 115$ hours 10 minutes when the jocular Conrad hopped, skipped and hummed across the surface. After joining him, Bean took the colour television camera to place it on a tripod, but the camera was pointed at the Sun and blacked out. The by now dwindling TV audiences switched off.

The first 3 hour 56 minute EVA on 19 November involved erecting the US flag and deploying the first ALSEP array of lunar experiments, one of which was powered by a radioisotope thermoelectric generator with a radioactive fuel source. The second

EVA on 20 November, lasting 3 hours 49 minutes, was highlighted by the visit to Surveyor, bits of which were cut off to be taken home for analysis. Conrad's fall in the lunar dust caused a "spacesuit might leak" scare, but from the antics of later moonwalkers, one wonders what the fuss was about.

The highly successful moonwalks over, after 31 hours 31 minutes on the Moon, Intrepid sailed for Yankee Clipper. The rendezvous and docking 3 hours 30 minutes later was watched live by TV audiences, who could even see Intrepid's crew in the windows of the LM and the little spurts of the RCS jets. Conrad and Bean removed their dusty spacesuits and crossed into Yankee Clipper naked, except for their headsets. Intrepid was sent crashing into the Moon and the reverberations from the impact were picked up by the ALSEP seismometer now on the surface.

Yankee Clipper broke anchor after 45 orbits and 88 hours 56 minutes over the Moon. The crew witnessed a spectacular solar eclipse on the way home and splashed down near USS *Hornet* at 15° south 165° west at $T + 10$ days 4 hours 36 minutes 25 seconds. Like the Apollo 11 crew, Conrad, Gordon and Bean had to live in the Apollo quarantine container for three weeks to ensure that no "moon bugs" came home with them.

Milestones

37th manned space flight
22nd US manned space flight
6th manned Apollo flight
6th Apollo CSM manned flight
4th Apollo LM manned flight
4th manned flight to and orbit of the Moon
2nd manned lunar landing and moonwalk
1st manned mission with two EVAs
1st manned spacecraft to spend a day on the Moon
1st manned mission to use radioisotope thermoelectric generators
8th US and 10th flight with EVA operations

APOLLO 13

Int. Designation	1970-029A
Launched	11 April 1970
Launch Site	Pad 39A, Kennedy Space Center, Florida
Landed	17 April 1970
Landing Site	Pacific Ocean
Launch Vehicle	Saturn V AS-508; spacecraft designations: CSM-109; LM-7
Duration	5 days 22 hrs 54 min 41 sec
Callsign	CSM – Odyssey; LM – Aquarius
Objective	Third manned lunar landing mission (H-2)

Flight Crew

LOVELL, James Arthur Jr., 42, USN, commander, 4th mission
Previous missions: Gemini 7 (1965); Gemini 12 (1966); Apollo 8 (1968)
SWIGERT, John Leonard "Jack" Jr., 39, command module pilot
HAISE, Fred Wallace Jr., 36, lunar module pilot

Flight Log

Command module pilot Thomas "Ken" Mattingly had the bad luck, two days before the flight of Apollo 13, to be declared not immune to the German measles that he had been exposed to by back-up LMP Charlie Duke. He was dropped and replaced by back-up Jack Swigert, who was put through his paces in the simulator to ensure his readiness and compatibility with the remaining prime crew members, James Lovell and Fred Haise. Lift-off seemed routine at 14:32 hrs local time, but the Saturn V burn lasted 44 seconds longer, because the four remaining engines of the S-II had to burn for an extra 34 seconds to make up for the loss of the fifth one, and the S-IVB had to burn for an additional ten seconds.

Initial orbit of 33.5° and 156 km (97 miles) apogee was achieved. The S-IVB ignited, the transposition and docking was successful and the stage was sent towards an impact on the Moon, big enough to be detected by the Apollo 12 seismometer. The television pictures were of high quality, but were not shown live by any network. Apollo 13 indeed seemed a milk run to the Moon, targeted for the Fra Mauro highlands. Then, at $T + 55$ hours 55 minutes 20 seconds on 13 April, oxygen tank No. 2 in the Service Module, which had undetected heater switches welded together due to an electrical malfunction in a pre-launch test, exploded 328,000 km (222,461 miles) from Earth.

The reaction of the crew was calm and stoic as they faced a lingering death in space. Power was going down fast in the Command Module. The only hope was to use the LM Aquarius, thankfully still attached, as it was the trans-lunar coast rather than

Apollo 13 crew (l to r) Haise, Swigert and Lovell, relieved to be back on Earth after a trying mission

the return journey. Aquarius's descent engine was used three times, $T + 61$ hours 30 minutes, for 30 seconds, to get Apollo 13 back on a lunar looping "free return" trajectory which would at least guarantee making landfall on Earth – somewhere, hopefully in the Indian Ocean: for 4 minutes 28 seconds to speed the return journey, at $T + 142$ hours 53 minutes; and for 15.4 seconds to fine-tune the trajectory. Rookies Haise and Swigert had strained at their windows to get a peek at the lunar far side during the lunar loop, which made them and Lovell the farthest travellers from Earth, at a distance of 397,848 km (247,223 miles).

Conditions on board were pitiful. It was extremely cold and the spacecraft was operating on the power equivalent of a single light bulb by the end of the mission. The crew, ably supported by thousands of engineers, scientists and fellow astronauts on the ground, even had to jury-rig an air conditioning unit to get rid of carbon dioxide. Aquarius was separated just before re-entry, followed by the Service Module, giving the incredulous crew their first view of the devastation that had been below them. Left with a little battery power, the Command Module Odyssey limped home to a splashdown at $T + 5$ days 22 hours 54 minutes 41 seconds, close to the USS *Iwo Jima* at 21° south 165°west. The shortest US three-person flight in history had captured the hearts of the world, and ended with a service of thanksgiving on the recovery ship.

The events of Apollo 13, as well as a tightening of the NASA budget, helped to seal the fate of future missions. Apollo 20 had already been axed in January 1970, and by September, Apollo 15 and 19 had been cancelled and the remaining missions renumbered to end with Apollo 17. The fear of losing a crew in space, or of their being stranded on the Moon with no hope of rescue, and the desire to move on to new programmes closer to Earth, together with the escalating cost of the war in southeast Asia and social unrest in the United States, all contributed to the end of the Apollo lunar programme.

Milestones

38th manned space flight
23rd US manned space flight
7th Apollo manned space flight
7th Apollo CSM manned flight
5th Apollo LM manned flight (docked only)
5th manned flight to the Moon
1st manned lunar loop flight
1st aborted lunar landing mission
1st flight by crewman on fourth mission
1st flight by crewman on second Moon mission

SOYUZ 9

Int. Designation	1970-041A
Launched	1 June 1970
Launch Site	Pad 1, Site 5, Baikonur Cosmodrome, Kazakhstan
Landed	19 June 1970
Landing Site	75 km west of Karaganda
Launch Vehicle	R7 (11A511); spacecraft serial number (7K-0K) #17
Duration	17 days 16 hrs 58 min 55 sec
Callsign	Sokol (Falcon)
Objective	Extended-duration Earth orbital mission (18 days)

Flight Crew

NIKOLAYEV, Andrian Grigoryevich, 40, Soviet Air Force, commander, 2nd mission
Previous mission: Vostok 3 (1962)
SEVASTYANOV, Vitaly Ivanovich, 34, civilian, flight engineer

Flight Log

The Soviet Union's first long-duration biomedical space mission, Soyuz 9, blasted away from Baikonur's Pad 1 at 00:00 hrs local time, the first manned launch at night, and was placed into a 51.6° orbit. During one of the five in-orbit changes, it reached a maximum altitude of 259 km (161 miles). In addition to medical experiments, including a torso harness which placed a load of about 40 kg (85 lb) on the cosmonaut's body, and other exercising equipment, the crew of Andrian Nikolayev and Vitaly Sevastyanov made several meteorological and geological observations in the relatively spacious Orbital Module. These were thwarted to a degree by a spacecraft window that had been smeared by the plume from an engine firing. These observations were compared with simultaneous observations by satellites, ships and aircraft. The crew also tested a new orientation system based on star-lock navigation, using the bright stars Canopus and Vega. After setting a new world space endurance record, Soyuz 9 came home to a "sniper accurate" landing in a ploughed field, 75 km (47 miles) west of Karaganda at $T + 17$ days 16 hours 58 minutes 50 seconds.

Doctors were keen to see whether the exercise regime and spaciousness of the Soyuz had allowed the cosmonauts to overcome what was thought to be difficult re-adaptation to gravity. However, as both cosmonauts were so busy they began avoiding the exercise programme and, despite a rebuke from mission control, never managed to return to the planned schedule. Both cosmonauts complained of feeling extraordinarily heavy and had to be carried out of the Soyuz 9 descent capsule. They were kept under close medical supervision for ten days, complaining of feeling twice their real weight, walking with difficulty, becoming flushed and breathing heavily.

Sevastyanov (left) and Nikolayev shown in the Soyuz 9 DM during their record-breaking 18-day space marathon

They even stumbled up stairs. Nikolayev was moved to express extreme pessimism about the possibilities of long-duration space flights. He need not have worried, because the reason for their discomfort was that Soyuz 9 had been in a partial, intermittent, gravity-inducing controlled spin throughout the mission, and it was this that had caused most of their problems. Their lack of regular daily exercise was also a factor in their physical condition upon return. However, this challenging and difficult mission was a wholly successful one, and an important stepping stone towards starting the space station programme the following year.

Milestones

39th manned space flight
16th Soviet manned space flight
8th Soyuz manned space flight
1st night launch of a manned space flight

5

The Second Decade: 1971–1980

APOLLO 14

Int. Designation	1971-008A
Launched	31 January 1971
Launch Site	Pad 39A, Kennedy Space Center, Florida
Landed	9 February 1971
Landing Site	Pacific Ocean
Launch Vehicle	Saturn V AS-509; spacecraft serial numbers: CSM-110; LM-8
Duration	9 days 0 hrs 1 min 57 sec
Callsign	CSM – Kittyhawk; LM – Antares
Objective	Third manned lunar landing (H-3); obtain photography of candidate exploration sites

Flight Crew

SHEPARD, Alan Bartlett Jr., 47, USN, commander, 2nd mission
Previous mission: Mercury–Redstone 3 (1961)
ROOSA, Stuart Allen, 37, USAF, command module pilot
MITCHELL, Edgar Dean, 40, USN, lunar module pilot

Flight Log

Apollo 14 was originally targeted for Taurus-Littrow but was diverted by the Apollo 13 abort to Fra Mauro. The mission was delayed for forty minutes by weather resembling the conditions for Apollo 12, and it surprised some observers to see the lift-off going ahead in very murky skies at 16:03 hrs local time. The parking orbit was 186 km (116 miles) and 32.4° inclination. The all-rookie crew in terms of orbital space flight, with only commander Alan Shepard having any space flight experience (a mere sub-orbital lob in 1961), were placed on their trans-lunar coast with no difficulty. CMP Stuart Roosa closed in for the transposition and docking but couldn't dock. A drama unfolded as Roosa tried six times, at last succeeding after 106 minutes. If the docking mechanism was faulty, the Moon landing would have to be cancelled. Once the Lunar Module Antares and Command Module Kitty Hawk were joined up, Shepard and his crew inspected the docking probe, but could not explain the earlier difficulty. NASA deliberated for a while before announcing that the Moon landing attempt would proceed.

Apollo 14 entered lunar orbit quietly, lowering its perilune to the lowest ever for the complete Apollo combination at just 16 km (10 miles). Manned by the steely-eyed Shepard and burly Edgar Mitchell, Antares separated and just before the PDI burn hit trouble. First, a short circuit in the LM computer abort switch meant the computer could not be persuaded not to abort the landing attempt. The crew just finished reprogramming themselves out of this problem when the landing radar failed. With an

Al Shepard proudly displays the US flag, standing on the Moon almost a decade after becoming America's first astronaut in space.

abort just seconds away, at which point the PDI could not bring them into the landing site, Mitchell's desperate switch flicking succeeded. Shepard brought Antares on to Fra Mauro with 40 seconds of fuel left, some 26.5 m (87 ft) off target, and at a tilt angle of 6°, at 30°40′27″ south 17°27′58″ west.

On 5 February, Shepard set foot on the Moon at $T + 114$ hours 31 minutes and was joined by Mitchell. The television camera was pointed at the sloping LM, with the sun very low in the sky, and it was not possible to follow the crew all the time as they set to work laying out the ALSEP instruments and collecting samples using the Modular Equipment Transporter. This was a lunar "wheelbarrow", which some of the less interested press reports suggested was to carry the aging Shepard when he got tired. The EVA lasted 4 hours 49 minutes and the second, on 6 February, lasted 4 hours 35 minutes and featured a thwarted and tetchy attempt to climb the 45 m (147.5 ft) high rim of Cone Crater. Although Shepard was convinced that they were nowhere near the rim and decided to abort the attempt, Mitchell had in fact been right in his assertion that they had nearly reached it. Shepard won the argument.

During the EVAs the crew had travelled about 2.72 km (2 miles) on foot, and at the end of the second EVA Shepard played his famous televised game of lunar golf, using a proper ball and a club made from the head of a 6-iron with an attachment to fit the handle of the contingency sample collector A passionate golfer, Shepard drew the ball out of his suit leg pocket and dropped it on the surface. Being limited in his mobility he sliced more lunar soil than ball in his first one handed "swing" barely moving the ball (something he was ribbed about back at the Astronaut Office after the mission). For the second attempt, a new ball was taken from the suit pocket and this time he hit it into a crater about 15 metres away. Not to be out done in this demonstration of lunar sports, Mitchell threw the staff from the solar wind composition experiment into the same crater. After 33 hours 39 minutes on the Moon, Antares lifted off and docked with Kitty Hawk, which itself leapt out of lunar orbit after a stay of 66 hours 39 minutes. The Command Module came home 6.4 km (4 miles) from the recovery ship USS *New Orleans* in the Pacific Ocean at 27° south 172° west at $T+9$ days 0 hours 1 minutes 57 seconds – the most accurate splashdown, 600 m (1950 ft) from the predicted target. The crew was the last to have to endure the quarantine container.

Milestones

40th manned space flight
24th US manned space flight
8th Apollo manned space flight
8th Apollo CSM manned flight
6th Apollo LM manned flight
6th manned flight to the Moon
5th manned flight to orbit the Moon
3rd manned Moon landing and walk
1st wheeled vehicle on the Moon
9th US and 11th flight with EVA operations

SOYUZ 10

Int. Designation	1971-034A
Launched	23 April 1971
Launch Site	Baikonur Cosmodrome, Kazakhstan
Landed	25 April 1971
Landing Site	120 km northwest of Karaganda
Launch Vehicle	R7 (11A511 – #25); spacecraft serial number (7K-T) #31
Duration	1 day 23 hrs 45 min 54 sec
Callsign	Granit (Granite)
Objective	Intended EO-1 resident crew for Salyut 1 (1971-032A)

Flight Crew

SHATALOV, Vladimir Aleksandrovich, 43, Soviet Air Force, commander, 3rd mission
Previous missions: Soyuz 4 (1969); Soyuz 8 (1969)
YELISEYEV, Aleksey Stanislavovich, 36, civilian, flight engineer, 3rd mission
Previous missions: Soyuz 5 (1969); Soyuz 8 (1969)
RUKAVISHNIKOV, Nikolay, Nikolayevich, 38, civilian, test engineer

Flight Log

On 19 April, the Soviets finally succeeded in placing a space station into orbit; the 18,900 kg (41,675 lb) Salyut. Originally to be called Zarya ("Dawn"), the name was changed when it was revealed that the Chinese had used the same name for one of their satellites. Salyut was chosen as a "salute" to the 10th anniversary of Gagarin's mission, but it was too late to repaint the side of the station, which still bore the named "Zarya". The first attempt at launching a crew to the station on 22 April was aborted at $T-1$ minute when one of the launch masts failed to retract. Three, three-person crews were prepared for resident stays on the station (known as EO in Russian), aimed at stealing some of the headlines from the Apollo lunar missions and in creating a the world's first station in space two years before America launched their Skylab workshop. However, just two days later all these plans seemed to be in doubt. Soyuz 10, crewed by Shatalov and Yeliseyev, the first Russians to make three space flights, plus new man Nikolay Rukavishnikov, took off on 23 April at 04:54 hrs local time and entered an orbit with an inclination of 51.6° and an apogee of 256 km (159 miles). About a day later, the rendezvous sequence – which also involved manoeuvres by Salyut 1 – ended with Soyuz 10 about 180 m (590 ft) away.

Shatalov jockeyed Soyuz (DOS-1) towards Salyut, and at 06:47 hrs Baikonur time docked with the space station. A 20- to 30-day stay aboard was in the offing and all seemed fine. The crew, however, never went on board. Shatalov had successfully soft-docked to Salyut but could not hard dock. With a 9 cm gap between them,

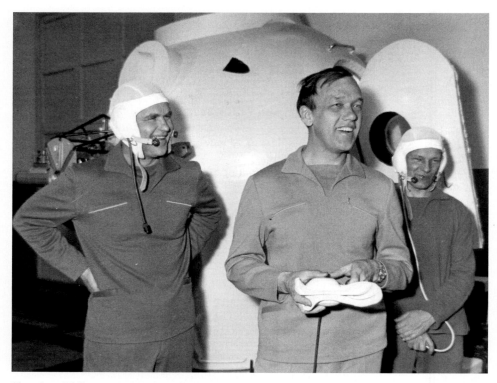

Shatalov, Yeliseyev and Rukavishnikov exit a Soyuz Volga simulator after a training session prior to their Soyuz 10 mission to Salyut

Shatalov tried firing the manoeuvring engines on Soyuz to push the spacecraft together, but to no avail. After six hours (four orbits), Shatalov was ordered to undock and try again. When he tried to do so, he found his Soyuz would not release from the Salyut, indicating a serious problem. The crew could have entered the OM and dismantled the docking device from the Soyuz side or jettisoned the OM, but either option would have rendered the single-ported Salyut useless for further operations. With onboard power supplies running out on the Soyuz (it would have been powered down during the occupation of the Salyut), Shatalov tried again and this time Soyuz slipped from her moorings, much to the relief of ground controllers and the crew. Other missions would be able to visit the Salyut. At 12:17 hrs Baikonur time, Soyuz 10 undocked and left Salyut uninhabited. The manned spacecraft flew in formation with Salyut for about 5 hours 30 minutes before making an emergency landing, at night for the first time and narrowly missing becoming the first Soviet splashdown missing a lake by a mere 50 metres, at $T + 1$ day 23 hours 45 minutes 54 seconds. It was left to Soyuz 11 to check in at the "space hotel".

On 10 May, the investigation team reported their findings. With the docking system lost in the re-entry of the OM, examination of flight hardware was impossible, but interpreting the data revealed that following the soft docking, the thrusters on Soyuz continued to fire for 30 seconds, causing a violent swing with a force equal to 160–200 kg on the docking mechanism shock absorbers. Ground tests found that the system could accept no more than 130 kg (60 per cent beyond design limits). This seemed to have buckled the docking system preventing the hard docking of the capture latches. Recommendations included strengthening the shock absorbers to twice the upper limit (260 kg) and additional controls for the commander to manually retract the probe.

Milestones

41st manned space flight
17th Soviet manned space flight
9th Soyuz manned space flight
1st Soyuz space ferry flight
1st manned space flight to land at night
1st manned space flight to be launched and to land at night

SOYUZ 11

Int. Designation	1971-053A
Launched	6 June 1971
Launch Site	Pad 1, Site 5, Baikonur Cosmodrome, Kazakhstan
Landed	30 June 1971
Landing Site	200 km east of Dzhezkazgan
Launch Vehicle	R7 (11A511); spacecraft serial number (7K-T) #32
Duration	23 days 18 hrs 21 min 43 sec
Callsign	Yantar (Amber)
Objective	Abbreviated occupation of first space station

Flight Crew

DOBROVOLSKY, Georgy Timofeyevich, 43, Soviet Air Force, commander
VOLKOV, Vladislav Nikolayevich, 35, civilian, flight engineer, 2nd mission
Previous mission: Soyuz 7 (1969)
PATSAYEV, Viktor Ivanovich, 38, civilian, test engineer

Flight Log

The Soyuz 11 prime crew comprised Aleksey Leonov, Valery Kubasov and rookie Pyotr Kolodin. A week before the launch a spot was discovered on Kubasov's lung and it was decided to replace the whole crew with their back-ups, Georgy Dobrovolsky, Vladislav Volkov and Viktor Patsayev, who took to the skies above Baikonur at 09:55 hrs. At 12:24 hrs the following day, the rendezvous was completed, and at a distance of about 150 m (492 ft), Dobrovolsky moved Soyuz 11 at about 0.9 m/sec (3 ft/sec) towards the docking port of Salyut 1. Would Soyuz 11 get a hard dock? Soyuz slowed to 0.3 m/sec (1 ft/sec) 60 m (196 ft) away and soft docking was completed at 12:49 hrs, with hard docking made at 12:55 hrs. The hatch was opened and the crew checked in at what Western press termed a "space hotel" for an expected six-week stay. The Soviets, however, had decided not to mount a 30-day flight but one of no more than 25 days. After surpassing the Soyuz 9 record of 18 days by the required ten per cent, they ended their mission on the 23rd day.

Manoeuvres on Salyut placed it at a maximum altitude of 282 km (175 miles) in the 51° orbit. The Soviets boasted that the size of the Soyuz/Salyut complex was 20.11 m (66 ft) long with workspace of 99.05 m^3 (3,500 ft^3) but at about 25 tonnes, it was far less than the combined Apollo modules. The crew conducted a highly successful mission, using telescopes to observe the stars, monitoring the weather, taking remote sensing photographs of the Earth, growing crops (including flax), hatching frogs' eggs, studying the atmosphere, conducting genetic experiments on fruit flies, using elasticated "Penguin" suits for exercising, and carrying out intensive biomedical work. On the surface Soyuz 11 was becoming the most successful mission

The ill-fated Soyuz 11 crew inside a Soyuz DM during training. From left, Dobrovolsky, Patsayev, Volkov

in Soviet history with TV pictures showing them clearly enjoying themselves, but there were moments of tension aboard the station. Personality clashes, probably due to the lack of previous flight experience for such a challenging mission (only Volkov had flown before), and their abbreviated training to make the flight they should only have supported, were contributing factors. The fear of an onboard fire on 16 June was triggered by the strong odour of smoke. Evacuating the station for Soyuz in case they had to make an emergency return to Earth, the atmosphere of Salyut was tested and found safe, with the crew returning to the station. The crew were also reluctant to follow their exercise programme, which was intended to maintain their physical condition to make the readaptation to gravity (and the stress of re-entry) much easier.

At 23:15 hrs Baikonur time on 29 June, the cosmonauts were back inside Soyuz 11, wearing their woollen flight suits and flight helmets. During preparation for undocking, a "hatch open" light indicated that the hatch between the DM and the OM was not sealed properly. Recycling the switch extinguished the light and the undocking was completed. Soyuz 11 undocked from Salyut thirteen minutes later, and retro-fire was completed at 03:35 hrs Baikonur time on 30 June. As programmed, the Soyuz separated into three parts: the Service Module, Descent Module and Orbital Module, to allow the Descent Module to continue its stable re-entry path. As this occurred, a pressure equalisation valve opened prematurely in the Descent Module and the crew's life-giving air began to be sucked into the vacuum of space. Patsayev and Dobrovolsky apparently tried to hand operate a pump to retain the cabin atmosphere but the crew succumbed to asphyxia. Soyuz 11, meanwhile, continued down automatically, landing safely at 04:17 hrs Baikonur time, just before sunrise, at $T + 23$ days 18 hours 21 minutes 43 seconds, the longest mission by far. Rescue teams rushed to the capsule to greet the crew but found three bodies slumped in their seats, as if sleeping peacefully. The Soviet space programme was grounded and one decision made immediately: to equip future Soviet cosmonauts with the obvious uniform – a spacesuit. Nine days after the loss of the Soyuz 11 cosmonauts, all crews assigned to Salyut were stood down. In October, the station re-entered the atmosphere, breaking up in a fiery descent.

Milestones

42nd manned space flight
18th Soviet manned space flight
10th Soyuz manned space flight
1st manned occupation of a space station
New duration record of 23 days 18 hours
1st birthday celebrated in space – (Patsayev's 38th on 19 June)
1st crew to die during entry phase

APOLLO 15

Int. Designation	1971-063A
Launched	26 July 1971
Launch Site	Pad 39A, Kennedy Space Center, Florida
Landed	7 August 1971
Landing Site	Pacific Ocean
Launch Vehicle	Saturn V AS-510; spacecraft serial numbers: CSM-112; LM-10; LRV-1
Duration	12 days 7 hrs 11 min 53 sec
Callsign	CSM – Endeavour; LM – Falcon
Objective	Fourth manned lunar landing mission (J1), first of three extended missions with expanded scientific objectives; first use of lunar roving vehicle

Flight Crew

SCOTT, David Randolph, 39, USAF, commander, 3rd mission
Previous missions: Gemini 8 (1966); Apollo 9 (1969)
WORDEN, Alfred Merrill, 39, USAF, command module pilot
IRWIN, James Benson, 41, USAF, lunar module pilot

Flight Log

A picture-perfect lift-off at 09:34 hrs began what was described by NASA as "the most complex and carefully planned scientific expedition in the history of exploration." From its 173 km (108 miles), 32.5° parking orbit, Apollo 15's S-IVB achieved not only TLI but also insertion into the required hybrid trajectory to the Moon. This was accompanied by three concerns: a potentially serious short-circuit in the SPS, which required a short test burst to ensure everything was functioning properly; a broken instrument cover in the Lunar Module which showered the interior with pieces of glass; and a leaking water pipe. These combined to cause a certain amount of tetchiness in the commander, Dave Scott.

After lunar orbit insertion, eventually lowered by 92.8 km (58 miles) to 12.8 km (8 miles), Command Module Endeavour separated from Lunar Module Falcon, and the great expedition reached the high point, with a steep 26° descent over the Appenine Mountains to a landing site close to the spectacular Hadley Rille. After poking his head out of the top of Falcon's docking port for a unique lunar stand-up EVA, Scott, playing the explorer part to perfection, set foot on Hadley Base on 31 July. He and his LMP, Jim Irwin, pulled the first lunar roving vehicle from the side of Falcon, and with the aid of lanyards, the rover unfurled. Soon the two crewmen were taking television viewers for spectacular rides to the edge of Hadley Rille, the TV camera on the rover being controlled by an engineer on the ground.

View of the Apollo 15 Hadley–Appenine landing area showing Hadley Rille

Three moonwalks (31 July, 1 and 2 August, lasting 6 hours 14 minutes, 6 hours 55 minutes and 4 hours 27 minutes respectively) and drives featured the deployment of the ALSEP array of instruments, drilling core samples up to 3 m (10 ft) deep, travelling a total of 27 km (17 miles) around the Moon and collecting 79 kg (174 lb) of lunar samples. The astronauts provided excellent descriptions of the surface geology and photographic documentation of the area. Towards the end of the final surface EVA, Scott proved the theory of Italian scientist, inventor and astronomer, Galileo Galilei (1564–1642) by dropping a hammer and a falcon feather. They fell and hit the surface at the same time in the one-sixth gravity and vacuum environment of the Moon, just as Galileo had predicted they would. Re-parking the Rover for the TV camera to record the lunar lift-off, Scott placed a plaque in the soil together with a statue of a "fallen astronaut". On the plaque were the names of eight astronauts and six cosmonauts (known at that time) who had died in while on active status.

With a spectacular finale seen live on television via the camera on the rover, after 66 hours 55 minutes on the Moon, with 17 hours 36 minutes of the period on the surface, Scott and Irwin lifted off in Falcon in a shower of multicoloured sparks. After docking with Endeavour, another Apollo 15 first was the deployment of the Particles and Fields Sub-satellite into lunar orbit, from the SIM bay on the side of the Service Module. This was packed with 16 scientific experiments, which kept CMP Al Worden busy operating while his colleagues were on the surface.

Endeavour broke out of lunar orbit after 74 orbits in 6 days 1 hour 18 minutes and sailed for home. During this return journey, Worden performed the first trans-Earth EVA, lasting 34 minutes, 318,400 km (197,854 miles) from Earth, retrieving experiments and samples from the side of the Service Module. Endeavour splashed down close to USS *Okinawa* at $T + 12$ days 7 hours 11 minutes 53 seconds, but at a speed of 33.6 kph (21 mph) rather than the planned 30.4 kph (19 mph), because one of the spacecraft's three parachutes failed to deploy. Apollo 15 was undoubtedly the maximum that human effort has achieved in exploration, though by some it is still rather sadly remembered more for the unauthorised commemorative first day covers carried during the mission than the science return or achievements by a hard working and dedicated crew.

Milestones

43rd manned space flight
25th US manned space flight
9th Apollo manned space flight
9th Apollo CSM manned flight
7th Apollo LM manned flight
7th manned flight to the Moon
6th manned flight to orbit the Moon
4th manned lunar landing and walk
1st two-day lunar stay
1st motorised vehicle on the Moon
1st manned mission with four EVAs and three moonwalks
1st manned space flight to deploy active sub-satellite
1st manned mission featuring trans-Earth spacewalk
1st televised lift-off from the Moon 10th US and
12th flight with EVA operations

APOLLO 16

Int. Designation	1972-031A
Launched	16 April 1972
Launch Site	Pad 39A, Kennedy Space Center, Florida
Landed	27 April 1972
Landing Site	Pacific Ocean
Launch Vehicle	Saturn V AS-511; spacecraft serial numbers: CSM-113; LM-11; LRV-2
Duration	11 days 1 hr 51 min 25 sec
Callsign	CSM – Casper; LM – Orion
Objective	Fifth manned lunar landing mission (J2)

Flight Crew

YOUNG, John Watts, 41, USN, commander, 4th mission
Previous missions: Gemini 3 (1965); Gemini 10 (1966); Apollo 10 (1969)
MATTINGLY, Thomas Kenneth II, 36, USN, command module pilot
DUKE, Charles Moss Jr., 36, USAF, lunar module pilot

Flight Log

Apollo 16 lifted off from KSC at 12:54 hrs local time and the public only pricked up their ears when Apollo 16 was in lunar orbit and the crew in the LM. Orion's landing was delayed because the back-up yaw gimbal servo loop on the CM Casper SPS engine failed and had to be restored before the CM – in 109 km by 19 km (68 miles by 12 miles) DOI orbit – could move into a high orbit with the assurance that the engine was workable for the rest of the mission. Given the go for landing after a frustrating and worrying 5 hours 43 minutes, astronauts John Young and Charlie Duke, piloting the heaviest Lunar Module to land on the Moon, came into Descartes Base like gangbusters.

Their high-spirited antics on the Moon later delighted the public at a time when walking on the Moon no longer attracted leading headlines. Frustratingly, Young, with his peripheral vision limited in the suit and unaware that his foot was snagged around a power cable, ruined a heat flow experiment by pulling the cable out of the instrument. The enthusiastic duo made three successful EVAs around their 8°69″ south 15°30″ east landing site, driving for 26 km (16 miles) and collecting 96.61 kg (218 lb) of moonrock. Young also drove the lunar rover at a record speed of 18 kph (11 mph) in a "lunar grand prix" demonstration of its ability in one-sixth gravity vacuum conditions. The moonwalks on 20–22 April lasted 7 hours 11 minutes, 7 hours 23 minutes and 5 hours 40 minutes. The third had been curtailed due to the late landing, but they had visited one of the largest boulders found on the Moon, which was dubbed "House Rock". In saluting the flag, Young leaped off the surface for the

Duke walks past the LRV during the first EVA of Apollo 16

cameras. When Duke tried to do this as well, out of the view of the camera, he forgot the mass of his backpack and came crashing down on his back on the surface, receiving a typically dry rebuke from Young. After the 20 hours of EVA and 71 hours 2 minutes on the Moon, Orion took off, again watched live on TV, to dock with Mattingly, who had flown the longest manned solo spaceflight of 81 hours 40 minutes. Another sub-satellite was deployed, and after 64 orbits in 125 hours 53 minutes, Casper lit its engine, one day earlier than planned, and headed home. *En route*, Mattingly made a 1 hour 24 minute EVA, before Casper landed about 2 km (1 mile) from USS *Ticonderoga* at $T + 11$ days 1 hr 51 minutes 25 seconds. They were on deck in 37 minutes, another record.

Milestones

44th manned space flight
26th US manned space flight
10th Apollo manned space flight
10th Apollo manned CSM flight
8th Apollo manned LM flight
2nd LRV operations
8th manned flight to the Moon
7th manned flight into lunar orbit
5th manned lunar landing and walk
11th US and 13th flight with EVA operations
2nd flight with trans-Earth EVA

APOLLO 17

Int. Designation	1972-096A
Launched	7 December 1972
Launch Site	Pad 39A, Kennedy Space Center, Florida
Landed	19 December 1972
Landing Site	Pacific Ocean
Launch Vehicle	Saturn V AS-512; spacecraft serial numbers: CSM-114; LM-12; LRV-3
Duration	12 days 13 hrs 51 min 59 sec
Callsign	CSM – America; LM – Challenger
Objective	Sixth and final Apollo lunar landing mission (J3)

Flight Crew

CERNAN, Eugene Andrew, 38, USN, commander, 3rd mission
Previous missions: Gemini 9 (1966); Apollo 10 (1969)
EVANS, Ronald Ellwin, 39, USN, command module pilot
SCHMITT, Harrison Hagen "Jack", 37, civilian, lunar module pilot

Flight Log

Because Apollo 17 became the final Moon landing mission as a result of budget cuts, LMP Joe Engle was dropped and replaced by the only mission trained geologist in the astronaut corps, Jack Schmitt, who had been down to fly Apollo 18. A computer malfunction delayed the first night launch in the US space programme by 2 hours 40 minutes to 12:33 hrs local time, when the Saturn V turned night into day amid a cataclysmic blast-off. Earth parking orbit was 178 km (111 miles) apogee, 32.5° and the lunar orbital DOI parameters were 25 km by 109 km (16 miles by 68 miles). A safe journey moonwards ended with Lunar Module Challenger making a perfect landing at the Taurus-Littrow landing site, at 20°10" north 30°45" east, with two minutes of fuel left.

A broken fender on the lunar rover made driving rather difficult, since it churned up piles of tacky moondust. The fender was ingeniously mended *in situ* and the rover featured in three highly successful moonwalks, highlighted by geologist Schmitt discovering orange soil. For a time, he and scientists on the ground believed, mistakenly, that this may have indicated recent volcanic activity and water on the Moon. TV pictures were spectacular and lunar surface activity ended with some ceremonial speeches by Cernan, marking the last steps on the Moon in the twentieth century.

Apollo 17 clocked up many firsts on the Moon – the longest EVA at 7 hours 37 minutes; longest EVA activity at 22 hours 5 minutes; longest distance travelled with the lunar rover at 33 km (21 miles); most samples collected during EVA 2; and furthest travelled from the LM at 7.3 km (5 miles), also during EVA 2. The times of EVAs 1

The Apollo lunar programme comes to an end as Apollo 17 spashes down

and 3 were 7 hours 12 minutes and 7 hours 16 minutes. Challenger lifted off and docked with Command Module America, which remained in lunar orbit for a record 147 hours 48 minutes. CMP Ron Evans made the customary trans-Earth EVA lasting 1 hour 6 minutes and the Apollo programme ended with a splashdown at $T + 12$ days 13 hours 51 minutes 59 seconds, near USS *Ticonderoga*, of the heaviest Command Module on landing. Another 110.22 kg (68 lb) of the Moon was on the Earth.

Milestones

45th manned space flight
27th US manned space flight
11th Apollo manned space flight
11th Apollo CSM manned flight
9th Apollo LM manned flight
9th manned flight to the Moon
8th manned flight into lunar orbit
6th manned lunar landing and walk
1st manned spacecraft to spend three days on the Moon
12th US and 14th flight with EVA operations
3rd flight with trans-Earth EVA

SKYLAB 2

Int. Designation	1973-032A
Launched	25 May 1973
Launch Site	Pad 39B, Kennedy Space Center, Florida
Landed	22 June 1973
Landing Site	Pacific Ocean
Launch Vehicle	Saturn 1B SA-206; spacecraft serial number CSM 116
Duration	28 days 0 hrs 49 min 49 sec
Callsign	Skylab
Objective	First Skylab resident crew (28 days)

Flight Crew

CONRAD, Charles "Pete" Jr., 42, USN, commander, 4th mission
Previous missions: Gemini 5 (1965); Gemini 11 (1966); Apollo 12 (1969)
KERWIN, Joseph Peter, 41, USN, science pilot
WEITZ, Paul Joseph, 40, USN, pilot

Flight Log

Early in the Apollo programme, when it became obvious to NASA that funding for a large space station to follow the Moon landings was unlikely to be forthcoming before the landing on the Moon had been achieved, the agency came up with an ingenious plan of using leftover Apollo hardware to build a smaller space station. The programme became known as the Apollo Applications Program, then in 1970, Skylab. By then it was clear that the agency could only support one Skylab space station with little prospect of the larger, 50-man space stations suggested only a few years before. The core space station, later named Skylab 1, was basically an empty S-IVB Saturn V third stage, internally converted on the ground into a manned orbital workshop. It was equipped with two large solar panels, a multiple docking adaptor, airlock and the Apollo Solar Telescope Mount (originally a modified LM), and equipped with four solar panels. The station would be launched unmanned and would house three, three-person crews for missions lasting 28, 56 and 56 days respectively. Inside and outside the 368 m^3 (13,000 ft^3) space station, these crews were to conduct the most detailed science programme ever attempted: 270 experiments in life sciences, solar physics, Earth observation, astrophysics, materials processing, engineering and technology.

The first step was to get Skylab 1 into orbit. The final two-stage Saturn V, AS-513, carried the 74,796 kg (164,925 lb) Skylab into space, but *en route* its micrometeoroid thermal shield was torn loose, together with one solar panel. Even worse, the other solar panel failed to deploy. Skylab (1973-027A) was the heaviest object in space but a useless one. Skylab 2, the manned flight of an Apollo Command and Service Module, launched on a Saturn 1B, was to have followed on 15 May, the day after Skylab 1, but

The Skylab Orbital Workshop in orbit, showing its one remaining solar panel and the parasol sunshield

was delayed until 25 May so that salvage procedures and tools could be devised. Equipment was being packed into Skylab 2 two hours before its launch at 09:05 hrs local time, from a uniquely configured Pad 39B, with a pedestal making the Saturn 1B tall enough to use some of the same pad and tower facilities as a Saturn 5.

The ebullient Conrad made a rendezvous with the crippled Skylab. The first salvage attempt was carried out by Paul Weitz, who stood on his seat in the Command Module, with science pilot Joe Kerwin desperately hanging on to his ankles. Weitz tried to pull the jammed solar panel free using a hooked pole, during a 37 minute stand-up EVA. This failed, then Skylab 2 failed to dock with the station. Finally, after eight attempts in two hours, Skylab 1 and 2 became one. The crew made Skylab, orbiting at 50°, 442 km (275 miles) apogee, habitable within four days by poking a parasol sun shield out of a small instrument airlock. On the twelfth day of the mission, 7 June, Conrad and Kerwin made the bravest and most hazardous EVA in history, lasting 3 hours 30 minutes. Using wirecutters, Conrad manually pulled out the errant solar panel, saving the mission and the whole programme.

The remarkable Skylab 2 mission lasted a total of 28 days 0 hours 49 minutes 49 seconds, during which time the astronauts completed 46 of the planed 55 experiments, working on them for 392 hours. Each astronaut performed one EVA, Weitz and Conrad having been outside on 19 June for 1 hr 44 minutes. Their exercise routine and lifestyle aboard enabled them to return to Earth feeling remarkably well, on board the recovery ship, USS *Ticonderoga*.

Milestones

46th manned space flight
28th US manned space flight
12th Apollo CSM manned space flight
1st US space station mission
New duration record – 28 days 0 hours
1st spacecraft salvage and repair mission
13th US and 15th flight with EVA operations
Conrad celebrates his 43rd birthday in space (2 June)

SKYLAB 3

Int. Designation	1973-050A
Launched	28 July 1973
Launch Site	Pad 39B, Kennedy Space Center, Florida
Landed	25 September 1973
Landing Site	Pacific Ocean
Launch Vehicle	Saturn 1B 207; spacecraft serial number CSM-117
Duration	59 days 11 hrs 9 min 4 sec
Callsign	Skylab
Objective	Second Skylab resident crew (59 days)

Flight Crew

BEAN, Alan LaVern, 41, USN, commander, 2nd mission
Previous mission: Apollo 12 (1969)
GARRIOTT, Owen Kay, 42, civilian, science pilot
LOUSMA, Jack Robert, 37, USMC, pilot

Flight Log

Skylab 3 was pressed into service earlier than planned because of worries about the integrity of the project, particularly its thermal protection qualities. It was launched at 07:11 hrs local time with the three astronauts, equipment and two spiders, Anita and Arabella, aboard. Docking was flawless and within hours the three crewmen Al Bean, Jack Lousma and Owen Garriott were floating inside the space station, 442 km (275 miles) up at an inclination of 50°. Space adaptation syndrome, or space sickness, hit them, particularly Lousma, and impaired their early activities.

When an RCS thruster quad on the Service Module of Skylab 3 was seen leaking and another quad showed signs of doing so, NASA put an emergency plan into action should the spacecraft be disabled. They would launch Skylab Rescue 1, with astronauts Vance Brand and Don Lind, to dock with the other port on Skylab and bring the three crew home, crammed together in the Command Module of the rescue vehicle. In the end, Brand and Lind's unique maiden space flight was not required.

The Skylab 3 crew endeared themselves to the ground, operating without histrionics and highly effectively. Bean and Lousma tested a prototype of a manned manoeuvring unit inside Skylab (as did an untrained Garriott, demonstrating how easy it was to "fly") and all three went for spacewalks to retrieve, observe and repair. On 6 August, Garriott and Lousma were outside for 6 hours 29 minutes, and on 24 August for 4 hours 30 minutes, while on 22 September Bean and Garriott made a 2 hour 45 minute EVA. They deployed two new parasols, and replaced gyros and nine other pieces of equipment.

This Apollo spacecraft carried the Skylab 3 crew, and two spiders, up to the station

Mission goals were exceeded by 50 per cent, with 305 man hours out of the total 1,081 hours of experimentation being spent on the Apollo Telescope Mount performing solar observations. The spiders spun webs in zero gravity. Altogether it was a rewarding flight, which ended at $T + 59$ days 11 hours 9 minutes 4 seconds, near the USS *New Orleans*, southwest of San Diego.

Milestones

47th manned space flight
29th US manned space flight
13th Apollo CSM manned space flight
1st manned space station re-occupation mission
1st manned space flight to exceed 50 days
New duration record – 59 days 11 hours
14th US and 16th flight with EVA operations

SOYUZ 12

Int. Designation	1973-067A
Launched	27 September 1973
Launch Site	Pad 1, Site 5, Baikonur Cosmodrome, Kazakhstan
Landed	29 September 1973
Landing Site	396 km southwest of Karaganda
Launch Vehicle	R7 (11A511); spacecraft serial number (7K-T) #37
Duration	1 day 23 hrs 15 min 32 sec
Callsign	Ural (Urals)
Objective	Manned test flight of Soyuz space station ferry without solar arrays

Flight Crew

LAZAREV, Vasily Grigoryevich, 45, Soviet Air Force, commander
MAKAROV, Oleg Grigoryevich, 40, civilian, flight engineer

Flight Log

After the Salyut 1/Soyuz 11 disaster, the Soviets redesigned the Soyuz vehicle and attempted to launch new Salyuts without much success. A new Soyuz ferry was designed to carry just two crewmen, not three, this time wearing the obviously necessary pressure suits, with additional life support systems. Another innovation was the removal of solar panels and the reliance on batteries to sustain the craft during a two-day independent flight and rendezvous and docking with a Salyut.

Cosmos 496 was a test flight of this new Soyuz ferry vehicle in June 1972 and the first manned space flights to a new Salyut 2 were readied, only for the intended Salyut 2 (DOS-2) to fail to orbit in July that year. A military version of a Salyut (Almaz-1), also called Salyut 2 (1973-017A), actually got into orbit in April 1973, only to break apart and decay. Then the following month the third attempted Salyut, which could have been called Salyut 3 (DOS-3), became Cosmos 557 (1973-026A) when it, too, failed. The Soyuz ferry was tested again as Cosmos 573 in June 1973.

Without a Salyut to fly to, the Soviets decided to fly the ferry vehicle anyway, as Soyuz 12, crewed by Vasily Lazarev and Oleg Makarov, who were originally to have stayed aboard a Salyut. Their two-day flight (restricted to prevent low battery power making it impossible to attempt a re-entry) was announced as such beforehand to prevent western news reports of a Soviet manned flight meeting a "premature" end. Launch from Baikonur at 17:18 hrs was followed by orbital manoeuvres in the 51° orbit, mimicking those that would have been made to reach a Salyut. Maximum altitude attained was 344 km (214 miles). Lazarev and Makarov were hardly stretched and Earth resources photography seemed the high point. The capsule landed about 396 km (246 miles) south west of Karaganda after a flight lasting only 1 day 23 hours

170   The Second Decade: 1971–1980

Makarov (left) and Lazarev discuss their preparations for their upcoming mission with Georgi Beregovoy

15 minutes 32 seconds. But at least the Soviets had launched men into space again, for the first time in 27 months, re-qualifying Soyuz for further operational use.

Milestones

48th manned space flight
19th Soviet manned space flight
11th Soyuz manned space flight
1st Soviet announcement of planned mission duration

SKYLAB 4

Int. Designation	1973-090A
Launched	16 November 1973
Launch Site	Pad 39B, Kennedy Space Center, Florida
Landed	8 February 1974
Landing Site	Pacific Ocean
Launch Vehicle	Saturn 1B 208; spacecraft serial number CSM-118
Duration	84 days 1 hr 15 min 37 sec
Callsign	Skylab
Objective	Third and final Skylab resident crew (84 days)

Flight Crew

CARR, Gerald Paul, 41, USMC, commander
GIBSON, Edward George, 37, civilian, science pilot
POGUE, William Reid, 43, USAF, pilot

Flight Log

This open-ended mission was extended from 56 to 84 days to make the best of the final mission to the station. The launch of Skylab 4 was also delayed for ten days because hairline cracks were found in the tail fins of the Saturn 1B booster, built in 1964. The delay was convenient in that it gave the astronauts an opportunity to observe Comet Kohoutek when it was at its "best". The comet, heralded as the Comet of the Century by some astronomers, proved to be an anticlimax. The first all-rookie American crew since Gemini 8 took off at 09:01 hrs into extremely clear skies and headed for a rendezvous with Skylab, in its 50°, 442 km (275 miles) orbit. Gerry Carr, Ed Gibson and Bill Pogue docked at the second attempt but before Pogue had time to enter Skylab, he vomited into a sick bag, which the crew dumped "secretly" into the space station's garbage storage system.

Data storage tapes recorded their conversations and when these were played back to the ground, it caused much ill feeling. The commander, Carr, apologised for his error and the crew got down to work, but not without elements of complaint, giving them the erroneous reputation of being almost mutinous. This resulted from totally frank and private comments they were invited to make about the Skylab systems and work regime by ground control being made known to some press, who unfairly labelled the crew as particularly difficult. They did insist on a reduced workload at one point early in the mission when they thought they needed a rest. This prompted the media to label them the first crew to "strike" in space, which again was not totally correct.

However, Skylab 4 was a very impressive mission – America's longest to date and for some time afterwards at 84 days. The crew conducted 56 experiments, 26 science

The Skylab 4 crew after recovery. (L to r) Gibson, Pogue and Carr

demonstrations, and studied the sun for 338 hours. They photographed and observed the elusive Comet Kohoutek during one of several spacewalks, on Christmas Day 1973, which gave Carr, with 15 hours 18 minutes, the most EVA experience on Skylab. Gibson and Pogue made a 6 hour 33 minute EVA on 22 November; Carr and Pogue's Christmas excursion lasted 7 hours 1 minute; Carr and Gibson went outside on 29 December for 3 hours 28 minutes; and again on 3 February for 5 hours 19 minutes.

Skylab was abandoned and Carr made a manual re-entry after noticing a misalignment of the Command Module, splashing down near USS *New Orleans*, the first US crew *not* to come home watched by live television. Their mission of 84 days 1 hour 15 minutes 31 seconds remained America's longest until the flight of Norman Thagard aboard the Russian Mir station in 1995 – some 21 years later. Plans for Skylab 5 were abandoned, as were the hopes of flying a Skylab B workshop in 1976. For some years, it was hoped that a Shuttle might have visited the station in the late 1970s, but an increase in solar activity increased the atmospheric drag on the station, and delays in the Shuttle programme meant this plan was not practical. Skylab made an ignominious exit from the scene, making a spectacular re-entry during orbit 34,981 over Australia in 1979, with some of its debris surviving re-entry. The debris footprint was about 65 km × 3,860 km, with debris found east and north west of the town of Perth, and the largest piece being 82 kg of aluminium – thought to be the door of one of the film vaults.

Milestones

49th manned space flight
30th US manned space flight
14th Apollo CSM manned space flight
New duration record – 84 days 1 hour
15th US and 17th flight with EVA operations
Pogue celebrates his 44th birthday in space (23 January)

SOYUZ 13

Int. Designation	1973-103A
Launched	18 December 1973
Launch Site	Pad 1, Site 5, Baikonur Cosmodrome, Kazakhstan
Landed	26 December 1973
Landing Site	198 km southwest of Karaganda
Launch Vehicle	R7 (11A511); spacecraft serial number (7K-T) #33
Duration	7 days 20 hrs 55 min 35 sec
Callsign	Kavkaz (Caucasus)
Objective	Astrophysical and biological research scientific solo Soyuz mission

Flight Crew

KLIMUK, Pyotr Illich, 31, Soviet Air Force, commander
LEBEDEV, Valentin Vitalyevich, 31, civilian, flight engineer

Flight Log

The first Soviet manned space flight dedicated to science, Soyuz 13, uniquely configured with an array of Orion 2 celestial telescopes at the front of the Orbital Module in place of a docking system, was launched at 16:35 hrs local time into a 51° orbit, which would have a maximum altitude of 256 km (159 miles). The Orion telescope mount and the Oasis 2 protein manufacturing unit on board Soyuz 13 should have been flown on a Salyut. Comet Kohoutek's appearance may also have been irresistible. This and the recent decision to dock a Soyuz with an American Apollo in 1975, seemed the *raison d'être* for flying an independent mission of what was, essentially, a Soyuz 10/11-type spacecraft on a useful mission to bolster US confidence in the Soyuz. However, it was the loss of the Salyut that initiated the Soviet decision to fly the solo Soyuz, rather than demonstrating their ability to the Americans after the Soyuz 11 tragedy. A Soyuz was specially built for the mission. The original prime crew consisted of Commander Lev Vorobyov and Flight Engineer Valeri Yazdovsky, with back-ups Pyotr Klimuk and Vitaly Sevastyanov (who was soon replaced when medical problems affected his flying status). An all-rookie team prepared for the mission, with Valentin Lebedev replacing Sevastyanov. However, just three or four days before launch, the prime crew were deemed incompatible for working together in space and were replaced by the back-ups. They flew the mission without a new back-up team assigned. This caused bitterness, as both original crewmembers were very principled men, often speaking their mind and making "enemies" of the very people who selected crews to fly, resulting in their removal so close to the launch date.

The young Soyuz 13 crew, Pyotr Klimuk and Valentin Lebedev, joined the Skylab 4 crew in space, on the first occasion that US and Soviets were in orbit together

Inside the cramped Soyuz 13 module, Klimuk (left) and Lebedev pose for one of the few on-orbit images taken and released in connection with the astrophysical solo Soyuz mission

– although they did not rendezvous or communicate with each other. The Orion 2 took interesting ultraviolet images of planetary nebulae close to the star Capella, down to a stellar magnitude of 13. The closed-loop Oasis system bred two types of bacteria in a test of the practicalities of food production in space colonies.

For the first time, the Soviets announced that a mission was at its half-way stage, but did not announce its end at $T + 7$ days 20 hours 55 minutes 35 seconds, 198 km (123 miles) southwest of Karaganda until an hour after space monitor Geoffrey Perry of the Kettering Grammar School's remarkable team.

Milestones

50th manned space flight
20th Soviet manned space flight
12th Soyuz manned space flight
1st time both US and Soviet spacemen in orbit at the same time

SOYUZ 14

Int. Designation	1974-051A
Launched	3 July 1974
Launch Site	Pad 1, Site 5, Baikonur Cosmodrome, Kazakhstan
Landed	19 July 1974
Landing Site	Southeast of Dzhezkazgan
Launch Vehicle	R7 (11A511); spacecraft serial number (7K-TA) #62
Duration	15 days 17 hrs 30 min 28 sec
Callsign	Berkut (Golden Eagle)
Objective	First Almaz military space station resident crew programme

Flight Crew

POPOVICH, Pavel Romanovich, 43, Soviet Air Force, commander, 2nd mission
Previous mission: Vostok 4 (1962)
ARTYUKHIN, Yuri Petrovich, 44, Soviet Air Force, flight engineer

Flight Log

On 24 June 1974, the replacement for the first Almaz military space station (designated Salyut 2 to mask its real purpose and which was never manned because it tumbled out of orbit in 1973), was launched by a Proton from Baikonur. As with Salyut 2, transmissions from the Salyut 3 (Almaz-2) space station as it manoeuvred itself into the operating orbit were on a military frequency, the first indications that manned military space operations were to follow. This seemed to be confirmed when Soyuz 14 was launched into orbit at 23:51 hrs local time from Baikonur, into a 51° orbit which would attain a maximum altitude of 273 km (170 miles).

On board were Vostok veteran Pavel Popovich, a blast from the past indeed, and military flight engineer, Lt-Col. Yuri Artyukhin. A day and two hours after launch, Popovich had docked at the rear port of the station, which had a recoverable capsule at its front. Salyut 3's main instrument appears to have been a 10 m (33 ft) focal length reconnaissance telescope. According to the many reports about the regime of the crew, Popovich and Artyukhin worked to a strict eight hourly routine: work for eight hours, relaxation and exercise, then sleep. They used special elasticated exercise suits, called Atlet and Penguin, harnessed to a treadmill to maintain their cardiovascular condition.

The crew's announced schedule of activities focused on medical experiments but it is clear that they were much busier conducting classified reconnaissance work. Equipment used included the Polimnon-2M, Rezeda-5, Levkoi-3 and Amak-3 medical units. There was a scare when a massive solar storm sent a worrying amount of radiation towards Salyut, threatening to abort the mission, whose planned 14 day duration was

Popovich (left) and Artyukhin wearing Sokol suits during a break in training

confirmed by the Soviets when they announced that it was at its half-way point. Soyuz 14 landed south east of Dzhezkazgan at $T + 15$ days 14 hours 30 minutes 28 seconds. Salyut 3, its recoverable capsule still attached, remained to receive another manned crew. The military nature of Soyuz 14 overshadowed the fact that it was the Soviets' first successful space station mission.

Milestones

51st manned space flight
21st Soviet manned space flight
13th Soyuz manned space flight
1st dedicated manned military space flight

SOYUZ 15

Int. Designation	1974-067A
Launched	26 August 1974
Launch Site	Pad 1, Site 5, Baikonur Cosmodrome, Kazakhstan
Landed	28 August 1974
Landing Site	48 km southwest of Tselininograd
Launch Vehicle	R7 (11A511); spacecraft serial number (7K-TA) #63
Duration	2 days 0 hrs 12 min 11 sec
Callsign	Dunay (Danube)
Objective	Second Salyut 3 resident crew programme

Flight Crew

SARAFANOV, Gennady Vasilyevich, 32, Soviet Air Force, commander
DEMIN, Lev, 48, Soviet Air Force, flight engineer

Flight Log

The second manned occupation of Salyut 3 was to be made by Soyuz 15, which made a routine exit from Baikonur at 00:58 hrs on 26 August. Docking was scheduled to be made on the sixteenth orbit about a day later. However, Soyuz 15 approached Salyut far too quickly, closing in at a speed of about 10 m/sec (33 ft/sec) due to excessive and uncontrollable burns, from a distance of about 48 m (157 ft) during what was planned to be a manual approach and docking. As a result it missed the station, and by the time docking was to have been achieved, Soyuz 15 was 112 km (70 miles) ahead of Salyut. The crew reported that the spacecraft's manoeuvring controls were operating in reverse, thus braking firings led to an increase in velocity and vice versa.

The automatic rendezvous system Igla failed, though this was not immediately acknowledged by engineers from NPO Energiya, who pushed the blame on to the crew. For some time there were suggestions that the failure was due to human error, although the cosmonauts reportedly made some attempt to re-establish a compatible orbit and approach though it is not clear whether they attempted manual docking. Indeed, the youthful Gennady Sarafanov and the elderly Lev Demin, at 48 the oldest man in space at that time, never made another flight. The battery power and propellant levels on the spacecraft approached the limits of the mission and the flight had to be abandoned with an emergency retro-fire out of its maximum altitude 236 km (147 miles), 5° inclination orbit, and a night landing 48 km (30 miles) southwest of Tselininograd at $T + 2$ days 0 hours 12 minutes 11 seconds.

The Soviets, presumably mindful of their responsibilities with the Soyuz–Apollo link planned for the following year, attempted to write off the Soyuz 15 mission as a unique manned test of an unmanned re-supply vessel, a test of a totally automatic docking which would have ended with the crew coming home after two days anyway.

Soyuz 15 prime crew, left Sarafanov, right Demin

It was even claimed that the night landing was a deliberate plan, too! As it turned out, the flight of Soyuz 15 would be far from unique. In 1999, it was finally revealed that the Igla system had failed and was issuing false commands to the orientation and manoeuvring system. Thus, when Soyuz 15 was only 350 metres from Salyut, Igla "thought" it was 20 km away and initiated a long range engine burn. Despite evidence to the contrary, the inexperience of the crew did not help their case and they were tarnished with an official reprimand, even though it was not their fault. Salyut 3 could have supported another crew but there were no Soyuz vehicles ready for a launch and on 23 September, its programme was completed with the recovery of a small capsule containing exposed film. On 24 January 1975, less than a month after Salyut 4 was safely in orbit and with a crew aboard, Salyut 3 was commanded to a destructive re-entry over the Pacific Ocean.

Milestones

52nd manned space flight
22nd Soviet manned space flight
14th Soyuz manned space flight
1st grandfather in space (Demin)

SOYUZ 16

Int. Designation	1974-096A
Launched	2 December 1974
Launch Site	Pad 1, Site 5, Baikonur Cosmodrome, Kazakhstan
Landed	8 December 1974
Landing Site	30 km north of Dzhezkazgan
Launch Vehicle	R7 (11A511U); spacecraft serial number (7K-TM) #73
Duration	5 days 22 hrs 23 min 35 sec
Callsign	Buran (Snowstorm)
Objective	Soviet ASTP manned test-flight

Flight Crew

FILIPCHENKO, Anatoly Vasilyevich, 46, Soviet Air Force, commander, 2nd mission
Previous mission: Soyuz 7 (1969)
RUKAVISHNIKOV, Nikolay Nikolayevich, 42, civilian, flight engineer, 2nd mission
Previous mission: Soyuz 10 (1971)

Flight Log

Following three years of discussions which featured a Soyuz docking with a Skylab, or an Apollo docking with a Salyut, the first international manned space mission was agreed. The climax of *détente* between the USA and the Soviet Union in 1972 was marked by the agreement between President Richard Nixon and Premier Leonid Brezhnev for a joint flight between Apollo and Soyuz spacecraft in July 1975. The Apollo would be fitted with a docking module with an androgynous docking ring adapter, which would mate with a similar docking ring adapter on the front of the Orbital Module of the Soyuz. The resulting Apollo–Soyuz Test Project, ASTP, was a remarkable example of international cooperation and a fusion of technology and communications. To ensure that they had everything working well, the Soviets conducted two unmanned tests of the newly configured Soyuz as Cosmos 638 and 672, and followed this up with a manned flight of Soyuz 16, which would make a simulated rendezvous and docking with an imaginary Apollo.

Despite the close cooperation, the US was caught by surprise when the Soviets announced, without warning, that they had launched Soyuz 16, with Anatoly Filipchenko and Nikolay Rukavishnikov, at 14:40 hrs local time from Baikonur, and that it was already in orbit. At first, the Soyuz was placed into a 51° orbit with an apogee of 291 km (181 miles), too high for a real ASTP mission. Soyuz, however, moved into a more compatible orbit for an evaluation of the Soyuz guidance and manoeuvring system.

Filipchenko inside the OM during the Soyuz 16 ASTP dress rehearsal mission

The docking system had an imitation Apollo docking ring fixed to it and the cosmonauts used this to simulate various docking modes. They also changed the cabin atmosphere to an oxygen–nitrogen mix as used in Apollo. Soyuz conducted further orbital manoeuvring tests, including a circularisation burn, and signalled to the Apollo team to conduct a mock Apollo launch and to start important tracking tests. All in all, it was pretty unspectacular but highly successful, giving the US more confidence in the Soviet system after its Salyut and Soyuz 15 failures.

Filipchenko and Rukavishnikov landed in the cold steppe land at $T + 5$ days 22 hours 23 minutes 35 seconds and were immediately wrapped in thick overcoats.

Milestones

53rd manned space flight
23rd Soviet manned space flight
15th Soyuz manned space flight
1st manned use of R7/Soyuz U (11A511U) launch vehicle

SOYUZ 17

Int. Designation	1975-001A
Launched	11 January 1975
Launch Site	Pad 1, Site 5, Baikonur Cosmodrome, Kazakhstan
Landed	9 February 1975
Landing Site	110 km northeast of Tselinograd
Launch Vehicle	R7 (11A511); spacecraft serial number (7K-T) #38
Duration	29 days 13 hrs 19 min 45 sec
Callsign	Zenit (Zenith)
Objective	First Salyut 4 resident crew programme

Flight Crew

GUBAREV, Aleksey Aleksandrovich, 42, Soviet Air Force, commander
GRECHKO, Georgy Mikhailovich, 42, civilian, flight engineer

Flight Log

The Soviet's next space station, the civilian (DOS-4) Salyut 4 (1974-104A), was launched on 26 December 1974. It was basically the same as Salyut 1, but instead of two pairs of solar panels it had three steerable panels mounted centrally. Soyuz 17, the first visitor to Salyut 4, with rookie cosmonauts Aleksey Gubarev and Georgy Grechko aboard, was launched at 02:43 hrs local time from Baikonur and entered a 51° orbit which had a maximum altitude of 354 km (220 miles). Docking with Salyut's front port took place about a day later. The crew – who had found a notice telling them to wipe their feet! – entered Salyut and freshened its air and powered it up.

Salyut 4, equipped with a new Delta automatic navigation system, was dedicated to science, carrying seven astronomical, eight medical and at least six other technological experiments. The crew was kept so busy that it was estimated that Grechko covered 4.8 weightless km (3 miles) a day moving from instrument to instrument. They used a bicycle ergonometer, operated a solar telescope, grew plants in a space garden, observed a supernova, and tried out different muscle loading suits. The crew was quiet and methodical and was described as the least demonstrative ever flown.

The flight was so quietly followed that it came as some surprise that when Soyuz 17 landed it had flown the longest Soviet space flight – $T + 29$ days 13 hours 19 minutes 45 seconds – and had exceeded the flight time of the first US Skylab mission. Gubarev and Grechko came through cloud only 240 m (787 ft) high and landed in gusts of wind of up to 70 kph (43 mph), 110 km (68 miles) northeast of Tselinograd.

184 The Second Decade: 1971–1980

Grechko (left) and Gubarev evaluating the restraint harness on the Salyut 4 mock-up at TsPK

Milestones

54th manned space flight
24th Soviet manned space flight
16th Soyuz manned space flight

SOYUZ 18-1

Int. Designation	None – failed to reach orbit
Launched	5 April 1975
Launch Site	Pad 1, Site 5, Baikonur Cosmodrome, Kazakhstan
Landed	5 April 1975
Landing Site	Southwest of Gorno-Altaisk, Western Siberia, 1,200 metres up on the slope of the Teremok-3 mountain
Launch Vehicle	R7 (11A511); spacecraft serial number (7K-T) #39
Duration	21 min 27 sec
Callsign	Ural (Urals)
Objective	Intended second Salyut 4 resident crew

Flight Crew

LAZAREV, Vasily Grigoryevich, 47, Soviet Air Force, commander, 2nd mission
Previous mission: Soyuz 12 (1973)
MAKAROV, Oleg Grigoryevich, 42, civilian, flight engineer, 2nd mission
Previous mission: Soyuz 12 (1973)

Flight Log

The task of the Soyuz 18 cosmonauts, Vasily Lazarev and Oleg Makarov, given a Salyut mission at last, was to extend the Soviet manned duration record to about 60 days. They were launched at 16:02 hrs local time from Baikonur. The strap-on boosters of the SL-4 shut down as planned, the core stage shut down and all seemed to be going well as the staging process began. The second stage ignited and spewed its exhaust through the lattice-like framework attaching the core and second stage. These stages were to be separated by firing two sets of six pyrotechnic latches, one set at the top and the other at the bottom of the lattice structure.

Three of the upper latches fired prematurely, partially separating the upper stage on one side of its circumference. The flight of Soyuz 18 was in deep trouble and the crew was powerless to do anything during this automatic phase of the mission. With separation incomplete the second stage was now dragging a spent core stage. Soyuz veered 10° off course before the abort system gyroscopes detected the lack of control. Soyuz itself fired its engine to separate from the strangely configured, errant rocket. The Orbital Module and Instrument Module were jettisoned and the Descent Module was positioned for its low-speed, high-G re-entry from a height of 192 km (119 miles). The crew experienced only 400 seconds of weightlessness instead of their planned 60 days!

The crew, being subjected to as much as 20.6-G, were extremely concerned that they were heading for a landing in China, as relations between the two countries

186 The Second Decade: 1971–1980

Lazarev and Makarov perform one of many Soviet pre-flight traditions. Unfortunately, these did not help them complete their mission as planned

were not good at the time. In the event, they missed the border by 320 km (199 miles), landing 1,574 km (978 miles) downrange from Baikonur, at $T + 21$ minutes 27 seconds – the longest sub-orbital manned space flight, the shortest Soviet manned space flight, and the shortest two-person manned space flight. The capsule apparently landed on a mountain and had been rolling down before its parachute became snarled on a tree.

Rescue teams spotted them 30 minutes later and they were recovered some time later that day. Detailed accounts of the launch abort have been released over the years, including recollections from the two cosmonauts themselves. However, it is doubtful whether any announcement would have been made at the time had it not been for the fact that ASTP was looming. As it was, the worried USA officials were told that Soyuz 18 was launched on an old booster and that the Soyuz intended for the joint mission would use the uprated version first used to launch a crew on Soyuz 16, the ASTP dress-rehearsal mission the previous December.

Milestones

55th manned space flight
25th Soviet manned space flight
17th Soyuz manned space flight (1st sub-orbital)
1st aborted launch and emergency landing

SOYUZ 18

Int. Designation	1975-044A
Launched	24 May 1975
Launch Site	Pad 1, Site 5, Baikonur Cosmodrome, Kazakhstan
Landed	26 July 1975
Landing Site	54 km northeast of Arkalyk
Launch Vehicle	R7 (11A511); spacecraft serial number (7K-T) #40
Duration	62 days 23 hrs 20 min 08 sec
Callsign	Kavkas (Caucasus)
Objective	Second (originally 3rd) Salyut 4 resident crew programme

Flight Crew

KLIMUK, Pyotr Ilyich, 32, Soviet Air Force, commander, 2nd mission
Previous mission: Soyuz 13 (1973)
SEVASTYANOV, Vitaly Ivanovich, 39, civilian, flight engineer, 2nd mission
Previous mission: Soyuz 9 (1970)

Flight Log

It didn't take the determined Soviet space team long to recover from the Soyuz 18-1 abort. Soyuz 18 was launched at 19:58 hrs local time with the original back-up crew, which duly docked with Salyut 4 in the standard one day. Cosmonauts Klimuk and Sevastyanov began a work regime which involved detailed work on one experiment before moving on to another. This single discipline approach was illustrated by the multi-spectral photography session lasting from 8 to 11 June. The crew, which reached a maximum altitude of 361 km (224 miles) during the mission, took over 2,000 remote-sensing photographs, and Sevastyanov claimed that as a result the whereabouts of even the smallest ore deposit in the Soviet Union was known.

The crew repaired a cosmic ray detector and took over 600 pictures of the sun with the OTS solar telescope. One of the crew's X-ray photographs indicated that the celestial object Cygnus X-1 was a black hole. Thirteen days were spent on astrophysics, six on technical and ten on medical experiments. The crew also spent ten days of relaxation and two days were taken up with packing and unpacking equipment, the Soviets estimated. On 15 July, there came a "symbolic rendezvous" with the ASTP crews, when the spacecraft came to within 320 km (199 miles) of each other. It has been reported that such were their struggles with the space station's environmental control system towards the end of the mission that it was impossible to see out of the windows and the walls of the station became mouldy.

There was considerable interest about how the crew would feel after the flight of over 60 days. Soyuz 18 came home to a televised landing 54 km (34 miles) northeast of Arkalyk at $T + 62$ days 23 hours 20 minutes 20 seconds, 59 days of which were spent

The second crew to reside aboard Salyut 4, left Sevastyanov and Klimuk

making 900 orbits aboard Salyut 4. Doctors wanted to carry the crew from the spacecraft but Klimuk emotionally insisted that the crew egress on their own. They both admitted that re-adaptation to gravity took some time. Indeed, Klimuk once awoke to see Sevastyanov sleeping in his terrestrial bed with his arms raised in the air as if floating in weightlessness.

Salyut 4 supported a further mission, the unmanned Soyuz 20 in November. This was used to test, in part, the planned Progress re-supply mission profile that would begin with Salyut 6 in 1978. The station re-entered the atmosphere in February 1977.

Milestones

56th manned space flight
26th Soviet manned space flight
18th Soyuz manned space flight
Sevastyanov celebrates his 40th birthday in space (8 July)
Klimuk celebrates his 33rd birthday in space (10 July)

SOYUZ 19 AND APOLLO 18 (ASTP)

Int. Designation	1975-065A (Soyuz 19) – 1975-066A (Apollo 18)
Launched	15 July 1975
Launch Site	Pad 1, Site 5, Baikonur Cosmodrome, Kazakhstan (Soyuz 19); Pad 39B, Kennedy Space Center, Florida (Apollo 18)
Landed	21 July 1975 (Soyuz 19); 24 July 1975 (Apollo 18)
Landing Site	86 km northeast of Arkalyk (Soyuz 19); Pacific Ocean, 432 km west of Hawaii (Apollo 18)
Launch Vehicle	R7 (11A511U); spacecraft serial number 7K-TM #75 (Soyuz 19); Saturn 1B SA-210; spacecraft serial number CSM-111; Docking Module-2 (Apollo 18)
Duration	5 days 22 hrs 30 min 51 sec (Soyuz 19); 9 days 1 hr 28 min 24 sec (Apollo 18)
Callsign	Soyuz/Apollo
Objective	First international manned space flight; docking of an American Apollo with a Russian Soyuz spacecraft in Earth orbit, with crew exchanges, returning to Earth in own vehicles

Flight Crew

LEONOV, Aleksey Arkhipovich, 41, Soviet Air Force, commander Soyuz 19, 2nd mission
Previous mission: Voskhod 2 (1965)
KUBASOV, Valery Nikolayevich, 40, civilian, flight engineer Soyuz 19, 2nd mission
Previous mission: Soyuz 6 (1969)
STAFFORD, Thomas Patten Jr., 44, USAF, commander Apollo 18, 4th mission
Previous missions: Gemini 6 (1965); Gemini 9 (1966); Apollo 10 (1969)
BRAND, Vance DeVoe, 44, civilian, command module pilot Apollo 18
SLAYTON, Donald Kent "Deke", 51, USAF, docking module pilot Apollo 18

Flight Log

After three years of extraordinary cooperation between the superpowers, involving reciprocal visits by teams of scientists and spacemen, development of a compatible docking system and flight planning, Apollo and Soyuz were made ready for the big link up, amid publicity comparable with one of the first Moon landings. First to go were Aleksey Leonov and Valery Kubasov aboard Soyuz 19 at 17:20 hrs local time at Baikonur, watched live on television while the reserve, fully fuelled Soyuz sat on a

It may not have been a lunar mission, but this launch of Apollo 18 was significant in its own right

sister pad about 30 km (19 miles) away. Soyuz entered an orbit with an inclination of 51.8° and a maximum altitude of 220 km (137 miles).

Sixteen thousand km (9,942 miles) away, on Pad 39B at the Kennedy Space Center, America's last manned space flight for almost six years was about to begin, seven and a half hours after the ascent of Soyuz 19. Apollo's crew, the veteran Stafford, the mature Brand and the positively aged former unflown Mercury astronaut Slayton, reached orbit and extracted a docking module from the S-IVB stage of the Saturn 1B. The race in space began with Soyuz relatively passive and Apollo doing most of the manoeuvring. At $T + 51$ hours 49 minutes ASTP mission time, Apollo

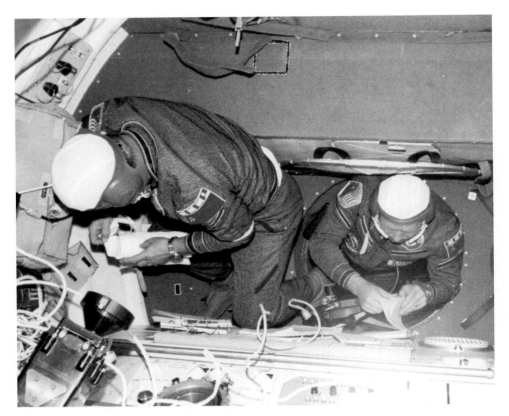

Kubasov and Leonov in the Soyuz 19 Orbital Module

and Soyuz docked together and a little later, with the connecting hatches open, Stafford and Leonov shook hands over Holland.

With the crews speaking their respective languages, visiting each other's spacecraft and exchanging TV conversations with President Gerald Ford and Premier Leonid Brezhnev, the mission took on the mantle of a space summit. At one point Brand, sitting in the Soyuz, spoke to Soviet TV viewers in Russian. The two spacecraft later separated for some joint experiments, one of which was to simulate a solar eclipse, and the rendezvous ended at $T + 102$ hours 16 minutes. During the series of crew transfers between the two spacecraft, Leonov spent 5 hours 43 minutes aboard Apollo, while Kubasov had been in the American spacecraft for 4 hour 47 minutes. Stafford had visited Soyuz for a total of 7 hours 10 minutes, Brand for 6 hours 30 minutes and Slayton for 1 hour 35 minutes.

Soyuz 19 landed at $T + 5$ days 22 hours 30 minutes 51 seconds, 86 km (53 miles) northeast of Arkalyk, watched live on TV. Apollo flew on, reaching a maximum altitude of 228 km (142 miles), almost as if savouring America's last days in space for as long as possible. The re-entry was textbook perfect, but as Apollo 18 descended towards USS *New Orleans* in the Pacific Ocean, 432 km (268 miles) west of Hawaii, Brand forgot to operate two switches to deploy the parachutes and shut down the RCS thrusters. When the drogue chute failed to deploy, Brand commanded it to do so manually, but the oscillations caused the still-armed thrusters to fire. Stafford shut them down but not before nitrogen tetroxide gas had boiled off and entered the cabin via a pressure fed valve. The astronauts began to cough and choke and Brand fell unconscious. He came round with the aid of oxygen from masks that were donned before splashdown at $T + 9$ days 1 hour 28 minutes 24 seconds. It was an unfortunate end to an historic and, it appears now, unique space flight.

Following the mission, there were enthusiastic plans for docking an American Shuttle with an improved (Salyut 6-class) space station in 1981, but relations between the two countries deteriorated with the Soviet invasion of Afghanistan and this contributed to the shelving of plans about future joint flights for almost 20 years.

Milestones

57th and 58th manned space flights
27th Soviet manned space flight
19th Soyuz manned space flight
31st US manned space flight
15th (and final) Apollo CSM manned space flight
1st joint US–Soviet manned space flight
1st time five people travel in "one" spacecraft
1st Soviet manned launch shown live on TV across the world
1st space traveller aged over 50

SOYUZ 21

Int. Designation	1976-064A
Launched	6 July 1976
Launch Site	Pad 1, Site 5, Baikonur Cosmodrome, Kazakhstan
Landed	24 August 1976
Landing Site	198 km southeast of Kokchetav
Launch Vehicle	R7 (11A511); spacecraft serial number (7K-TA) #41
Duration	49 days 6 hrs 23 min 32 sec
Callsign	Baikal (Baikal)
Objective	First Salyut 5 (Almaz) resident crew programme

Flight Crew

VOLYNOV, Boris Valentinovich, 41, Soviet Air Force, commander, 2nd mission
Previous mission: Soyuz 5 (1969)
ZHOLOBOV, Vitaly Mikhailovich, 39, Soviet Air Force, flight engineer

Flight Log

Another military Almaz space station, designated Salyut 5 (Almaz-3), was launched on 22 June 1976 (1976-057A), together with its conical, recoverable capsule, to continue operations which began with Salyut 3 in 1974. It also operated on frequencies used by reconnaissance satellites and previous military Salyuts. The Soyuz 21 ferry vehicle ascended from sunny Baikonur at 17:09 hrs local time, and one day later a smooth docking was completed. Maximum altitude achieved by Soyuz 21 was 274 km (170 miles) during the mission at 51.6°.

Salyut 5 was crammed with over 20 pieces of scientific equipment for science–technological experiments, biological investigations, astronomical and Earth observations, medical checks and technical evaluations. Apart from classified military work, the crew of veteran Boris Volynov and the first moustached cosmonaut, Vitaly Zholobov, were clearly going to have their hands full on the projected 60-day plus mission. The behaviour and breeding of fish was studied using an aquarium – another space first – and the cosmonauts monitored aerosol and industrial pollution.

They also grew crystals in microgravity and studied the behaviour of liquids as a prelude to in-orbit refuelling by tanker craft. Equipment was even used to harden and solder metals. It is interesting to note that the equipment used for Earth resources was unnamed, unlike all the other scientific experiments, clearly pointing to the reconnaissance nature of the mission, with its quick response being aided by the teleprinter on board. Soviet reports honed in on the science equipment on board but did little to describe the cosmonauts' other activities, which included monitoring the extensive air and sea operations around Siberia during Operation Sevier.

Inspecting the business end of their launch vehicle at Baikonur are Soyuz 21 cosmonauts Volynov (left) and Zholobov

Seemingly out of the blue to western observers, the end of the mission was announced just 12 hours before the crew was back on Earth. Volynov and Zholobov had evacuated the space station when an odour in Salyut's atmosphere became so acrid as to be unbearable. First time space flyer Zholobov's health had been deteriorating for some time – even prescribed medication seemed not to work – and this initiated reports of him suffering sensory deprivation due to prolonged isolation. But Volynov was also suffering (to a lesser extent) from what was assumed to be nitric acid fumes leaking from the propellant system of the Salyut. The cosmonauts landed at night 198 km (123 miles) southeast of Kokchetav at $T + 49$ days 6 hours 23 minutes 32 seconds. The next crew got ready with new equipment – gas masks.

Milestones

59th manned space flight
28th Soviet manned space flight
20th Soyuz manned space flight

SOYUZ 22

Int. Designation	1976-096A
Launched	15 September 1976
Launch Site	Site 1, Pad 5, Baikonur Cosmodrome, Kazakhstan
Landed	23 September 1976
Landing Site	148 km northwest of Tselinograd
Launch Vehicle	R7 (11A511U); spacecraft serial number (7K-TM) #74
Duration	7 days 21 hrs 52 min 17 sec
Callsign	Yastreb (Hawk)
Objective	In-flight testing and evaluation of space station-designated scientific equipment on a solo Soyuz mission

Flight Crew

BYKOVSKY, Valery Fyodorovich, 42, Soviet Air Force, commander, 2nd mission
Previous mission: Vostok 5 (1963)
AKSENOV, Vladimir Viktorovich, 41, civilian, flight engineer

Flight Log

In July 1975, Soyuz spacecraft serial number 74 had been prepared as a second back-up vehicle for the ASTP programme. The first Soyuz back-up vehicle, serial number 76, had been fuelled and sitting on Pad 31 in case the primary Soyuz 19 mission failed. As it had to be launched within 75 days of being fuelled, it was dismantled and sent back to NPO Energiya for short-term storage. The DM was recycled to be used on Soyuz 31 but the OM and SM were used in a display with a mock-up DM at the Energiya museum. Spacecraft 74, being unfuelled, had a longer shelf life and was available for its own mission. This independent Soyuz spacecraft, rather than a ferry, was then rather ingeniously and cost-effectively converted into a specialised Earth observatory, manned by cosmonauts Valery Bykovsky and Vladimir Aksenov. It was launched into a unique 64.5° inclination orbit at 14:48 hrs local time at Baikonur. The docking system on the front of the Orbital Module had been replaced with the East German Karl Zeiss MKF-6 multi-spectral camera. This was loaded with film and operated from within the Orbital Module in much the same way as the Orion astronomical equipment on board the Soyuz 13 spacecraft.

From its 296 km (184 miles) maximum altitude orbit, Soyuz 22 made special observations of the Soviet Union and East Germany and may not have been able to resist some military reconnaissance of the coast of Norway where a NATO exercise was being conducted. The MKF camera was able to take six photographs simultaneously in the visible and infrared bands, providing a stereo imaging capability which would be useful to agricultural experts, cartographers, geologists and hydrol-

Aksenov recording data during the Soyuz 22 mission

ogists. The camera took pictures in 164 km (102 miles) swathes with a resolution of 28 m (92 ft). The Soviets claim that the resulting pictures helped find the best route for the construction of a new railway, identified optimum timber and production sites, and mapped tidal zones to assist in the siting of tidal power stations. Images also assessed mineral prospects in the continental shelf, harvest projections, land reclamation possibilities and atmospheric pollution monitoring. In all, the MKF took 2,400 images covering 30 special targets.

Bykovsky and Aksenov came home 148 km (92 miles) northwest of Tselinograd at $T + 7$ days 21 hours 52 minutes 17 seconds. Soyuz 22 was the final solo flight of a Soyuz spacecraft. All the future flights were associated with space stations, and though it seems other solo flights were expected, no such long-term programme was planned.

Milestones

60th manned space flight
29th Soviet manned space flight
21st Soyuz manned space flight
Final solo "scientific" Soyuz mission

SOYUZ 23

Int. Designation	1976-100A
Launched	14 October 1976
Launch Site	Pad 1, Site 5, Baikonur Cosmodrome, Kazakhstan
Landed	16 October 1976
Landing Site	Lake Tengiz 194 km southwest of Tselinograd
Launch Vehicle	R7 (11A511); spacecraft serial number (7K-T) #65
Duration	2 days 0 hrs 6 min 35 sec
Callsign	Radon (Radon)
Objective	Intended second Salyut 5 resident crew

Flight Crew

ZUDOV, Vyacheslav Dmitriyevich, 34, Soviet Air Force, commander
ROZHDESTVENSKY, Valery Ilyich, 37, Soviet Air Force, flight engineer

Flight Log

Like Apollo 13 six years before, Soyuz 23 seemed to have its gremlins. Firstly, the cosmonaut transfer bus broke down on the way to the launch pad. During the ascent, high winds caused the R7 to deviate from its intended flight path to such a degree that it almost triggered a launch abort. This resulted in a lower orbit than planned. Then the Igla system failed and a challenging recovery awaited the rookie crew.

Armed with gas masks to prevent possible asphyxiation by Salyut 5's acrid atmosphere, rookie cosmonauts Vyacheslav Zudov and Valery Rozhdestvensky took off in the Soyuz 23 ferry at 22:40 hrs local time from Baikonur for a routine flight to the space station. The flight was to become an arduous and much shorter one than the planned 14-day mission. The *bête noir* of the Salyut programme reared its ugly head again, as Soyuz 23 approached and failed to dock because of a failure with the automatic system. The rendezvous radar electronics had failed. The ferry, without solar panels and running on just batteries, had a limited lifetime in its 275 km (171 miles), 51.6° orbit. The happy-go-lucky crew, as they had been described before the launch, shut down as many systems as possible and floated around cold and disappointed until the next convenient re-entry pass for a landing in the main recovery area. The Soviets had failed to dock with their space stations seven times in eleven attempts.

Following a curt Soviet announcement that the visit to Salyut 5 had been cancelled, Soyuz 23 fired the retro-rockets and came home. Though de-orbit burn and re-entry occurred as programmed, the high winds in the recovery area resulted in a 121 km overshoot from the planned landing near Arkalyk and they descended in a thick fog, with temperatures at −22°C. The 32 km (20 miles) wide Lake Tengiz was situated in the recovery area, 194 km (121 miles) southwest of Tselinograd, and the

The only Russian crew to splash down at the end of their mission. Zudov (left) and Rozhdestvensky

unlucky duo, expecting a hard ground impact, headed straight for it, making the first splashdown of sorts in the Soviet space programme, at $T + 2$ days 0 hours 6 minutes 35 seconds. As the lake was covered with ice, which was broken by the Soyuz descent capsule, the recovery, in temperatures of $-20°C$ in the fog and in darkness, was very difficult and not without "a certain amount of heroism," said the Soviets later. The true extent of the danger and difficult recovery did not emerge for some years. The capsule rapidly cooled in the freezing waters and the cosmonauts took off their thin Sokol pressure suits, instead donning their warmer flight suits before breaking into some of their emergency rations. To conserve energy they stopped moving and talking, but this hampered efforts to find them, as the signal beacon was obscured by the fog and corrosion activated the reserve parachute which rapidly filled with water. Fortunately, as the lake was shallow, it did not drag the capsule beneath the water line, allowing the equalisation valve to supply additional oxygen to the dwindling supplies on board. The capsule was found by chance and in extreme conditions, with a lack of adequate equipment meaning that the recovery had to wait until the next morning. It was snowing and several attempts to reach the capsule were thwarted. The crew were in contact with their rescuers, but by the next day the antenna had frozen, and rescue teams thought that they would find a dead crew. The weather improved slightly, however and divers communicated with the crew, but were unable to attach a lifting cable. The only way to rescue the crew was by towing the capsule to finally retrieve a cold and exhausted crew. The rookies of Soyuz 23 never flew again.

Milestones

61st manned space flight
30th Soviet manned space flight
22nd Soyuz manned space flight
1st Soyuz manned splash down
Final manned use of R7 11A511

SOYUZ 24

Int. Designation	1977-008A
Launched	7 February 1977
Launch Site	Pad 1, Site 5, Baikonur Cosmodrome, Kazakhstan
Landed	25 February 1977
Landing Site	36 km northeast of Arkalyk
Launch Vehicle	R7 (11A511U); spacecraft serial number: (7K-TA) #66
Duration	17 days 17 hrs 25 min 58 sec
Callsign	Terek (Terek)
Objective	Second (originally third) Salyut 5 resident crew programme

Flight Crew

GORBATKO, Viktor Vasilyevich, 42, Soviet Air Force, commander, 2nd mission
Previous mission: Soyuz 7 (1969)
GLAZKOV, Yuri Nikolayevich, 37, Soviet Air Force, flight engineer

Flight Log

Baikonur was treated to yet another fireworks display at 23:12 hrs local time when another Soyuz departed at night. Soyuz 24 made four orbital manoeuvres and a day later got to within 1,500 m (4,921 ft) of Salyut 5 at a speed of 2 m/sec (6.5 ft/sec). When Igla again failed, Commander Viktor Gorbatko took control at 80 m (262 ft) distance, and with Salyut illuminated by Soyuz 24's searchlight, he made a smooth docking. After a sleep period inside the Soyuz and wearing breathing apparatus, they entered the space station, and to their surprise found conditions comfortable. There was no longer an acrid odour. A full operational mission lasting 17 days, in a 51.6° orbit and at a maximum altitude of 281 km (175 miles), was on the agenda.

The cosmonauts conducted many classified military experiments and had much success with the science equipment. Soldering tests were successful but the casting of metals was not. The cosmonauts conducted crystal growth experiments, observed fungi and fish roe development in microgravity, photographed the sun, carried out Earth resources photography and made a study of glacial precipitation based on observations made by the Soyuz 21 crew. Continuing investigations into the effects of weightlessness on the human body, the crew conducted several biological and medical tests, including regular electrocardiogram sessions and studies into space sickness. A unique event occurred on 21 February when much of the spacecraft's atmosphere was purged, using air in onboard containers.

The highly successful mission ended quietly at $T + 17$ days 17 hours 25 minutes 58 seconds, outside the planned recovery zone, in snow, high winds, low cloud and sub-zero temperatures, 36 km (22 miles) northeast of Arkalyk. After leaving the

The final military Soyuz crew. Glazkov and Gorbatko

capsule to await the rescue team, the cosmonauts found it too cold outside and re-entered the capsule. The rescue teams arrived an hour after landing. It was from the experiences on several Soyuz off nominal landings (notably the Soyuz abort, Soyuz 23 and Soyuz 24) that improvements to survival gear were authorised for future missions. This was the final military Soyuz flight and the last flight of a manned Almaz station. Though a further station launch was planned for the early 1980s, it was cancelled. Salyut 5 finally re-entered the atmosphere in August 1977.

Milestones

62nd manned space flight
31st Soviet manned space flight
23rd Soyuz manned space flight
Final Soyuz "Almaz" military station mission

SOYUZ 25

Int. Designation	1977-099A
Launched	9 October 1977
Launch Site	Pad 1, Site 5, Baikonur Cosmodrome, Kazakhstan
Landed	11 October 1977
Landing Site	184 km from Tselinograd
Launch Vehicle	R7 (11A511U); spacecraft serial number: (7K-T) #42
Duration	2 days 0 hrs 44 min 45 sec
Callsign	Foton (Photon)
Objective	First intended Salyut 6 resident crew programme

Flight Crew

KOVALENOK, Vladimir Vasilyevich, 35, Soviet Air Force, commander
RYUMIN, Valery Viktorovich, 38, civilian, flight engineer

Flight Log

The Soviet Union started its 20th anniversary celebrations of Sputnik 1 with the launch of the Salyut 6 (DOS 5-1) space station on 29 September. Salyut 6 (1977-097A) was very similar to Salyut 4 but significantly upgraded, equipped with more experiments, two docking ports instead of one allowing for re-supply missions, and an operational EVA system. Rhetoric was emotional as the Soyuz 25 crew, rookies Vladimir Kovalenok and Valery Ryumin, prepared to start a new era in space operations, the third decade of Soviet space flight. Their launch at 07:40 hrs local time was unimpressive since the rocket almost immediately disappeared into low cloud.

Kovalenok and Ryumin, heading for a stay of 90 days or more in space, made the routine rendezvous with Salyut 6 and things looked good. Soyuz 25, which reached a maximum altitude of 349 km (217 miles) during the mission, at 51.6°, made contact but it was only a soft dock. Radio silence fell over the mission. The cosmonauts made three more desperate attempts to get a hard dock but ran out of time before the critical limit in the ferry vehicle's power capability was reached. The Soviets announced that the much heralded mission had ended prematurely and ignominiously. There are conflicting reports on the docking operations; some indicate the orientation of the station was incorrect, while other suggest that the docking mechanism failed. Whatever the reason, once again a crew could not enter a station.

The disappointed crew – the last all-rookie crew in Soviet history – came home, 184 km (114 miles) from Tselinograd at $T + 2$ days 0 hours 44 minutes 45 seconds, after the eighth docking failure in thirteen Soviet attempts. There were concerns that the docking port had been damaged, perhaps by imperfections in crew control, since Kovalenok and Ryumin did not get the usual full honours on their return, although

Studying hard for a mission they could not complete, Ryumin (left) and Kovalenok were the intended first crew to Salyut 6

they did return to space. If the port had been damaged it did not matter, because for the first time there was an alternative port on a Salyut.

Milestones

63rd manned space flight
32nd Soviet manned space flight
24th Soyuz manned space flight

SOYUZ 26

Int. Designation	1977-113A
Launched	10 December 1977
Launch Site	Pad 1, Site 5, Baikonur Cosmodrome, Kazakhstan
Landed	16 March 1978 (aboard Soyuz 27)
Landing Site	264 km west of Tselinograd
Launch Vehicle	SL-4
Duration	96 days 10 hrs 0 min 7 sec
Callsign	Tamyr (Tamyr)
Objective	Amended 1st resident crew programme to Salyut 6

Flight Crew

ROMANENKO, Yuri Viktorovich, 33, Soviet Air Force, commander
GRECHKO Georgy Mikhailovich, 45, civilian, flight engineer, 2nd mission
Previous mission: Soyuz 17 (1975)

Flight Log

Soyuz 26 was originally planned as the second long-duration mission to Salyut 6 with rookies Yuri Romanenko and Aleksandr Ivanchenkov, who had been paired since 1973. After the Soyuz 25 abort, the Soviets decided not to fly all-rookie crews again and Romanenko and Ivanchenkov were split. Romanenko was to command what was now to be the first long-duration mission, with Georgy Grechko as the new flight engineer. Grechko had been drafted in from the intended third crew as he was a Salyut designer and would be able to evaluate any fix to the forward docking port that may have been damaged by Soyuz 25.

Soyuz 26 got underway at 06:19 hrs local time at Baikonur, and just over two days later Romanenko made a manual docking at the rear port of Salyut. Had the forward port been damaged by the Soyuz 25 attempt, the second port on Salyut 6 had saved the day. Romanenko and Grechko floated into the new space station to sample the delights of several improvements since Salyut 4, in their 356 km (221 miles), 51.6° orbit.

The station was equipped with a temperature regulation and water regeneration system, external TV cameras, EVA handrails, airlock, sun sensor, waste ejection airlock, dust filter, running track, shower, ion sensor and the MKF multi-spectral camera, whose housing took most of the room in the aft pressurised section. The rear docking port replaced the Salyut main propulsion system engine chamber which was divided into two either side of the port. Salyut 6 was equipped internally with several scientific experiments, including a smelting furnace for crystal growth and metal-lurgical processing brought up later by cargo vehicles. Much emphasis was placed on medical experiments since this mission was to exceed all previous ones in duration.

Romanenko and Grechko near the main control panels of Salyut 6 during their record-breaking mission

On 19 December, the cosmonauts' routine work was interrupted by the first EVA in nine years, when Grechko, wearing a new semi-rigid spacesuit called Orlan with a portable life support system in a back pack (which also acted as the suit's "back door" entry), emerged to inspect possible damage to the forward docking port. Grechko did not find any damage during the 20 minute EVA. He had practiced his EVA inside a huge water simulator, called the Hydrolab, at Star City. During the 88 minutes of depressurisation, the space-suited Romanenko, keen to look outside as far as possible, floated free. For several years after the flight, Grechko maintained a "story" that his instinctive grab saved Romanenko from becoming the first person to make an independent EVA and meet a lingering death in space. The commander, however, insisted that he was always attached by a small safety line and more recently Grechko has admitted that it was his "big joke" and that Romanenko was never in danger of floating away from the station.

Relations between the two were not always good. They had been thrown together with six weeks' notice for the mission and the rookie commander had a much older and more experienced person as his second in command and flight engineer. Romanenko spent a good part of the mission with a raging toothache which he insisted be kept secret in case the mission be shortened. Wrapped with a scarf around his painful jaw, the miserable Romanenko swallowed aspirin after aspirin in the hope of a cure.

After seeing in the New Year, the crew became the first to receive visitors when Soyuz 27 arrived on 11 January. Before the Soyuz 25 abort, the first visiting crew would have comprised two rookie cosmonauts, Vladimir Dzhanibekov and Pyotr Kolodin on a test of the dual docking port capability. The docking failure saw Oleg Makarov replace Kolodin. Their stay of eight days ended with their return in Soyuz 26, leaving a fresh craft for the resident crew. Another major milestone in the Soviet space programme occurred on 20 January 1978, when the unmanned Progress 1 tanker was launched. It made an automatic docking at the rear of Salyut 6, carrying compressed air, food, water, films, letters, parcels and scientific equipment. It also carried propellants and conducted the first in-flight refuelling in space, replenishing Salyut 6's tanks. Progress, based on the Soyuz ferry but with no Descent Module, also acted as a waste disposal unit. Crammed with rubbish by the cosmonauts, it undocked from Salyut 6 on 6 February and made a commanded re-entry, but not before conducting a rendezvous test from a distance of 16 km (10 miles).

Romanenko and Grechko got to work on some of the equipment that had been delivered by Progress 1. This included a solar electric furnace capable of temperatures of 1,000°C which was used to study the diffusion processes in molten metals and the interaction of solid and liquid metals in weightlessness. Another piece of equipment, called Medusa, was used to assess the effects of radiation on amino acids and other biological building blocks. Other equipment such as air purification filters and lithium hydroxide canisters were also fitted into the Salyut's environmental control system. The next highlight was the arrival in March of the Soyuz 28 crew, with the Czechoslovakian visitor, Vladimir Remek and his commander Aleksey Gubarev, who arrived as the crew was breaking the 84-day duration record held by the Skylab 4 crew since 1974.

After the brief Soyuz 28 visit, the Soyuz 26 mission ended at $T + 96$ days 10 hours 0 minutes 7 seconds, aboard the Soyuz 27 descent capsule, 264 km (164 miles) west of Tselinograd. Grechko learned that his father had died during the flight, a fact that Romanenko had known but was asked to keep from his flight engineer. Re-adaptation to 1-G took a long time but the cosmonauts were nonetheless fit and well. The first visit to Salyut 6 had been an unqualified success and a major milestone in the Soviet programme. That Romanenko had been part of it was later deemed prophetic.

Milestones

64th manned space flight
33rd Soviet manned space flight
25th Soyuz manned space flight
New duration record – 96 days 10 hours
1st crew to receive visiting crew
1st in-flight refuelling in space
3rd Soviet and 18th flight with EVA operations

SOYUZ 27

Int. Designation	1978-003A
Launched	10 January 1978
Launch Site	Pad 1, Site 5, Baikonur Cosmodrome, Kazakhstan
Landed	16 January 1978 (aboard Soyuz 26)
Landing Site	307 km west of Tselinograd
Launch Vehicle	R7 (11A511U); spacecraft serial number (7K-T) #44
Duration	5 days 22 hrs 58 min 58 sec
Callsign	Pamir (Pamirs)
Objective	First exchange of resident crew Soyuz craft for fresher vehicle; structural tests of two docking ports on same space station

Flight Crew

DZHANIBEKOV, Vladimir Aleksandrovich, 35, Soviet Air Force, commander
MAKAROV, Oleg Grigoryevich, 45, civilian, flight engineer, 3rd mission
Previous missions: Soyuz 12 (1973); Soyuz 18-1 (1975)

Flight Log

This all-Soviet visiting mission was designed to evaluate the structural strength of the new second docking port with another Soyuz already docked to the station. In light of the Soyuz 25 failure, it also provided a good opportunity to confirm what the Soyuz 26 cosmonauts had discovered during their EVA; that the front port of Salyut was able to receive visiting craft. With the two cosmonauts aboard their Soyuz 27 ferry, the Salyut 6 space station flew over Baikonur and was 17 minutes further away when the SL-4 booster came to life at 17:26 hrs local time.

Before manually docking, Dzhanibekov flew around Salyut 6 to inspect its suspect port carefully before berthing. Apart from the obvious firsts, four people were on board a space station at one time and the space station was now 37 m (121 ft) long and weighed 32 tonnes. The first space visitors, reaching 352 km (219 miles) in the 51.6° orbit, had brought with them the Soviet–French Cyton biology experiment, and also conducted what was called the Resonance Experiment, which involved "jumping" on the floor, restrained by bungee cords, to measure the stresses exerted on the space station and the two docked vehicles. Dzhanibekov, an expert in electrical engineering, also gave the Salyut station the once-over.

After their short stay, the visitors packed Soyuz 26 with equipment, samples and other packages, exchanged their custom couches, and headed home. They landed at $T + 5$ days 22 hours 58 minutes 58 seconds, 307 km (191 miles) west of Tselinograd.

210 The Second Decade: 1971–1980

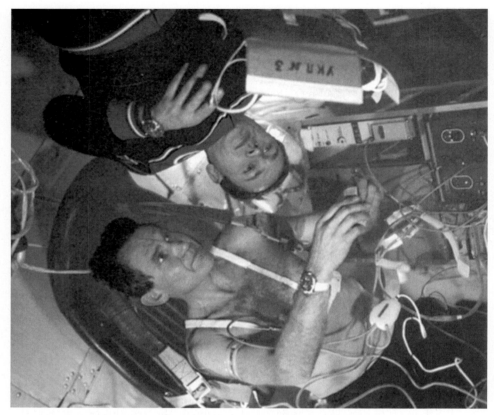

Dzhanibekov and Makarov conducting medical tests during their week on the station for the first ferry exchange mission

Milestones

65th manned space flight
34th Soviet manned space flight
26th Soyuz manned space flight
1st manned re-supply mission
1st manned dual docking
1st full crew to land in a different spacecraft

SOYUZ 28

Int. Designation	1978-023A
Launched	2 March 1978
Launch Site	Pad 1, Site 5, Baikonur Cosmodrome, Kazakhstan
Landed	10 March 1978
Landing Site	307 km west of Tselinograd
Launch Vehicle	R7 (11A511U); spacecraft serial number (7K-T) #45
Duration	7 days 22 hrs 16 min
Callsign	Zenit (Zenith)
Objective	First international crew – Czech visiting mission to Salyut 6

Flight Crew

GUBAREV, Aleksey Aleksandrovich, 45, Soviet Air Force, commander, 2nd mission
Previous mission: Soyuz 17 (1975)
REMEK, Vladimir, 29, Czech Army Air Force, cosmonaut researcher

Flight Log

The Soviet Union decided to fly guest cosmonauts from the Eastern Block "Interkosmos space science programme" on space station visits in 1976, at the same time that the USA announced that it was to fly West Europeans on the Space Shuttle. Three countries, Poland, East Germany and Czechoslovakia, were chosen to supply the first candidates for these missions, and Czechoslovakia's Vladimir Remek came out of the political hat first. He was teamed with Aleksey Gubarev for the mission, which began with a night launch at 20:28 hrs local time. The docking the following day was shown live on television via a camera on Soyuz 28. Once linked together, the two visiting cosmonauts came through the Salyut 6 entry hatch to a welcome so long-lasting and exuberant that ground controllers told the foursome to sober up, although nothing more powerful than cherry juice was drunk.

Remek got down to work with some national science experiments, most of which were conducted using onboard Soviet equipment. He used the Splav furnace to conduct the Morava experiment to study the growth of super pure crystals in microgravity and the possibility of obtaining semi-conductor optical materials. The Extinctica experiment observed the change in the brightness of stars when viewed through the atmosphere. The Chorella experiment studied the growth of algae cultures in a nutrient medium, and an oxymeter was used to study the concentrations of oxygen in human tissue in weightlessness.

The flight took place with enough national rhetoric to cover 50 missions, but the irony was that ten years later the Soviets admitted that they were worthless propaganda missions. Remek's was the first of many international missions which after a

Remek (at rear) working in Salyut 6 with his commander Gubarev (left). Salyut commander Romanenko is on the right

while became rather monotonous other than to the country involved at the time or to ardent space watchers. What they did achieve was access to a lot of unique experiments prepared by specialists outside of the Soviet Union, expanding the research programme of the station. For nine Soviet cosmonauts however, it cost them perhaps their only chance of a seat into space after years of waiting and training, to foreign "part-time" cosmonauts with only months of preparation. Soyuz 28, which reached 353 km (219 miles) during the 51.6° mission, undocked and landed at $T + 7$ days 22 hours 16 minutes, 307 km (191 miles) west of Tselinograd, where a fleet of jets and helicopters were waiting.

Milestones

66th manned space flight
35th Soviet manned space flight
27th Soyuz manned space flight
1st manned space flight by non-Soviet, non American
1st Interkosmos mission
1st manned space flight by a Czechoslovakian

SOYUZ 29

Int. Designation	1978-061A
Launched	15 June 1978
Launch Site	Pad 1, Site 5, Baikonur Cosmodrome, Kazakhstan
Landed	2 November 1978 (aboard Soyuz 31)
Landing Site	307 km west of Tselinograd
Launch Vehicle	R7 (11A511U), spacecraft serial number (7K-T) #46
Duration	139 days 14 hrs 47 min 32 sec
Callsign	Foton (Photon)
Objective	Second Salyut 6 resident crew programme

Flight Crew

KOVALENOK, Vladimir Vasilyevich, 36, Soviet Air Force, commander, 2nd mission
Previous mission: Soyuz 25 (1977)
IVANCHENKOV, Aleksandr Sergeyevich, 37, flight engineer

Flight Log

Yet another night launch, at 01:17 hrs Baikonur time on 16 June, dispatched Vladimir Kovalenok and Aleksandr Ivanchenkov towards a rendezvous and docking with the vacant Salyut 6 space station after a one day solo flight in Soyuz 29. The two entered the Salyut to find a welcoming letter from Romanenko and Grechko and soon got to work with some repair jobs. With a balky ventilator repaired and the airlock given the once-over, Kovalenok and Ivanchenkov embarked on a three day production cycle using the Splav furnace. Before the flight, the cosmonauts had been given special instruction in Earth resources observation and potential strategic reconnaissance, flying a TU-134 at a height of 9,200 m (30,180 ft).

During their Salyut mission, some 36 times higher up at 366 km (227 miles) in the 51.6° orbit, the cosmonauts eventually took over 18,000 photographs using the onboard MKF-6m multi-spectral camera and a new topographical camera called the KT-140. To assess their ability to identify objects on the ground, an area near Rostow was specially laid out with grains, grasses and other types of vegetation. Following the standard Soviet procedures, Kovalenok and Ivanchenkov concentrated on one experiment at a time using a scheduled programme, before moving on to another. These tasks also included work on a new furnace called Kristall which was used to assess glass fusing techniques for the semi-conductor industry. This was delivered with Progress 2 later in the mission.

The cosmonauts had a busy time in space, receiving manned spacecraft and unloading several unmanned tankers. The first visitor was Soyuz 30 on 29 June, with a Polish guest cosmonaut. This visit was followed by Progress 2, on 9 July, which

Kovalyonok (with Ivanchenkov reflected in visor) stands on the Salyut during the first operational Salyut EVA.

remained docked for 25 days and automatically transferred 600 kg (1,323 lb) of fuel and oxidiser. On 29 July, 45 days after their launch, the two exited Salyut for an EVA which lasted 2 hours 5 minutes. Ivanchenkov, the lead EVA cosmonaut, retrieved several materials samples from the side of Salyut, including rubbers, polymers and biopolymers, which would be studied on the ground by scientists to assess the effects of exposure to space conditions. After completion of the assigned EVA tasks, Kovalenok requested an extension to the televised spacewalk, just to give the two cosmonauts a chance to sightsee and relax after being cooped up for so long.

By 8 August, the manned occupation of Salyut 6 had totalled 171 days, exceeding the total manned occupation of America's Skylab. On 10 August, Progress 3 arrived bearing food, water, oxygen, new processing equipment and Kovalenok's guitar. Progress 3 remained attached for 12 days. The rear docking port was again left vacant, this time to receive Soyuz 31 on 28 August, with an East German guest. The guest crew left in Soyuz 29, leaving the newer Soyuz attached to the rear of the station, On 21 September, the Soyuz 29 crew became the first space travellers to remain in space for 100 days, and seven days later, Ivanchenkov was celebrating his birthday in orbit.

In a new space first, Kovalenok and Ivanchenkov entered Soyuz 31, undocked from the rear and re-docked at the front in preparation to receive a further Progress delivery. Actually, Salyut 6 did the flying, reorienting itself so that Soyuz faced the front port, rather than the rear port from which it had undocked. As a precaution, in case the re-docking failed, the Soviets planned this unique operation during a standard landing window. Progress 4 duly followed on 4 October, carrying a cargo which included fur boots – either in preparation for the cosmonauts' winter landing or because Salyut was becoming a cold home. On 9 October, the Soviets announced that the mission would end after operations with Progress 4 had been completed. The tanker's final operational task was to fire its engine, to raise the Salyut 6 orbit in preparation for a vacant period before the arrival of a new long duration crew. Progress undocked for its standard destructive re-entry on 24 October.

Soyuz 31, with the Soyuz 29 cosmonauts aboard, landed on 2 November, 180 km (112 miles) southeast of Dzhezkazgan after a mission lasting 139 days 14 hours 47 minutes 32 seconds. The cosmonauts' re-adaptation to 1-G after this record period of weightlessness was considered very good, and this was the result of extensive and intensive exercise while in orbit.

Milestones

67th manned space flight
36th Soviet manned space flight
28th Soyuz manned space flight
1st manned space flight to exceed 100 days
New duration record – 139 days 14 hrs
1st manned spacecraft transfer between docking ports
4th Soviet and 19th flight with EVA operations
Ivanchenkov celebrates 38th birthday in space (28 September)

SOYUZ 30

Int. Designation	1978-065A
Launched	27 June 1978
Launch Site	Pad 1, Site 5, Baikonur Cosmodrome, Kazakhstan
Landed	5 July 1978
Landing Site	300 km west of Tselinograd
Launch Vehicle	R7 (11A511U); spacecraft serial number (7K-T) #67
Duration	7 days 22 hrs 2 min 59 sec
Callsign	Kavkaz (Caucasus)
Objective	Polish Salyut 6 visiting mission

Flight Crew

KLIMUK, Pyotr Ilyich, 35, Soviet Air Force, commander, 3rd mission
Previous missions: Soyuz 13 (1973); Soyuz 18 (1975)
HERMASZEWSKI, Miroslaw, 36, Polish Air Force, cosmonaut researcher

Flight Log

At 20:27 hrs local time at Baikonur, Soyuz 30 thundered into the skies, heading for Salyut 6 and its resident cosmonauts Kovalenok and Ivanchenkov. On board were the seasoned and youthful veteran Pyotr Klimuk and his Polish cosmonaut researcher, Miroslaw Hermaszewski. Their spacecraft docked with Salyut 26 hours 19 minutes later. During the brief mission, which reached 343 km (213 miles) in the 51.6° orbit, materials processing was high on the list of Polish priorities.

The Serena experiment was used to attempt to make usable cadmium–tellurium–mercury semi-conductor material, the most sensitive known director of infrared radiation, apparently worth thousands of pounds per gram. The Relaks experiment was dedicated to finding the best position in weightlessness in which to relax. The Zierna programme of Earth resources photography was somewhat thwarted when Hermaszewski's homeland remained hidden below clouds. Then there was the Kardiolider experiment to study the cardiovascular system and the health experiment named Zdrowie. The final intensely scientific experiment was Smak, an investigation of why some foods which tasted delicious on Earth tasted of sawdust in weightlessness.

The Pole's mission ended in a maize field near Rostov, 300 km (186 miles) west of Tselinograd at $T + 7$ days 22 hours 2 minutes 59 seconds. A unique aspect of the mission was the Polish cosmonaut's candid reflection of his sometimes pensive thoughts during the fiery re-entry.

Polish cosmonaut Hermaszewski turns "upside down" for the camera on Salyut 6, next to commander Klimuk and Salyut 6 FE Ivanchenkov.

Milestones

68th manned space flight
37th Soviet manned space flight
29th Soyuz manned space flight
2nd Interkosmos mission
1st manned space flight by a Polish cosmonaut

SOYUZ 31

Int. Designation	1978-081A
Launched	26 August 1978
Launch Site	Pad 1, Site 5, Baikonur Cosmodrome, Kazakhstan
Landed	3 September 1978 (aboard Soyuz 29)
Landing Site	140 km southeast of Dzhezkazgan
Launch Vehicle	R7 (11A511U); spacecraft serial number (7K-T) #47
Duration	7 days 20 hrs 49 min 4 sec
Callsign	Yastreb (Hawk)
Objective	East German Salyut 6 visiting mission; Soyuz ferry exchange mission

Flight Crew

BYKOVSKY, Valery Fyodorovich, 44, Soviet Air Force, commander, 3rd mission
Previous missions: Vostok 5 (1963); Soyuz 22 (1976)
JAEHN, Sigmund, 41, East German Air Force, cosmonaut researcher

Flight Log

Another science-packed international mission blasted off from Baikonur at 19:51 hrs local time with commander Valery Bykovsky, on his third space flight – 15 years after his first – and a cosmonaut researcher from East Germany, Sigmund Jaehn. An emotional, bear-hugging and kissing welcome followed the docking with Salyut 6 one day after launch, and Jaehn got down to work.

His main tool was the East German Karl Zeiss MKF-6M multi-spectral Earth resources camera, which was also used to photograph a major military exercise in eastern Europe. The East German experiments included one called Rech which involved crew members repeating a series of numbers during the flight to see if their speech levels changed. There was another, called Audio, which was to "study subtle nuances of sound", that is, a hearing test. Other work included operations using the Splav furnace to study crystal growth.

After seven days aboard Salyut, during which they reached 355 km (221 miles) altitude at 51.6°, Bykovsky and Jaehn left residents Kovalenok and Ivanchenkov with their fresh spacecraft and landed inside Soyuz 29 – the first spacecraft switch involving an international crew – at $T + 7$ days 20 hours 49 minutes 4 seconds, 140 km (87 miles) southeast of Dzhezkazgan.

The Soyuz 29 and 31 crews pose for the camera inside Salyut 6. L to r Bykovsky, Kovalyonok, Jähn and Ivanchenkov. National flags, medallions, formal portraits and pennants adorn the station, the typical accoutrements of the ceremonial activities during visiting Interkosmos missions.

Milestones

69th manned space flight
38th Soviet manned space flight
30th Soyuz manned space flight
3rd Interkosmos mission
1st manned space flight by an East German

SOYUZ 32

Int. Designation	1979-018A
Launched	25 February 1979
Launch Site	Pad 31, Site 6, Baikonur Cosmodrome, Kazakhstan
Landed	19 August 1979 (aboard Soyuz 34)
Landing Site	211 km southeast of Dzhezkazgan
Launch Vehicle	R7 (11A511U); spacecraft serial number (7K-T) #48
Duration	175 days 0 hrs 37 min 37 sec
Callsign	Proton (Proton)
Objective	Third Salyut 6 resident crew programme

Flight Crew

LYAKHOV, Vladimir Afanasevich, 37, Soviet Air Force, commander
RYUMIN, Valery Viktorovich, 39, civilian, flight engineer, 2nd mission
Previous mission: Soyuz 25 (1977)

Flight Log

The original plan for Salyut 6 was to include three long-duration missions, by Soyuz 25, 26 and 28, with visits by Soyuz 27, 29 and 30. The Soyuz 25 failure resulted in a reschedule and at first an anticipated reduction from three to two long-duration missions. By February 1979, however, Salyut's condition was assessed to be worth a try at the third long-duration mission, by Soyuz 32, during which two further international missions would also be attempted. Soyuz 32, crewed by veteran Valery Ryumin and rookie commander Vladimir Lyakhov, took off in extremely overcast and murky skies at 16:54 hrs local time. Their record residency aboard Salyut 6, which they entered the following day, was destined to be a long, lonely one.

The crew immediately checked over the station and found it in good condition, apart from the confirmation that there was a potentially serious problem with the station's fuel system. A membrane in a liquid fuel and gaseous nitrogen gas line had warped. Orders were given to launch Progress 5 to start the restocking process. This arrived on 12 March, carrying, among other things, a TV monitor to permit the first two-way visual communications with the ground and an improved Kristall furnace. Before Progress 5 started refuelling, the tank connected to the line which had warped was ingeniously emptied by spinning the space station. This fuel was then transferred into another tank.

Each day, the crew embarked on a strenuous two-and-a-half-hour period of exercise to ward off the ill-effects of weightlessness which would be felt on landing back in 1-G. They took the first Soviet shower in space and by the end of March had completed nearly 40 repairs on the station, as well as their scientific work. All was ready for the arrival of Soyuz 33 with a Bulgarian visitor on 11 April. When this

Lyakhov and Ryumin in the transfer compartment of the Salyut 6 1-G trainer at TsPK.

mission failed to dock and had to make an emergency return to Earth, the following Hungarian mission, too, was cancelled, and the resident crew, which reached a maximum altitude of 406 km (252 miles) in the 51.6° orbit, resigned themselves to unmanned visits by further Progress tankers. The sixth of these arrived on 13 May.

An unusual step was then taken to ensure that Lyakhov and Ryumin had a fresh ferry vehicle. Soyuz 32 was replaced by Soyuz 34 (1979-049A), which was launched unmanned on 6 June. This docked at the rear and after Soyuz 32 returned unmanned, the crew took Soyuz 34 from the back to the front, in preparation for the arrival of Progress 7. Launched on 28 June, this new tanker delivered a radio telescope called the KRT-10, designed to monitor pulsars and other celestial phenomena. It was pushed through the aft docking port before Progress departed, and as the tanker undocked was automatically unfurled to a diameter of 10 m (33 ft), a record for a space dish antenna.

It may have jammed against some of the station's appendages and not have unfurled correctly however, because after work with the KRT-10, the mission was thrown into a frenzy when the cosmonauts could not discard it in order to clear the aft docking port for another possible Progress launch to replenish the station later. Thanks to the intensive repair work by the cosmonauts, the station was being considered for several more missions, rather than being de-orbited. Therefore, Ryumin, assisted by Lyakhov, made an unscheduled and highly risky spacewalk on 15 August, reminiscent of the Skylab 2 EVA in 1973, to free the KRT using wire cutters. Their 1 hour 23 minute effort was successful.

The crew became the first to land in a spacecraft, Soyuz 34, that had been launched unmanned, at $T + 175$ days 0 hours 37 minutes 37 seconds, 211 km (131 miles) south east of Dzhezkazgan. They had to be carried from the capsule and placed in reclining chairs. Although recovered fit and well after four days, Lyakhov and Ryumin at first had difficulty speaking properly and felt the effects of re-adaptation quite starkly; blankets felt like chainmail and beds like boards, they said.

Milestones

70th manned space flight
39th Soviet manned space flight
31st Soyuz manned space flight (Soyuz 32 up only)
33rd Soyuz manned space flight (Soyuz 34 down only)
New duration record – 175 days 0 hours
1st manned space flight to land in spacecraft launched unmanned
5th Soviet and 20th flight with EVA operations
Lyakhov celebrates 39th birthday in space (28 Jul)
Ryumin celebrates 40th birthday in space (16 Aug)

SOYUZ 33

Int. Designation	1979-029A
Launched	10 April 1979
Launch Site	Pad 31, Site 6, Baikonur Cosmodrome, Kazakhstan
Landed	12 April 1979
Landing Site	318 km southeast of Dzhezkazgan
Launch Vehicle	R7 (11A511U); spacecraft serial number (7K-T) #49
Duration	1 day 23 hrs 1 min 6 sec
Callsign	Saturny (Saturn)
Objective	Bulgarian Interkosmos visiting mission programme

Flight Crew

RUKAVISHNIKOV, Nikolay Nikolayevich, 46, civilian, commander, 3rd mission
Previous missions: Soyuz 10 (1971); Soyuz 16 (1974)
IVANOV, Georgy Ivan, 38, Bulgarian Air Force, cosmonaut researcher

Flight Log

Conditions at the Baikonur Cosmodrome at 22:34 hrs local time on 10 April were described as the worst ever for a Soviet launch, as the SL-4 booster was committed to lift-off in winds of 40 kph (25 mph). Another international mission was under way, this time with the moustached Bulgarian Georgy Ivanov alongside the first Soviet civilian commander, Nikolay Rukavishnikov, who also happened to be the first non-pilot to command a spacecraft.

Soyuz 33 began its orbital manoeuvres for the rendezvous with Salyut 6, reaching a maximum altitude of 261 km (162 miles) at 51.6°, but Rukavishnikov was not happy with the main propulsion system. Instinctively, he knew that something was not quite right and this was confirmed on the control panel, with readings indicating lower chamber pressures than normal. One of the final manoeuvres for the rendezvous was a 6-second burst of the main engine, which fired erratically for 3 seconds and then shut down. The docking was off. The fault was later traced to the gas generator feeding the turbo-pump in the spacecraft's main engine.

While the failure and routine return to Earth of the powerless Soyuz ferry were reported matter-of-factly by the Soviet news agencies, controllers and cosmonauts were very worried about the Soyuz 33 crew's safety. The back-up propulsion unit would have to be used for the retro-burn and had never been used before on a manned mission. In addition, it was only capable of one continuous burn, rather than a two-phase retro-burn normally used to induce a stress-reduced, 3-G lifting re-entry. The ballistic re-entry would therefore involve much higher g-forces. The nominal back-up engine burn time was 188 seconds, but there was no guarantee that it would perform as

The Bulgarian Interkosmos crew during training, with Ivanov seated behind commander Rukavishnikov

advertised. There was a facility to allow the crew to start it manually if it cut off after burning for at least 90 seconds, but even so, the re-entry would be extremely inaccurate.

Anything shorter would have meant them being stranded in orbit. A longer burn than nominal could place an intolerable G-force on the spacecraft and crew. Ivanov was told to make sure he tucked his moustache inside his helmet! Soyuz 33's back-up engine fired for 213 seconds which was longer than normal and Rukavishnikov had to manually shut it down. For 530 seconds, the crew was subjected to G-forces between 10-G and a peak of 15-G. Rukavishnikov later commented that as the Soyuz DM descended it was like being inside the flames of a blowtorch. Externally, temperatures reached 3,000°C and the crew experienced a lot of vibration and noise. Pinned to their seats by the G forces, both reported difficulty in breathing but were still able to talk to each other. The capsule was glowing red as it landed beneath darkening skies 318 km (198 miles) southeast of Dzhezkazgan at $T + 1$ day 23 hours 1 minute 6 seconds.

Unlike NASA, there are no copies of the post-flight reports of the Soviet accident investigation boards publicly available. Accounts of what happened are left to personal recollections, sometimes years after the event, making for conflicting accounts of the facts. The official RKK Energiya history and the memoirs of Flight Director Alexei Yeliseyev record that the engine failed to provide enough thrust and the deceleration impulse was less than planned. Soviet space analyst and historian Bart Hendrickx has suggested that perhaps the automatic systems on the Soyuz tried to compensate for the inadequate firing of the engine by commanding a longer burn than planned and the crew, unaware of this, simply shut it down. Landing close to the planned recovery site was pure coincidence according to Yeliseyev. Due to the insufficient deceleration, Soyuz 33 took longer to enter the atmosphere, but as a result of the longer engine burn, a steeper angle compensated for the increase descent time.

Milestones

71st manned space flight
40th Soviet manned space flight
32nd Soyuz manned space flight
1st manned space flight commanded by non-pilot
1st civilian engineer-cosmonaut to command a Soviet mission
1st manned space flight by a Bulgarian
4th Interkosmos mission

SOYUZ 35

Int. Designation	1980-027A
Launched	9 April 1980
Launch Site	Pad 31, Site 6, Baikonur Cosmodrome, Kazakhstan
Landed	11 October 1980 (in Soyuz 37)
Landing Site	180 km southeast of Dzhezkazgan
Launch Vehicle	R7 (11A511U); spacecraft serial number (7K-T) #51
Duration	184 days 20 hrs 11 min 35 sec
Callsign	Dneiper (Dneiper)
Objective	Fourth Salyut 6 resident crew programme

Flight Crew

POPOV, Leonid Ivanovich, 34, Soviet Air Force, commander
RYUMIN, Valery Viktorovich, 40, civilian, flight engineer, 3rd mission
Previous missions: Soyuz 25 (1977); Soyuz 32 (1979)

Flight Log

Assigned to this next long-duration mission to Salyut 6 were commander Leonid Popov and flight engineer Valentin Lebedev, who injured his knee shortly before lift-off while exercising on a trampoline. Mission officials decided to adhere to the new rule that at least one flight experienced crewman should fly a mission and chose a certain Valery Ryumin as Lebedev's replacement. So, eight months after coming home from the Soyuz 32 marathon, the tall, burly Ryumin was off again. Soyuz 35 lifted off into a decidedly murky sky at 19:38 hrs local time on 9 April. Within 26 hours, Soyuz 35 was safely nestled at Salyut 6 and Ryumin entered the station to read a welcoming letter written by himself before he last left. Before Soyuz 35 had arrived, Salyut 6 had received two visits from unmanned spacecraft; the new Soyuz T1 ferry vehicle on an automatic test and a regular tanker visit from Progress 8, which was used to boost Salyut's orbit and was there ready for unloading when the cosmonauts arrived. The rookie Popov went about his work in such a frenzy of enthusiasm that the experienced Ryumin told him to slow down – they were going to be there for a while.

After unloading Progress 8, the cosmonauts discarded it and almost immediately received Progress 9 in order to take on more equipment, including a new motor for a biogravity centrifuge. Progress 9 was also used for the first time to unload water into Salyut 6. Ryumin keenly tended his Oasis space garden and proudly displayed one of its products, a huge cucumber, which he later revealed was a plastic one.

The cosmonauts played host to several visiting crews and also unloaded two more Progress tankers. Soyuz 36 arrived with a Hungarian cosmonaut researcher in May, two Soviet cosmonauts arrived in Soyuz T2 in June, Progress 10 arrived the same month, and Soyuz 37 in July, with a crew including a cosmonaut researcher from

Popov (seated) watches rescue teams remove Ryumin from the DM of Soyuz 35 at the end of the 185-day mission.

Vietnam. A Cuban followed in Soyuz 38 during September, and finally that month Progress 11 arrived and stayed docked for a record 70 days, well after Popov and Ryumin had returned to Earth. These regular visits meant that the resident crew estimated that they spent 25 per cent of their time in Salyut unloading equipment and cosmonauts. One major exercise during a hiatus in the visits during August was a dedicated period of Earth resources photography of targets in the Soviet Union, using the MKF-6M and KT-140 cameras. The approximately 4,500 images covered a total of 96 million km^2 (37,056,000 $miles^2$).

During the mission Ryumin put on 3.1 kg (7 lb), which was an unusual statistic for a space traveller. He and Popov had reached 365 km (227 miles) maximum altitude during the 51.6° orbit. They boarded the fresh Soyuz 37 and came home 180 km (112 miles) southeast of Dzhezkazgan at $T + 184$ days 20 hours 11 minutes 35 seconds, remarkably fresh and fit after a new record duration mission, which was not officially recognised as such by the IAF since it only exceeded the previous mission by nine days, not the required ten per cent. Nonetheless, Ryumin, with three flights under his belt, had clocked up almost a year's space experience, and declared that he was ready to fly to Mars.

Milestones

72nd manned space flight
41st Soviet manned space flight
34th manned Soyuz space flight
Ryumin celebrates his 41st birthday in space (16 August)
1st person to celebrate two consecutive birthdays in space (Ryumin)
Popov celebrates his 35th birthday in space (31 August)

SOYUZ 36

Int. Designation	1980-041A
Launched	26 May 1980
Launch Site	Pad 31, Site 6, Baikonur Cosmodrome, Kazakhstan
Landed	3 June 1980 (in Soyuz 35)
Landing Site	140 km southeast of Dzhezkazgan
Launch Vehicle	R7 (11A511U); spacecraft serial number (7K-T) #52
Duration	7 days 20 hrs 45 min 44 sec
Callsign	Orion (Orion)
Objective	Hungarian Interkosmos visiting mission programme; Soyuz ferry exchange mission

Flight Crew

KUBASOV, Valery Nikolayevich, 45, civilian, commander, 3rd mission
Previous missions: Soyuz 6 (1969); Soyuz 19 ASTP (1975)
FARKAS, Bertalan, 30, Hungarian Air Force, cosmonaut researcher

Flight Log

Regarded with much hullabaloo in Hungary but not elsewhere, Soyuz 36, delayed from June 1979, lifted off from Baikonur at 00:21 hrs local time on 27 May, carrying cosmonaut researcher Bertalan Farkas and his Soviet commander Valery Kubasov. Farkas arrived inside Salyut 6 bearing succulent gifts of goulash, pâté de foie gras, fried pork and jellied tongue, and soon got down to work with his national experiments called Daignost, Balaton, Interferon, Dose, Opros, Andio, Oxymeter, Biosphere, Refraction, Zarya, Bealuca, Eotvos and Ilkeminator.

Farkas, who adapted to weightlessness much faster than his flight experienced colleague Kubasov, helped to process gallium arsenide crystals with chromium, assessed the misalignment of instruments due to long exposure to the space environment, made a photographic geomorphological map of the Carpathian Basin, studied hearing and human motor response in weightlessness and attempted to study the behaviour of the cancer fighting drug Interferon in weightlessness. The Interferon experiment was divided into three separate studies: to assess the effect of weightlessness in Interferon production in the lymphatic system on the human body, study changes in Interferon samples in weightlessness, and assess the effect of Interferon on blood samples taken before launch and after landing.

The mission, which reached 355 km (221 miles) in the 51.6° orbit, ended with a night landing at $T + 7$ days 20 hours 45 minutes 44 seconds in Soyuz 35, 140 km (87 miles) southeast of Dzhezkazgan. The fresher Soyuz 36 was left for the resident crew in case of the need for an emergency return.

The Second Decade: 1971–1980

Kubasov and Farkas complete a typical operation during Soyuz ferry exchanges, relocating the personal seat liners and Sokol suits from one Soyuz DM to the other

Milestones

73rd manned space flight
42nd Soviet manned space flight
35th Soyuz manned space flight
1st manned space flight by a Hungarian
5th Interkosmos mission

SOYUZ T2

Int. Designation	1980-045A
Launched	5 June 1980
Launch Site	Pad 1, Site 5, Baikonur Cosmodrome, Kazakhstan
Landed	9 June 1980
Landing Site	198 km from Dzhezkazgan
Launch Vehicle	R7 (11A511U); spacecraft serial number: (7K-ST) #07L
Duration	3 days 22 hrs 19 min 30 sec
Callsign	Yupiter (Jupiter)
Objective	Manned test flight of new Soyuz T ferry

Flight Crew

MALYSHEV, Yuri Vasilyevich, 38, Soviet Air Force, commander
AKSENOV, Vladimir Viktorovich, 45, civilian, flight engineer, 2nd mission
Previous mission: Soyuz 22 (1976)

Flight Log

After the unmanned Soyuz T1 test flight in 1979, the cautious Soviets decided to test the new vehicle again on a short mission with a crew aboard. Wearing new style space-suits, Yuri Malyshev and Vladimir Aksenov lifted off from Baikonur at 20:19 hrs local time, heading for a totally automatic rendezvous and docking with Salyut 6. The hands-off docking became a hands-on one from 210 m (689 ft) because the avionics aboard the new ship failed. Malyshev and Aksenov, who reached a maximum altitude of 349 km (217 miles) at 51.6°, spent a brief time with Salyut 6 hosts Ryumin and Popov before heading home, departing the station 2 days 17 hours 22 minutes after docking.

En route, they jettisoned the Orbital Module before rather than after retro-fire to save fuel and, according to some western analysts, possibly to demonstrate that an Orbital Module could be left attached to Salyut to act as a special laboratory. Soyuz T2 came home at T + 3 days 22 hours 19 minutes 30 seconds, 198 km (123 miles) from Dzhezkazgan, with black soot covering its windows after the fiery re-entry and with the aid of larger deceleration rockets to aid touchdown.

Milestones

74th manned space flight
43rd Soviet manned space flight
36th Soyuz manned space flight
1st Soyuz T manned space flight

A successful test flight. Soyuz T2 cosmonauts Aksyonov (left) and Malyshev back on Earth after evaluating the capabilities of the improved Soyuz ferry vehicle.

SOYUZ 37

Int. Designation	1980-064A
Launched	23 July 1980
Launch Site	Pad 1, Site 5, Baikonur Cosmodrome, Kazakhstan
Landed	31 July 1980 (in Soyuz 36)
Landing Site	179 km southeast of Dzhezkazgan
Launch Vehicle	R7 (11A511U); spacecraft serial number (7K-T) #53
Duration	7 days 20 hrs 42 min
Callsign	Terek (Terek)
Objective	Vietnamese Interkosmos visiting mission programme; Soyuz ferry exchange mission

Flight Crew

GORBATKO, Viktor Vasilyevich, 45, Soviet Air Force, commander, 3rd mission
Previous missions: Soyuz 7 (1969); Soyuz 24 (1977)
TUAN, Pham, 33, Vietnamese Air Force, cosmonaut researcher

Flight Log

The Interkosmos international manned missions to Salyut 6 were accompanied by much propaganda, most of which rubbed up many western observers the wrong way. This was particularly true for the flight of Soyuz 37 with cosmonaut researcher Pham Tuan, hailed as the only pilot to have shot down an American B-52 during the Vietnam War, a claim refuted by the USA. One of his tasks during the flight, according to the Soviets, was to study the effects on the Vietnamese countryside, plants and forests of the enormous amounts of defoliants and fire bombs dropped during the conflict.

Western sports journalists reporting from the Moscow Olympics – much boycotted as a result of the Soviet invasion of Afghanistan – were told to applaud when told of the launch, which was the usual spectacular event, since it again took place at night, 00:33 hrs local time at Baikonur. Twenty-six hours later, Tuan and his commander Viktor Gorbatko had docked with Salyut 6 to give more company to the residents, Ryumin and Popov.

Tuan's adaptation to weightlessness was rather uncomfortable as he suffered a headache and loss of appetite, but he soon picked up to start his busy schedule of highly scientific tasks, such as taking photos of Vietnam with the MKF-6M camera to reveal details of tidal flooding, silting at river mouths and other hydrological features. The flight ended with a landing in Soyuz 36, at $T + 7$ days 20 hours 42 minutes, some 179 km (111 miles) southeast of Dzhezkazgan. Maximum altitude reached during the $51.6°$ mission was 351 km (218 miles).

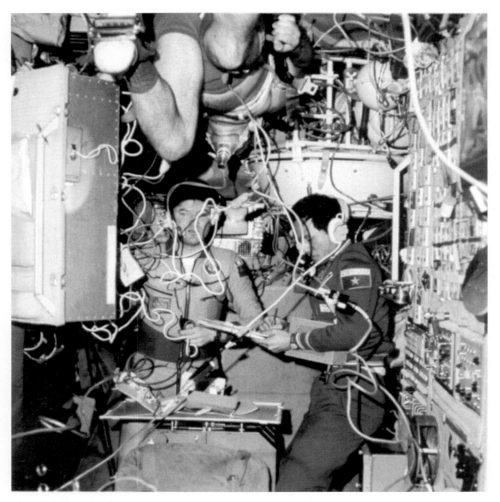

Pham Tuan (right) records experiment data during his week aboard Salyut 6

Milestones

75th manned space flight
44th Soviet manned space flight
37th Soyuz manned space flight
36th (original) Soyuz manned space flight
1st manned space flight by a Vietnamese
6th Interkosmos mission

SOYUZ 38

Int. Designation	1980-075A
Launched	18 September 1980
Launch Site	Pad 1, Site 5, Baikonur Cosmodrome, Kazakhstan
Landed	26 September 1980
Landing Site	17 km southeast of Dzhezkazgan
Launch Vehicle	R7 (11A511U); spacecraft serial number (7K-T) #54
Duration	7 days 20 hrs 43 min 24 sec
Callsign	Tamyr (Tamyr)
Objective	Cuban Interkosmos visiting mission programme

Flight Crew

ROMANENKO, Yuri Viktorovich, 36, Soviet Air Force, commander, 2nd mission
Previous mission: Soyuz 26 (1977)
TAMAYO-MENDEZ, Arnaldo, 38, Cuban Air Force, cosmonaut researcher

Flight Log

Even the most enthusiastic space watcher, except the few in Cuba, would have found the latest Interkosmos mission little more than a monotonous repeat of those that had gone before. Cuban Arnaldo Tamayo-Mendez studied the crystallisation of sucrose in weightlessness for the benefit of his country's sugar industry on his Soyuz 38 mission, with commander Yuri Romanenko. This began at 01:11 hrs local time at the Baikonur Cosmodrome and, as usual, 26 hours later, Soyuz 38 homed in on Salyut 6 with its occupants Ryumin and Popov.

Other experiments being conducted by Mendez included one in which he wore heavy overshoes, to assess ways of preventing muscular deterioration in space. One unique point about the mission, apart from its Cuban connection, was that at its end it touched down within 2.9 km (2 miles) of the predicted landing point in the most accurate landing in the Soviet space programme to date. Flight time was 7 days 20 hours 43 minutes 24 seconds and the mission achieved a maximum altitude of 350 km (217 miles) in the 51.6° orbit.

Milestones

76th manned space flight
45th Soviet manned space flight
38th Soyuz manned space flight

236   The Second Decade: 1971–1980

37th (original) Soyuz manned space flight
1st manned space flight by a Cuban
7th Interkosmos mission

(L to r) Ryumin, Tamayo-Mendez, Romanenko and Popov aboard Salyut 6

SOYUZ T3

Int. Designation	1980-094A
Launched	27 November 1980
Launch Site	Pad 1, Site 5, Baikonur Cosmodrome, Kazakhstan
Landed	10 December 1980
Landing Site	129 km southeast of Dzhezkazgan
Launch Vehicle	R7 (11A511U); spacecraft serial number #08L
Duration	12 days 19 hrs 7 min 42 sec
Callsign	Mayak (Lighthouse)
Objective	Salyut 6 maintenance mission; Soyuz T three-crew test flight

Flight Crew

KIZIM, Leonid Denisovich, 39, Soviet Air Force, commander
MAKAROV, Oleg Grigoryevich, 47, civilian, flight engineer, 4th mission
Previous missions: Soyuz 12 (1973); Soyuz 18-1 (1975); Soyuz 27 (1978)
STREKALOV, Gennady Mikhailovich, 40, civilian, research engineer

Flight Log

The first three-crew Soviet manned space flight since the Soyuz 11 disaster of 1971, Soyuz T3 was also the first maintenance mission, designed to overhaul the aging Salyut 6 space station to prolong its already long life sufficiently to accommodate a further three missions, two of which were international flights. Lift-off came at 19:18 hrs local time at Baikonur, carrying rookie commander Leonid Kizim and research engineer Gennady Strekalov, along with the first Soviet space four-timer, Oleg Makarov. Unlike Soyuz T2, this flight made a totally automatic docking with Salyut 6 and a hectic two-week repair mission began.

In addition to the busy schedule of repairs, the crew did conduct two science experiments; to study the behaviour of crystals in weightlessness, extensively using the onboard furnaces; and the use of a laser-based holographic system to photograph the dissolution of a salt crystal. Kizim and his colleagues overhauled the hydraulic system on board Salyut, installing a new hydraulic unit and pumps. They also replaced a programming device in the control system, fitted a new transducer to a compressor in the in-orbit refuelling system, and replaced electronic components in the communications system.

While the crew was still on board, the Progress 11 tanker's main propulsion system was fired to raise the station's orbit. The tanker was then discarded, followed by the crew's entry into Soyuz T3 and an automatic retro-fire and re-entry to a landing at $T + 12$ days 19 hours 7 minutes 42 seconds, 129 km (80 miles) southeast of Dzhezkazgan. Maximum altitude achieved during the 51.6° mission was 398 km (247 miles). Salyut was ready for another long-duration visit and in preparation

The safe recovery of the Soyuz T3 crew, the first three-person Soviet space crew since 1971. L to r: Makarov, Kizim, Strekalov

for it, another Progress was dispatched from Baikonur, to await a new crew. With Soyuz T3 featuring a three-person crew, it proved the greater flexibility of the new spacecraft and opened up the possibility of crewing a Salyut with three residents instead of two, offering greater productivity. However, the Soviets had revealed that on future Soyuz T missions, the crew composition would vary and would not always include a cosmonaut researcher (who had no piloting functions) in the third seat. On occasion, the third seat would be occupied by additional supplies or science experiments.

Milestones

77th manned space flight
46th Soviet manned space flight
39th Soyuz manned space flight
2nd Soyuz manned space flight
1st Soviet three-person crew since 1971

6

The Third Decade: 1981–1990

SOYUZ T4

Int. Designation	1981-023A
Launched	12 March 1981
Launch Site	Pad 1, Site 5, Baikonur Cosmodrome, Kazakhstan
Landed	26 May 1981
Landing Site	124 km east of Dzhezkazgan
Launch Vehicle	R7 (11A511U); spacecraft serial number (7K-ST) #10L
Duration	74 days 17 hrs 37 min 23 sec
Callsign	Foton (Photon)
Objective	Fifth Salyut 6 resident crew; first use of Soyuz T for resident crew delivery and support

Flight Crew

KOVALENOK, Vladimir Vasilyevich, 39, Soviet Air Force, commander, 3rd mission
Previous missions: Soyuz 25 (1977); Soyuz 29 (1978)
SAVINYKH, Viktor Petrovich, 41, civilian, flight engineer

Flight Log

The long-duration mission of Soyuz T4 was not originally intended as such but was more the result of rescheduling after the Soyuz 33 docking abort cancelled some international Interkosmos flights, leading to a situation where a Mongolian and a Romanian were still to visit Salyut. The station had fortunately been given a stay of execution by the Soyuz T3 mission. Soyuz T4, set for a mission that would last long enough to accommodate the two Interkosmos missions and verify a new Soyuz spacecraft for a period of extended docking with Salyut, took off at midnight from Baikonur, the second such launch since Soyuz 9. On board were Vladimir Kovalenok, who was already rather familiar with Salyut 6, and his flight engineer, Viktor Savinykh. Though he was a rookie, Savinykh nevertheless had the unique statistic of being both the hundredth person and the fiftieth Soviet to enter space. As the two visiting crews had not trained on Soyuz T and were not qualified to return in one, the resident crew would not have an exchange of vehicle to support an extended-duration mission, so no attempt would be made to exceed the space flight endurance record on this mission.

After docking with Salyut, the crew unpacked Progress 12 and finished off some refurbishment work, including repairs to a battery unit and a condensation unit on the thermal control system. This was necessary because only one solar panel was generating enough power, and as a result excessive condensation was forming inside the station. Kovalenok and Savinykh then discarded Progress 12, the last unmanned tanker to berth at Salyut 6, and prepared for their first visit, on 23 March, by the

End of an era. The T4 crew's return to Earth brought Salyut 6 operations to a close. The crew is seen here displaying the national emblems of the communist countries whose representatives visited the station during the Soviet and Interkosmos missions between 1977 and 1981

Mongolian mission of Soyuz 39. After the brief visit, the crew then changed the docking unit on Soyuz T4, possibly to demonstrate the space rescue capability. By taking the docking unit out of Soyuz, it was possible to dock another Soyuz with it, in an operation that may have been prompted by the near disaster on Soyuz 33 when the two crewmen could have been left stranded in space. Another international mission followed on 15 May, this time by a Romanian, before Kovalenok and Savinykh mothballed Salyut 6 for the last time and headed home.

Soyuz T4 landed at $T + 74$ days 17 hours 37 minutes 23 seconds, 124 km (77 miles) east of Dzhezkazgan. Maximum altitude achieved during the mission was 374 km (232 miles), at $51.6°$. The mission of Salyut 6 was not over, however, as a new module called Cosmos 1267, the same size as Salyut, docked with it on 19 June, remaining in orbit until the whole combination was de-orbited in July 1982. The Cosmos 1267 Heavy Cosmos module, flight tested as Cosmos 929, was launched by a Proton booster on 25 April. It performed extensive orbital manoeuvres and even dispatched an unmanned re-entry capsule back to Earth for recovery on 15 May. Further orbital manoeuvres were made by the joint craft and it became evident that future Salyuts and other generation space stations would be enlarged by the addition of other Heavy Cosmos derivatives. Salyut 6 was not manned again after the end of the T4 mission.

It re-entered the atmosphere in July 1982, once Salyut 7 had been successfully placed in orbit to replace it. Salyut 6 had been an outstanding success for the Soviet Union, at a time when all bar three US astronauts were grounded between the end of Apollo and the beginning of the Shuttle program.

Milestones

78th manned space flight
47th Soviet manned space flight
40th Soyuz manned space flight
3rd Soyuz T manned space flight
100th person in space

SOYUZ 39

Int. Designation	1981-029A
Launched	22 March 1981
Launch Site	Pad 1, Site 5, Baikonur Cosmodrome, Kazakhstan
Landed	30 March 1981
Landing Site	169 km southeast of Dzhezkazgan
Launch Vehicle	R7 (11A511U); spacecraft serial number (7K-T) #55
Duration	7 days 20 hrs 42 min 3 sec
Callsign	Pamir (Pamirs)
Objective	Mongolian Salyut 6 visiting mission programme

Flight Crew

DZHANIBEKOV, Vladimir Aleksandrovich, 38, Soviet Air Force, commander, 2nd mission
Previous mission: Soyuz 27 (1978)
GURRAGCHA, Jugderdemidyin, 33, Mongolian People's Army, cosmonaut researcher

Flight Log

During the 1960s, the name of a certain Nigerian leader used to terrify newscasters. The name of the next spaceman could also have caused apoplexy in newsrooms around the world had he not been the one-hundred-and-first and a Mongolian. Jugderdemidyin Gurragcha and his commander Vladimir Dzhanibekov were launched at 19:59 hrs local time from the Baikonur Cosmodrome and a day later were inside Salyut 6 with residents Kovalenok and Savinykh. The experiments were mainly medically oriented, but also included Gurragcha's photography of his homeland to conduct an Earth resources survey of oil, gas and mineral deposits, and the use of a visual polarising analyser to assess the effects of prolonged exposure to space on the station's portholes.

Maximum altitude achieved during the 51.6° mission was 355 km (221 miles). Gurragcha may have been one of the few space travellers to have reacted violently to weightlessness. Only one photo of him aboard Salyut 6 has ever been released, but he seemed in good spirits after landing in fog and drizzle at $T + 7$ days 20 hours 42 minutes 3 seconds, 169 km (105 miles) southeast of Dzhezkazgan.

Milestones

79th manned space flight
48th Soviet manned space flight

41st Soyuz manned space flight
38th (original) Soyuz manned space flight
1st flight by a Mongolian
8th Interkosmos flight

Dzhanibekov (right) and Gurragcha in the Salyut Hall at TsPK during training for their mission to Salyut 6

STS-1

Int. Designation	1981-034A
Launched	12 April 1981
Launch Site	Pad 39A, Kennedy Space Center, Florida
Landed	14 April 1981
Landing Site	Runway 23, Edwards Air Force Base, California
Launch Vehicle	OV-102 Columbia/ET-2/SRB A07; A08/SSME #1 2007; #2 2006; #3 2005
Duration	2 days 6 hrs 20 min 53 sec
Callsign	Columbia
Objective	First manned orbital test flight (OFT-1) of Shuttle system

Flight Crew

YOUNG, John Watts, 50, USN, commander, 5th mission
Previous missions: Gemini 3 (1965); Gemini 10 (1966); Apollo 10 (1969); Apollo 16 (1972)
CRIPPEN, Robert Laurel, 43, USN, pilot

Flight Log

The build up to this momentous space mission for the US programme was painfully slow. A budget lower than that afforded to Apollo for a space system five times more technically demanding resulted in inevitable glitches at almost every turn. The first space flight by the Space Shuttle was originally scheduled for 1978, but in fact all that happened was that the first four space crews were rather optimistically named for missions that would start the following year. Thus, veteran John Young and rookie Bob "Crip" Crippen began what was to become one of the longest periods of training ever, ending with a lift-off in 1981. Coincidentally, for such a major space milestone, the launch would be on the twentieth anniversary of the first manned space flight by Yuri Gagarin.

That the launch had been scrubbed at $T - 36$ min, by a computer synchronisation glitch two days before, which had been dubbed by the media assembled at the Kennedy Space Center as a "fiasco", is indicative of the reputation of the Shuttle. The USA had been through a period of several major technical disasters, including Three Mile Island, and there were many cynics expecting to be reporting another from the Kennedy Space Center on a maiden flight being manned for the first time. There is no doubting the heroism of the crew, who had only the dubious opportunity of ejection seats available to them for an early bail-out.

The cataclysmic blast-off occurred at 07:00 hrs local time, causing unpredicted over pressurisation of the orbiter and a potential collision with the launch tower, followed almost immediately by the roll programme which alarmed already nervous

STS-1 comes home, with a T-38 chase plane in attendance, at the end of the first full Shuttle Mission profile

spectators with its brute force. Thrust was five per cent higher than anticipated, leading to a steeper, "heads down" climb to orbit. The solid rocket boosters were ejected at $T + 2$ minutes 11 seconds and the three main engines cut off at $T + 8$ minutes. The Space Shuttle Columbia was in initial orbit and was then boosted by four burns of the orbiter's own propulsion system. Inclination was 40.3° and maximum altitude 232 km (144 miles).

With Columbia flying "upside-down" with its back facing the Earth, the payload bay doors were opened, exposing a vast interior which was empty for this test flight. TV cameras also showed that some heatshield tiles were missing from the rear of the orbiter and much was made of this in the popular press. They were not critical tiles, but all the same if they were missing, could other more critical tiles on the orbiter's underside be loose or lost? The crew would find out after their thirty-sixth orbit, when after an almost flawless orbital workout by the jubilant Young and Crippen, the OMS engines initiated the 2 minute 27 second long retro-fire burn.

The Mach 25 re-entry, during which some tiles were exposed to 1,260°C, was accompanied by the usual radio blackout. Then, at Mach 10 and 57.3 km (36 miles), the happy Young reported that all was well. He proceeded to bring Columbia in like an airliner, landing on the dry lake bed runway 23, at Edwards Air Force Base, with main gear touchdown at $T + 2$ days 6 hours 20 minutes 32 seconds. Routine space flight with airliner-like landings seemed to have begun. Fifty Shuttle flights a year were being predicted.

Milestones

80th manned space flight
32nd US manned space flight
1st Shuttle mission
1st flight of Columbia
1st manned space flight in a reusable spacecraft
1st manned space flight on previously untested spacecraft
1st manned space flight to be boosted by solid propellants
1st flight by crewman on fifth space mission
1st flight to end with conventional runway landing

SOYUZ 40

Int. Designation	1981-042A
Launched	15 May 1981
Launch Site	Pad 1, Site 5, Baikonur Cosmodrome, Kazakhstan
Landed	22 May 1981
Landing Site	224 km southeast of Dzhezkazgan
Launch Vehicle	R7 (11A511U); spacecraft serial number (7K-T) #56
Duration	7 days 20 hrs 41 min 52 sec
Callsign	Dneiper (Dneiper)
Objective	Romanian Salyut 6 Interkosmos visiting mission

Flight Crew

POPOV, Leonid Ivanovich, 35, Soviet Air Force, commander, 2nd mission
Previous mission: Soyuz 35 (1980)
PRUNARIU, Dumitru Dorin, 28, Romanian Army Air Force, cosmonaut researcher

Flight Log

The final Interkosmos mission involving a cosmonaut researcher from a Soviet bloc country, Soyuz 40, was also the last of this Soyuz model. Crewed by Leonid Popov and Dumitru Prunariu, the mission got under way at 23:17 hrs from Baikonur, followed by the docking with Salyut 6 a day later and greetings from residents Kovalenok and Savinykh. Experiments on board included those to study the Earth's upper atmosphere and changes in its magnetic field. The mission ended at $T + 7$ days 20 hours 41 minutes 52 seconds, 224 km (139 miles) southeast of Dzhezkazgan. Maximum altitude during the 51.6° mission was 374 km (232 miles).

When Soyuz T4 returned later, Salyut 6 had received 16 cosmonaut crews and 15 unmanned spacecraft in three-and-a-half years. No fewer than 35 dockings had been made with it and Salyut 6 was occupied for 676 days. Some 13,000 photographs of the Earth had been taken and 1,310 experiments operated a remarkable record.

Milestones

81st manned space flight
49th Soviet manned space flight
42nd Soyuz manned space flight
39th (original) Soyuz manned space flight
Final flight of original Soyuz variant
1st manned space flight by a Romanian
9th and final Interkosmos mission

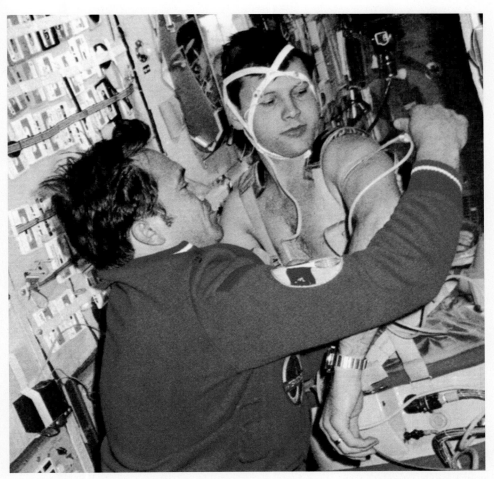

Prunariu (right) wears the Chibis lower body negative pressure garment aboard Salyut 6, assisted by Popov

STS-2

Int. Designation	1981-111A
Launched	12 November 1981
Launch Site	Pad 39A, Kennedy Space Center, Florida
Landed	14 November 1981
Landing Site	Runway 23, Edwards Air Force Base, California
Launch Vehicle	OV-102 Columbia/ET-3/SRB A09; A10/SSME #1 2007; #2 2006; #3 2005
Duration	2 days 6 hrs 13 min 13 sec
Callsign	Columbia
Objective	Second Orbital Test Flight (OFT-2); tests of the Remote Manipulator System (RMS)

Flight Crew

ENGLE, Joseph Henry, 49, USAF, commander
TRULY, Richard Harrison, 44, USN, pilot

Flight Log

As Columbia was being prepared for its second mission, originally scheduled for 10 September then pushed to 30 September, with 1,000 new tiles, new RCS components, an OMS nozzle and fuel cells, nitrogen tetroxide leaked from a ruptured fuel line during propellant loading, damaging another 240 tiles on its nose. The launch was postponed to 4 November. Things looked good but the launch was delayed for 2 hours 40 minutes and finally scrubbed, holding at $T - 31$ seconds, when an oil flush on one of the APUs failed and a computer malfunctioned. Both had to be replaced before a new launch attempt, eight days later.

The first flight of a re-used manned spacecraft began at 10:10 hrs local time and astronauts Joe Engle and Richard Truly (marking a space first by being launched on his birthday) were at last on their way, into a 38° inclination orbit which would have a maximum altitude of 219 km (136 miles). A five day flight was on the agenda, with a complement of science experiments and the testing of the Remote Manipulator System robot arm, being carried for the first time. An increase in the alkaline level of the electrolyte in one of Columbia's three fuel cells spoiled the day for Engle and Truly, who were told to cram five days work into two, as mission rules dictated.

The RMS did not work quite as well as planned since it suffered a back-up drive failure, but TV cameras at its end gave interesting views of the payload bay and other parts of the Shuttle. Engle did not get the chance to try out the Shuttle EVA suit inside the airlock. Mission scientists were pleased with the results from the science experiments, particularly the multi-spectral imaging radiometer and the Shuttle Imaging Radar. Columbia came home on the dry lake bed runway 23 at Edwards Air Force

The launch of STS-2 sees Columbia become the first manned spacecraft to return to orbit

Base, touching down at a speed of 361 kph (224 mph), at $T + 2$ days 6 hours 13 minutes 13 seconds, main gear touchdown time.

Milestones

82nd manned space flight
33rd US manned space flight
2nd Shuttle flight
2nd flight of Columbia
1st manned space flight by reused spacecraft
Truly celebrates his 44th birthday by being launched into space (12 November)

STS-3

Int. Designation	1982-022A
Launched	22 March 1982
Launch Site	Pad 39A, Kennedy Space Center, Florida
Landed	30 March 1982
Landing Site	Runway 17, Northrup Strip, White Sands, New Mexico
Launch Vehicle	OV-102 Columbia/ET-4/SRB A11; A12/SSME #1 2007; #2 2006; #3 2005
Duration	8 days 0 hrs 4 min 45 sec
Callsign	Columbia
Objective	Third Orbital Test Flight (OFT-3)

Flight Crew

LOUSMA, Jack Robert, 46, USMC, commander, 2nd mission
Previous mission: Skylab 3 (1973)
FULLERTON, Charles Gordon "Gordo", 45, USAF, pilot

Flight Log

When Columbia returned to the Kennedy Space Center after STS-2, it was scheduled to be launched again on 22 March 1982. It was launched into murky skies, watched by one of the largest crowds since the moonshots, at 11:00 hrs local time. The first two minutes on the SRBs were enough for Lousma to describe the experience as a real barnburner, during which the vibrations caused the loss of 37 tiles from the nose and rear. His attention was diverted by an overheating APU which had to be shut down and when he got into his 38° inclination orbit, he became sick, repeating his experience of Skylab 3.

Lousma and his balding rookie pilot Gordon Fullerton started work on a hectic schedule of test flying and science. The RMS was to be tested heavily, moving two payloads around but not actually deploying them. The failure of TV cameras on the RMS, however, meant the cancellation of testing with the heaviest payload, although some operations were permitted with the Plasma Diagnosis Package. Other niggling failures, including the much-publicised toilet, were rather over-emphasised in the media, giving STS-3 a reputation it did not necessarily deserve.

Columbia was given long hot and cold soaks, pointing in the same direction for up to 80 hours, exposing it to temperatures of between −66°C and +93°C. One of these cold soaks froze a fitment on one of the payload bay doors which refused to close properly. The mission, which reached a maximum altitude of 204 km (127 miles), was to last seven days and to end at White Sands for a change, because the runway at Edwards Air Force Base was waterlogged. Just 40 minutes before retro-fire, Columbia was waived off by high winds and given a day's extension. When she finally came home

STS-3 lands at White Sands, New Mexico

to the Northrup Strip's runway at $T + 8$ days 0 hours 4 minutes 45 seconds, Lousma caused a scare by looking as though he was trying to take off again, at a record Shuttle landing speed of 404 kph (251 mph), when he over-corrected what he thought was excessive nose pitch down rate.

Milestones

83rd manned space flight
34th US manned space flight
3rd Shuttle flight
3rd flight of Columbia

SOYUZ T5

Int. Designation	1982-042A
Launched	13 May 1982
Launch Site	Pad 1, Site 5, Baikonur Cosmodrome, Kazakhstan
Landed	10 December 1982 (in Soyuz T7)
Landing Site	150 km southeast of Dzhezkazgan
Launch Vehicle	R7 (11A511U); spacecraft serial number (7K-ST) #11L
Duration	211 days 9 hrs 4 min 32 sec
Callsign	Elbrus (Elbrus)
Objective	First Salyut 7 resident crew programme

Flight Crew

BEREZOVOY, Anatoly Nikolayevich, 40, Soviet Air Force, commander
LEBEDEV, Valentin Vitalyevich, 40, civilian, flight engineer, 2nd mission
Previous mission: Soyuz 13 (1973)

Flight Log

With Salyut 6 and its Heavy Cosmos module orbiting somewhat uselessly, on 19 April 1982 the Soviets launched Salyut 7 (DOS 5-2/1982-033A), similar to Salyut 6 although its interior was fitted with an eye to décor. It was equipped with the Salyut 6-type MKF and Kate telescopes and a new large X-ray telescope for astronomy. Improved medical and physical exercise machines were incorporated, and on the outside the space station had extra handholds to improve EVA productivity. The three solar panels, too, were fitted with an attachment that could hold new, secondary sets of panels. The primary docking port was equipped to accommodate the Heavy Cosmos class modules comfortably and safely, and there were also new portholes.

The first crew to inhabit Salyut 7 was launched at 15:58 hrs local time on 13 May. It comprised rookie commander Anatoly Berezovoy and the experienced flight engineer Valentin Lebedev, a nit-picking duo who were soon to build up such a bad relationship that they only spoke to each other when necessary during the first 200-day long mission in history, which, no doubt fortunately for them, included the visit of a French cosmonaut and the first lady in space since Valentina Tereshkova. Soon after boarding, Berezovoy and Lebedev hand-deployed a small Iskra communications satellite from an airlock, the first such deployment by the Soviets and the first from a space station. Progress 13 then arrived on 25 May, to stock up the station for the long-duration medical and science mission.

The cosmonauts operated cameras, the new telescope, a Kristall materials processing furnace, a star sensor and the Oasis plant growing cabinet. The first visit occurred on 25 June, when French cosmonaut Jean-Loup Chrétien and two Soviets came aboard in Soyuz T6 for a short stay. Another Progress, No.14, arrived on 12 July

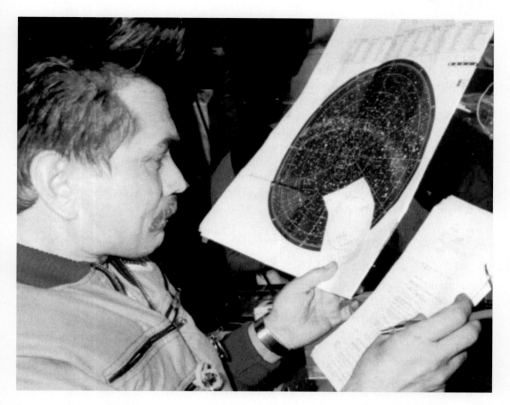

Where off Earth are we? Berezovoy consults the star charts during the long, 211-day mission

bearing more cargo, water and fuel. The first Salyut 7 spacewalk was made on 30 July, with the cosmonauts spending 2 hours 33 minutes outside retrieving samples that had been exposed to space and replacing some science equipment. Lebedev was the prime EVA crewman, with his commander supporting activities with a television camera to provide some live pictures for the folks at home. The flight engineer also conducted some space assembly tests under the code name Pamyat, in which joints between girders were made and assessed.

By 20 August, this altogether highly successful mission was receiving Svetlana Savitskaya and two male colleagues from Soyuz T7, and afterwards two more Progress tankers, 15 and 16, came to roost. The cosmonauts, who reached a maximum altitude of 374 km (232 miles) during the $51.6°$ mission, had even launched another Iskra communications satellite. A manned crew changeover was expected later in the year but never came. Apparently it was decided to bring Berezovoy and Lebedev home earlier than anticipated, before the New Year rather than after.

They had a rough return, coming back at $T + 211$ days 9 hours 4 minutes 32 seconds aboard the fresh Soyuz T7, which landed hard, turned over and rolled down a slope. Lebedev ended up on top of his commander. The weather conditions were so awful – thick fog, heavy snow and temperatures of $-18°C$ – that helicopters could not reach them for a day. The pale, tired and drawn duo had to wait 20 minutes for a ground team to reach them and ended up spending the night in the back of a truck! When the helicopter did arrive, it crash-landed and the second vehicle had to be talked down by the commander of the first.

The cosmonauts had lost several pounds in weight, their red blood counts were reduced, and their pulse rates and blood pressure were high. Indeed, Berezovoy and Lebedev were reported to be still suffering from a space hangover by mid-January. On 2 March 1983, the Soviets launched another Heavy Cosmos module, Cosmos 1443. This was similar to the Cosmos 1267 module attached to Salyut 6, with a re-entry capsule at the front. Cosmos 1443 docked with Salyut 7 on 10 March, in preparation for a new manned occupation.

Milestones

84th manned space flight
50th Soviet manned space flight
43rd Soyuz manned space flight
4th Soyuz T manned mission
1st "operational" Soyuz T flight
1st manned space flight over 200 days
New duration record – 211 days 9 hours
6th Soviet and 21st flight with EVA operations

SOYUZ T6

Int. Designation	1982-063A
Launched	24 June 1982
Launch Site	Pad 1, Site 5, Baikonur Cosmodrome, Kazakhstan
Landed	2 July 1982
Landing Site	65 km northeast of Arkalyk
Launch Vehicle	R7 (11A511U); spacecraft serial number (7K-ST) #09L
Duration	7 days 21 hrs 50 min 52 sec
Callsign	Pamir (Pamirs)
Objective	First (French) International visiting crew to Salyut 7

Flight Crew

DZHANIBEKOV, Vladimir Aleksandrovich, 40, Soviet Air Force, commander, 3rd mission
Previous missions: Soyuz 27 (1978); Soyuz 39 (1981)
IVANCHENKOV, Aleksandr Sergeyevich, 41, civilian, flight engineer, 2nd mission
Previous mission: Soyuz 29 (1978)
CHRÉTIEN, Jean-Loup, 44, French Air Force, cosmonaut researcher

Flight Log

The highlight to France's long-term cooperation with the Soviet Union in space was the decision in 1980 to fly a national cosmonaut. However, the cooperation between the chosen man, Jean-Loup Chrétien and the chosen commander, Yuri Malyshev, in 1981, was not very smooth, leading to Malyshev's replacement by Vladimir Dzhanibekov, with flight engineer Aleksandr Ivanchenkov making up the numbers. The highly qualified Chrétien had been forbidden by Malyshev to touch anything during simulations and was so frustrated that he took a pillow along with him for one simulation at Star City and went to sleep during the session, much to Malyshev's exasperation.

Relations improved with a new commander in the seat, and at 22:29 hrs local time on 24 June, Soyuz T6 ascended, watched by French officials from a stand some 1,800 m (5,905 ft) away. Before the mission the prime crew and back-up crews had drawn lots to decide which emergency situations they would cope with during final simulator training. Chrétien and his colleagues came out with automatic docking failure, which was repeated in space when the spacecraft's computer failed, necessitating a manual docking by Dzhanibekov. Once the Soyuz trio had joined Berezovoy and Lebedev, the experiments began.

The Soviets thought that working with the French was more like the real thing. The experiments were more technically sophisticated and useful compared with

Chrétien works with Ivanchenkov during their visit to Salyut 7. Note the fur-lined boots being worn to keep their feet warm

some of those carried on earlier Interkosmos missions. These included the French Echograph heart monitor, which was left on Salyut 7 after Chrétien's departure. During his stay aboard Salyut, during which he reached 360 km (224 miles) in the 51.6° orbit, and with the US Space Shuttle Columbia also in orbit, seven men were in space for the first time since 1969.

Soyuz T6's successful mission ended in fine weather near Arkalyk at $T + 7$ days 21 hours 50 minutes 52 seconds, with Chrétien highly impressed with the dynamics of re-entry, rather than his launch. Later, he was to criticise the Soviet planners for cramming too much into the work schedule and to remark that throughout the mission he never fully acclimatised to weightlessness.

Milestones

24 June 1982
85th manned space flight
51st Soviet manned space flight
44th Soyuz manned space flight
5th Soyuz T manned space flight
1st Soyuz international mission
1st manned space flight by a Frenchman
1st manned space flight by a West European

STS-4

Int. Designation	1982-065A
Launched	27 June 1982
Launch Site	Pad 39A, Kennedy Space Center, Florida
Landed	4 July 1982
Landing Site	Runway 22, Edwards Air Force Base, California
Launch Vehicle	OV-102 Columbia/ET-5/SRB A13; A14/SSME #1 2007; #2 2006; #3 2005
Duration	7 days 1 hr 9 min 31 sec
Callsign	Columbia
Objective	Fourth and final orbital flight test (OFT-4); first DoD classified payload

Flight Crew

MATTINGLY, Thomas Kenneth II, 46, USN, commander, 2nd mission
Previous mission: Apollo 16 (1972)
HARTSFIELD, Henry Warren "Hank" Jr., 48, USAF, pilot

Flight Log

The first military payload to fly aboard a US manned spacecraft was designated DoD-82-1. Not much detail was released and because of this secrecy, the STS-4 mission marked a change in media relations. The openness of NASA was restricted by the Department of Defense. Conversations with the crew would be classified for most of the mission and photographs taken during it would be limited to those that did not show any classified hardware. STS-4, which was the first US mission to be flown by astronauts without a back-up crew, was not entirely classified because apart from the range of science and declassified payloads, the DoD-82-1 was known to be the Cirris cryogenic infrared radiance instrument to obtain spectral data on the exhausts of vehicles powered by rocket and air breathing engines, and an ultraviolet horizon scanner. Cirris would not perform well, because its lens cap didn't come off!

The first on-time Shuttle launch, at 11:00 hrs local time, was handled extremely matter-of-factly by young Mark Hess, the NASA press officer, making his first launch commentary. Commander Ken Mattingly and his sidekick Hank Hartsfield sailed into 28.5° inclination orbit, the lowest for a manned space flight but one that would become fairly usual for a Shuttle mission, with a maximum altitude during the mission of 275 km (127 miles). This was still 7 km (4 miles) shorter than planned after the heavier than planned launch weight, caused by water under the heat shield tiles which had collected after a thunderstorm days before launch, and which resulted in an increased SSME burn time of 3 seconds and several OMS burns. In addition, the

The STS-4 crew is greeted by President and Mrs Reagan after completing their mission on America's 206th birthday

two SRBs were lost in the Atlantic rather than recovered as planned, as a result of parachute failures.

The first US commercial payload in space, more than nine experiments from Utah University crammed inside Getaway Special (GAS) canisters in the payload bay, began operating together with over 20 others packed aboard the busy Columbia orbiter. The mission seemed to have been a spectacular success, despite the Cirris lens cap saga, which Mattingly tried to knock off with the RMS and even suggested that he make a spacewalk to rectify. He did try out the EVA suit in the airlock as planned, however. President Reagan was waiting at Edwards Air Force Base to greet the returning crew, which landed on the concrete runway 22 at a speed of 374 kph (232 mph), at main gear touchdown time of 7 days 1 hour 9 minutes 40 seconds. The Independence Day celebrations seemed complete amid the patriotic fervour but were left a little damp by the President's lacklustre support for a space station. The Shuttle was rather too enthusiastically declared "operational" as from its next flight.

Milestones

86th manned space flight
35th US manned space flight
4th Shuttle flight
4th flight of Columbia
1st US manned military space flight
1st US manned space flight without a back-up crew
1st manned space flight to carry an official commercial payload

SOYUZ T7

Int. Designation	1982-080A
Launched	19 August 1982
Launch Site	Baikonur Cosmodrome, Kazakhstan
Landed	27 August 1982 (in Soyuz T5)
Landing Site	112 km northeast of Arkalyk
Launch Vehicle	R7 (11A511U); spacecraft serial number (7K-ST) #12L
Duration	7 days 21 hrs 52 min 24 sec
Callsign	Dnieper (Dnieper)
Objective	All-Soviet visiting mission to Salyut 7; Soyuz exchange mission

Flight Crew

POPOV, Leonid Ivanovich, 36, Soviet Air Force, commander, 3rd mission
Previous missions: Soyuz 35 (1980); Soyuz 40 (1981)
SEREBROV, Aleksandr Aleksandrovich, 38, civilian, flight engineer
SAVITSKAYA, Svetlana Yevgenyevna, 34, civilian, research engineer

Flight Log

A Soyuz with a difference lit up the Baikonur skies at 23:12 hrs local time on 19 August, when a crew of three lifted off for a visiting mission to Salyut 7. This crew included the first female in space for 19 years, since the first, Valentina Tereshkova, was launched. While Tereshkova's mission was mere propaganda, the inclusion of Svetlana Savitskaya, bona fide test pilot and a world aerobatic champion, seemed logical and acceptable – except that she just happened to beat the first American female, Sally Ride, into space.

Amid much ballyhoo and publicity, as well as live TV coverage, Savitskaya and her two seemingly anonymous male colleagues docked with Salyut about 25 hours after launch. The Salyut 7 resident, Valentin Lebedev, gave her an apron and told her to start work. Savitskaya's main task was not to do the washing up, but to operate a series of life sciences experiments to study the cardiovascular system, motion sickness and eye movement. She also operated an electrophoresis experiment to separate cells. Popov, Serebrov and Savitskaya landed in Soyuz T5 at $T + 7$ days 21 hours 52 minutes 24 seconds, 112 km (70 miles) northeast of Arkalyk. Maximum altitude reached during the 51.6° mission was 315 km (196 miles).

Milestones

87th manned space flight
52nd Soviet manned space flight

45th Soyuz manned space flight
6th Soyuz T manned space flight
1st manned space flight by mixed female and male crew

Berezovoy and Savitskaya in Salyut 7

STS-5

Int. Designation	1982-110A
Launched	11 November 1982
Launch Site	Pad 39A, Kennedy Space Center, Florida
Landed	16 November 1982
Landing Site	Runway 22, Edwards Air Force Base, California
Launch Vehicle	OV-102 Columbia/ET-6/SRB A15; A16/SSME #1 2007; #2 2006; #3 2005
Duration	5 days 2 hrs 14 min 26 sec
Callsign	Columbia
Objective	First "operational" Shuttle mission – commercial satellite deployment mission

Flight Crew

BRAND, Vance DeVoe, 51, civilian, commander, 2nd mission
Previous mission: Apollo 18 ASTP (1975)
OVERMYER, Robert Franklyn, 46, USMC, pilot
ALLEN, Joseph Percival, 45, civilian, mission specialist 1
LENOIR, William Benjamin, 43, civilian, mission specialist 2

Flight Log

The news of the death of Soviet premier Leonid Brezhnev, events in Poland, and a British spy scandal served to overshadow this unique space flight, which began at 07:19 hrs local time at the Kennedy Space Center. Commander Vance Brand, pilot Bob Overmyer and mission specialist Bill Lenoir (evaluating the MS2/Flight Engineer role for ascent), were seated in the flight deck, while the other mission specialist, Joe Allen, was seated below in the mid-deck, which also served as the kitchen and toilet. Columbia was still fitted with ejection seats for the commander and pilot but they were not armed. The crew was the first from America not to have any means of escape in the event of a launch accident and were also the first to fly in flight overalls, and oxygen-fed helmets, in case of cabin depressurisation.

After MECO and two OMS burns, Columbia was in its 256 km (159 miles) maximum altitude 28.4° inclination orbit. At $T + 7$ hours 58 minutes 35 seconds into the mission, the crew dispatched the communications satellite SBS from its spin table in the payload bay, on the first commercial manned trucking mission, earning for NASA a cool $12 million. The satellite's own Pam D upper stage fired later, to place it into a geostationary transfer orbit where it would normally have been placed by a conventional expendable launch vehicle. Another satellite, Canada's Anik 3, was launched later and the crew proudly displayed an "Ace Trucking Company – We Deliver" sign to TV cameras.

One of the commercial satellite deployment operations during STS-5

There were disappointments, however. First Overmyer was space sick, vomiting at $T + 6$ hours and continuing to feel queasy. Lenoir felt less sick, describing his symptoms as a "wet belch". The astronauts were prescribed drugs and were also angry that their illness was publicised, possibly to the detriment of their careers. In future, NASA decided, if an astronaut was sick it would remain a confidential matter. The first Shuttle spacewalk by Allen and Lenoir was delayed by a day, and then never took place at all because both astronauts experienced spacesuit problems on the brink of opening the airlock door. Lenoir's primary oxygen pressure regulator failed and Allen's fan assembly sounded like a motorboat. Allen, now seated in the flight deck (evaluating the FE role for entry), took pictures during re-entry, which was like being inside a blast furnace, he said.

Columbia was aiming for a lake bed landing at Edwards Air Force Base but was diverted to the concrete runway 22 because the "dry" lake was rather wet. Main gear touchdown came at $T + 5$ days 2 hours 14 minutes 26 seconds, the longest four-crew space flight.

Milestones

88th manned space flight
36th US manned space flight
5th Shuttle flight
5th flight of Columbia
1st flight with four crew members
1st flight of mission specialists
1st manned space flight to deploy commercial satellites
1st flight with cancelled EVA operations
1st launch and landing by crew member not seated in cockpit
1st US flight with no emergency crew escape
1st US flight by crew without spacesuits
1st US flight to carry engineers

STS-6

Int. Designation	1983-026A
Launched	4 April 1983
Launch Site	Pad 39A, Kennedy Space Center, Florida
Landed	9 April 1983
Landing Site	Runway 22, Edwards Air Force Base, California
Launch Vehicle	OV-099 Challenger/ET-8/SRB A17; A18/SSME #1 2017; #2 2015; #3 2012
Duration	5 days 0 hrs 23 min 42 sec
Callsign	Challenger
Objective	Maiden flight of OV-099 (Challenger); EVA demonstration; deployment of first TDRS

Flight Crew

WEITZ, Paul Joseph, 50, civilian, commander, 2nd mission
Previous mission: Skylab 2 (1973)
BOBKO, Karol Joseph, 45, USAF, pilot
MUSGRAVE, Franklin Story, 47, civilian, mission specialist 1
PETERSON, Donald Herod, 49, civilian, mission specialist 2

Flight Log

The first Challenger orbiter mission was originally due to have taken place on 27 January 1983 but was delayed by a series of potentially disastrous engine problems which first came to light after Challenger's Flight Readiness Firing on 18 December 1982. Engineers detected an abnormal level of gaseous hydrogen. A second FRF was scheduled for 25 January and the TDRS payload was removed from Challenger. The hydrogen leak was detected again and this time was traced to a 2 cm ($\frac{3}{4}$ in) crack in the No.1 main engine combustion chamber coolant outlet manifold. Engine 1 was ordered to be replaced. TDRS was replaced, only to be slightly damaged by fine salt sea spray after a severe storm. It was back inside Challenger's cargo bay by 19 March.

Worse was to follow. The replacement engine 1 was found to be faulty and had to be replaced itself, then an inspection of the No.2 and 3 engines revealed hairline cracks which had to be repaired. Challenger sat engineless on the pad. At last, on 4 April 1983, at the comparatively late hour of 18:30 KSC time, Challenger ascended flawlessly into clear blue skies, the only anomaly being the annoying deposition of some black soot on Challenger's windows at SRB separation.

The rookie crew (called the F Troop after a TV programme and the fact that they were the sixth Shuttle crew) proceeded to achieve the main objective – to deploy NASA's first $100 million communications station in space, TDRS, on only the second IUS solid propellant two-stage upper stage flown. This was duly deployed

The first Shuttle EVA demonstration was conducted during STS-6

from its tilt table and was later injected into geostationary transfer orbit. A second stage failure stranded the satellite, however, and through no fault of its own the Shuttle was tarred with the same brush by some of the press. TDRS was eventually nudged into its planned geostationary orbit by careful firing of its own thrusters over a period of 58 days.

Maximum altitude reached by Challenger in the 28.4° orbit was 248 km (154 miles). On 8 April, Story Musgrave (EV1) and Donald Peterson (EV2) made the delayed EVA that was planned for STS-5, lasting 4 hours 17 minutes, to check out the Shuttle spacesuit and practice making space repairs, featuring in some spectacular TV. It was also the first US EVA since Skylab 4, nine years earlier. Later, Challenger came home to Edwards Air Force Base, landing on runway 22 at $T + 5$ days 0 hours 23 minutes 42 seconds, the shortest four-crew space flight.

Milestones

89th manned space flight
37th US manned space flight
6th Shuttle flight
1st flight of Challenger
16th US and 22nd flight with EVA operations
1st Shuttle-based EVA
1st TDRS deployment mission

SOYUZ T8

Int. Designation	1983-035A
Launched	20 April 1983
Launch Site	Pad 1, Site 5, Baikonur Cosmodrome, Kazakhstan
Landed	22 April 1983
Landing Site	96 km northeast of Arkalyk
Launch Vehicle	R7 (11A511U); spacecraft serial number (7K-ST) #13L
Duration	2 days 0 hrs 17 min 48 sec
Callsign	Okean (Ocean)
Objective	Second Salyut 7 resident crew programme

Flight Crew

TITOV, Vladimir Georgyevich, 36, Soviet Air Force, commander
STREKALOV, Gennady Mikhailovich, 43, civilian, flight engineer 1, 2nd mission
Previous mission: Soyuz T3 (1980)
SEREBROV, Aleksandr Aleksandrovich, 39, civilian, research engineer, 2nd mission
Previous mission: Soyuz T7 (1982)

Flight Log

Salyut 7 remained empty during the Russian winter of 1982–3 and was joined by the unmanned Cosmos 1443 module in March. Trained to work aboard Salyut and the new module were Vladimir Titov, Aleksandr Serebrov, making the first successive national manned space flight, and Gennady Strekalov. Their attempt to dock with Salyut, however, was doomed very soon after lift-off, at 19:11 hrs local time from Baikonur, when the payload shroud tore away Soyuz T's rendezvous radar antenna which only partially deployed. The crew used the RCS thrusters to try to shake the antenna free but to no avail. In trying to hide the serious problem, these engine firings were reported as tests of the attitude control system.

Although mission rules would normally dictate a return to Earth, the rookie commander Titov got permission to try a visual rendezvous and attempted docking using radar readings from the ground. The docking was perceived as having a low success probability by the ground controllers. It could have been a complete disaster, for Soyuz T8 flew past Salyut 7 at great speed, missing a catastrophic collision by 160 m (525 ft). Titov had made an optically guided approach to Salyut's rear docking port after a 50 second rocket burn. The seventh space station flight had to be aborted not because of lack of power but because propellant reserves were not high enough to try again. The difficulty in guiding the Soyuz T to the station becomes more apparent when it was later revealed by Titov that he had not trained for a fully manual docking

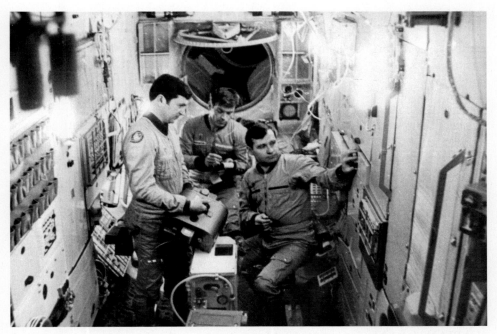

Soyuz T8 crew during a training session in the Salyut 7 mock-up, something they did not put into practice in space. L to r Titov, Serebrov, Strekalov

approach and was unsure of his depth perception through the spacecraft periscope as he attempted a difficult manoeuvre.

The crew, which would have been the first three-man long duration crew since Soyuz 11, came home 96 km (60 miles) northeast of Arkalyk at $T + 2$ days 0 hours 17 minutes 48 seconds. Maximum altitude reached in the 51.6° orbit was 300 km (186 miles).

Milestones

90th manned space flight
53rd Soviet manned space flight
46th Soyuz manned space flight
7th Soyuz T manned space flight
1st space flight by crewman on successive national missions

STS-7

Int. Designation	1983-026A
Launched	18 June 1983
Launch Site	Pad 39A, Kennedy Space Center, Florida
Landed	24 June 1983
Landing Site	Runway 15, Edwards Air Force Base, California
Launch Vehicle	OV-099 Challenger/ET-7/SRB A51; A52/SSME #1 2017; #2 2015; #3 2012
Duration	6 days 2 hrs 23 min 59 sec
Callsign	Challenger
Objective	Commercial satellite deployment mission; space adaptation medical investigations

Flight Crew

CRIPPEN, Robert Laurel, 45, USN, commander, 2nd mission
Previous mission: STS-1 (1981)
HAUCK, Frederick "Rick" Hamilton, 42, USN, pilot
FABIAN, John McCreary, 44, USAF, mission specialist 1
RIDE, Sally Kristen, 32, civilian, mission specialist 2
THAGARD, Norman Earl, 39, civilian, mission specialist 3

Flight Log

By contrast to the Soviet reaction to the flight of Svetlana Savitskaya in 1982, the US launch of Sally Ride was played down as much as possible by NASA and by the lady herself, not with total success. The on-time lift-off occurred at 07:33 hrs local time and after MECO, two OMS burns were required to carry Challenger to its operational 28.45° orbit with a maximum altitude of 272 km (169 miles). Crippen, the first person to fly the Shuttle for a second time, described the launch as a bit smoother than he remembered on STS-1.

The first commercial satellite payload was delivered into orbit at $T + 9$ hours 29 minutes, with an accuracy estimated at within 457 m (1,500 ft) of the target point and within 0.085° of the required pointing vector. Canada's Anik 2C later made its way into geostationary orbit. The following day, satellite number two, India's Palapa, was safely deployed. With the commercial trucking mission over, the crew got down to the third satellite deployment, that of the West German SPAS free flier, using the Remote Manipulator System (RMS) arm operated by John Fabian. Almost immediately, Fabian grabbed the satellite, demonstrating the first satellite retrieval. SPAS was released again and Crippen moved Challenger 300 m (984 ft) away and performed a series of station-keeping manoeuvres. Cameras on SPAS, meanwhile, took spec-

Clockwise from top left: Crippen, Hauck, Fabian, Thagard and Ride, the crew of STS-7

tacular photos of Challenger in space, with the RMS arm conveniently cocked in the shape of a number 7.

Six science experiments were on board and these operated for nine-and-a-quarter hours autonomously before the free flier was retrieved, this time by Sally Ride. The third mission specialist, the flight doctor Norman Thagard, who had been added to the crew to study space adaptation syndrome, even had a go at using the RMS. A complement of onboard experiments was operated by the crew, including the first demonstration of the Shuttle's Ku-band rendezvous radar system and a reduction in cabin pressure from 760 mm (3 in) to 527 mm (2 in) for 30 hours to investigate the possibility of eliminating the required three-and-a-half hours pre-breathing period for EVA astronauts.

The high point of the mission was to be Challenger's return to the Kennedy Space Center, the first such return to launch site in history. Bad weather thwarted the attempt and Crippen was diverted to Edwards Air Force Base to land on runway 15. His request for a two day orbital extension was turned down because of concerns over one of the APUs on Challenger. Mission time was $T + 6$ days 2 hours 23 minutes 59 seconds.

Milestones

91st manned space flight
38th US manned space flight
7th Shuttle flight
2nd flight of Challenger
1st flight with five crew
1st US female in space

SOYUZ T9

Int. Designation	1983-062A
Launched	27 June 1983
Launch Site	Pad 1, Site 5, Baikonur Cosmodrome, Kazakhstan
Landed	23 November 1983
Landing Site	160 km east of Dzhezkazgan
Launch Vehicle	R7 (11A511U); spacecraft serial number (7K-ST) #14L
Duration	149 days 10 hrs 46 min 1 sec
Callsign	Proton (Proton)
Objective	Second Salyut 7 resident crew programme (revised)

Flight Crew

LYAKHOV, Vladimir Afanasevich, 42, Soviet Air Force, commander, 2nd mission
Previous mission: Soyuz 32 (1979)
ALEKSANDROV, Aleksandr Pavlovich, 40, civilian, flight engineer

Flight Log

Following the docking failure of Soyuz T8, the next crew were assigned to complete most of the tasks planned for the previous one. However, Titov and Strekalov had conducted extensive EVA training which the T9 crew had not, so the plan was to launch Soyuz T10 with Titov and Strekalov aboard to take over from the T9 crew and conduct the extensive EVAs they had trained for.

Soyuz T9, with a crew of two rather than the expected three (due to additional propellant load), took off from Baikonur at 15:12 hrs local time, and just over a day later docked at the rear of Salyut 7 to start a mission that would, according to mission controller Valery Ryumin, be shorter than Soyuz T5's 211 days. They almost did not make it as, for the first time since Soyuz 1, one of the twin solar panels on Soyuz failed to deploy (although this did not prevent the docking with the Salyut), a fact not revealed for 20 years. The crew, Vladimir Lyakhov and rookie flight engineer Aleksandr Aleksandrov, were the first to operate using a Heavy Cosmos module, No.1443, attached to the front of Salyut 7. This two-part spacecraft contained a 1.5 m^3 (50 ft^3) habitable module, an Instrument Module, and a descent capsule capable of returning 500 kg (1,103 lb) to Earth. The module was equipped with 38 m^2 (40 ft^2) of solar panels, providing 3 kW of electricity.

Lyakhov and Aleksandrov got down to work producing virus cells and conducting Earth resources surveys, saving Soviet citizens from disaster by warning of the formation of a lake from a melting glacier which threatened to flood several towns beneath. While the crew were inside Soyuz T9 conducting a mock evacuation exercise,

With a traditional traveller's gift of bread and salt (as well as flowers), the T9 crew relax after their recovery from the mission. Lyakhov is on the left, Alexandrov on the right

one of Salyut's 14 mm ($\frac{1}{2}$ in) thick windows was pitted to a depth of 4 mm (0.16 in) by the impact of an unidentified object.

Cosmos 1443 separated from Salyut 7 on 14 August and, while flying autonomously, returned its descent capsule containing film and some equipment. It landed 100 km (62 miles) southeast of Arkalyk on 23 August. The major part of Cosmos was destroyed during re-entry on 19 September. Soyuz T9 had been flown from the back of Salyut to the front to prepare for the arrival of Progress 17 on 19 August. Progress left on 17 September, leaving the port free for the Soyuz T10 crew, who were to have been launched on 27 September to help with repairs, including an EVA to correct solar panel problems and add additional panels to increase the electrical supply on board the station.

By this time, Salyut 7 was in pretty bad shape, propellant leaks leaving the station with little manoeuvrability. Salyut's back-up main engine was also crippled and a solar panel failure had reduced solar power. A major incident occurred on 9 September during the refuelling operations by Progress 17. A Salyut fuelling line used to feed oxidiser from the Progress to the Salyut ruptured. With only half of the 32 thrusters working, it seemed likely Salyut would have to be abandoned, but a decision was made to work around the problem and let the current mission continue while options for repair were evaluated. After the Soyuz T10 crew failed to reach orbit following the first on-the-pad launch abort in history, rumours spread in the west that Lyakhov and Aleksandrov were stranded in space, particularly as the Soyuz T9 ferry was exceeding its 115-day lifetime, according to the rumours, which created sensational press stories.

The flight continued, and a Progress ferry craft was launched to Salyut, on 21 October, carrying new solar panels, fuel and equipment. It also provided a means of propulsion for the crippled station. The crew even made two spacewalks on 1 and

3 November, lasting 2 hours 50 minutes and 2 hours 55 minutes respectively, to erect new solar panels, while cosmonauts Leonid Kizim and Vladimir Solovyov carried out a simulated EVA at the same time in the neutral buoyancy tank at Star City. The two cosmonauts on Salyut had not trained to perform such complicated EVAs and struggled to complete the tasks, as reflected in the durations of each spacewalk. The tasks had originally been planned to be completed during one EVA, but were spread over two EVAs due to the cosmonauts' inexperience. First-time space explorer Alexandrov was amazed by the whole experience of EVA and at one point casually discarded a small unwanted item into space to see what happened. This earned him a rebuke from Mission Control, who feared confusing the station's stellar orientation system into "thinking" that the light refection from the object might be a star. Progress separated on 13 November and the so-called doomed cosmonauts made an unheralded landing on 23 November, at $T + 149$ days 10 hours 46 minutes. Maximum altitude reached in the $51.6°$ orbit was 354 km (220 miles). The unexpected extension to the mission had gave rise to concerns over the reliability of Soyuz T in supporting a crew after such a long time in space. Soyuz T9 proved such fears were unfounded, however, and the recovery occurred without incident, giving great confidence for longer Soyuz T-supported station residences.

Milestones

92nd manned space flight
54th Soviet manned space flight
47th manned Soyuz space flight
8th manned Soyuz T space flight
7th Soviet and 23rd flight with EVA operations
Lyakhov celebrates his 42nd birthday in space (20 July)

STS-8

Int. Designation	1983-089A
Launched	30 August 1983
Launch Site	Pad 39A, Kennedy Space Center, Florida
Landed	5 September 1983
Landing Site	Runway 22, Edwards Air Force Base, California
Launch Vehicle	OV-099 Challenger/ET-9/SRB A53; A54/SSME #1 2017; #2 2015; #3 2012
Duration	6 days 1 hr 8 min 43 sec
Callsign	Challenger
Objective	Satellite deployment mission; RMS load evaluation tests; space adaptation medical investigations

Flight Crew

TRULY, Richard Harrison, 45, USN, commander, 2nd mission
Previous mission: STS-2 (1981)
BRANDENSTEIN, Daniel Charles, 40, USN, pilot
GARDNER, Dale Allan, 34, USN, mission specialist 1
BLUFORD, Guion Stewart, 40, USAF, mission specialist 2
THORNTON, William Edgar, 54, civilian, mission specialist 3

Flight Log

An awe-inspiring lift-off from the Kennedy Space Center at 02:32 hrs local time, the first night launch in the Shuttle programme and only the second in US manned space flight history, was seen within a radius of 720 km (447 miles), but was lucky to have been given the go-ahead. Lightning had struck the launch tower hours before the launch and rain swept conditions delayed it for 17 minutes until mission controllers felt that they had found a hole in the weather and that conditions would be good enough for Challenger to actually make it back to the KSC in one piece following any return to launch site abort.

Already delayed from 4 August due to technical problems including an in-orbit check of TDRS-1, Challenger headed through the clouds as a fuzzy orange halo while the moisture-laden air reflected and amplified the sound, making it the noisiest affair. Inside the Shuttle, the visual effects were both spectacular and a bit frightening. During the SRB burn, unbeknown to NASA at the time, ablative material on one of the SRB nozzles, designed to burn through to 4 cm ($1\frac{1}{2}$ in) in the 3,200°C temperatures, actually burned through to just 1.3 cm ($\frac{1}{2}$ in). Complete burn through, NASA discovered later, could have caused side-thrusting exhaust to put Challenger out of control. The problem delayed the next mission, STS-9, which had its SRBs replaced as a precautionary measure.

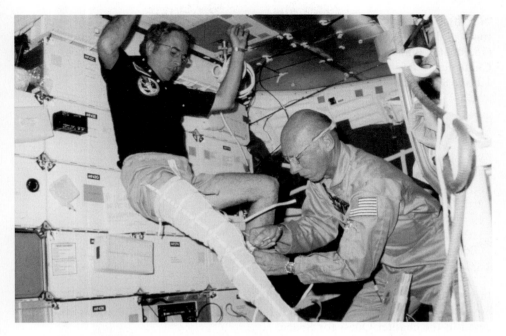

Dr. Bill's clinic on STS-8. His "patient" is commander Dick Truly

Once on orbit, with an inclination of 28.4° and a maximum altitude of 265 km (165 miles), the commercial objectives of STS-8 were achieved quickly when India's Insat 1A was deployed from the payload bay, with a slight clinking sound which may have been the result of it knocking against a Shuttle fixture. This probably caused the damage to the satellite that was discovered later when one of the solar panels would not deploy properly at first. Challenger was originally to have carried TDRS-2 but this was replaced by Insat and a 3,383 kg (7,458 lb) Payload Test Article, brought forward from STS-11 and shaped like a giant dumb-bell. This was unberthed and held in various positions to evaluate the performance of the RMS in handling heavy payloads.

Challenger appraised the use of the TDRS-1 satellite to communicate with mission control and the new link enabled the first in-flight press conference since ASTP to be staged and excellent TV coverage to be beamed to Earth. Communications during re-entry via TDRS were not possible, however, due to a computer fault. Commander Truly concentrated on a programme of the most detailed Earth photography since Skylab, while the other astronauts concentrated on their specialities, including Bluford's operation of the McDonnell Douglas electrophoresis machine to process living cells for the first time. Meanwhile, the oldest man in space, Dr. Bill Thornton, aged 54 – who was only added to STS-8 at a late stage to study space motion sickness – operated "Dr. Bill's Clinic", learning more in an hour than all the previous years he had put in on the Earth, he said.

After a smooth re-entry, during which Truly performed a series of hypersonic turns and banks, the crew got their first site of Edwards at Mach 2 and 22,860 m (75,000 ft), illuminated by the six xenon lights of runway 22, which greeted Challenger's first US night landing in manned space flight history, at $T + 6$ days 1 hour 8 minutes 43 seconds.

Milestones

93rd manned space flight
39th US manned space flight
8th Shuttle mission
3rd flight of Challenger
1st US manned space flight to end at night
1st African American space traveller
Oldest first time space traveller (Thornton), aged 54

In between the flights of STS-8 and STS-9, the Soviet Union attempted to launch Soyuz T10. The mission was aborted following a launch pad fire and is covered in detail in the chapter Quest for Space.

STS-9

Int. Designation	1983-116A
Launched	28 November 1983
Launch Site	Pad 39A, Kennedy Space Center, Florida
Landed	8 December 1983
Landing Site	Runway 17, Edwards Air Force Base, California
Launch Vehicle	OV-102 Columbia/ET-11/SRB A55; A60/SSME #1 2011; #2 2018; #3 2019
Duration	10 days 7 hrs 47 min 23 sec
Callsign	Columbia
Objective	First flight of European-built Spacelab pressurised scientific laboratory (Spacelab 1)

Flight Crew

YOUNG, John Watts Jr., 53, USN, commander, 6th mission
Previous missions: Gemini 3 (1965); Gemini 10 (1966); Apollo 10 (1969); Apollo 16 (1972), STS-1 (1981)
SHAW, Brewster Hopkinson Jr., 38, USAF, pilot
GARRIOTT, Owen Kay, 53, civilian, mission specialist 1, 2nd mission
Previous mission: Skylab 3 (1973)
PARKER, Robert Allan Ridley, 46, civilian, mission specialist 2
MERBOLD, Ulf, 42, civilian, payload specialist 1
LICHTENBERG, Byron Kurt, 35, civilian, payload specialist 2

Flight Log

This mission, originally scheduled for 30 September, was delayed for a month to allow time for Columbia's SRB nozzles to be replaced following the STS-8 launch scare. The Shuttle rose from Pad 39A at 16:00 hrs local time at the Kennedy Space Center, carrying a unique crew of six, including the apparently ageless veteran of space flight, John Young, in the commander's seat. Columbia performed a startling 140° roll programme to place it on an azimuth heading for a 57° inclination orbit – the highest achieved on a US manned space flight. As a result, the Shuttle could be seen passing over Britain just after entering its orbit, which would have a maximum altitude of 216 km (134 miles).

On the schedule was a nine-day intensively scientific mission aboard the first Spacelab payload bay-mounted laboratory. This was built by the European Space Agency as a result of an agreement signed with NASA ten years earlier, at a cost of $850 million. The nine-day mission, which featured the first time that the principal investigators could talk directly to experiments in space via the TDRS-1 relay satellite, was judged a phenomenal success. Altogether, 73 separate investigations were

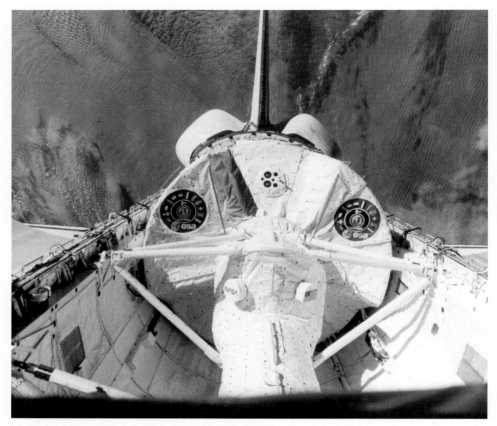

The Spacelab 1 Module in Columbia's payload bay. In the foreground is the access tunnel from the Shuttle mid-deck

completed in the fields of astronomy, physics, atmospheric physics, Earth observations, life sciences, material sciences, space plasma physics and technology. The crew split into two shifts which lasted twelve hours, sometimes more like eighteen hours, so much so that mission controllers extended the mission. The Red Shift comprised Young, Parker and Merbold (representing ESA), while the Blue Shift included Shaw, Garriott and Lichtenberg (representing the American science community).

Strapped in their seats and ready to come home, the crew got an unplanned nine hour flight extension when a major computer failure stopped the OMS retro-burn just before it started. Computer 1 failed, then computer 2, which had taken over operations instantly, also shut down. Young and his pilot Brewster Shaw sweated over troubleshooting the problem, and with everything crossed, fired the OMS engines for another try. The computer fault had been due to contaminated integrated circuits. All went well and Columbia homed in on Edwards Air Force Base's dry lake bed runway 17, landing at $T + 10$ days 7 hours 47 minutes 23 seconds, the longest six-crew mission. Two minutes later, the troubled Columbia caused another scare by catching fire, leaking hydrazine from an APU being the cause. As the pilot and commander winged their way to Houston, the four other crewmen took part in intensive life sciences investigations, which included another dose of zero-G in a KC-135 aircraft.

Milestones

94th manned space flight
40th US manned space flight
9th Shuttle mission
6th flight of Columbia
1st Spacelab Long Module mission
1st flight with six crew members
1st flight to carry a West German
1st flight with crew member on sixth mission
1st US flight with non-American crew member
1st flight with payload specialist crew members

STS 41-B

Int. Designation	1984-011A
Launched	3 February 1984
Launch	Site Pad 39A, Kennedy Space Center, Florida
Landed	11 February 1984
Landing Site	Runway 15, Kennedy Space Center, Florida
Launch Vehicle	OV-099 Challenger/ET-10/SRB A57; A58/SSME #1 2109; #2 2015; #3 2012
Duration	7 days 23 hrs 15 min 55 sec
Callsign	Challenger
Objective	Satellite deployment mission; first tests of Manned Manoeuvring Unit (MMU)

Flight Crew

BRAND, Vance DeVoe, 52, civilian, commander, 3rd mission
Previous missions: Apollo 18 ASTP (1975); STS-5 (1982)
GIBSON, Robert Lee "Hoot", 37, USN, pilot
McNAIR, Ronald Erwin, 33, civilian, mission specialist 1
STEWART, Robert Lee, 41, US Army, mission specialist 2
McCANDLESS, Bruce II, 46, USN, mission specialist 3

Flight Log

The first flight of the manned manoeuvring unit apart, this mission offered one rather infamous distinction. Was it STS-10, STS-11 or STS 41-B? It was originally planned as STS-11 but when a military flight, STS-10, was delayed for what would turn out to be a year, STS-11 moved up one slot, becoming what would logically be called STS-10. Instead, it confusingly retained the STS-11 designation, and then NASA confused the numbering system even further by introducing an extraordinary system of nomenclature that would soon have most non-specialist space followers in a real pickle.

The "4" in STS 41-B represented the US fiscal year 1984. The "1" represented the Kennedy Space Center launch site (and a "2" would have referred to Vandenberg Air Force Base in California, where Shuttle missions were expected to launch from beginning in 1986). The "B" stood for the second flight of the 1984 fiscal year. Adding to the confusion was the fact that the mission was the first of the actual 1984 year and that STS-9 was sometimes termed as STS 41-A. Despite these diversions, most attention was focused on the fact that it was to perform one of the last "firsts" in manned space flight – an independent EVA. NASA's Manned Manoeuvring Unit (MMU) was to be operated by Bruce McCandless, who had helped to develop it and who had waited eighteen years for a space flight.

Alone in space, Bruce McCandless becomes the first person to fly an untethered EVA using the MMU

Lift-off – delayed from 29 January by APU problems – took place on time at 08:00 hrs local time, the only anomaly being the failure of one of the parachutes on each SRB. Orbit was 28.45° inclination and would reach a maximum altitude of 281 km (175 miles). Six-and-a-half orbits later came the first of two deployments of similar Hughes-built satellites. Westar 6 spun out of Challenger's cargo bay first. Later, its PAM-D perigee motor shut down early, stranding Westar. The failure was widely blamed on the Shuttle. The press had another field day when a small instrumented rendezvous balloon target burst on deployment and they went into anti-Shuttle overkill when the second main satellite, Palapa B2, was also inexplicably stranded in orbit by an identical upper stage failure. Because the wrist joint on the RMS failed, the SPAS free flier would not be deployed either.

Astronauts McCandless (EV1) and Robert Stewart (EV2) saved the flight from ignominy when they emerged on 7 February for the first EVA, which featured McCandless's solo MMU flight. The jocular astronaut flew as far as 100 m (328 ft) from Challenger, as did Stewart when he tried the MMU later. McCandless likened flying the MMU to flying a helicopter at Mach 25. A second MMU unit inside the payload bay was tried out by McCandless on the second EVA the following day. The EVAs lasted 5 hours 55 minutes and 6 hours 17 minutes respectively, and McCandless and Stewart had operated the MMUs for 4 hours 42 minutes. (Unit 2 made two sorties lasting 1 hour 3 minutes, and Unit 3 three flights lasting 3 hours 39 minutes.)

The next highlight of the mission of mixed fortune was the first landing back at base, on the Kennedy Space Center Shuttle runway, just 6.4 km (4 miles) from its take-off point. Commander Vance Brand was surprised as he flew over the KSC that the autopilot had taken Challenger to 15,000 m (49,200 ft) and that he was far too high, even for a steep Shuttle descent. All went well, however, despite more than a hint of ground fog and the first bird strike for the Shuttle. Main gear touchdown on runway 15, designated for approaches from the north, came at $T + 7$ days 23 hours 15 minutes 55 seconds.

Milestones

95th manned space flight
41st US manned space flight
10th Shuttle mission
4th flight of Challenger
1st independent EVA using manned manoeuvring unit
1st manned space flight to land at launch site
17th US and 24th flight with EVA operations

SOYUZ T10

Int. Designation	1984-014A
Launched	8 February 1984
Launch Site	Pad 31, Site 6, Baikonur Cosmodrome, Kazakhstan
Landed	2 October 1984 (aboard Soyuz T11)
Landing Site	145 km northeast of Dzhezkazgan
Launch Vehicle	R7 (11A511U); spacecraft serial number (7K-ST) #15L
Duration	236 days 22 hrs 49 min 4 sec
Callsign	Mayak (Lighthouse)
Objective	Third Salyut 7 resident crew; extensive EVA repair programme

Flight Crew

KIZIM, Leonid Denisovich, 42, Soviet Air Force, commander, 2nd mission
Previous mission: Soyuz T3 (1980)
SOLOVYOV, Vladimir Alekseyevich, 37, civilian, flight engineer
ATKOV, Oleg Yuryevich, 34, civilian, cosmonaut researcher

Flight Log

There could not have been much wrong with Salyut 7 that couldn't be righted, for the next Soviet space mission would be one of sheer endurance. In the cosmonaut researcher's seat in Soyuz T10, too, was a cardiologist, Oleg Atkov, who had designed a portable ultrasound cardiograph which he would use to monitor crew health throughout the flight. Soyuz lifted off at 17:07 hrs local time from the Baikonur Cosmodrome on 8 February and docked with Salyut the following day. During the mission, the crew would reach a maximum altitude of 375 km (233 miles) in the 51.6° orbit and two of them would achieve new heights in EVA experience. The EVAs took place much later in the mission, after another Shuttle (STS 41-C) had conducted some unique EVA operations of its own during April.

Soon after boarding Salyut, Progress 19 arrived, providing all-important consumables for the long mission, the duration of which was not announced by the Soviets. On 4 April, Leonid Kizim, Vladimir Solovyov and Dr. Atkov were visited by two Soviet and one Indian cosmonaut, and with five Americans aboard the Space Shuttle Challenger also in orbit, eleven people were in space for the first time. The Indian international crew returned to Earth aboard Soyuz T10, leaving the fresh Soyuz T11 attached to Salyut. Progress 20 arrived on 20 April, with a cargo of tools and equipment to enable Kizim and Solovyov to perform EVAs to repair the errant main propulsion unit on the space station.

The first of the record six EVAs started on 23 April (after STS 41-C had been launched) and lasted 4 hours 15 minutes. The cosmonauts prepared the exterior for

During six EVAs, the T10 crew successfully repaired and restored the station to operational use

their series of sorties, including the erection of a work base with the necessary equipment. Three days later, they were at the propulsion end of Salyut starting the repair work during an EVA that would last 4 hours 56 minutes. EVAs on 29 April and 3 May, both lasting 2 hours 45 minutes, completed the repair work at that end of Salyut. In quick succession came the undocking of Progress 20 and the arrival of Progress 21, carrying new solar arrays for the cosmonauts to erect on the outside of Salyut during the first fifth expedition EVA in history, lasting 3 hours 5 minutes. Over 24 m^2 (78 ft^2) of solar arrays had been added to Salyut. The repaired propulsion system was replenished with propellant from the newly docked Progress 21, which also carried additional equipment, including more for Atkov's space surgery. When the next Progress, No.22, undocked, it left the rear port free to receive the unique crew of Soyuz T12, who arrived on 18 July and left on 29 July.

To the surprise of observers, Kizim and Solovyov made a record sixth EVA on 8 August, lasting 5 hours, to conduct further and unrehearsed repairs to the propulsion unit, having been given detailed instructions from the ground. Another Progress, No.23, arrived later in August and in early September, the three cosmonauts became the space endurance record holders, beating Soyuz T5's 211 days. Another month in space was still to follow, however, and the return came at $T + 236$ days 22 hours 49 minutes aboard Soyuz T11, which is the longest three-crew manned space flight. Atkov estimated that he had spent 87 days on medical work while the others had spent 22 hours on EVAs on what was a very productive mission. The crew looked frail and pale lying in reclining chairs close to the capsule after landing, but were nonetheless in good health.

Milestones

96th manned space flight
55th Soviet manned space flight
48th Soyuz manned space flight
9th Soyuz T manned space flight
New duration record – 236 days 22 hours
1st manned space flight to feature five and six EVAs
8th Soviet and 25th flight with EVA operation
Atkov celebrates his 35th birthday in space (9 May)
Kizim celebrates his 43rd birthday in space (5 Aug)

SOYUZ T11

Int. Designation	1984-032A
Launched	3 April 1984
Launch Site	Pad 31, Site 6, Baikonur Cosmodrome, Kazakhstan
Landed	11 April 1984 (in Soyuz T10)
Landing Site	46 km from Arkalyk
Launch Vehicle	R7 (11A511U); spacecraft serial number (7K-ST) #17L
Duration	7 days 21 hrs 40 min 6 sec
Callsign	Yupiter (Jupiter)
Objective	Indian international Salyut 7 visiting mission; Soyuz T exchange mission

Flight Crew

MALYSHEV, Yuri Vasilyevich, 42, Soviet Air Force, commander, 2nd mission
Previous mission: Soyuz T2 (1980)
STREKALOV, Gennady Mikhailovich, 42, civilian, flight engineer, 4th mission
Previous missions: Soyuz T3 (1980); Soyuz T8 (1983); Soyuz T10-1 (1983)
SHARMA, Rakesh, 35, Indian Air Force, cosmonaut researcher

Flight Log

The next and eleventh Interkosmos spaceman, Rakesh Sharma, came from India, a country which, like France, had already had close ties with the Soviet Union in the field of unmanned space flight. The Soyuz T11 mission began at 18:09 hrs local time at the Baikonur Cosmodrome on 3 April and docking with Salyut, which was to house a record six crew, came 25 hours 20 minutes later. India's science programme included detailed Earth resources photography, weightlessness adaptation studies – with Sharma floating in yoga positions – and the possibility of making amorphous metals in space. Sharma and his colleagues, Yuri Malyshev and Gennady Strekalov, came home in Soyuz T10, 46 km (29 miles) from Arkalyk at $T + 7$ days 21 hours 40 minutes. Maximum altitude during the 51.6° flight was 298 km (185 miles). The original FE for this mission had been Rukavishnikov, but he failed his medicals and was replaced by Strekalov. This was a bitter disappointment for Rukavishnikov who had trained for years to work aboard a Salyut space station only to be thwarted several times. In 1971 he was on the Soyuz 10 crew which failed to enter Salyut 1. Then he was assigned to ASTP, flying the dress-rehearsal mission Soyuz 16 in 1974 instead of receiving an assignment to Salyut 4. Finally, in 1979, he failed to dock with Salyut 6 in Soyuz 33. Sadly, he would never make it inside a space station, and did not return to space.

Demonstrating the power of weightlessness, rather than yoga, Sharma and Malyshev "hold up" Salyut commander Kizim

Milestones

97th manned space flight
56th Soviet manned space flight
49th Soyuz manned space flight
10th Soyuz T manned space flight
1st flight by an Indian
2nd Soyuz international mission

STS 41-C

Int. Designation	1984-034A
Launched	6 April 1984
Launch Site	Pad 39A, Kennedy Space Center, Florida
Landed	13 April 1984
Landing Site	Runway 17, Edwards Air Force Base, California
Launch Vehicle	OV-099 Challenger/ET-10/SRB BI-012/SSME #1 2109; #2 2020; #3 2012
Duration	6 days 23 hrs 40 min 7 sec
Callsign	Challenger
Objective	Repair and re-deployment of Solar Max; deployment of Long Duration Exposure Facility (LDEF)

Flight Crew

CRIPPEN, Robert Laurel, 46, USN, commander, 3rd mission
Previous missions: STS-1 (1981); STS-7 (1983)
SCOBEE, Francis Richard "Dick", 44, USAF, pilot
HART, Terry Jonathan, 37, civilian, mission specialist 1
NELSON, George Driver, 33, civilian, mission specialist 2
VAN HOFTEN, James Douglas Adrianus, 39, civilian, mission specialist 3

Flight Log

Space Shuttle flights were seemingly becoming more and more audacious mission by mission. STS 41-C was to retrieve, repair and redeploy the Solar Maximum Mission, or Solar Max, a science satellite that had been launched in 1980 but had blown some fuses, spoiling its fine pointing ability and rendering it useless. It was a tough assignment for the Challenger crew, led by the inspiring Bob "Mr. Shuttle" Crippen. Challenger took off two days late from the Kennedy Space Center at 08:58 hrs local time. For the first time, the Shuttle made a direct insertion into 28.45° orbit, to conserve valuable OMS propellant for the intensive manoeuvres required for Solar Max rendezvous. Lower than anticipated SRB performance almost resulted in the use of the OMS engines to achieve the planned orbit, which would have cancelled the Solar Max portion of the flight. The crew reached a maximum altitude of 435 km (270 miles) during the mission.

Before Solar Max could be retrieved, Challenger's payload bay had to be emptied of a rather unique cylindrical satellite which almost filled it completely. This was the Long Duration Exposure Facility, or LDEF, which was a 12-sided craft with 57 materials experiments mounted on the outside. LDEF was scheduled to be retrieved in 1985 to enable scientists to assess the effects of exposure to the space environment on

"Ox" Van Hoften test-flies an MMU in the payload bay of Challenger during STS 41-C

the different materials. Its deployment over the Kennedy Space Center, using the RMS, was a spectacular sight.

Crippen, his pilot Dick Scobee, and Challenger's rendezvous radar, star trackers and computers, got to work on the Solar Max rendezvous, which was achieved effortlessly on 8 April. The rotating spacecraft was about 54 m (177 ft) away as mission specialist George Nelson (EV1), wearing an MMU, flew from Challenger. Attached to his chest was a T-pad docking device with which he was to attach himself to Solar Max, to steady it for an RMS grapple. But try as he may, while a soft-docking was achieved, Nelson could not make a hard dock with the satellite. In an effort to steady it manually, Nelson undocked and tried to hold the solar panels. Solar Max went even more out of control and the mission seemed doomed. The unhappy Nelson was recalled to Challenger.

Its solar panels now pointing away from the Sun, Solar Max was losing power, but ground engineers managed to bring it under some sort of control, so that Challenger, its mission extended one day, could attempt a direct RMS grab. Mission specialist Terry Hart's one and only chance had to succeed and his deft handling worked. Solar Max was captured. During a record 7 hour 16 minute EVA on 11 April, Nelson and James van Hoften, who had assisted during the first 2 hour 57 minute EVA, repaired Solar Max, which was later redeployed to continue its mission. The repair to the satellite's electronics and attitude control system was a great demonstration of what a crew and the Shuttle could do. Van Hoften (EV2) was allowed a little go on the MMU, clocking up 28 minutes on Unit 3, compared with Nelson's 42 minutes on Unit 2. It was during the STS 41-C EVAs that Nelson experienced a spacewalker's worst nightmare (apart from a punctured pressure suit) – a minor urine contamination problem – in other words his waste collection device leaked. Fortunately, the liquid coolant garment (LCG) soaked up some of the liquid like a sponge, and though some helmet fogging was experienced, post-flight inspection revealed that no urine had entered the helmet. The fogging was a result of turning down the LCG after the astronaut felt cold as the urine was soaked up. The real danger of inhaling a small globule of any liquid in an EVA suit is that it could become trapped in the throat. The possibility of an astronaut being drowned by less than a teaspoon of liquid is a very real one. When Nelson returned to the airlock and crew compartment, the aroma of six hours perspiration in the suit coupled with the soaking urine reminded his crew members of "the inside of a toilet that had not been cleaned." The smell was so bad that Nelson's colleagues threatened to put him back outside for the remainder of the mission!

The icing on the cake was to be a landing back at the Kennedy Space Center but, as was the case with Crippen's STS-7 mission, bad weather thwarted the attempt. Challenger came home after a one-orbit extension, on runway 17 at Edwards Air Force Base at $T + 6$ days 23 hours 40 minutes 7 seconds. "Mr. Shuttle" headed straight for the simulators for his next mission. Solar Max re-entered in 1990.

Milestones

98th manned space flight
42nd US manned space flight
11th Shuttle mission
5th flight of Challenger
1st satellite retrieval, in-orbit repair and redeployment
1st astronaut docking with satellite
19th US and 26th flight with EVA operations

SOYUZ T12

Int. Designation	1984-073A
Launched	17 July 1984
Launch Site	Pad 31, Site 5, Baikonur Cosmodrome, Kazakhstan
Landed	29 July 1984
Landing Site	140 km southeast of Dzhezkazgan
Launch Vehicle	R7 (11A511U2); spacecraft serial number (7K-ST) #18L
Duration	11 days 19 hrs 14 min 36 sec
Callsign	Pamir (Pamirs)
Objective	All-Soviet Salyut 6 visiting mission; on-orbit instruction activities; tests of new EVA equipment

Flight Crew

DZHANIBEKOV, Vladimir Aleksandrovich, 42, Soviet Air Force, commander, 4th mission
Previous missions: Soyuz 27 (1978); Soyuz 39 (1981); Soyuz T6 (1982)
SAVITSKAYA, Svetlana Yevgenyevna, 35, civilian, flight engineer, 2nd mission
Previous mission: Soyuz T7 (1982)
VOLK, Igor Petrovich, 47, civilian, research engineer

Flight Log

Lift-off of this Soyuz with a difference came at 23:41 hrs local time at Baikonur. The commander was making his fourth flight, the flight engineer was the first woman to make two missions and the cosmonaut researcher was a Buran space shuttle pilot on a familiarisation trip. To cap it all, the commander and flight engineer made an EVA on the mission, the first by a female and the first by a man and a woman together. All these statistical firsts seemed to be linked to the fact that in three months' time the US was to launch a Shuttle to perform all these facts and feats.

Soyuz T12 was not altogether a Khrushchev-style propaganda mission, since Vladimir Dzhanibekov, the commander, was giving Salyut 7 a once-over and training the resident crew in the updated repair techniques necessary to keep it operational and which had been developed since their launch. Docking with Salyut occurred on 18 July and seven days later, Dzhanibekov and Svetlana Savitskaya started a 3 hour 55 minute EVA, during which both operated welding equipment. A 30 kg (66 lb) portable electron beam welder was carried outside, together with the control panel, transformer and metal samples. Savitskaya started the cutting, soldering and welding tests and was followed by her commander.

At the end of the spacewalk, the cosmonauts retrieved some samples from the outside of Salyut. The rest of the mission included experiments with the French Cytos 3 biological unit. The visiting mission was a long one compared with those that went

On board Salyut 7, clockwise from bottom left: Dzhanibekov, Savitskaya, Solovyov, Volk, Atkov and Kizim

before and ended at $T + 11$ days 19 hours 14 minutes 36 seconds. Maximum altitude during the 51.6° mission was 372 km (231 miles).

As Buran was still a state secret, the activities of Volk were quite vague. Volk had trained for a flight to Salyut 7 with Kizim and Solovyov but his mission had been delayed due to the problems Salyut was experiencing. After completing the T12 mission and shortly after landing, Volk piloted both the Tupolev 154 and MiG 25 with adapted Buran control systems in order to test the effects of a 12-day space flight on his piloting skills. This was a simulation of what a Buran pilot would have to experience upon returning the shuttle from orbit, something the Americans had been doing for the previous three years.

Milestones

99th manned space flight
57th Soviet manned space flight
50th Soyuz manned space flight
11th Soyuz T manned space flight
1st space flight by female on second mission
1st EVA by female
1st male–female EVA
9th Soviet and 27th flight with EVA operations
Final manned space flight from Pad 31, Site 6
First manned flight of new Soyuz uprated booster (11A511U2) – Soyuz U2

STS 41-D

Int. Designation	1984-093A
Launched	30 August 1984
Launch Site	Pad 39A, Kennedy Space Center, Florida
Landed	5 September 1984
Landing Site	Runway 17, Edwards Air Force Base, California
Launch Vehicle	OV-103 Discovery/ET-12/SRB BI-013/SSME #1 2109; #2 2018; #3 2021
Duration	6 days 0 hrs 56 min 4 sec
Callsign	Discovery
Objective	Maiden flight of OV-103 (Discovery); satellite deployment mission

Flight Crew

HARTSFIELD, Henry Warren "Hank", 50, USAF, commander, 2nd mission
Previous mission: STS-4 (1982)
COATS, Michael Lloyd, 38, USN, pilot
MULLANE, Richard Michael, 38, USAF, mission specialist 1
HAWLEY, Steven Alan, 32, civilian, mission specialist 2
RESNIK, Judith Arlene, 35, civilian, mission specialist 3
WALKER, Charles David, 36, civilian, payload specialist

Flight Log

Scheduled for 25 June 1984, the first flight of the new orbiter Discovery was to clock up another first in space, with the transportation of the first fare-paying payload specialist. McDonnell Douglas paid NASA about $30,000 to fly Charles Walker to operate the Continuous Flow Electrophoresis System experiment, which he had helped to develop. CFES had flown before but McDonnell Douglas and NASA, both seeing the commercial possibilities, felt that CFES was on the verge of producing biological materials which could form the basis of a huge pharmaceutical space processing industry.

Walker also became the first payload specialist to make a late entry into a flight crew, which otherwise consisted of NASA career astronauts from the Group 8 cadre, which Walker himself had failed to join, and who had waited much longer to be assigned to a mission. The first launch attempt, already delayed three days, was stopped at $T - 6$ min by a fault in the fifth general purpose computer, and one day later, 2.6 seconds into main engine start for Discovery on Pad 39A on 26 June, engines three and two had ignited when both shut down with an alarming metallic graunching sound after the SSME 3 main fuel valve actuator channel A failed. This was the first launch pad abort in the Shuttle programme and the first in the US since

The first launch pad abort of the Shuttle programme was on 26 June 1984, for mission STS 41-D

Gemini 6 in December 1965. The abort was followed by a scare when residual hydrogen gas caught fire outside the Shuttle, but in the end the crew made a graceful exit after closing out the flight deck instrumentation.

As a result of the cancellation, missions 41-D and 41-F were merged and 41-D took on more cargo. All was ready again on 29 August but launch was cancelled due to computer software problems before the flight crew got aboard. The next day, it was delayed by a further 6 minutes 50 seconds by intruding aircraft. Finally, at 07:07 hrs local time, Discovery was airborne heading for its 28.45° orbit and a maximum altitude of 286 km (178 miles). Three large communications satellites were successfully dispatched by the gleeful crew – a Shuttle first – and a 31 m (102 ft) solar sail was unfurled from the cargo bay. The longest structure erected in space, the Shuttle Power Extension Package prototype provided 250 watts of additional electrical power.

Walker, spending most of his time in the lower mid-deck, beavered away with CFES – which he had to repair – but its hormone biological material which was to be tested for its treatment for diabetes proved to be contaminated after the flight. Venting water from the fuel cell system had caused a large chunk of ice on the outside of Discovery, which the crew removed using the RMS thus making Hawley and Mullane's unique contingency EVA unnecessary. The highly successful mission ended at $T + 6$ days 0 hours 56 minutes 4 seconds on Edwards runway 17, the shortest six-crew space flight.

Milestones

100th manned space flight
43rd US manned space flight
12th Shuttle flight
1st flight of Discovery
1st Shuttle pad abort
1st flight with a commercial payload specialist

STS 41-G

Int. Designation	1984-108A
Launched	4 October 1984
Launch Site	Pad 39A, Kennedy Space Center, Florida
Landed	13 October 1984
Landing Site	Runway 33 North, Kennedy Space Center, Florida
Launch Vehicle	OV-099 Challenger/ET-15/SRB A63; A64/SSME #1 2023; #2 2020; #3 2021
Duration	8 days 5 hrs 23 min 38 sec
Callsign	Challenger
Objective	Satellite deployment mission; Space Imaging Radar experiments; satellite refuelling demonstration

Flight Crew

CRIPPEN, Robert Laurel, 47, USN, commander, 4th mission
Previous missions: STS-1 (1981); STS-7 (1983); STS 41-C (1984)
MCBRIDE, Jon Andrew, 41, USN, pilot
RIDE, Sally Kristen, 33, civilian, mission specialist 1, 2nd mission
Previous mission: STS-7 (1983)
SULLIVAN, Kathryn Dwyer, 32, civilian, mission specialist 2
LEESTMA, David Cornell, 35, USN, mission specialist 3
SCULLY-POWER, Paul Desmond, 40, USN, payload specialist 1
GARNEAU, Marc, 35, Canadian Navy, payload specialist 2

Flight Log

Bob Crippen, the first astronaut to fly four Shuttle missions, was specially selected to command this mission to evaluate the effectiveness of flying two missions close together (he had commanded STS 41-C six months before). The main reason was to determine the feasibility of recycling complete Shuttle flight crews to minimise training time and free up limited simulators and resources as Shuttle flight rates increased. Delayed from 1 October, the launch at 07:03 hrs on 4 October was early enough to create spectacular colour schemes as Challenger punched a hole in the cloud-filled sky as it headed for its 57° orbit and a maximum altitude of 304 km (189 miles). The scientific mission was almost thrown into disarray immediately when the satellite part of the Earth Radiation Budget Experiment project, ERBE, misbehaved prior to deployment. Computer software errors and the failure of the satellite's solar panels combined to foil the mission specialists, until Ride got hold of ERBE with the RMS robot arm and shook it. The panel unfolded and the satellite was deployed to begin its work.

Clockwise from top left, the STS 41-G crew of McBride, Garneau, Leestma, Ride, Sullivan, Crippen and Scully-Power heads for the launch pad

Then a Shuttle Imaging Radar, SIR-B, antenna panel failed to deploy and Challenger's Ku-band antenna failed to lock into position, making it impossible to send SIR data real-time. The crew performed some electronic troubleshooting, locking the antenna into one position, enabling about 40 per cent of SIR data to be relayed in real-time. These images were of such clarity that many were impounded temporarily by the Department of Defense. During their 3 hour 27 minute EVA on 11 October, Leestma (EV1) and Sullivan (EV2) practised an in-orbit refuelling technique, preparing the transfer of highly dangerous hydrazine propellant between two containers before entering the orbiter to monitor events from relative safety.

The first Canadian in space, Marc Garneau, operated a suite of ten experiments, labelled CANEX (Canadian Experiments), that focused on space technology, Earth and space sciences. Australian born, and US naturalized oceanographer Paul Scully-Power was flying for the US Navy to conduct real-time observations of ocean phenomena from space.

After a re-entry over the east coast of the USA for the first time, Challenger made a 384 kph (239 mph) landing at runway 33 North at the Kennedy Space Center, with main gear touchdown at 8 days 5 hours 23 minutes 33 seconds, and a 3,220 m (10,564 ft) 59-second rollout. After the mission, the longest by seven crew, it was found that Challenger could have met with disaster during re-entry. Over 4,000 heatshield tiles were found to be loose, their adhesive weakened by a new injection waterproofing technique.

Milestones

101st manned space flight
44th US manned space flight
13th Shuttle mission
6th flight of Challenger
1st flight with seven crew members
1st manned space flight by a Canadian
1st flight with two female crew members
1st US male–female EVA
19th US and 28th flight with EVA operations

STS 51-A

Int. Designation	1984-113A
Launched	8 November 1984
Launch Site	Pad 39A, Kennedy Space Center, Florida
Landed	16 November 1984
Landing Site	Runway 15 North, Kennedy Space Center, Florida
Launch Vehicle	OV-103 Discovery/ET-16/SRB A65; A66/SSME #1 2109; #2 2018; #3 2012
Duration	7 days 23 hrs 44 min 56 sec
Callsign	Discovery
Objective	Satellite deployment and retrieval mission

Flight Crew

HAUCK, Frederick Hamilton "Rick", 43, USN, commander, 2nd mission
Previous mission: STS-7 (1983)
WALKER, David Mathieson, 40, USN, pilot
ALLEN, Joseph Percival, 47, civilian, mission specialist 1, 2nd mission
Previous mission: STS-5 (1982)
FISHER, Anna Lee Tingle, 35, civilian, mission specialist 2
GARDNER, Dale Allan, 36, USN, mission specialist 3, 2nd mission
Previous mission: STS-8 (1983)

Flight Log

The loss of the Westar and Palapa communications satellites in useless orbits during the Shuttle STS 41-B mission was unkindly blamed on the Shuttle, which had an opportunity to make spectacular amends the following November. Lloyds of London had paid out about $270 million for the loss of the satellites, and in the hope of recouping some of that loss it invested a further $15 million to mount the most ambitious Shuttle mission yet: to retrieve the satellites, return them to Earth, refurbish them, re-sell them and re-launch them into orbit, generating perhaps $90 million and reducing the overall insurance loss.

STS 51-A was to have been just a routine deployment mission for its five crew, including Joe Allen (EV1) and Dale Gardner (EV2), who as the EVA crewmen devised the rescue plan with satellite contractors. Discovery's take-off was delayed by a day when the count was stopped due to high winds at altitude, but all went well the following day at 07:15 hrs local time when STS 51-A ascended into clear blue skies, treating observers to a fine view of SRB separation. The Anik and Leasat communications satellites were routinely deployed on the second and third days, and eventually found themselves on station in geostationary orbit.

Joe Allen retrieves the Palapa satellite using the MMU and the "stinger" device

On 12 November, Discovery made a rendezvous to within 10 m (33 ft) of the first stranded satellite, Palapa, which had been nudged into a lower orbit by remote control to facilitate the rescue. The Shuttle performed a record 16 manoeuvres for the rendezvous in the 28° orbit, which reached a maximum altitude of 312 km (194 miles). Joe Allen, floating inside his spacesuit, propelled himself across to Palapa using an MMU, and with the aid of a docking rod called a "stinger" joined up with the satellite's apogee motor nozzle. Allen and his catch were themselves then snared by RMS, deftly operated by Anna Fisher. Meanwhile, inside the payload bay, Gardner cut away at the satellite's antenna to ensure that it could fit inside the payload bay in a special frame. The frame didn't work, however, and Allen had to hold the 544 kg (1,200 lb) satellite steady over his head for 77 minutes while Gardner fixed a contingency adapter. The procedure worked so well that it was decided that when

Gardner rescued Westar the following day, the same method would be used again, with Allen again playing Charles Atlas.

The two EVAs lasted 6 hours 13 minutes and 6 hours 1 minutes, with Allen and Gardner clocking up 2 hours 22 minutes and 1 hour 40 minutes MMU flying time. It had been a brilliant demonstration of human abilities in space and the unique capability of the Shuttle system. Discovery came home to runway 15 at the KSC, at $T + 7$ days 23 hours 44 minutes 56 seconds and the crew later received the Lloyds Silver Medal for the first salvage in space. After much effort, Lloyds managed to dispose of the refurbished satellites. Westar became Asiasat and was scheduled to be launched by a Chinese Long March 3 in 1990. Palapa was re-sold to Indonesia and was also launched in 1990, by a Delta 2 from Cape Canaveral.

Milestones

102nd manned space flight
45th US manned space flight
14th Shuttle mission
2nd flight of Discovery
1st retrieval of satellites and their return to Earth
20th US and 29th flight with EVA operations
Gardner celebrates his 36th birthday with the launch of STS 51-A (8 November)

STS 51-C

Int. Designation	1985-010A
Launched	24 January 1985
Launch Site	Pad 39A, Kennedy Space Center, Florida
Landed	27 January 1985
Landing Site	Runway 15 North, Kennedy Space Center, Florida
Launch Vehicle	OV-103 Discovery/ET-14/SRB BI-015/SSME #1 2109; #2 2018; #3 2012
Duration	3 days 1 hr 23 min 23 sec
Callsign	Discovery
Objective	First classified dedicated DoD shuttle mission

Flight Crew

MATTINGLY, Thomas Kenneth, 48, USN, commander, 3rd mission
Previous missions: Apollo 16 (1972); STS-4 (1982)
SHRIVER, Loren James, 40 USAF, pilot
ONIZUKA, Ellison Shoji, 38, USAF, mission specialist 1
BUCHLI, James Frederick, 39, USMC, mission specialist 2
PAYTON, Gary Eugene, 36, USAF, payload specialist

Flight Log

This mission was originally to have been designated STS-10 and to have flown in January 1984. When it finally got to the launch pad as STS 51-C (originally designated STS-20), it was assigned to orbiter Discovery, rather than Challenger, which was suffering a rather disturbing tile problem. The assignment of Discovery to the mission caused a disruption to the 1985 Shuttle schedules. Although a classified military mission, press leaks resulted in most people knowing full well what the STS 51-C crew would be doing: deploying a geostationary electronic monitoring satellite on an IUS upper stage. It was also unique in that it carried the first military specialist passenger, Gary Payton, from a cadre of US Air Force Manned Space Flight Engineers, most of whom it was anticipated at the time would fly on later military Shuttle missions.

The launch of STS 51-C was delayed by one day by the coldest weather in memory at the KSC and was quite spectacular a day later at 14:50 hrs. The countdown had not been announced until $T - 9$ minutes under new rules for military launches, although observers could tell it was in progress a lot earlier by seeing wisps of liquid oxygen coming off the ET. Discovery entered a 28.4° inclination orbit with a maximum altitude of 341 km (212 miles). Trouble with the IUS on the STS-6 mission had been the main reason for this mission's delay, and after the Acquacade ELINT (electronic

USAF astronauts Onizuka (left) and Shriver give the thumbs up during the classified STS 51-C mission

signals intelligence satellite) had been deployed, the IUS misbehaved again, its first stage thrust shortfall being made up by thruster firings.

Discovery also carried a blood flow experiment but little else was officially reported about the mission, which ended after the shortest five-crew flight of just $T + 3$ days 1 hour 23 minutes 13 seconds on runway 15 at the KSC.

Milestones

103rd manned space flight
46th US manned space flight
15th Shuttle mission
3rd flight of Discovery
1st US classified manned military mission
1st flight of a Military Spaceflight Engineer (MSE – Payton)

STS 51-D

Int. Designation	1985-028A
Launched	12 April 1985
Launch Site	Pad 39A, Kennedy Space Center, Florida
Landed	19 April 1985
Landing Site	Runway 33, Kennedy Space Center, Florida
Launch Vehicle	OV-103 Discovery/ET-18/SRB BI-018/SSME #1 2109; #2 2018; #3 2012
Duration	6 days 23 hrs 55 min 23 sec
Callsign	Discovery
Objective	Satellite deployment mission

Flight Crew

BOBKO, Karol Joseph, 47, USAF, commander, 2nd mission
Previous mission: STS-6 (1983)
WILLIAMS, Donald Edward, 42, USN, pilot
GRIGGS, Stanley David, civilian, mission specialist 1
HOFFMAN, Jeffrey Alan, 40, civilian, mission specialist 2
SEDDON, Margaret Rhea, 37, civilian, mission specialist 3
GARN, Edwin Jacob "Jake", 52, US Senator, payload specialist 1
WALKER, Charles David, 36, civilian, payload specialist 2, 2nd mission
Previous mission: STS 41-D (1984)

Flight Log

This mission was originally designated STS-14 or STS 41-F. After the STS 41-D abort, much of the 41-F cargo was incorporated into a new flight attempt and the mission was re-designated as STS 51-E with a TDRS satellite as the primary payload. This flight was to use the original 41-F crew and the orbiter Challenger, plus two unique payload specialists, Frenchman Patrick Baudry and US Senator Jake Garn, the first space passenger–observer. The fated 41-F/51-E mission was again cancelled, this time because a fault was found in the TDRS satellite, due for launch in February. Challenger was rolled back to the VAB to be configured for a later mission, while 51-E and its crew took on the planned 51-D mission mantle, ousting that crew and now assigned both new payloads and a new orbiter, Discovery.

In the ensuing mammoth crew reshuffle for 1985 flights, Baudry was replaced by an original 51-D McDonnell Douglas payload specialist, Charlie Walker, making a unique second space flight. The new launch date was set as 12 April 1985, but when it arrived it was so dark and gloomy that observers were resigned to a launch scrub as the count was inevitably held for 55 minutes, following a short hold due to a stray ship in the SRB splashdown zone. With just 55 seconds of the launch window remaining, the

The crew of STS 51-D display the "fly swatter" devices they fabricated to activate the Leasat satellite

go-ahead was given to proceed with the count, surprising most observers including astronaut John Young, who was reporting rain drops on the window of the Shuttle training aircraft prowling the skies over the launch pad. Discovery disappeared into thick clouds seconds after lifting off in gloom at 08:59 hrs local time.

The routine deployment of Anik was followed by that of Leasat. Deployment from the payload bay should have activated a spring on the satellite to initiate spin-up and antenna deployment, but clearly this had not happened and yet another Shuttle-deployed satellite was in deep trouble. A contingency EVA was suggested, during which Jeff Hoffman and David Griggs would manually deploy the spring by pulling an arming pin on the side of the satellite while Discovery performed an extremely close station-keeping manoeuvre. This was deemed far too risky and instead the crew manufactured a "fly swatter" device using on-board materials, which could be placed on the end of the RMS during an EVA so that the robot arm could pull the pin.

Hoffmann (EV1) and Griggs (EV2) did their job during a 3 hour 10 minute EVA on 16 April, and it was left to Rhea Seddon operating the RMS to try to pull the pin on Leasat as Discovery closed in. The attempt was useless and Leasat was left stranded. Observers noted that Jake Garn was missing from most of the in-flight TV broadcast and assumed correctly that the senator was having a rather uncomfortable time in the mid-deck getting used to weightlessness. His payload specialist colleague, Charlie Walker, busied himself operating CFES for a second time.

Discovery made the fourth consecutive landing at the Kennedy Space Center on runway 33 at $T + 6$ days 23 hours 55 minutes 23 seconds, damaging its brakes and bursting a tyre as commander Karol Bobko tried to compensate for crosswinds. Maximum altitude of the 28° orbit was 401 km (249 miles).

Milestones

104th manned space flight
47th US manned space flight
16th Shuttle mission
4th flight of Discovery
1st flight with unscheduled EVA
1st flight of a political observer
1st re-flight of a payload specialist
21st US and 30th flight with EVA operations

STS 51-B

Int. Designation	1985-034A
Launched	29 April 1985
Launch Site	Pad 39A, Kennedy Space Center, Florida
Landed	6 May 1985
Landing Site	Runway 17, Edwards Air Force Base, California
Launch Vehicle	OV-099 Challenger/ET-17/SRB BI-016/SSME #1 2023; #2 2020; #3 2021
Duration	7 days 0 hrs 8 min 46 sec
Callsign	Challenger
Objective	Spacelab 3 research programme

Flight Crew

OVERMYER, Robert Franklyn, 48, USMC, commander, 2nd mission
Previous mission: STS-5 (1982)
GREGORY, Frederick Drew, 44, USAF, pilot
LIND, Don Leslie, 54, civilian, mission specialist 1
THAGARD, Norman Earl, 41, civilian, mission specialist 2
THORNTON, William Edgar, 56, civilian, mission specialist 3
WANG, Taylor G., 44, civilian, payload specialist 1
VAN DEN BERG, Lodewijk, 53, civilian, payload specialist 2

Flight Log

Space Shuttle activities were building up to a frenetic pace by April 1985. Discovery was dispatched on mission 51-D, Challenger rolled out to the now vacant Pad 39A for 51-B, and the new orbiter Atlantis arrived at the KSC in preparation for its first mission later that year. It was all looking rather routine stuff, especially when 51-B finally got off the ground – 17 days after 51-D – with a seven-man crew that included three people over 50, as if to emphasise the apparent routine nature of manned space flight. NASA was pushing the system and time was running out. Spacelab 2 featured the Instrument Pointing System and a pallet-only development flight. It was delayed so much due to preparing the IPS that Spacelab 3 flew before it, adding to the confusing Shuttle identification sequence. Research on Spacelab 3, considered to be the first operational mission of the long series, focused on five disciplines: materials science, life sciences, fluid mechanics, atmospheric physics and astronomy. The flight featured 15 primary experiments, of which 14 were considered successful. The crew worked in two shifts: Gold (Gregory, Thagard, Van Den Berg) and Silver (Overmyer, Lind, Thornton, Wang).

Challenger lifted off just 2 minutes 18 seconds later than anticipated, after a liquid oxygen drain back had to be manually commanded, at 12:02 hrs local time.

(L to r) The STS 51-B crew of Gregory, Overmyer, Lind, Thagard, Thornton, Wang, and van den Berg. Note the different coloured shirts, denoting the two-shift operations

Apart from an overheating APU which had to be shut down, the launch was smooth and Challenger, in its 57° orbit which would reach a maximum altitude of 308 km (191 miles), was placed into a tail down, nose up gravity gradient attitude, vital for the array of mainly microgravity processing experiments to be operated inside the Spacelab 3 laboratory. The two payload specialists, Lodewijk van den Berg and Taylor Wang, both naturalised American citizens, operated their own crystal growth and fluid physics experiments, the latter only after spending days getting it to work following an electrical fault that almost spoiled years of hard work.

Also on board Challenger – another Shuttle first – were two monkeys and 24 rats, to help with the study of space adaptation syndrome, SAS, under the guidance of doctors Norman Thagard and William Thornton. The performance of the Animal Holding Facility left much to be desired and the astronauts spent a lot of time clearing up floating droppings. Two small research satellites were to be deployed from GAS canisters in the payload bay, but one failed to get away. Science astronaut Don Lind, having waited a record 19 years to get into space, marvelled at the sight of the aurora borealis from space.

The highly esoteric science mission, which went over most people's heads, was extremely successful and ended with a long rollout on the Edwards Air Force Base desert runway 17, and with the heaviest cargo to return from space – 14,198 kg (31,307 lb) – at $T+7$ days 0 hours 8 minutes 46 seconds. Further landings at the KSC had been banned after the 51-D landing incident.

Milestones

105th manned space flight
48th US manned space flight
17th Shuttle flight
7th flight of Challenger
Thornton retains oldest person in space record (56)
2nd Spacelab Long Module mission

SOYUZ T13

Int. Designation	1985-043A
Launched	6 June 1985
Launch Site	Pad 1, Site 5, Baikonur Cosmodrome, Kazakhstan
Landed	26 September 1985
Landing Site	220 km northeast of Dzhezkazgan
Launch Vehicle	R7 (11A511U2) spacecraft serial number (7K-ST) # 19L
Duration	112 days 3 hrs 12 min 6 sec (Dzhanibekov)
	168 days 3 hrs 51 min 0 sec (Savinykh – returned in Soyuz T14)
Callsign	Pamir (Pamirs)
Objective	Salyut 7 rescue and recovery mission

Flight Crew

DZHANIBEKOV, Vladimir Aleksandrovich, 43, Soviet Air Force, commander, 5th mission
Previous missions: Soyuz 27 (1978); Soyuz 39 (1981); Soyuz T6 (1982); Soyuz T12 (1984)
SAVINYKH, Viktor Petrovich, 43, civilian, flight engineer, 2nd mission
Previous mission: Soyuz T4 (1981)

Flight Log

After the return of the long-duration Soyuz T10 trio from the Salyut 7 space station in late 1984, another three manned missions were planned for 1985/1986 to continue manned operations. Indeed, the crews had been selected and were in training. Soyuz T13 would be launched in May, crewed by Vladimir Vasyutin, Viktor Savinykh and Aleksandr Volkov to complete a six-month mission. They would be visited by Soyuz T14, carrying an all-female crew on a two-week mission in November. This crew would comprise commander Svetlana Savitskaya on her third mission, along with Yekaterina Ivanova and Yelena Dobrokvashina. The Soyuz T15 flight would then launch at the end of 1985 to complete the Salyut 7 programme by the spring of 1986, when a new station would have been launched. Soyuz T15 was to have been crewed by Viktor Viktorenko, Alexandr Alexandrov and Yevgeny Salei.

Then, in early 1985, came the blow that Salyut 7 was effectively dead in space. Contact had been lost and the space station was out of control. Systems were freezing and observers expected that it would never be manned again. There was concern that it would make a Skylab-like uncontrolled re-entry. It was decided to send the most qualified veteran cosmonaut (Vladimir Dzhanibekov, with four previous space flights to his credit) to the station along with veteran Salyut 6 FE Savinykh to see if they could restore the station to operational use. Dzhanibekov had already proven his

Savinykh and Dzhanibekov wearing thermals during the early occupation of the stricken space station

docking skills in past missions, and all his knowledge would be required on this demanding space flight. If the two men could restore the station sufficiently, then the Soyuz T14 crew would be sent to Salyut to resume their intended programme. Flying along with the T14 crew would be Salyut veteran Georgi Grechko, on a short science-orientated mission. He would return with Dzhanibekov after a few days on the Salyut.

The fate of the other crews would be decided if Salyut could be restored. The launch of Soyuz T13 at 12:40 hrs local time from Baikonur on 6 June to conduct "joint work" with Salyut 7 caught western observers by surprise. The two cosmonauts, Vladimir Dzhanibekov – the first Soviet to make five space flights – and flight engineer Viktor Savinykh, proceeded to fly one of the bravest and most remarkable space missions ever.

Soyuz T13 arrived at Salyut two days later, flying all around to check the condition of the exterior and finally docking at the front port with the aid of a new laser ranging device. The crew donned oxygen masks and lots of woolly clothing and entered the freezing space station. Salyut was stabilised so that its solar panels were pointing at the Sun for long enough periods to re-start some form of electrical power. The station's life support system was fixed thanks to expert repairs by the two crewmen, who often retreated to Soyuz to warm their bodies. Communications directly from the station were restored and by late June, Salyut 7 was declared to be in an operational state.

Progress 24 arrived on 23 June with additional repair and replenishment supplies, and the crew even got to work on some experiments dedicated to Earth observation, while Progress loaded the Salyut propulsion system with fuel. Confounding experts who regarded the Soyuz T13 mission as finished, the Soviets launched a Heavy Cosmos module, rather than the Progress-class spacecraft that it was first suspected to be by analysts. Designated Cosmos 1669, it docked with Salyut on 21 July, enabling even more fruitful work to be conducted by the remarkable crew, which even went on EVA on 2 August to place two small solar arrays on the third large array on Salyut. The walk lasted about 5 hours.

Cosmos 1669 undocked on 28 August, making a controlled re-entry two days later, and the T13 crew went on conducting a mission as ordinary and routine as a normal residency. The 51.6° mission reached a maximum altitude of 359 km (223 miles). Soyuz T14 was launched and joined them on 18 September. Dzhanibekov and one of the T14 crew, the burly Georgy Grechko, undocked on 25 September, flew a day's autonomous mission and came home – the first individual space travellers who were launched separately but landed together – at T13 flight time of $T + 112$ days 3 hours 12 minutes. Savinykh, meanwhile, remained on board Salyut 7 to attempt the longest manned space flight in history, only to be thwarted by his new commander's illness.

Milestones

106th manned space flight
58th Soviet manned space flight
51st Soyuz mission
12th Soyuz T mission
1st reactivation of a dead space station
10th Soviet and 31st flight with EVA activities

STS 51-G

Int. Designation   1985-048A
Launched   17 June 1985
Launch Site   Pad 39A, Kennedy Space Center, Florida
Landed   24 June 1985
Landing Site   Runway 23, Edwards Air Force Base, California
Launch Vehicle   OV-103 Discovery/ET-20/SRB BI-019/SSME #1 2109; #2 2018; #3 2012
Duration   7 days 1 hr 38 min 52 sec
Callsign   Discovery
Objective   Satellite deployment mission

Flight Crew

BRANDENSTEIN, Daniel Charles, 42, USN, commander, 2nd mission
Previous mission: STS-8 (1983)
CREIGHTON, John Oliver, 42, USN, pilot
FABIAN, John McCreary, 43, USAF, mission specialist 1, 2nd mission
Previous mission: STS-7 (1983)
NAGEL, Steven Ray, 38, USAF, mission specialist 2
LUCID, Shannon Wells, 42, civilian, mission specialist 3
BAUDRY, Patrick, 39, French Air Force, payload specialist 1
AL-SAUD, Prince Sultan Salman Abdul Aziz, 28, civilian, payload specialist 2

Flight Log

The smoothest Space Shuttle to date, STS 51-G, with the orbiter Discovery in tow, made a majestic, on-time lift-off at 07:33 hrs local time from Pad 39A, carrying a cargo of three large communications satellites and a crew of seven which for the first time included passengers (or, more correctly, payload specialists) from two other countries, France (CNES) and Saudi Arabia. Three days later, the Mexican satellite, Morelos, Saudia Arabia's Arabsat, and the USA's Telstar were safely deployed *en route* to geostationary orbit, with the aid of PAM-D stages.

Another satellite payload, called SPARTAN 1, was deployed for an autonomous flight to conduct X-ray observations of the Milky Way, before it was retrieved by the RMS. 51-G also conducted the first manned Strategic Defense Initiative (SDI, or "Star Wars")-related tests, attempting, eventually successfully, to reflect a laser beam directed at the Shuttle from Hawaii back to Earth via a small mirror mounted on the orbiter's mid-deck side hatch window. French crew member Baudry completed a programme of biomedical experiments similar to those flown by his colleague Jean-Loup Chrétien aboard Salyut 7 the previous year. Baudry had been Chrétien's back-up on that mission before completing an abbreviated Shuttle payload specialist

The multi-national STS 51-G crew. L to r: Al-Saud, Creighton, Nagel, Lucid, Fabian, Baudry and Brandenstein

training programme in America. The experiments included studies in physiology, biology, materials processing, and astronomy. Al-Saud took photographs of his homeland, participated in several experiments (include assisting Baudry in his programme) and continued his religious commitments, fulfilling his Muslim customs as well as he could. He admitted that he could not totally "bend down" while floating, due to the tendency to cause space sickness; and facing Mecca created its own problems when he was orbiting Earth every 90 minutes.

Discovery came home to Edwards Air Force Base's runway 23, making the shortest rollout so far, of 2,265 m (7,431 ft), only for its main landing gear to sink partially in the wet lake bed. It had to be rather ignominiously righted using a plank of wood. Flight time was $T + 7$ days 1 hour 38 minutes 52 seconds. Orbital inclination was 28.45° and maximum altitude was 334 km (208 miles).

Milestones

107th manned space flight
49th US manned space flight
18th Shuttle flight
5th flight of Discovery
1st flight by crew from three nations
1st flight by a Saudi Arabian
1st royalty in space (Al-Saud)
1st nation (France) to make space flights with both the USA and Russia

STS 51-F

Int. Designation	1985-063A
Launched	29 July 1985
Launch Site	Pad 39A, Kennedy Space Center, Florida
Landed	6 August 1985
Landing Site	Runway 23, Edwards Air Force Base, California
Launch Vehicle	OV-099 Challenger/ET-19/SRB BI-017/SSME #1 2023; #2 2020; #3 2021
Duration	7 days 22 hrs 45 min 26 sec
Callsign	Challenger
Objective	Spacelab 2 research programme; verification of Spacelab Igloo/pallet configuration

Flight Crew

FULLERTON, Charles Gordon, 48, USAF, commander, 2nd mission
Previous mission: STS-3 (1982)
BRIDGES, Roy Dunbard Jr., 42, USAF, pilot
HENIZE, Karl Gordon, 58, civilian, mission specialist 1
MUSGRAVE, Franklin Story, 49, civilian, mission specialist 2
ENGLAND, Anthony Wayne, 43, civilian, mission specialist 3
ACTON, Loren Wilbur, 49, civilian, payload specialist 1
BARTOE, John-David, 40, civilian, payload specialist 2

Flight Log

The highly esoteric astronomical observation Spacelab 2 science payload had been proposed a decade before and the mission itself had been in preparation for over five years and, at last, 51-F was ready. On 12 July, on Pad 39A all three main engines were up and running with three seconds to go before lift-off, when the hydrogen chamber coolant valve on engine 2 failed to close. A computer ordered a redundant command to be made but mission rules dictated a launch pad abort, called an RCLS abort. As an abort had already occurred on STS 41-D, the event was not greeted with so much drama, although to the crew, which included Karl Henize (who was waiting to be the new oldest man in space after an 18-year wait for a flight), it was a bitter disappointment.

It was also a disappointment to the scientists, because a later launch in less than perfect lighting conditions, as there would be more moon shine, would degrade three of the thirteen primary experiments, mounted on pallets in the payload bay. Challenger tried again on 29 July, and until $T + 4$ minutes 55 seconds, the launch went well, after a 1 hour 37 minute hold due to incorrect telemetry. The crew had a close eye on the centre engine which was apparently overheating early after lift-off and watched

Humorous crew photo by the STS 51-F crew. Clockwise from top: Acton, England, Fullerton, Bartoe, Musgrave, Bridges and Henize

helplessly as an indicator showed it had shut down, just 33 seconds after Challenger would have had to have performed a transatlantic abort, possibly ditching into the sea and sinking. As it was, an abort-to-orbit was ordered, and at the flick of a switch commander Fullerton initiated a new flight programme. Later, another engine appeared to be overheating and if the crew had not been told to inhibit what was suspected to have been an over-zealous sensor, a worse abort would have resulted.

Challenger struggled into a 49.5° orbit and its OMS engines placed it at a peak altitude of 276 km (171 miles), much lower than had been planned for the Spacelab mission. The use of the OMS engines also restricted flight experiments using the free-flying Plasma Diagnostic Package payload. Nonetheless, the Shuttle had at least reached orbit, in which the busy crew worked in two 12-hour shifts on the vast array of science experiments. These included the Instrument Pointing System (IPS), which did not quite achieve its advertised 1 arc-second of pointing accuracy.

Despite the abort-to-orbit situation, the significant mission re-planning allowed the flight to be deemed a success. As well as verification of the IPS, the flight also featured part of the modular Spacelab system called the Igloo, which was placed at the front of the three-pallet "train" in the payload bay and designed to provide in-flight support to the instruments that were installed on each of the pallets. The main objective of STS 51-F was to verify the Igloo, pallet and IPS system and its interface with the orbiter, in addition to monitoring the immediate environment around the orbiter, which may or may not interfere in the gathering of scientific data. The experiment programme for the flight encompassed life sciences, plasma physics, astronomy, high-energy astrophysics, solar physics, atmospheric physics, and technology research. This latter category included an evaluation of new beverage containers by the crew. Coke and Pepsi cans, with specially adapted mouth dispensers to both retain the carbonated drinks' condition and prevent it leaking into the cabin, were evaluated. The system worked but the taste did not, and according to the astronauts who tried the drinks, having extra gas in the digestive system when in space is not the most enjoyable feeling or experience.

Challenger's performance was so good that an extra day was offered to the crew, which turned it down and came home to runway 23 at Edwards Air Force Base, at $T + 7$ days 22 hours 45 minutes 26 seconds.

Milestones

108th manned space flight
50th US manned space flight
19th Shuttle flight
8th flight of Challenger
1st Spacelab pallet only science mission
Henize becomes oldest person to fly in space (58)

STS 51-I

Int. Designation	1985-076A
Launched	27 August 1985
Launch Site	Pad 39A, Kennedy Space Center, Florida
Landed	3 September 1985
Landing Site	Runway 30, Edwards Air Force Base, California
Launch Vehicle	OV-103 Discovery/ET-21/SRB BI-020/SSME #1 2109; #2 2018; #3 2012
Duration	7 days 2 hrs 17 min 42 sec
Callsign	Discovery
Objective	Satellite deployment and repair mission

Flight Crew

ENGLE, Joseph Henry, 53, USAF, commander, 2nd mission
Previous mission: STS-2 (1981)
COVEY, Richard Oswalt, 39, USAF, pilot
VAN HOFTEN, James Douglas Adrianus, 41, civilian, mission specialist 1, 2nd mission
Previous mission: STS 41-C (1984)
LOUNGE, John Michael "Mike", 39, civilian, mission specialist 2
FISHER, William Frederick, 39, civilian, mission specialist 3

Flight Log

Leasat 3 had been in low-Earth orbit and inoperable since its fated deployment from the Shuttle 51-D the previous April, when Discovery took off to rescue it. Such an audacious mission would not have been imagined had it not been for the confidence engendered in the Shuttle programme by the 51-A dual satellite retrieval in November 1984, and it turned the otherwise routine 51-I mission into a spectacular one. Discovery's lift-off was delayed twice, once on 24 August by predicted bad weather at T − 5 minutes, to nobody's surprise, and again the following day, this time at T − 9 minutes because of a computer glitch. On 27 August, the mission got away in spectacular style at 06:58 hrs local time, after a 3 minute 1 second hold, piercing a low cloud deck that threatened a thunderstorm. Mission managers insisted that launch criteria had not been broken.

Once in 28° inclination orbit, which would reach a maximum altitude of 387 km (241 miles), the crew encountered a problem when a sunshield failed to close over the Aussat satellite in the payload bay. To prevent overheating, the satellite was deployed the same day, instead of on day two, shortening the mission by one day. Another satellite, ASC, was also deployed, this time as planned on day one, making a first for

"Ox" van Hoften hand-launches the Leasat satellite after its repair

the Shuttle with two deployments in one day. A third satellite, a sister Leasat, was also successfully deployed but ironically failed in orbit later.

The EVA operations with the errant Leasat 3 were planned to last just one sortie, but it was decided to divide the exercise into two EVAs when the RMS, which was crucial to the job, did not respond properly to computer commands and had to be operated manually by time-consuming switching. During the first EVA on 1 September, "Ox" van Hoften (EV1), the biggest astronaut, grabbed Leasat manually and held it in place while former surgeon Bill Fisher (EV2) attached a grapple bar so that it could be linked to the RMS end effector. The astronauts then conducted a by-pass operation of a failed electrical sequencer and powered up the satellite.

The remarkable EVA, lasting 7 hours 20 minutes, ended with Leasat poised over the payload bay for the night. The following day, 2 September, a 4 hour 31 minute EVA completed the repair job and Leasat was pushed away by "Ox", who then took on a Charles Atlas pose as he stood on the end of the RMS. After 51-I had landed, Leasat 3 was safely on station in geostationary orbit and operating perfectly, testimony to the crew's and the Shuttle's ability in space. Discovery came home to Edwards Air Force Base's runway 30 at $T + 7$ days 2 hours 17 minutes 42 seconds.

Milestones

109th manned space flight
51st US manned space flight
20th Shuttle flight
6th flight of Discovery
22nd US and 32nd flight with EVA operations

SOYUZ T14

Int. Designation	1985-081A
Launched	17 September 1985
Launch Site	Pad 1, Site 5, Baikonur Cosmodrome, Kazakhstan
Landed	21 November 1985
Landing Site	180 km southeast of Dzhezkazgan
Launch Vehicle	R7 (11A511U2); spacecraft serial number (7K-ST) #20L
Duration	64 days 21 hrs 52 min 8 sec (Vasyutin and Volkov)
	8 days 21 hrs 13 min 6 sec (Grechko, returned in Soyuz T13)
Callsign	Cheget (Cheget)
Objective	Salyut 7 fourth resident crew (revised) programme

Flight Crew

VASYUTIN, Vladimir Vladimirovich, 33, Soviet Air Force, commander
GRECHKO, Georgy Mikhailovich, 53, civilian, flight engineer, 3rd mission
Previous missions: Soyuz 17 (1975); Soyuz 26 (1977)
VOLKOV, Aleksandr Aleksandrovich, 37, Soviet Air Force, cosmonaut researcher

Flight Log

Following the remarkable Soyuz T13 repair job to Salyut 7, a new long-duration resident crew was launched in T14 at 18:39 hrs local time from Baikonur. Aiming to stay up for 179 days were Vladimir Vasyutin and Aleksandr Volkov, the original T13 long-duration crew. Also on board was Georgy Grechko, the first Soviet cosmonaut over 50 to fly, who was due to return to Earth with T13 commander, Dzhanibekov, while T13 flight engineer Savinykh – Vasyutin and Volkov's originally planned partner – would remain to clock up a record 282 days in space. EVA operations by Vasyutin and Volkov were also on the schedule.

All this seemed quite possible and likely when, after the return of T13, the new residents got down to routine space operations, with a new module in tow. Cosmos 1686, which had docked on 2 October, was full of equipment, experiments, telescopes and possibly visual reconnaissance equipment, to be operated by the unique military cosmonaut researcher Volkov. The revised plan for Salyut 7 now saw one further mission, the all-female crew, visiting the station, with no further long-duration flights launched. The Cheget crew would complete the Salyut 7 experiment programme before they came home in March 1986 , leaving the way clear to launch the next station, expected to be called Salyut 8 and thought in the west to be of an upgraded design.

All smiles at the start of the T14 mission as the T13 crew prepares to depart. L to r Vasyutin, Grechko, Savinykh, Volkov and Dzhanibekov

Meanwhile, however, flight commander Vasyutin was displaying disturbing medical symptoms. He first developed a minor ailment which was treated with antibiotics but was never cured fully. This reduced the commander's workload and he felt that he was not contributing to the mission. He could not sleep and he lost his appetite. This made him depressed and anxious and he was a bag of nerves. At first this was kept between the crew, but by the end of October, it soon became evident that his condition need the assistance of medical specialists on the ground. Western observers noted that the normal reports of "the crew were in good health" were being omitted in the regular progress reports of the flight. The orbital dynamics of the flight meant favourable return conditions were not available until the middle of November, so the crew remained in orbit.

By 17 November, Vasyutin's condition had deteriorated to the point where he was in acute pain and doctors decided he was too ill to continue with the flight. More secret medical conferences with the ground followed and it was decided to end the mission, with Savinykh taking charge of the re-entry and landing. The mission ended at $T + 64$ days 21 hours 52 minutes, with Savinykh having clocked up 168 days and the early-exiting Grechko 8 days. Maximum altitude during the $51.6°$ mission was 359 km (223 miles).

Immediately after the landing, Vasyutin was examined by medical members of the recovery team and was flown to Moscow for hospital treatment. The suspicion of appendicitis was soon dismissed as Vasyutin was found to have an infection of his prostrate that had inflamed and caused a fever. It was revealed that the condition had manifested before the mission, but the cosmonaut had kept the illness from the doctors. As a result, even more stringent medical standards were imposed on the cosmonaut team, which saw several veteran cosmonauts fail their next medical and forced to leave the team, some without flying. Vasyutin left the cosmonaut team in February 1986 and returned to his air force career.

Milestones

110th manned space flight
59th Soviet manned space flight
52nd Soyuz manned space flight
13th manned Soyuz T space flight
1st partial resident crew exchange
1st flight curtailed due to crew illness
First person to fly to three different space stations (Grechko – Salyut 4, Salyut 6, Salyut 7)

STS 51-J

Int. Designation	1985-092A
Launched	3 October 1985
Launch Site	Pad 39A, Kennedy Space Center, Florida
Landed	7 October 1985
Landing Site	Runway 23, Edwards Air Force Base, California
Launch Vehicle	OV-104 Atlantis/ET-25/SRB BI-021/SSME #1 2011; #2 2019; #3 2017
Duration	4 days 1 hr 44 min 38 sec
Callsign	Atlantis
Objective	2nd classified DoD Shuttle mission

Flight Crew

BOBKO, Karol Joseph, 48, USAF, commander, 3rd mission
Previous missions: STS-6 (1983); STS 51-D (1985)
GRABE, Ronald John, 40, USAF, pilot
HILMERS, David Carl, 35, USMC, mission specialist 1
STEWART, Robert Lee, 43, US Army, mission specialist 2, 2nd mission
Previous mission: STS 41-B (1984)
PAILES, William, 33, USAF, payload specialist 1

Flight Log

The maiden flight of the Atlantis orbiter began in spectacular style from Pad 39A at 11:15 hrs local time, but the first anyone was to have known about the mission was nine minutes earlier, when the ground launch sequencer started the final countdown. Mission 51-J was a Department of Defense flight and is one of the most anonymous in Shuttle history because of its classification. It is thought to have deployed two DSCS communications satellites into orbit aboard an IUS upper stage. According to data revealed by the North American Air Defense Command, NORAD, Atlantis reached a record 512 km (318 miles) altitude in the 28.5° orbit.

Also on board was an experiment called Bios, which studied the damage to biological samples by high-energy cosmic rays. The mission also marked the end of the brief career of the USA Air Force Manned Space Engineer corps, whose William Pailes was the second and last to fly. At one time, one or two representatives from the MSE corps were to have flown every DoD mission. After the Challenger accident the next year, these already limited opportunities disappeared altogether.

Atlantis made a longer than usual return from its high orbit, landing on runway 23 at Edwards Air Force Base at $T + 4$ days 1 hour 44 minutes 38 seconds.

This Earth image is one of the few released for the classified STS 51-J mission

Milestones

111th manned space flight
52nd US manned space flight
21st Shuttle mission
1st flight of Atlantis

STS 61-A

Int. Designation	1985-104A
Launched	30 October 1985
Launch Site	Pad 39A, Kennedy Space Center, Florida
Landed	6 November 1985
Landing Site	Runway 17, Edwards Air Force Base, California
Launch Vehicle	OV-099 Challenger/ET-24/SRB BI-022/SSME #1 2023; #2 2020; #3 2021
Duration	7 days 0 hrs 44 min 53 sec
Callsign	Challenger
Objective	Spacelab D1 research programme

Flight Crew

HARTSFIELD, Henry Warren "Hank", 51, USAF, commander, 3rd mission
Previous missions: STS-4 (1982); STS 41-D (1984)
NAGEL, Steven Ray, 39, USAF, pilot, 2nd mission
Previous mission: STS 51-D (1985)
DUNBAR, Bonnie Jean, 36, civilian, mission specialist 1
BUCHLI, James Frederick, 40, USMC, mission specialist 2, 2nd mission
Previous mission: STS 51-C (1985)
BLUFORD, Guion Stewart, 41, USAF, mission specialist 3, 2nd mission
Previous mission: STS-8 (1983)
MESSERSCHMID, Ernst Willi, civilian, payload specialist 1
FURRER, Reinhard, 44, civilian, payload specialist 2
OCKELS, Wubbo, 39, civilian, payload specialist 3

Flight Log

The STS 61-A mission carrying a Spacelab Long Module was chartered by West Germany for $175 million, contributing most of the 76 scientific experiments and two payload specialists – who preferred to be called payload scientists – to the seven-day expedition. The first flight by eight crew members included five NASA astronauts, two Germans and the first space-flying Dutchman, Wubbo Ockels. Pilot Steve Nagel, former mission specialist of 51-G, was flying again after only 128 days since his previous mission, a record turnaround. The Spacelab experiment operations were controlled by the West German DFVLR centre, near Munich, via the TDRS 1 and Intelsat satellites. Lift-off came at 12:00 hrs from Pad 39A and Challenger rolled on to its launch azimuth in dramatic fashion, heading towards its 57° inclination orbit, which would have a highest point of 288 km (179 miles).

A few technical problems, including communications, RCS thruster and fuel cell anomalies, delayed the entry into Spacelab by over three hours, but soon a 24-hour

Payload Specialist Reinhard Furrer participates in medical experiments during Spacelab D1

round-the-clock regime of experimental work began, with the crew split into two shifts. They were aided when required by commander Hank Hartsfield and payload specialist Ockels, who overworked early in the mission and was ordered to rest. A unique experiment was the Space Sled, which was designed to investigate the reactions and adaptation of the human balance and orientation functions. It was moved backwards and forwards along a 7 m (23 ft) long track in the module. The Spacelab D1 programme included experiments in basic and applied microgravity research in materials science, life sciences and technology, communications and navigation.

Another first was achieved at the end of the 7 day 0 hour 44 minute 51 second mission, on runway 17 at Edwards Air Force Base, when Hartsfield conducted a computer-controlled nosewheel steering test, deliberately steering up to 10 m (33 ft) off the centre line, to gain data on ways of eliminating excessive brake and tyre wear, such as that suffered by Discovery at the end of the 51-D Kennedy landing the previous April. The Spacelab D1 mission was considered a great success, so much so that West Germany booked a repeat mission for five years time (which actually flew eight years later).

Milestones

112th manned space flight
53rd US manned space flight
22nd Shuttle flight
9th flight of Challenger
3rd Spacelab Long Module mission
1st flight with eight crew members
1st flight by a Dutchman
1st commercially leased manned space flight
1st flight by two West Germans
1st US flight to be controlled outside the USA

STS 61-B

Int. Designation	1985-109A
Launched	27 November 1985
Launch Site	Pad 39A, Kennedy Space Center, Florida
Landed	3 December 1985
Landing Site	Runway 22, Edwards Air Force Base, California
Launch Vehicle	OV-104 Atlantis/ET-22/SRB BI-023/SSME #1 2011; #2 2019; #3 2017
Duration	6 days 21 hrs 4 min 49 sec
Callsign	Atlantis
Objective	Satellite deployment; EVA construction demonstration mission

Flight Crew

SHAW, Brewster Hopkinson Jr., 40, USAF, commander, 2nd mission
Previous mission: STS-9 (1983)
O'CONNOR, Bryan Daniel, 38, USMC, pilot
ROSS, Jerry Lynn, 37, USAF, mission specialist 1
CLEAVE, Mary Louise, 38, civilian, mission specialist 2
SPRING, Sherwood Clark "Woody", 41, US Army, mission specialist 3
WALKER, Charles David, 37, civilian, payload specialist 1, 3rd mission
Previous missions: STS 41-D (1984); STS 51-D (1985)
NERI VELA, Rudolpho, 33, civilian, payload specialist 2

Flight Log

Such was the apparently routine nature of Space Shuttle flights by November 1985 that the 61-B mission's extraordinary EVA operations were left unheralded. The flight got off to a spectacular start at 19:29 hrs local time at the KSC, the third night launch in the US manned space programme and the second by the Shuttle. Unlike Challenger's ascent into thunder clouds on STS-8, Atlantis began the 61-B mission in skies so clear that the ascent could be seen over 640 km (398 miles) away. Riding the mid-deck were two payload specialists with a difference, Mexico's Rudolpho Neri Vela, flying courtesy of his country's booking of the Shuttle to deploy the Morelos national communications satellite – and who was to become the last international passenger on the Shuttle – and McDonnell Douglas's Charlie Walker, who was flying for the third time – more than any of the professional NASA crew. Indeed, by the end of the mission, Walker had clocked up more Shuttle flight experience than all the NASA astronauts, except Crippen and Hartsfield.

Atlantis reached a 28° inclination orbit and a maximum height of 334 km (208 miles) during the mission, which included the routine deployments of Morelos,

Ross and Spring construct the EASE-ACCESS hardware in the payload bay of Atlantis

Aussat and a Satcom Ku-band satellite, and the remarkable EASE-ACCESS EVA experiments. These were performed by astronauts Jerry Ross (EV1) and Sherwood Spring (EV2), who erected a series of truss frames in a rehearsal of proposed space station construction procedures. The photography of the two EVAs on 1 and 3 December was splendid, one showing Spring standing at the end of a 13.7 m (45 ft) long tower, erected over the payload bay. The EVAs lasted 5 hours 34 minutes and 6 hours 46 minutes.

Probably the best Shuttle flight in the pre-Challenger era of the programme, 61-B came home to Edwards Air Force Base's runway 22 at $T + 6$ days 21 hours 4 minutes 49 seconds, after a mission shortened by one orbit because of concerns over landing lighting conditions.

Milestones

113th manned space flight
54th US manned space flight
23rd Shuttle flight
2nd flight of Atlantis
1st manned space flight by a Mexican
23rd US and 33rd flight with EVA operations

STS 61-C

Int. Designation	1986-003A
Launched	12 January 1986
Launch Site	Pad 39A, Kennedy Space Center, Florida
Landed	18 January 1986
Landing Site	Runway 22, Edwards Air Force Base, California
Launch Vehicle	OV-102 Columbia/ET-30/SRB BI-024/SSME #1 2015; #2 2018; #3 2109
Duration	6 days 2 hrs 3 min 51 sec
Callsign	Columbia
Objective	Satellite deployment mission

Flight Crew

GIBSON, Robert Lee "Hoot", 39, USN, commander, 2nd mission
Previous mission: STS 41-B (1984)
BOLDEN, Charles Frank Jr., 39, USMC, pilot
NELSON, George Driver, 35, civilian, mission specialist 1, 2nd mission
Previous mission: STS 41-C (1984)
HAWLEY, Steven Alan, 34, civilian, mission specialist 2, 2nd mission
Previous mission: STS 41-D (1984)
CHANG-DIAZ, Franklin Ramon, 35, civilian, mission specialist 3
CENKER, Robert J., 37, civilian, payload specialist 1
NELSON, C. William, 43, US Congressman, payload specialist 2

Flight Log

This mission was dubbed "Mission Impossible" by the press and thoroughly earned its nickname, making eight attempts to launch, five of them when the crew was aboard. Columbia, making its first space flight since STS-9 in 1983 and after a planned period of modification, was on the pad on 18 December 1985 for its first attempt, which was routinely postponed for a day because of the longer time it was taking to close out the aft compartment of the orbiter. The launch reached $T - 14$ seconds on 19 December, when a back-up hydraulic unit on one of the SRBs failed to activate on time.

With Christmas coming up and with the orbiter Challenger already poised for a lift-off on sister pad 39B, NASA felt comfortable about delaying "Mission Impossible" until 6 January. That launch was scrubbed at $T - 31$ seconds after 6,350 kg (14,000 lb) of liquid oxygen had been accidentally drained from the external tank by a careless but overworked controller. The following day, the crew climbed aboard resigned to another launch scrub, at $T - 9$ minutes, such were the appalling weather conditions. Before the crew had got aboard, all seemed to augur well for

Payload specialist and US Congressman Bill Nelson exercises on the treadmill in the mid-deck of Columbia

9 January, but the countdown again reached $T - 9$ minutes only for a sensor to break on the launch pad and stop the count. After another delay due to bad weather and with the crew on board for the fifth time (and, thanks to his 41-D experience, with Steven Hawley on board a Shuttle poised for lift-off for a record eighth time), Columbia at last made a dramatic dawn exit from the Kennedy Space Center on 12 January, leaving a dark contrail in the dark blue sky with a pinpoint of light at its end as it headed for space. A hydraulic failure on one of the SSMEs just after lift-off momentarily raised fears of an RTLS abort.

With commander Hoot Gibson and his crew of two veterans and four rookies, including Congressman Bill Nelson (who served on a Senate committee which controlled NASA's purse strings) flying as a political observer, Columbia set to work. This did not seem too hard, since only one satellite, Satcom Ku 1, needed to be deployed. A special camera designed to photograph Halley's Comet failed, much to the chagrin of astronomer George Nelson (no relation to Bill, on the first flight to include crew members with the same surname). Another passenger was the second industrial payload specialist, RCA's Robert Cenker. NASA planned to fly several more from other companies, and even a citizen's representative, a US teacher, during 1986.

The "Mission Impossible" tag stuck until the end of the mission too, since NASA, wanting Columbia home at the Kennedy Space Center for the first landing there since 51-D for a turnaround of the 61-E Shuttle mission to observe Halley's Comet, delayed the landing twice, by which time the weather was so bad that it was diverted to Edwards. Columbia landed at $T + 6$ days 2 hours 3 minutes 51 seconds – the shortest seven-crew flight – adding days to the turnaround time. Maximum altitude achieved in the 28° inclination orbit was 296 km (184 miles).

Milestones

114th manned space flight
55th US manned space flight
24th Shuttle flight
7th flight of Columbia
1st manned space flight carrying two crew with same surname

STS 51-L

Int. Designation	None – failed to reach orbit, lost in launch phase explosion
Launched	28 January 1986
Launch Site	Pad 39B, Kennedy Space Center, Florida
Landed	N/A
Landing Site	N/A
Launch Vehicle	OV-099 Challenger/ET-26/SRB BI-026/SSME #1 2023; #2 2020; #3 2021
Duration	1 min 13 sec
Callsign	Challenger
Objective	Planned deployment of TDRS-B; observations of Comet Halley, Teacher in Space programme

Flight Crew

SCOBEE, Francis Richard "Dick", 47, USAF, commander, 2nd mission
Previous mission: STS 41-C (1984)
SMITH, Michael John, 40, USN, pilot
ONIZUKA, Ellison Shoji, 39, USAF, mission specialist 1, 2nd mission
Previous mission: STS 51-C (1985)
RESNIK, Judith Arlene, 40, civilian, mission specialist 2, 2nd mission
Previous mission: STS 41-D (1984)
McNAIR, Ronald Erwin, 35, civilian, mission specialist 3, 2nd mission
Previous mission: STS 41-B (1984)
JARVIS, Gregory Bruce, 41, civilian, payload specialist 1
McAULIFFE, Sharon Christa Corrigan, 37, civilian, payload specialist 2

Flight Log

The apparently routine nature of Space Shuttle missions and the transportation of industrial payload specialists, foreign passengers and political observers resulted in an inevitable call for a US citizen to fly. At first, a journalist-in-space programme was suggested, but President Reagan decided that the first citizen in space would be a teacher. A nationwide competition resulted in the choice of Christa McAuliffe, while at the same time, a journalist-in-space nationwide competition reached a stage where 100 had been short-listed for a flight to take place later in 1986.

McAuliffe was assigned to Challenger mission STS 51-L, along with an industrial payload specialist from Hughes, Gregory Jarvis, whose bad luck it had been to have already been assigned to and bumped off three missions in 1985. STS 51-L was delayed from December 1985 to 22 January 1986 and was routinely delayed further due to operational difficulties on 23 and 24 January. By 25 January, the launch window was to have been pushed into the morning, requiring reprogramming of computers, so

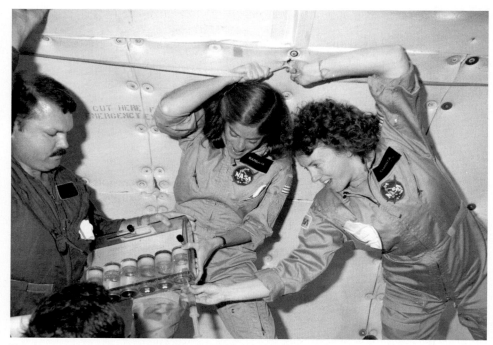

Teachers in Space candidates Barbara Morgan (centre) and Christa McAuliffe (right) practice working on experiments for STS 51-L aboard NASA's zero-G aircraft

another delay ensued. The following day, 26 January, seemed acceptable, but officials decided to cancel the attempt because of predicted bad weather, which actually turned out to be fine. Vice President George Bush had expressed an interest in seeing the Challenger launch but would not have been able to get to the Kennedy Space Center until 27 January (which was also the 19th anniversary of the Apollo 1 fire), he told NASA. In the end, he did not come for the new launch attempt on that date, which was cancelled when the handle of the orbiter hatch would not close properly. When that problem had been fixed, crosswinds at the KSC runway in the event of an RTLS abort were unacceptable.

The crew exited, prepared for an attempt the following day, and were aboard Challenger again at 08:36 hrs on a bright but very cold day, with ice having been reported on the launch pad. Liquid hydrogen tanking delays caused by hardware interface problems further delayed the launch two hours, until 11:38 hrs. As the SSMEs built up to full thrust as usual, the Shuttle "stack" thrusted forward laterally, only to be "twanged" back at SRB ignition. This manoeuvre places an excessive load on the lower attach rings of each booster where they are linked to the external tank. In the case of 51-L, the excessive loads caused a small rupture in the casing of the right-hand SRB, which immediately caused a sideways spurt of flame that could not be spotted within the intense brilliance of the main SRB exhaust at lift-off.

Mission specialist Judy Resnik immediately noticed that the temperature of the liquid hydrogen being fed from the external tank to the SSMEs was high. Commander Dick Scobee tried to look sideways and backwards out of his window to see if he could spot the reason why. As Challenger completed its roll programme, it was obvious to the crew and observers north and south of the pad that the Shuttle was "fishtailing". Loss of thrust in the right-hand SRB was being compensated for by the thrust vector control in the left-hand booster. Challenger continued its rocky and wild ride upwards, startling the pilot Mike Smith. Scobee and Smith realised that there was a serious problem and called up the pressure reading on the right-hand SRB. Meanwhile, the rupture in the SRB casing, which was shedding debris throughout the launch and causing an extremely messy contrail, in stark contrast to previous launches, caused the double O-rings to fail in one of the joints of the SRB and flame was later seen emanating from it.

When Scobee and Smith saw the pressure reading on the SRB they must have known that they were in trouble. Meanwhile, Houston had given Challenger the "go at throttle up" call, indicating that all was well, and no doubt a somewhat puzzled Scobee gave the last public utterance from the doomed ship, acknowledging the call. Leaking hydrogen from the now-breached external tank was igniting and a glow could be seen in the TV shots of Challenger, which did not show what was about to happen in a matter of milliseconds on its other side. At $T + 73$ seconds, pilot Smith uttered the final recorded words, "Uh-oh," as the right-hand booster broke from its mooring, broke through the wing of Challenger as the upper attachment broke, and ruptured the external tank.

Flying at Mach 1.92 at an altitude of 14,020 m (45,997 ft), Challenger was enveloped in the explosion, and broke apart under severe aerodynamic loads. The cabin section of Challenger emerged intact from the cataclysm, imposing a tolerably brief period of 20-G forces on the crew, as the two SRBs continued their flight independently before being blown up some seconds later. The cabin may have depressurised, rendering the astronauts unconscious in seconds, but there is evidence to suggest that it did not happen immediately, as at some stage, Resnik activated both Scobee and Smith's emergency compressed air packs. These would not have been much use at high altitude. The crew was alive but not necessarily conscious as the cabin began a two-and-a-half minute plunge into the Atlantic Ocean, breaking up on impact and instantly killing all seven crew members, Scobee, Smith, Resnik and Ellison Onizuka in the flight deck and Ronald McNair, Jarvis and McAuliffe in the mid-deck. Along with several pieces of debris, the shattered crew cabin and parts of the remains of the crew were recovered later, along with the crew cockpit recorder, which revealed the conversation during the launch.

NASA said that the crew was totally unaware that anything was wrong, a conclusion reached by the Rogers commission which investigated the accident and, despite a mountain of evidence, decided that the only cause of the accident was failed O-rings, damaged by the low temperatures before launch. Analyses of the accident by several engineers were refuted by NASA, though many suggestions were later incorporated quietly into the programme, including strengthening the attach ring system and testing it thoroughly during the Shuttle recovery programme. Partly in respect to the crew, STS 51-L is categorised by NASA as a "space mission in progress" when it failed and is counted as a "mission" for the crew with a 1 minute 13 second "duration", the point when contact was lost with the crew, even though it did not reach the official recognised altitude to be called a true "space flight".

Several important redesigns were made to the whole system, including the O-rings and seals between the segments of the SRBs, the official cause of the disaster. The spectre of "Challenger" continued to haunt NASA and was brought back to the headlines in February 2003 when a second Shuttle and crew were lost, leading to the decision to finally retire the Shuttle when its obligation to the International Space Station programme had been completed.

Milestones

25th Shuttle mission
10th and final flight of Challenger
2nd (intended) TDRS deployment mission
1st attempted manned space flight to launch and not reach space
1st loss of manned space vehicle during launch
1st US in-flight fatalities in space programme

SOYUZ T15

Int. Designation	1986-022A
Launched	13 March 1986
Launch Site	Baikonur Cosmodrome, Kazakhstan
Landed	16 July 1986
Landing Site	55 km northeast of Arkalyk
Launch Vehicle	R7 (11A511U2); spacecraft serial number (7K-ST) #21L
Duration	125 days 0 hrs 0 min 56 sec
Callsign	Mayak (Lighthouse)
Objective	First Mir resident (EO) programme; completion of work at Salyut 7

Flight Crew

KIZIM, Leonid Denisovich, 44, Soviet Air Force, commander, 3rd mission
Previous missions: Soyuz T3 (1982); Soyuz T10 (1984)
SOLOVYOV, Vladimir Alekseyevich, 39, civilian, flight engineer, 2nd mission
Previous mission: Soyuz T10 (1984)

Flight Log

As a result of the evacuation of Salyut 7 by the T14 crew, some of the experiment results had not been returned and some of the research programmes not completed. The sole remaining Soyuz T vehicle was ready to fly to the station (using the refurbished DM from the September 1983 T10-1 pad abort), and as the replacement Soyuz TM was not yet ready, this was the only vehicle available to the Soviets to continue their presence in space during 1986. In addition, the next space station was ready for launch, but its planned add-on modules were not. Therefore, it was decided to send the T15 crew to the core module of the new station first, then revisit Salyut 7 to complete the work there, and finally return to the new station to complete their programme in preparation for fully operational activities from 1987. On 20 February 1986, 22 days after the American manned space programme's tragic Space Shuttle loss and the beginning of what was to be a lengthy period on the ground, the Soviets launched the new space station, called Mir (Peace) and not Salyut 8 as many Western observers had expected. The ballyhoo surrounding the launch seemed to indicate that this was the expected giant station, launched aboard the first heavy lift Saturn V-type booster. In fact, it soon became obvious that Mir (1986-017A/DOS-7) was merely a beefed up Salyut, with one major difference: a docking pod at the front which would take five separate spacecraft. With the rear docking port, a sixth could be added, making it possible to erect the first modular space station.

Mir was to be fully operational by 1990, said the Soviets, with four large add-on modules mounted radially on the front pod. Mir weighed 20.9 tonnes and measured

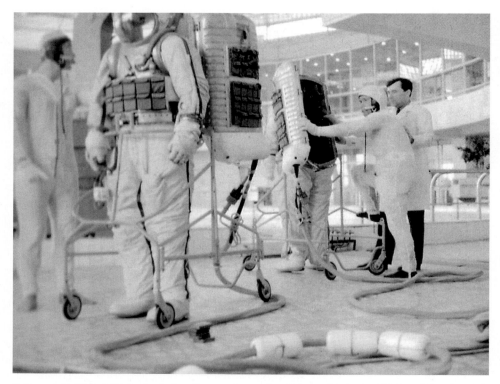

Solovyov (left) and Kizim (entering the rear door of an Orlan suit) during EVA training for their historic dual space station mission

13.13 m (43 ft), with a maximum diameter of 4.15 m (13.5 ft), and comprised a transfer compartment and two work compartments, with a service propulsion system at the rear. It had two solar panels with a total span of 29.73 m (97.5 ft). The first crew to visit in space were Leonid Kizim and Vladimir Solovyov, who were trained for a mission to Salyut. They were the first all-Soviet crew to fly an exclusively Soviet mission that was named before lift-off, the time of which was also given and seen live on TV, another first for an all-Soviet crew mission. Launch was at 18:33 hrs local time from Baikonur on 13 March, with docking occurring two days later.

Their basic job was to ready Mir for occupation by Mir-trained crews, when add-on modules had been attached. The crew also became the first to use a relay satellite, Cosmos 1700, to conduct real-time conversations with mission control at Kaliningrad. Two Progress tankers, Nos. 25 and 26, arrived in April and May to stock up the station, which the crew then surprisingly left on 5 May. But this was not to return home. Instead, they made a unique rendezvous and docking with Salyut 7. While the cosmonauts were aboard Salyut 7, an unmanned version of the new Soyuz TM spacecraft docked to Mir on 23 May, while Progress 26 was still attached. It returned home on 29 May.

During their stay aboard Salyut 7, the cosmonauts completed much of the work that had been intended for the Soyuz T14 crew, including two EVAs to erect a structure in space 50 m (164 ft) long that resembled the Shuttle 61-B EASE-ACCESS exercises. The walk on 28 May lasted 3 hours 50 minutes and the one on 31 May, 4 hours 40 minutes, making them the first spacemen to make seven and eight EVAs. Kizim and Solovyov ended their second stay aboard Salyut 7 on 25 June when they undocked and, again surprising some observers, returned to Mir rather than coming home, clocking up yet another first in a flight with so many such facts and feats for the statistician to record.

Salyut 7-Cosmos 1686 was placed into a 475 km (295 miles) storage orbit. Salyut 7 remained in space, unoccupied, for the next five years, finally re-entering in February 1991. The cosmonauts, with their spacecraft crammed with as much equipment from Salyut as possible, docked with Mir on 26 June. Work continued aboard Mir before it was announced – again another first – that they were returning home on 16 July and would be covered live on TV. Touchdown came at $T + 125$ days 0 hours 1 minute. During the mission, Kizim had become the first person to clock up a year's space experience, bringing his personal total at the end of the mission to 374 days. Maximum altitude achieved during the 51° mission was 349 km (217 miles).

Milestones

115th manned space flight
60th Soviet manned space flight
53rd Soyuz manned space flight
14th Soyuz T manned space flight
14th and last Soyuz T mission
1st flight by crew making seventh and eighth EVAs
1st manned docking with two separate vehicles
1st manned visit to two space stations
1st manned re-visit to space station
1st crewman to amass one year of space experience
1st Soviet use of a data relay satellite on manned space flight
11th Soviet and 34th flight with EVA operations

SOYUZ TM2

Int. Designation	1987-013A
Launched	5 February 1987
Launch Site	Pad 1, Site 5, Baikonur Cosmodrome, Kazakhstan
Landed	29 December 1987
Landing Site	80 km from Arkalyk
Launch Vehicle	R7 (11A511U2); spacecraft serial number (7K-M) #052
Duration	326 days 11 hrs 37 min 57 sec (Romanenko landed in TM3)
	174 days 3 hrs 25 min 56 sec (Laveikin landed in TM-2)
Callsign	Tamyr (Tamyr)
Objective	Mir EO-2 resident programme

Flight Crew

ROMANENKO, Yuri Viktorovich, 42, Soviet Air Force, commander, 3rd mission
Previous missions: Soyuz 26 (1977); Soyuz 38 (1980)
LAVEIKIN, Aleksandr Ivanovich, 36, civilian, flight engineer

Flight Log

The first crew to be fully operational aboard Mir was due to be Vladimir Titov and Aleksandr Serebrov, but their places aboard Soyuz TM2 were taken by the veteran Yuri Romanenko and rookie Aleksandr Laveikin, apparently days before the launch and possibly due to illness. The spectacular night launch was shown live on TV from Baikonur, at 01:38 hrs local time. The new spacecraft, Soyuz TM2 (Soyuz TM, the first of the series, was an unmanned test flight that docked to Mir in May 1986) took two days to reach Mir. The crew docked safely at the front and their first job was to unload Progress 27 from the rear port, which had docked on 18 January in preparation for their mission.

Laveikin became one of the few space explorers whose adaptation to weightlessness is "painful", as it was described by officials. Another Progress, No.28, arrived on 5 March, armed to the hilt with an array of scientific instruments to make the crew as busy and productive as possible. The equipment included the Korund crystal growth machine which was ultimately used to make useful electronic crystals, and the KATE 140 remote-sensing camera. Another device already aboard Mir, called Pion M, was used for extensive materials processing experiments.

The first add-on module was expected to be launched, and to prepare for it, Progress 28 undocked. Kvant, a module armed with an array of astrophysics experiments and with an externally mounted electrophoresis device attached, was launched by a Proton on 1 April, at 05:06 hrs local time. It weighed as much as Mir but after its planned docking it was to shed its propulsion module and would weigh 11 tonnes. Its

Romanenko and Laveikin review training plans for their residency on the Mir station

docking was a dramatic affair to say the least. On 5 April, Kvant failed to brake at a distance of 200 m (656 ft) and, according to Romanenko, sped dangerously by Mir at a distance of just 30 m (98 ft). Another docking attempt was made on 9 April, with Mir aiding the manoeuvres, but Kvant's probe got stuck and a hard-dock was not achieved. Something was blocking the docking port.

What that something was would be discovered during an unscheduled EVA by Romanenko and Laveikin on 11 April, during which they had to crawl through a tightly fitting hatch at the front of Mir and then scramble over the station to get to the rear. The cosmonauts saw that a white sheet of plastic was blocking the port, having been left there by Progress 28 which had inadvertently been launched with it attached after a technician failed to remove it. It was removed from the docking port and the hard docking was completed. The cosmonauts ended a highly successful 3 hour 40 minute EVA. They later entered Kvant to place it in operational status while its propulsion module was jettisoned – but not de-orbited due to lack of fuel – to expose a docking port for new Progress tankers, the next of which arrived on 21 April.

Progress 29, the third craft attached to the Mir core, was primarily used to load propellant into Mir's propulsion system, ingeniously through lines in Kvant. Another Progress was to arrive on 21 May, while Kvant was being used to make the first of its extremely useful observations of the Universe using telescopes, one of which was provided by the UK. A second EVA, lasting 1 hour 53 minutes, was made on 12 June, to begin the erection outside Mir of a two-part array of solar panels, 10 m (33 ft) high, which would increase Mir's electrical power. The second EVA, on June 16, lasted 3 hours 5 minutes.

The first international and visiting mission to Mir, Soyuz TM3, docked on 24 July, carrying a rookie commander, a Syrian pilot and a veteran, Aleksandr Aleksandrov, who was told a month before his scheduled eight-day trip that he was to replace Laveikin, whose heart was giving doctors cause for concern. Soyuz TM2 returned with Laveikin, having clocked up 174 days, leaving Aleksandrov with Romanenko, who exceeded the Soyuz T10 endurance record on 30 September. During their residency, Progress 31 (5 August) and 32 (26 September) were used to replenish the station with fuel, propellant and supplies, while on 31 August the crew were ordered to don their suits and get into the fresh TM3 for an emergency return to Earth. This exercise was not completed since it was, unbeknown to the crew at that time, a safety test.

During another unique test on 10 November, Progress 32 was undocked and re-docked to the rear of Kvant, perhaps to prepare for the introduction in later Mir operations of free-flying materials processing labs, conducting experiments autonomously of the type that Romanenko and Aleksandrov were in Mir. Yet another Progress, No. 33, replaced No. 32 on 23 November, with more esoteric processing equipment to test. Maximum altitude achieved during the 51° mission was 370 km (230 miles). On 23 December, Soyuz TM4 arrived, with the next long duration duo, Vladimir Titov and rookie Musa Manarov, as well as a Shuttle test pilot whom Romanenko had never met, called Anatoly Levchenko.

TM3 returned to Earth, with Romanenko, Aleksandrov and Levchenko all having clocked up different flight times and all of whom had been launched separately in the first landing of its kind in space history. Romanenko's condition after the 326 day 11 hour 37 minute mission originally launched in TM2 was remarkably good, although for precautionary reasons he was stretchered from the spacecraft. Ridiculous rumours persisted in some popular western newspapers that he was a "wreck".

Milestones

116th manned space flight
61st Soviet manned space flight
54th Soyuz manned space flight
1st Soyuz TM manned space flight
1st manned space flight over 300 days
1st landing with three crewmen launched separately
1st flight with crew members meeting for the first time
12th Soviet and 35th flight with EVA operations
Romanenko sets new space endurance record 326 days 11 hours
Laveikin celebrates his 36th birthday in space (21 Apr)
Romanenko celebrates his 43rd birthday in space (1 Aug)

SOYUZ TM3

Int. Designation	1987-063A
Launched	22 July 1987
Launch Site	Pad 1, Site 5, Baikonur Cosmodrome, Kazakhstan
Landed	30 July 1987
Landing Site	140 km NE of Arkalyk
Launch Vehicle	R7 (11A511U2); spacecraft serial number (7K-M) #53
Duration	7 days 23 hrs 4 min 55 sec (Viktorenko and Faris – returned in TM2)
	160 days 7 hrs 16 min 58 sec (Aleksandrov – returned in TM3)
Callsign	Vityaz (Viking)
Objective	Syrian visiting mission; partial resident crew exchange

Flight Crew

VIKTORENKO, Aleksandr Stepanovich, 40, Soviet Air Force, commander
ALEKSANDROV, Aleksandr Pavlovich, 44, civilian, flight engineer, 2nd mission
Previous mission: Soyuz T9 (1983)
FARIS, Mohammed, 36, Syrian Air Force, cosmonaut researcher

Flight Log

Accompanied by the now customary live television coverage, Soyuz TM3 blasted of from Baikonur at 07:59 hrs local time into clear blue skies *en route* for the Mir space station, carrying the first visiting crew to see the new complex in which Yuri Romanenko and Aleksandr Laveikin had been residing since the previous February. TM3 carried a cosmonaut from Syria, the nineteenth nation to have a man in space, plus commander Aleksandr Viktorenko and flight engineer Aleksandr Aleksandrov. The latter was told a month before lift-off that he was to remain aboard Mir until the end of the year, replacing Laveikin whose electrocardiograph readings were causing some concern.

Two days later, TM3 docked at the rear of Mir at the Kvant port, and soon five people were floating around inside. The Syrian, Mohammed Faris, had the usual complement of national experiments to conduct, including electrophoresis, crystal growth, remote sensing – during rare passes over Syria – and an ionospheric monitoring programme. The first international mission to return with a replacement crewman, Faris and Viktorenko entered the old TM2 with Laveikin and landed at $T + 7$ days 23 hours 4 minutes 5 seconds, 140 km (87 miles) northeast of Arkalyk, just 2 km (1 mile) from some settlement buildings, much to the relief of observers.

The prime and back-up crews of Soyuz TM3. At rear (l to r) Viktorenko and A. Solovyov. At front (l to r) Faris, Habib, Aleksandrov and Savinykh

Laveikin seemed pale but cheerful after his 174-day flight and doctors later declared him fit for another flight, while Syria celebrated. Maximum altitude during the 51° mission was 359 km (223 miles).

Milestones

117th manned space flight
62nd Soviet manned space flight
55th Soyuz manned space flight
2nd Soyuz TM manned space flight
3rd Soyuz international mission
1st flight by a Syrian

SOYUZ TM4

Int. Designation	1987-104A
Launched	21 December 1987
Launch Site	Pad 1, Site 5, Baikonur Cosmodrome, Kazakhstan
Landed	21 December 1988
Landing Site	180 km southeast of Dzhezkazgan
Launch Vehicle	R7 (11A511U2); spacecraft serial number (7K-M) #54
Duration	365 days 22 hrs 38 min 57 sec (Titov and Manarov)
	7 days 21 hrs 58 min 12 sec (Levchenko)
Callsign	Okean (Ocean)
Objective	Mir EO-3 research programme; Russian Buran pilot visiting mission

Flight Crew

TITOV, Vladimir Georgyevich, 41, Soviet Air Force, commander, 3rd mission
Previous missions: Soyuz T8 (1983); Soyuz T10-1 (1983)
MANAROV, Musa Khriramovich, 36, civilian, flight engineer
LEVCHENKO, Anatoly Semenovich, 46, civilian, cosmonaut researcher

Flight Log

In September 1985, the initial Mir resident crews were formed. The first crew would be Vladimir Titov and Alexander Serebrov; the second, Romanenko and Manarov; and the third, Alexander Volkov and Sergei Yemelyonov. When the Soyuz T15 crew were launched to Mir, the training of these crews was extended, and delays in the construction of modules also delayed their missions into 1987/1988. Then Manarov became ill and was replaced by Laveikin. Serebrov also fell ill and so Crew 2 replaced Crew 1 for the Soyuz TM2 mission flown in 1987. Manarov was restored to flight status and replaced Serebrov, joining Titov to form a new Crew 3. Yemelyanov was also removed from flight status during 1987 for medical reasons and was replaced first by Alexander Kaleri, then Sergei Krikalev, who was paired with Volkov as crew 4.

The Soyuz TM4/EO-3 crew would remain in space for a year, and would be joined for the ascent to Mir by one of the Buran shuttle pilots, who would return with the EO-2 crew after about a week in space. Like Igor Volk, a crewman on Soyuz T12 in 1984, another of this band of future shuttle pilots, Anatoly Levchenko, was to be given eight days space experience, at the end of which he would immediately get aboard a shuttle training aircraft to make simulated landings, to ensure that weightlessness did not impair his faculties. The first flight of the Soviet shuttle was clearly quite close, observers thought.

The new TM4 trio were launched from Baikonur at 17:18 hrs local time on 21 December and within two days were aboard Mir, with Romanenko and

The Soyuz TM4 crew of Titov, Levchenko (standing) and Manarov (front)

Aleksandrov, who were meeting Levchenko for the first time. Five days later, Romanenko, Aleksandrov and Levchenko were landing aboard Soyuz TM3, a unique crew indeed, having each been launched separately and each having clocked up different flight times. With the changeover complete, Titov and the rookie flight engineer Manarov began their long, historic, but quietly received mission by placing Soyuz TM4 at the front of Mir to prepare for the arrival of their first Progress support ship, No.34. This duly arrived on 23 January 1988 with two tonnes of scientific and other equipment. Progress also transferred propellant into Mir's depleting tanks.

Titov and Manarov took 4 hours 25 minutes exercise outside Mir on 26 February, when they replaced a faulty part of the new solar panel that had been erected by Romanenko and Laveikin. Progress 34 was replaced by No.35 on 25 March and then No.36 on 15 May, as the mission proceeded routinely, unheralded and unwatched by most of the world.

The cosmonauts were due to make an EVA in late May to repair faulty instruments on the Kvant astrophysics laboratory, but their work was delayed until after the departure of a new visiting crew aboard Soyuz TM5. Bulgaria's second cosmonaut (confusingly also called Aleksandr Aleksandrov) arrived with commander Anatoly Solovyov and flight engineer Viktor Savinykh on 9 June and remained for eight days, rather than the usual six for a Mir visiting mission. When they had left, Titov and Manarov moved the fresh TM5 craft to the front of Mir, making room for another Progress, and prepared for their next spacewalk which they made on 30 June.

During the 5 hour EVA, the cosmonauts struggled to replace a detector unit on Kvant because of a faulty tool, and had to be recalled while new tools were prepared and launched on the next Progress, No.37, which arrived on 20 July. The EVA, however, was delayed until after the departure of the next visiting crew aboard Soyuz TM6, which replaced Progress 37. This arrived at Mir on 31 August, carrying Vladimir Lyakhov, an Afghan cosmonaut researcher and, getting a taste of space at last, Doctor Valery Polyakov, who was to stay with Titov and Manarov throughout the remainder of their stay to monitor the medical condition of the core crew as they approached the end of their long pioneering mission, and then remain himself with the next resident crew. Lyakhov and the Afghan, Abdul Mohmand, left in Soyuz TM5 on 5 September, straight into a one-day drama which could have ended with them being stranded in orbit. Progress 38 duly arrived at Mir on 12 September.

On 20 October, the cosmonauts made their much-delayed EVA, after as much preparation and discussion from the ground as possible. The 4 hour 20 minute effort was a success, with Kvant's TTM telescope fully operational again. Titov and Manarov broke Romanenko's duration record on 11 November. By this time it had been announced that French cosmonaut Jean-Loup Chrétien would be launched on 21 November with two Soviets, on a longer than usual visiting mission, returning with Titov and Manarov on their 365th day in space. This launch of Soyuz TM7 was delayed to 26 November and reached Mir two days later.

With Chrétien were Aleksandr Volkov, with whom he made an EVA, and flight engineer Sergei Krikalev. He and Volkov were to be Polyakov's new companions until the following April, said the Soviets. The comings and goings from Mir were by now getting rather complicated to follow.

The long awaited return of the new space record holders, Titov and Manarov, with Chrétien, took place aboard Soyuz TM6 on 21 December after a one-orbit delay before retro-fire because of a computer software problem. The Orbital Module remained with the craft until after retro-fire to avoid problems that were experienced with TM5. The crew flew straight to Star City rather than to Baikonur because of an outbreak of hepatitis at the launch facility. Both cosmonauts recovered well after their 365 day 22 hour 39 minute mission, the longest manned space flight to that point. Chrétien clocked up 25 days. The Soviets then announced that, for the time being, long duration flights would be restricted to about five months.

Milestones

118th manned space flight
63rd Soviet manned space flight
56th Soyuz manned space flight
3rd Soyuz TM manned space flight
1st manned space flight to last one year
Titov and Manarov set new endurance record of 365 days 22 hours
13th Soviet and 36th flight with EVA operations
Titov celebrates his 41st birthday in space (1 Jan)
Manarov celebrates his 37th birthday in space (22 Mar)

SOYUZ TM5

Int. Designation	1988-048A
Launched	7 June 1988
Launch Site	Pad 1, Site 5, Baikonur Cosmodrome, Kazakhstan
Landed	17 June 1988 (in TM4)
Landing Site	160 km southeast of Dzhezkazgan
Launch Vehicle	R7 (11A511U2); spacecraft serial number (7K-M) #55
Duration	9 days 20 hrs 9 min 19 sec
Callsign	Rodnik (Spring)
Objective	Bulgarian visiting mission; Soyuz ferry exchange mission

Flight Crew

SOLOVYOV, Anatoly Yakovlovich, 40, Soviet Air Force, commander
SAVINYKH, Viktor Petrovich, 48, civilian, flight engineer, 3rd mission
Previous missions: Soyuz T4 (1981); Soyuz T13 (1985)
ALEKSANDROV, Aleksandr Panayatov, 37, Bulgarian Air Force, cosmonaut researcher

Flight Log

Bulgaria's first cosmonaut, Georgy Ivanov, became the only international space traveller not to reach a Soviet space station when Soyuz 33 malfunctioned in 1979. To make amends, the Soviets offered Bulgaria another try, even flying Ivanov's back-up, Aleksandr Aleksandrov. This caused confusion among the non-specialist press covering the mission because the other, Soviet, Aleksandr Aleksandrov had recently returned from Mir himself. The Soviets also provided a flight of ten days to make up for the five days that Ivanov lost when he had to come home early.

The Bulgarian flight was well prepared and packed with 40 science experiments, many of which were left on Mir for further use. Of all the international missions thus far, this seemed to be the most fruitful nationally. Launch in Soyuz TM5, with rookie commander Anatoly Solovyov and veteran flight engineer Viktor Savinykh, came at 20:03 hrs local time, and the docking followed smoothly on 9 June. Aleksandrov's array of experiments included eight for Earth remote sensing and manufacturing units to produce composite alloys in weightlessness. The mission ended inside Soyuz TM4 at $T + 9$ days 20 hours 10 minutes in a lake bed – fortunately bone dry – in 42° temperatures.

Milestones

119th manned space flight
64th Soviet manned space flight

57th Soyuz manned space flight
4th Soyuz TM manned space flight
4th Soyuz international space flight
1st Soviet flight carrying second foreign crewman from the same country (Bulgaria)

The crew for the second Bulgarian space flight. (l to r) Savinykh, Solovyov and Aleksandrov

SOYUZ TM6

Int. Designation	1988-075A
Launched	29 August 1988
Launch Site	Baikonur Cosmodrome, Kazakhstan
Landed	7 September 1988 (in TM5)
Landing Site	160 km southeast of Dzhezkazgan
Launch Vehicle	R7 (11A511U2); spacecraft serial number (7K-M) #56
Duration	8 days 20 hrs 26 min 27 sec (Lyakhov and Mohmand)
	240 days 22 hrs 34 min 47 sec (Polyakov – returned in TM7)
Callsign	Proton (Proton)
Objective	Afghan visiting mission; Soyuz TM ferry exchange mission

Flight Crew

LYAKHOV, Vladimir Afanasyevich, 47, Soviet Air Force, commander, 3rd mission
Previous missions: Soyuz 32 (1979); Soyuz T9 (1983)
POLYAKOV, Valery, 46, civilian, flight engineer
MOHMAND, Abdul Ahad, 29, Afghan Air Force, cosmonaut researcher

Flight Log

Soyuz TM6 was launched routinely from Baikonur at 07:23 hrs local time, watched live on television. Inside the spacecraft were commander Vladimir Lyakhov, Doctor Valery Polyakov and Afghan pilot, Abdul Mohmand who, considering the Soviet–Afghan war at the time, featured in the most incongruous and unabashed political Soviet international mission, the last "free" ride for a foreigner. After Mohmand's statutory six-day stay aboard Mir, he left the space station aboard TM5, with his commander, leaving Polyakov to ride out Titov and Manarov's marathon stay.

The Orbital Module was jettisoned and TM5's computer began the retro-fire sequence. The prime infra horizon sensors attempted to orientate the spacecraft correctly, but because the Sun had just set, became confused. The retro-engine was primed and ready to ignite when, just 30 seconds before ignition, the confused computer vetoed the firing. The puzzled Lyakhov was reporting the incident and musing over what to do next when the sensors suddenly indicated that orientation was indeed correct and fired the engines, seven minutes later than planned. This would have resulted in a landing far off course, possibly in Manchuria. Lyakhov routinely shut the engine down after six seconds.

Reasonably unconcerned, Soviet officials decided to attempt the retro-fire two orbits later using the back-up flight computer, thus plunging the mission into dangerous chaos. TM5 was the back-up spacecraft for the Bulgarian mission and the back-up computer still held the software for it. TM5's inertial measurement unit was being

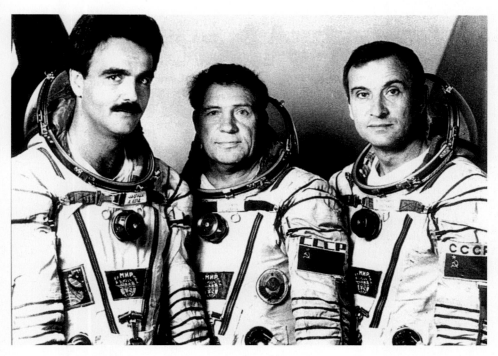

Portrait of the Soyuz TM6 crew. L to r: Mohmand, Lyakhov and Polyakov

used to orbit the spacecraft this time when suddenly the computer, programmed to perform a rendezvous manoeuvre burn instead, lit the engine for six seconds and shut it down, as programmed. Lyakhov, anxious to get home and unaware of the cause of the problem, manually re-ignited the engine and it fired for sixty seconds before the navigation computer sensed an attitude error and shut it down again.

In a normal retro-burn, the engine should have fired for 230 seconds and the crew was extremely fortunate that the engine shut down when it did and not later because it would have placed TM5 into an extremely low orbit. Had they been unable to orientate TM5 and get the confused computer to fire the engine, the cosmonauts would have been forced to separate the descent craft in preparation for a natural re-entry which would have imposed abnormal loads on it. The crew would have burned up. Officials delayed the retro-fire for 24 hours and there was a danger that the crew could be stranded in space, to die when the limited environmental control system of the descent capsule ceased to work. There were emergency rations, but Lyakhov was more concerned about the rather basic lavatorial facilities. The toilet had been discarded with the OM, so they had to rely on plastic sleeves of the sleeping bag Mohmand had brought from Mir.

The back-up computer was reprogrammed with the correct retro-fire data to attempt a re-entry towards the prime recovery zone, and everything worked perfectly. Mission time was 8 days 20 hours 27 minutes. The extremely relieved Lyakhov suggested that in future the OM should not be jettisoned until after retro-fire, even though this would be costly in propellant, while the extremely unruffled Mohmand took it all in his stride, stating the crew were more tense than afraid. The Soviets admitted a lack of vigilance which had resulted from a sequence of successes. Lyakhov and Mohmand got away with it and the relatively good safety record for manned space flight remained.

Milestones

120th manned space flight
65th Soviet manned space flight
58th Soyuz manned space flight
5th Soyuz TM manned space flight
5th Soyuz international space flight
1st flight of an Afghan
1st failure of retro-rocket during burn

STS-26

Int. Designation	1988-091A
Launched	29 September 1988
Launch Site	Pad 39B, Kennedy Space Center, Florida
Landed	3 October 1988
Landing Site	Runway 17, Edwards Air Force Base, California
Launch Vehicle	OV-103 Discovery/ET-28/SRB BI-029/SSME #1 2019; #2 2022; #3 2028
Duration	4 days 1 hr 0 min 11 sec
Callsign	Discovery
Objective	Return-to-Flight mission; TDRS-C deployment

Flight Crew

HAUCK, Frederick Hamilton "Rick", 47, USN, commander, 3rd mission
Previous missions: STS-7 (1983); STS 51-A (1984)
COVEY, Richard Oswalt, 42, USAF, pilot, 2nd mission
Previous mission: STS 51-I (1985)
LOUNGE, John Michael, 38, civilian, mission specialist 1, 2nd mission
Previous mission: STS 51-I (1985)
HILMERS, David Carl, 38, USMC, mission specialist 2, 2nd mission
Previous mission: STS 51-J (1985)
NELSON, George Driver, 38, civilian, mission specialist 3, 3rd mission
Previous missions: STS 41-C (1984); STS 51-C (1985)

Flight Log

Following the release of the findings of the Rogers Commission into the Challenger disaster in June 1986, NASA was directed to follow nine major recommendations to improve the safety and management of the Space Shuttle programme. The path to recovery was a tortuous one. At first, a re-launch in late 1987 seemed a possibility, or early 1988, or June 1988. Discovery finally and patriotically made it to the pad on 4 July. A successful launch and flight of the Space Shuttle was considered crucial. It was to be the most important manned mission of the US space programme. A failure of any kind could have spelled the death knell of the programme and NASA knew it. No.chances were being taken; so much so that many experienced space watchers reckoned that a few abortive countdowns were going to be unavoidable and once Discovery did take off, it would be an anticlimax, perhaps what NASA wanted.

True to form, as the all-veteran crew of STS-26 – a Shuttle first – left the crew quarters on 29 September, looking like astronauts again wearing high-altitude pressure suits, the chances of launching that day were put at 50-50, mainly because the winds at high altitude were not strong enough. The flight computer was programmed

Return to flight. The launch of STS-26 was the start of America's journey back to space

to expect stronger seasonal winds. It was re-programmed during holds caused by other niggling problems and the count stood at $T-9$ minutes for 1 hour 38 minutes. The go for launch was suddenly given and people realised that perhaps Discovery was going to get off first time after all. Things went well until it was announced that the count would hold at $T-31$ seconds because a problem had been experienced. This proved to be an erroneous switch and at 11:37 hrs local time, on Challenger's Pad 39B, America returned to space with a smooth lift-off and ascent.

Concern was caused by the sight of flames around the SRBs just before burn out but these were caused by the SSME exhaust being sucked into an aerodynamically low pressure area of the Shuttle stack as it rose at Mach 4. It was all so smooth that observers did indeed feel the anticlimax, a tribute to the launch team under former astronaut Bob Crippen. The STS-26 mission continued on its winning way, performing an OMS burn to circularise the 29.45° orbit at 284 km (176 miles), and deploying the TDRS-C satellite on its IUS upper stage.

The crew conducted several science experiments, practiced donning and doffing the ascent/descent suit to see how quickly it could be done in an emergency, and experienced, for a short while, uncomfortably high cabin temperatures of 29°C due to ice blocking a cooling duct. On day four, the crew made a moving tribute to the Challenger Seven, covered live on TV. The flight was also a re-qualification of Discovery within the Return-to-Flight programme.

Only the landing remained. The de-orbit and re-entry were routine and Discovery came home in triumph, to a rapturous welcome from observers at Edwards Air Force Base, including Vice President George Bush, landing on runway 17 at $T+4$ days 1 hour 0 minutes 11 seconds. The Shuttle was poised for routine operations again but the difference was that even NASA admitted that things could go wrong again, something that before Challenger would have seemed a sacrilege, such was the apparent ease and safety of the system.

Milestones

121st manned space flight
56th US manned space flight
26th Shuttle flight
7th flight of Discovery
3rd TDRS deployment mission

SOYUZ TM7

Int. Designation	1988-104A
Launched	26 November 1988
Launch Site	Pad 1, Site 5, Baikonur Cosmodrome, Kazakhstan
Landed	26 April 1989
Landing Site	140 km northeast of Dzhezkazgan
Launch Vehicle	R7 (11A511U2); spacecraft serial number (7K-M) #57
Duration	151 days 11 hrs 8 min 23 sec (Volkov and Krikalev)
	24 days 18 hrs 7 min 25 sec (Chrétien)
Callsign	Donbass (Donbass)
Objective	Mir EO-4 research programme; French Aragatz visiting mission

Flight Crew

VOLKOV, Aleksandr Aleksandrovich, 40, Soviet Air Force, commander, 2nd mission
Previous mission: Soyuz T14 (1985)
KRIKALEV, Sergei Konstantinovich, 30, civilian, flight engineer
CHRÉTIEN, Jean-Loup, 50, French Air Force, cosmonaut researcher, 2nd mission
Previous mission: Soyuz T6 (1982)

Flight Log

France's close ties with the Soviet space programme produced beneficial results, none more so than the Soyuz TM7 mission in which the highest ranking spaceperson, Brigadier General Jean-Loup Chrétien would make his second flight on a Soviet spacecraft and be the first non-US and non-Soviet spaceman to walk in space. His 30-day mission would also be considerably longer than the usual seven-day jaunts by foreigners. The longer flight was provided in return for the supply of much French scientific equipment for use by the Soviet crews on Mir, but it was the last to be provided free by the Soviets; the next Frenchman had to pay $12 million.

France's President Mitterand scored a spectacular own goal before the mission by insisting on going to Baikonur to watch the launch, which would therefore have to be delayed four days to 26 November, reducing Chrétien's time in space. Mitterand winged his way in and out of Baikonur on a Concorde with an entourage of such high number and rank that Baikonur's modest hospitality facilities and traditional pre-launch pomp and circumstance became unmanageable. The result was a chaotic crew walk out in which Mitterand and other officials were bundled about by hordes

French cosmonaut Chrétien (left) joins his Soviet colleagues Volkov (centre) and Krikalev for the Soyuz TM7 crew photo

of eager bystanders and press, as crew commander Aleksandr Volkov tried to make his traditional speech of dedication of the mission to General Kerim Kerimimov, the president of the state commission.

The launch, the 301st from Pad 1 at Baikonur, was spectacularly routine, with the Soyuz booster that had only been rolled out to the fog-bound pad two days previously lighting up the sky at 20:49 hrs local time. Once aboard Mir, after the two-day rendezvous flight, Chrétien, Volkov and the impassive young flight engineer Sergei Krikalev, got to work with Titov, Manarov and Polyakov, the high point of which was Chrétien's EVA with Volkov on 9 December, three days earlier than planned originally. During the 5 hour 57 minute EVA, Chrétien and Volkov deployed an experiment called ERA, provided by France, which comprised folded carbon fibre tubes that could be unfurled to form a cube structure in a test of erectable space structures. The $8 million experiment seemed doomed to failure when it could not be commanded to unfurl and engineers considered jettisoning it. Volkov saved the day – he admitted later – by giving it a hefty kick with his space boot. Both spacemen were utterly exhausted by their efforts.

The fruitful French mission ended with Chrétien returning to Earth with the record-breakers Titov and Manarov on 21 December, leaving Volkov, Krikalev and the doctor Polyakov to remain until 27 April, to be replaced by the TM8 crew. The fresh Soyuz TM7 craft was moved to the front and Progress 39 linked up on 27 December with New Year supplies.

Delays in launching new modules to Mir meant that this crew, like previous ones on Mir, were rather limited in what experiments they could conduct, most of which seemed to focus on the astrophysics telescopes on the Kvant module and Polyakov's surgery. The crew also spent much of their time repairing balky equipment, particularly environmental control systems which were misbehaving so badly that some electrical equipment was covered in condensation. The module delays and these niggling equipment problems raised concerns over whether Mir, three years old on 20 February 1989, would ever see out its operational life before being declared operational with all four modules.

But life went on. Progress 40 replaced No.39 on 12 February, delivering pickled cucumbers by request. A planned EVA by Volkov and Krikalev was cancelled and there were suggestions that Polyakov might remain on Mir with the next crew. When Progress 40 departed on 3 March it remained close to Mir for the cosmonauts to observe a unique experiment in which the unmanned tanker deployed two folding structures, which were unfurled from it by heating electrical wires in its body in a space structures test. Progress 40 was destroyed during a controlled re-entry two days later and was replaced by Progress 41 on 18 March.

Meanwhile, on the ground, cosmonauts Aleksandr Viktorenko and Aleksandr Balandin, the latter having replaced Aleksandr Serebrov because of the delays in the launches of the new modules which Serebrov had been trained specifically to operate, were ready to launch on 19 April onboard Soyuz TM8, to replace the crew of TM7 which was to come home with Polyakov on 27 April. Then, on 12 April, the Soviets sprang a surprise, saying that the Soyuz TM7 crew would leave Mir empty for several months. Flying another crew when the new modules were not ready for launch seemed wasteful and leaving Mir empty would save money. So Volkov and Krikalev clocked up a TM7 flight time of 151 days 11 hours 10 minutes, landing on 27 April northeast of Dzhezkazgan, the prime recovery zone, as planned. Polyakov had clocked up 240 days 22 hours 36 minutes flight time, the fourth longest individual space mission.

Milestones

122nd manned space flight
66th Soviet manned space flight
59th Soyuz manned space flight
6th Soyuz TM manned space flight
6th Soyuz international mission
1st non-Soviet, non-US crewman to make two space flights
1st non-Soviet, non-US crewman to perform EVA
14th Soviet and 37th flight with EVA operations
Volkov celebrates his 41st and Polyakov his 47th birthday (27 Apr) on the day both returned to Earth on TM7

STS-27

Int. Designation	1988-106A
Launched	2 December 1988
Launch Site	Pad 39B, Kennedy Space Center, Florida
Landed 6	December 1988
Landing Site	Runway 17, Edwards Air Force Base, California
Launch Vehicle	OV-104 Atlantis/ET-23/SRB BI-030/SSME #1 2027; #2 2030; #3 2029
Duration	4 days 9 hrs 5 min 35 sec
Callsign	Atlantis
Objective	3rd classified DoD Shuttle mission

Flight Crew

GIBSON, Robert Lee "Hoot", 42, USN, commander, 3rd mission
Previous missions: STS 41-B (1984); STS 61-C (1986)
GARDNER, Guy Spence, 40, USAF, pilot
MULLANE, Richard Michael, 43, USAF, mission specialist 1, 2nd mission
Previous mission: STS 41-D (1984)
ROSS, Jerry Lynn, 40, USAF, mission specialist 2, 2nd mission
Previous mission: STS 61-B (1985)
SHEPHERD, William McMichael, 39, USN, mission specialist 3

Flight Log

STS 62-A, the first manned polar orbiting space flight, was to have been launched from Vandenberg Air Force Base in California in 1986. The flight was cancelled and the Vandenberg pad mothballed after the Challenger disaster. It re-emerged as STS-27, with a new commander, Hoot Gibson, replacing Bob Crippen, and a new mission specialist, William Shepherd, replacing Dale Gardner. Orbiter Atlantis was equipped with an enormous electronic intelligence and digital imaging reconnaissance satellite which was to be placed into a 57° inclination orbit, the highest inclination permitted by a Shuttle from the KSC.

The first launch attempt was called off on 1 December with the crew aboard at $T-9$ minutes due to high winds at altitude. There was a minor delay the following day, before the spectacular take-off at 09:30 hrs local time, with the Shuttle making a dramatic, sloping lateral movement away from the pad as it performed a 140° roll programme and headed up the east coast of the USA. Debris from the top of one of the SRBs broke away and severely damaged some of the underside of Atlantis, as the crew would see after they landed. Once in orbit the official communications, which began at $T-9$ minutes, ended for this military mission.

L to r: Gardner, Gibson and Shepard at work during STS-27

According to analysis and ground observations, the giant Lacrosse was deployed from the payload bay by the RMS and inspected. Its 45 m (147 ft) span solar panels were supposed to unfurl but did not at first. If an EVA was required to free the panels it was not announced but the panels were freed, possibly by an RMS-induced shake, and the deployment followed at about $T + 7$ hours into the mission. Little information about it was released, except that a gallon of water had leaked in the cockpit. STS-27 marked the third time that eleven people were in space at once, with six cosmonauts on board the Mir space station at the same time.

The flight ended at $T + 4$ days 9 hours 5 minutes 35 seconds on runway 17 at Edwards Air Force Base after a northerly approach from its high-inclination orbit, only the second afternoon landing in the Shuttle programme. The crew busied themselves examining the underside of the orbiter, which had suffered extensive damage to its heatshield tiles, resulting in the need to replace 707 of them, the greatest tile loss on the programme. Despite this, STS-27 had qualified Atlantis for the Return-to-Flight programme.

Milestones

123rd manned space flight
57th US manned space flight
27th Shuttle mission
3rd flight of Atlantis

STS-29

Int. Designation	1989-021A
Launched	13 March 1989
Launch Site	Pad 39B, Kennedy Space Center, Florida
Landed	18 March 1989
Landing Site	Runway 22, Edwards Air Force Base, California
Launch Vehicle	OV-103 Discovery/ET-36/SRB BI-031/SSME #1 2031; #2 2022; #3 2028
Duration	4 days 23 hrs 38 min 50 sec
Callsign	Discovery
Objective	TDRS-D deployment mission

Flight Crew

COATS, Michael Lloyd, 43, USN, commander, 2nd mission
Previous mission: STS 41-D (1984)
BLAHA, John Elmer, 46, USAF, pilot
SPRINGER, Robert Clyde, 46, USMC, mission specialist 1
BUCHLI, James Frederick, 43, USMC, mission specialist 2, 3rd mission
Previous missions: STS 51-C (1985); STS 61-A (1985)
BAGIAN, James Philip, 37, civilian, mission specialist 3

Flight Log

With the Shuttle up and rolling again, it did not take NASA long to establish an ambitious launch schedule for 1989, which would see seven launches starting with STS-29 on 14 February. The mission was soon put back to 23 February and was delayed again to at least mid-March. As a precautionary measure, the high-pressure oxidiser turbopumps on all three SSMEs on Discovery – by now already on Pad 39B – were replaced because stress corrosion cracks had been discovered on an SSME used for STS-27.

There was concern that the delay might cancel the mission altogether because STS-30 Atlantis needed to be on the pad by 23 March to meet the first day of its planetary launch window for the Magellan spacecraft. The work was completed satisfactorily and a date of 11 March set. Then a master events computer failure caused another delay to 13 March. It did not look too hopeful on this day either, as Discovery and the launch pad were draped in ground fog and there were concerns about conditions on the Kennedy runway should there be a return to launch site abort.

The count proceeded and was held at $T - 9$ minutes for 1 hour 50 minutes before Discovery rose majestically into the bright sunlit skies at 09:57 hrs local time – accompanied by the first female launch commentary of a US manned launch by NASA's Lisa Malone – into its 28.45° inclination orbit which would reach a maximum

This view of the flight deck of Atlantis shows MS2 Buchli (left) and commander Coats (centre) at work. Pilot Blaha is hidden from view by his seat (right)

altitude of 283 km (176 miles). For the first time, photographs were captured of the external tank, looking like a fat cigar before being stubbed out during re-entry. In space at last were the original crew of STS 61-H, which was to have flown in June 1986 carrying Britain's would-be first man in space, Sq. Ldr. Nigel Wood, had it not been for the Challenger disaster. The only newcomer to the crew was James Bagian who replaced Anna Fisher from the original 61-H mission.

At $T + 6$ hours 13 minutes, the primary payload of the mission, TDRS-D, was deployed, while concerns were raised about pressure surges in the fuel cells of the orbiter. The rest of the mission went swimmingly, concentrating on an array of science experiments and photography using the large format IMAX camera. The weather was so smooth at Edwards Air Force Base that a proposed crosswind landing test could not be performed, so commander Mike Coats aimed for runway 22 and a braking test on the concrete. Mission time was $T + 4$ days 23 hours 38 minutes 52 seconds. Discovery was in great shape and STS-29 proved to be one of the smoothest missions ever.

Milestones

124th manned space flight
58th US manned space flight
28th Shuttle mission
8th flight of Discovery
4th TDRS deployment mission

STS-30

Int. Designation	1989-033A
Launched	4 May 1989
Launch Site	Pad 39B, Kennedy Space Center, Florida
Landed	8 May 1989
Landing Site	Runway 22, Edwards Air Force Base, California
Launch Vehicle	OV-104 Atlantis/ET-29/SRB BI-027/SSME #1 2027; #2 2030; #3 2029
Duration	4 days 0 hrs 56 min 27 sec
Callsign	Atlantis
Objective	Magellan Venus probe deployment mission

Flight Crew

WALKER, David Mathieson, 44, USN, commander, 2nd mission
Previous mission: STS 51-A (1984)
GRABE, Ronald John, 43, USAF, pilot, 2nd mission
Previous mission: STS 51-J (1985)
LEE, Mark Charles, 36, USAF, mission specialist 1
THAGARD, Norman Earl, 45, civilian, mission specialist 2, 3rd mission
Previous missions: STS-7 (1983); STS 51-B (1985)
CLEAVE, Mary Louise, 42, civilian, mission specialist 3, 2nd mission
Previous mission: STS 61-B (1985)

Flight Log

The radar mapping satellite, Magellan, destined to explore the planet Venus, had been scheduled for a Shuttle launch on STS-72 on 6 April 1988, atop a Centaur-G Prime liquid-fuelled upper stage. The Challenger disaster intervened and not only delayed the mission but also meant that, for new safety reasons, the Centaur would not be carried on Shuttle. With a replacement upper stage, the less powerful IUS, Magellan was scheduled for launch between 28 April and 23 May, preferably as early as possible, but still would not reach Venus for over 16 months.

Assigned to STS-30 Atlantis, Magellan got to within 31 seconds of launching on the first launch day window, but a fault in one of the SSME fuel recirculation pumps stopped the count. Another problem was discovered later and a leaking fuel pump had to be replaced. A new launch date of 4 May was set, but Atlantis only got away with 5 minutes of the 64 minute window remaining, at 14:47 hrs local time and following a hold at $T - 5$ minutes until the only cloud in the area – sitting over the KSC runway that would have been used for an RTLS abort – cleared and crosswinds died down.

Deployment of Magellan and the IUS from the payload bay of Atlantis

The Shuttle performed the first ascent of its type, called an inertially targeted profile, to place Atlantis at exactly the right point for Magellan's deployment. This did, however, result in an initial orbit of just 6.4 by 136 km (85 miles) before two OMS burns. The deployment occurred flawlessly at $T + 6$ hours 18 minutes, only after much in-orbit checking. Magellan's solar panels deployed and the IUS stages fired, placing it *en route*. STS-30, in 28.85° orbit with a maximum altitude of 283 km (176 miles), got down to a routine job of experiments, tests and rest. But the routine was broken when the third general purpose computer on board failed and had to be replaced by the crew – a first for the Shuttle. Once Magellan had been deployed, the crew occupied themselves with a range of mid-deck experiments.

Atlantis was aiming to make a crosswinds landing at Edwards Air Force Base at the end of its mission and such were the conditions there that it was only when the orbiter was at Mach 15 that the final choice of runway, the hard No.22, was made. Mission time was $T + 4$ days 0 hours 57 minutes 9 seconds.

For Magellan, the interplanetary cruise lasted from 4 May 1989 to 10 August 1990, when it entered orbit around the shrouded planet. The first mapping cycle was completed between 15 September 1990 and 15 September 1991. In an orbit with a period of 3.25 hours and an inclination of 86°, the radar mapped the surface for 37.2 minutes per orbit. In total 98 per cent of the surface of Venus was radar-mapped by Magellan, and 95 per cent of the gravity data from the planet recorded. Magellan's extended mission lasted between 15 September and 11 October 1994. Split into cycles, the extended mission focused on: (Cycle 2) Imaging the south pole region and gaps from cycle 1; (Cycle 3) filling remaining gaps and collecting stereo imagery; (Cycle 4) measuring Venus' gravitational field; (Cycle 5) aerobraking to circularise the orbit and global gravity measurements; (Cycle 6) collecting high-resolution gravity data, conducting radio science experiments and a windmill experiment to observe the behaviour of molecules in the upper atmosphere. Experiment data collection was completed on 11 October 1995. Magellan was programmed to complete a destructive entry into the Venusian atmosphere on 12 October 1994, ending a highly successful mission and closing the flight operations chapter on STS-30, over five years after the launch from Kennedy Space Center.

Milestones

125th manned space flight
59th US manned space flight
1st deployment of planetary spacecraft on manned space flight
29th Shuttle flight
4th flight of Atlantis
1st Shuttle planetary deployment mission

STS-28

Int. Designation	1989-061A
Launched	8 August 1989
Launch Site	Pad 39B, Kennedy Space Center, Florida
Landed	13 August 1989
Landing Site	Runway 17, Edwards Air Force Base, California
Launch Vehicle	OV-102 Columbia/ET-31/SRB BI-028/SSME #1 2019; #2 2022; #3 2028
Duration	5 days 1 hr 0 min 9 sec
Callsign	Columbia
Objective	Fourth classified DoD Shuttle mission

Flight Crew

SHAW, Brewster Hopkinson Jr., 44, USAF, commander, 3rd mission
Previous missions: STS-9 (1983); STS 61-B (1985)
RICHARDS, Richard Noel, 42, USN, pilot
ADAMSON, James C., 43, US Army, mission specialist 2
LEESTMA, David Cornell, 40, USN, mission specialist 1, 2nd mission
Previous mission: STS 41-G (1984)
BROWN, Mark Neil, 38, USAF, mission specialist 3

Flight Log

In 1985, NASA announced the crew of a Department of Defense mission, STS 61-N, the last crew to be named before the Challenger disaster, which of course cancelled the mission. It eventually became known as STS-28, with one crew replacement. STS-28, by rights, should have taken off in late 1988 or early 1989, but the main reason for the delay to August 1989 was that the refurbishment of the oldest orbiter, Columbia, was taking much longer than anticipated. Columbia was being brought up to the standard of the later orbiters and even in early 1989 was short of 2,400 heatshield tiles. Other processing difficulties and parts shortages meant that the launch schedule had to be changed, with STS-33 moving to November to accommodate STS-28.

Columbia eventually reached the pad on 14 July and was ready to go at 07:57 hrs local time on 8 August. Fog on the Shuttle runway at KSC took time to lift, so Columbia remained Earthbound for 40 minutes, leaving the pad at 08:37 hrs, heading up the eastern seaboard of the USA into its 57° inclination orbit. On its fifth orbit, the crew deployed the advanced reconnaissance satellite KH-12 (USA-40). A smaller satellite was deployed the next day and the crew conducted classified experiments and one unclassified experiment on radiation monitoring before coming home to Edwards Air Force Base runway 17 at $T + 5$ days 1 hour 0 minutes 9 seconds, just after sunrise.

The STS-28 crew poses in the now-familiar "starburst" formation. Clockwise from top left: Shaw, Adamson, Leestma, Brown and Richards

Later, it was reported that the KH-12 was spinning out of control in orbit and was inoperable. Apparently, however, the satellite was brought under control. The classified nature of the mission was lifted later, when a photo of the crew in orbit was released. This was the final re-qualification mission of the Shuttle fleet as part of the Return-to-Flight programme.

Milestones

126th manned space flight
60th US manned space flight
30th Shuttle mission
8th flight of Columbia
4th Shuttle DoD mission

SOYUZ TM8

Int. Designation	1989-071A
Launched	6 September 1989
Launch Site	Pad 1, Site 5, Baikonur Cosmodrome, Kazakhstan
Landed	19 February 1990
Landing Site	55 km northeast of Arkalyk
Launch Vehicle	R7 (11A511U2); spacecraft serial number (7K-M) #58
Duration	166 days 6 hrs 58 min 16 sec
Callsign	Vityaz (Viking)
Objective	Mir EO-8 research programme

Flight Crew

VIKTORENKO, Aleksandr Stepanovich, 42, Soviet Air Force, commander, 2nd mission
Previous mission: Soyuz TM3 (1987)
SEREBROV, Aleksandr Aleksandrovich, 45, civilian, flight engineer, 3rd mission
Previous missions: Soyuz T7 (1982); Soyuz T8 (1983)

Flight Log

The Soviet Union's premier space station, Mir, which was to have been permanently manned from February 1987 when the TM2 crew boarded her, remained empty for four months during 1989 as engineers readied the much-delayed add-on modules for launch so that future crews would have something more to do than use the Kvant telescopes and operate within the relatively cramped quarters of the Mir core module and Kvant. On 23 August 1989, there was a sure sign that Mir was being readied for another crew when a Progress was launched to dock with the station two days later. This was the first of a new series, called Progress M, a basic uprated version that would, according to the Soviets, carry a small re-entry vehicle to return 150 kg (331 lb) of samples in the future.

Soyuz TM8, which should have been launched the previous April, was readied on Pad 1, daubed for the first time with advertisements, and launched at 03:38 hrs local time at Baikonur on 6 September, docking manually after a minor malfunction two days later. The commander was, as expected, Aleksandr Viktorenko, but his flight engineer was not Aleksandr Balandin of the TM8 that should have launched in April, but Aleksandr Serebrov, in space at last and sure that his new modules would be launched.

The first of these modules was to carry the first Soviet manned manoeuvring unit which Serebrov helped to develop. The launch of the first module, designated "D", was due in October and the second, designated "T" later in the year. The first got off

The crew of Soyuz TM8: Victorenko (left) and Serebrov

the ground much later than planned in 1989, but the second was delayed again, first to January then to Spring 1990, by which time another crew were scheduled to replace the TM8 crew, who were due to return to Earth on 19 February 1990. The delay to the "D" module from October to November was due to problems with checking it out for launch when several faults were found, a fact that was announced self-critically by Soviet officials.

The 19,565 kg (43,141 lb) module was finally launched on 26 November and was called Kvant 2. It gave a performance in space rather reminiscent of its predecessor in 1987. First, one of its two solar panels failed to deploy and there were suggestions that an EVA would be required to pull it out when the module arrived at Mir. Then the first docking attempt was called off at a distance of 20 km (12 miles) from the station because Kvant 2 was not in the right orbit. A subsequent attempt on 6 December was successful. The docking took place at the front port, replacing Progress M which had made the first front port unmanned tanker docking while TM8 was at the rear.

Kvant 2 was then ingeniously moved to a side port at the front by a simple crane-like device. TM8 moved back to the front after a crewed 20 minute fly around and was replaced at the rear by Progress M2 on 20 December, which for the first time carried a commercial US experiment to be operated by Soviet cosmonauts. Kvant 2 was fitted with a wide EVA airlock, the MMU, scientific equipment, a remote-sensing camera and life support systems, including a shower.

On 9 January, Viktorenko and Serebrov made an EVA lasting 2 hours 56 minutes, using the Mir core module airlock, and moving as far as the rear of Kvant to install two 80 kg (176 lb) star sensors. A second EVA on 11 January, lasting 2 hours 54 minutes, made alterations to the docking port on Mir to receive the "T" module which was due in March. They also retrieved some materials left outside by the TM7 EVA cosmonauts. A third walk was completed on 26 January, and was the first to use the new wider airlock on Kvant 2. During the 3 hour 2 minute EVA, Viktorenko and Serebrov assembled a magnetic device on the outside of Kvant 2 on which the Icarus MMU could be placed during later planned spacewalks. The first of these came on 1 February, when Serebrov became the first Soviet to operate an MMU – which was tethered to Mir – moving as far as 33 m (108 ft) from the station. This 4 hour 59 minute EVA was followed on 5 February by a 3 hour 45 minute effort, during which Viktorenko also had a go on the MMU, extending the distance from Mir to 45 m (147 ft). Icarus weighed 220 kg (485 lb) and was powered by 32 compressed air thrusters, 16 of which were primary thrusters. Maximum speed available was 30 m/sec (98 ft/sec).

After cramming in the five spacewalks, which lasted a total of over 17 hours, the cosmonauts were joined by Anatoly Solovyov and Aleksandr Balandin in Soyuz TM9 on 13 February, and prepared to come home. They landed in biting winds and temperatures of $-30°C$ 55 km (34 miles) northeast of Arkalyk at $T + 166$ days 6 hours 58 minutes, to a Soviet Union and Soviet bloc that had changed dramatically in such a relatively short time. This was the seventh longest manned space flight.

Milestones

127th manned space flight
67th Soviet manned space flight
60th Soyuz manned space flight
7th Soyuz TM manned space flight
15th Soviet and 38th flight with EVA operations
1st Soviet manned flight to operate commercial US experiments
1st Soviet test of MMU (tethered)
Serebrov celebrates his 46th birthday in space (15 Feb)

STS-34

Int. Designation	1989-084A
Launched	18 October 1989
Launch Site	Pad 39B, Kennedy Space Center, Florida
Landed	23 October 1989
Landing Site	Runway 23, Edwards Air Force Base, California
Launch Vehicle	OV-104 Atlantis/ET-27/SRB BI-032/SSME #1 2027; #2 2030; #3 2029
Duration	4 days 23 hrs 39 min 21 sec
Callsign	Atlantis
Objective	Galileo Jupiter probe deployment mission

Flight Crew

WILLIAMS, Donald Edward, 47, USN, commander, 2nd mission
Previous mission: STS 51-D (1985)
MCCULLEY, Michael James, 46, USN, pilot
LUCID, Shannon Wells, 46, civilian, mission specialist 1, 2nd mission
Previous mission: STS 51-G (1985)
CHANG-DIAZ, Franklin Ramon, 39, civilian, mission specialist 2, 2nd mission
Previous mission: STS 61-C (1986)
BAKER, Ellen Schulman, 36, civilian, mission specialist 3

Flight Log

During an extraordinary week in May 1986, two Space Shuttles were to have taken off from both launch pads at the KSC, carrying their Ulysses and Galileo deep space probes, both atop liquid-fuelled Centaur G Prime upper stages. The prospect caused such concern among the astronauts, even before the Challenger disaster, that the flights were dubbed the "Death Star" missions after the *Star Wars* films. The Challenger accident put paid to such ambitious NASA flight rate plans and resulted in the elimination of the Centaur stage from Shuttles. Galileo and Ulysses were to fly the less powerful solid-propellant IUS upper stages and, instead of taking off six days apart, were to be launched in October 1989 and October 1990 respectively.

Galileo was assigned to the orbiter Atlantis and mission STS-34 with a launch window starting on 12 October and extending well into November if necessary. NASA aimed hopefully for 12 October, having successfully avoided returning Atlantis to the VAB to escape Hurricane Hugo by a whisker. The astronaut crew, including Shannon Lucid and Ellen Baker, the first two-female crew to be launched since October 1984, arrived in good spirits at the KSC on 9 October. They were not in such good spirits flying back to Houston the following day after the launch was cancelled due to a fault in SSME No.2's main events controller, which had to be replaced.

Galileo is deployed from Atlantis atop its IUS to start its long journey to Jupiter

Atlantis was rescheduled for 17 October and a tight 24 minute planetary launch window. While the crew stared into blue skies, the RTLS Shuttle runway was threatened by a thundercloud and the transatlantic abort sites were also suffering inclement weather. Commander Don Williams took his crew back to their quarters and an attempt the following day seemed destined for more of the same, such were the forecasts. A launch hold was ordered because the orbiter computers had to be reprogrammed for a potential TAL landing at a secondary site due to bad weather at the first, while at the KSC, the weather was perfect for the launch at 12:53 hrs local time, boosted by the first SRBs to use components from previous Shuttle missions since the Challenger disaster.

Once Atlantis had reached its unique 34.30° inclination, with a maximum altitude of 286 km (178 miles), Pad 39B was almost obliterated by thunderstorms. At $T + 6$ hours 21 minutes, Galileo began its journey to Jupiter, leaving the payload bay on its IUS. It would go backwards to Jupiter via Venus and two Earth fly-bys, one as close as 900 km (559 miles) in December 1990, taking six years to reach its destination.

The flawless departure from the Shuttle, like Magellan's the previous April, did much to boost NASA's morale, particularly in light of the criticism of flying these payloads on the Shuttle rather than an ELV and the particular storm in a teacup which brewed over the fact that environmentalists had been concerned about Galileo's radioactive RTG power system. Such was the ridiculous media hype that local residents near the Cape were led to believe that, had Atlantis gone the same way as Challenger, the effect would have been a nuclear explosion. Adding to the drama was the fact that the IUS control room at a US Air Force facility in Sunnyvale, California was hit by an earthquake just before launch and personnel, some of whom had lost their houses and belongings, were controlling the IUS amid dust and more than a little rubble.

The crew then got to work with film-making using the IMAX camera, medical experiments under the watchful eyes of Dr. Baker, and remote-sensing photography from the unique orbit. Atlantis made a picture perfect touchdown on the dry lake bed runway 23 at Edwards Air Force Base at $T + 4$ days 23 hours 39 minutes 24 seconds. By the time Galileo soared past Venus in February 1990, two Shuttle flights later, STS-34 was but a statistic on the flight manifest.

Galileo had a long and troubled journey to Jupiter and at times it looked as though the spacecraft might not succeed at all thanks to difficulties with unfurling the antenna. Finally, on 7 December 1995, the probe entered Jovian orbit. The probe had flown past Venus on 10 February 1990, Earth and the Moon on 8 December 1990, and the asteroid Gaspra on 29 October 1991, before returning to the vicinity of Earth and the Moon for a second time on 8 December 1992, passing the asteroid Ida on 28 August 1993, and taking historic images of Comet Shoemaker–Levy striking Jupiter during July 1994. The probe was released on 13 July 1995 and entered the upper atmospheric cloud layers of the planet on 7 December 1995, the same day that the main spacecraft entered orbit. The primary mission (orbits 1–11) ran from January 1996 through December 1997. This was followed by the Europa phase between December 1997 and December 1999 (orbits 12–25), and finally an extended mission phase. A decade after leaving Earth, the Galileo mission continued to rewrite the text books on the largest planet in our solar system. Mostly forgotten was the short mission of STS-34, another mission that blended unmanned and manned space exploration into one programme.

Milestones

128th manned space flight
61st US manned space flight
31st Shuttle mission
5th flight of Atlantis
2nd Shuttle planetary deployment mission

STS-33

Int. Designation	1989-090A
Launched	22 November 1989
Launch Site	Pad 39B, Kennedy Space Center, Florida
Landed	27 November 1989
Landing Site	Runway 04, Edwards Air Force Base, California
Launch Vehicle	OV-103 Discovery/ET-38/SRB BI-034/SSME #1 2011; #2 2031; #3 2107
Duration	5 days 0 hrs 6 min 48 sec
Callsign	Discovery
Objective	5th classified DoD Shuttle mission

Flight Crew

GREGORY, Frederick Drew, 48, USAF, commander, 2nd mission
Previous mission: STS 51-B (1985)
BLAHA, John Elmer, 47, USAF, pilot, 2nd mission
Previous mission: STS-29 (1989)
CARTER Jr., Manley Lanier "Sonny", 42, USN, mission specialist 1
MUSGRAVE, Franklin Story, 54, civilian, mission specialist 2, 3rd mission
Previous missions: STS-6 (1983); STS 51-F (1985)
THORNTON, Kathryn Cordell Ryan, 37, mission specialist 3

Flight Log

With STS-33 flying after STS-34 but before STS-32, it was understandably difficult to keep track of how many Space Shuttles had been launched by November 1989. STS-33 Discovery was the thirty-second Space Shuttle mission but only the thirty-first to reach space. It was due to launch in August 1989 but had to make way for the delayed STS-28 and for the STS-34 planetary launch window. It was also the first Shuttle to fly with a crewman replacing one who had died. Pilot David Griggs had been killed in the crash of an aerobatics plane in June 1989 and was replaced on the mission by John Blaha, recently returned from STS-29, who was thus making a second flight in a record seven months.

The mission was unusual in that while it was a classified military affair, it carried two civilian crew persons, mission specialists Kathryn Thornton on her first flight, and the veteran Story Musgrave on his third (though both had previous experience in classified roles. Thornton had worked with the Army Foreign Science and Technology Center before being selected for astronaut training and Musgrave had served in the USMC in the 1950s). The fact that the third mission specialist, Manley Carter, was a doctor like Musgrave and that Thornton was a nuclear physicist indicated that several biomedical–radiation crew experiments were on the schedule after the deployment of

Carter and Thornton display a slogan for Astronaut Group 10, "The Maggots". This was the unofficial nickname for the group which came from their self-professed love of food

the main payload. This was a SIGINT electronic signals intelligence satellite, ELINT, deployed in the 28° inclination, 561 km (249 miles) apogee orbit.

Discovery was raring to go on the first attempt and was held for just 90 seconds at $T-5$ minutes before making a spectacular departure from Pad 39B at 19:23 hrs local time, turning night into day and the quiet peace of the Cape's lagoons into a frightening cacophony. The SIGINT was deployed on orbit seven. The flight was due to last four days but was extended for almost a day by excessive winds at Edwards, where it was to have made the first night landing since the Challenger accident. Discovery was waived off again by one orbit and after its long re-entry from high orbit was diverted from the concrete runway to runway 04 at Edwards, landing at $T+5$ days 0 hours 6 minutes 48 seconds.

Milestones

129th manned space flight
62nd US manned space flight
1st military manned flight with "civilian" and female crew
32nd Shuttle mission
9th flight of Discovery
5th classified DoD Shuttle mission

STS-32

Int. Designation	1990-002A
Launched	9 January 1990
Launch Site	Pad 39A, Kennedy Space Center, Florida
Landed 20	January 1990
Landing Site	Runway 22, Edwards Air Force Base, California
Launch Vehicle	OV-102 Columbia/ET-32/SRB BI-035/SSME #1 2024; #2 2022; #3 2028
Duration	10 days 21 hrs 0 min 36 sec
Callsign	Columbia
Objective	Satellite deployment and LDEF retrieval mission

Flight Crew

BRANDENSTEIN, Daniel Charles, 46, USN, commander, 3rd mission
Previous missions: STS-8 (1983); STS 51-G (1985)
WETHERBEE, James Donald, 37, USN, pilot
DUNBAR, Bonnie Jean, 40, civilian, mission specialist 1, 2nd mission
Previous mission: STS 61-A (1985)
IVINS, Marsha Sue, 38, civilian, mission specialist 2
LOW, George David, 33, civilian, mission specialist 3

Flight Log

When the Long Duration Exposure Facility (LDEF) was deployed in 1984, the plan was that it would be retrieved the following year. The NASA Space Shuttle manifest got itself into a real pickle under pressure from all directions and had to push the LDEF retrieval mission into September 1986. That would have been flight STS 61-L, commanded by Don Williams, piloted by Mike Smith – who was also assigned to 51-L Challenger – and with mission specialists Bonnie Dunbar, James Bagian and Manley Carter. After the Shuttle programme had recovered from the Challenger accident, the LDEF retrieval mission was assigned to STS-32 with the lone survivor from 61-L, Bonnie Dunbar. The commander of what was going to be one of the more high-profile Shuttle missions was the new chief of the astronauts, Dan Brandenstein.

STS-32 was subject to several delays, partly due to the longer time in getting the orbiter Columbia spaceworthy. Eventually, Columbia was rolled out to Pad 39A just after the launch of STS-33 and would be the first Shuttle to take off from this refurbished pad since STS 61-C in January 1986. It was set for a mammoth ten-day mission, starting on 18 December and taking in a Christmas in space, but problems bringing the new pad on line for launches meant a delay first to 21 December, then for three weeks to 8 January. NASA felt it prudent to give the launch and support teams a full holiday.

STS-32 retrieves LDEF after almost six years in space

As the crew left their quarters on 8 January, they knew they would be coming back the same day because the weather gave them less than a ten per cent chance of taking off. Going through a full countdown to $T-5$ minutes, however, provided a good opportunity to give Pad 39A a full workout. The following day, Columbia took off at 07:35 hrs local time, featuring in one of the most beautiful lift-offs of a Shuttle, making a direct insertion burn to 28.5° orbit. On day two, the Shuttle's major payload on the upward journey, Syncom IV, or Leasat 5, was deployed, and Columbia sailed on towards its dramatic meeting with the LDEF. There was a serious water leak on the third day, involving the collection of two gallons of water globules.

The complicated LDEF rendezvous was completed on the fourth day, 12 January, when Columbia flew towards, over and down to the facility, with its payload bay doors opening towards the Earth, waiting to receive. While Brandenstein deftly manoeuvred the Shuttle as it had never been manoeuvred before, Dunbar got ready with the RMS robot arm, which she was operating using a monitor showing scenes from the TV camera at its end. Brandenstein stopped all motion and, as rehearsed hundreds of times, Dunbar made the great space catch. As pilot Jim Wetherbee flew Columbia belly first, the LDEF was manoeuvred into several positions while the other mission specialists, David Low and Marsha Ivins, took close up photographs of every part, just in case the LDEF could not be safely secured in the payload bay and had to be left in space. Following the style of the mission, LDEF was berthed in the payload bay later, after 2,093 days autonomous flying in space, pitted, torn and worn. Columbia continued on its winning way, with the crew busying themselves with an array of science experiments, a range of medical experiments under the Extended-Duration Orbiter Medical Programme (EDOMP) and Dunbar getting the news that her husband (Ronald Sega) had been selected for astronaut training.

The landing on the ninth mission day was called off by a failure of one of the suite of five computers on board, and as a result, Columbia returned to Edwards Air Force Base on runway 22 at night, and after a Shuttle-record mission lasting 10 days 21 hours 0 minutes 36 seconds – the longest five-crew space flight, and with the heaviest landing weight of 103,572 kg (228,376 lb). STS-32 was probably the most complicated space flying mission and certainly the most successful and rewarding, as scientists pored over the LDEF to see how its time in space had affected its array of different materials.

Milestones

130th manned space flight
63rd US manned space flight
33rd Shuttle mission
9th flight of Columbia
Brandenstein celebrates his 47th birthday in space (17 Jan)

SOYUZ TM9

Int. Designation	1990-014A
Launched	11 February 1990
Launch Site	Pad 1, Site 5, Baikonur Cosmodrome, Kazakhstan
Landed	9 August 1990
Landing Site	70 km from Arkalyk
Launch Vehicle	R7 (11A511U2); spacecraft serial number (7K-M) #60
Duration	179 days 1 hrs 17 min 57 sec
Callsign	Rodnik (Spring)
Objective	Mir EO-6 research programme

Flight Crew

SOLOVYOV, Anatoly Yakovlovich, 42, Soviet Air Force, commander, 2nd mission
Previous mission: Soyuz TM5 (1988)
BALANDIN, Aleksandr Nikolayevich, 36, civilian, flight engineer

Flight Log

What was planned as a now-standard five month residency aboard the Mir complex began at 07:16 hrs local time at Baikonur on 11 February, when Soyuz TM9 lifted off, watched by US astronaut guests Dan Brandenstein, Paul Weitz, Ron Grabe and Jerry Ross. Docking was completed two days later and, yet again, was a manual affair, with the automatic approach malfunctioning at the last moment. The TM9 cosmonauts, Anatoly Solovyov and Aleksandr Balandin, joined Aleksandrs Viktorenko and Serebrov for the traditional handover period. The TM9 residency began officially on 19 February and was due to last until 30 July, following the 22 July launch of Soyuz TM10.

The TM9 crew were expected to receive the second large add-on module, Kristall, in April and begin an intensive programme of materials processing, so that they could return to Earth with 100 kg (221 lb) of space products to make a profit of 25 million roubles from the 80 million rouble space flight. Thus, the space flight could be seen as actually contributing to the economy and not as wasteful and extravagant as it was regarded by much of the Soviet public.

As Soyuz TM9 approached Mir, TV pictures, seen on the national news, revealed that the thermal insulation blankets around the flight cabin had become unclipped. The Soviets routinely announced that at some time during the mission the crew would have to make an unscheduled spacewalk to clip them back on. No fuss was made of the event. After settling into the routine of life aboard Mir, TM8 cosmonauts Viktorenko and Serebrov left them to it, and the routine continued with the docking of the Progress M3 supply ship on 3 March.

Solovyov (right) and Balandin reviewing EVA equipment and hardware during training

The mission proceeded very quietly, but the scheduled launch date for Kristall passed before the Soviets announced that the new module had been delayed yet again, this time until June. The crew which had trained especially to operate Kristall would only have about two months to do so, rather than the planned four months. Progress 42, the last of the original spacecraft first launched in 1978, docked to Mir on 8 May and later in the month, the most bizarre case of inaccurate and distorted media hype of the space age occurred when *Aviation Week* magazine "discovered" the already three-month-old story of the unclipped insulation, leading the western press to print stories of the cosmonauts being stranded in space. If there had been any danger, the Soviets would have launched an unmanned replacement ferry immediately, rather like they did with Soyuz 34 which replaced Soyuz 32 during the Salyut 6 mission of 1979. The delayed Kristall was at last launched on 31 May but at first failed to dock when a computer fouled up during the final approach. It finally moored at Mir on 10 June.

Because of the delay to the launch of Kristall, the Soviets decided to extend the TM9 mission from 29 July to 9 August and to delay the launch of the replacement TM10 from 22 July to 1 August. On 1 July, Solovyov and Balandin made a 7 hour EVA to clip back the loose insulation on their TM9 ferry. They used the Kvant 2 airlock and while exiting, opened the outer hatch before the airlock had fully depressurised. It flew open with such a force that it almost came off its hinges. Not surprisingly, after their tortuous record-breaking EVA, scrambling over the Soyuz and successfully re-clipping only two of the three insulation panels, the cosmonauts couldn't close the hatch properly and were forced to depressurise the rest of Kvant to gain entry to Mir. Another spacewalk, lasting three hours on 26 July, closed the hatch but did not completely seal it.

It would be left for the TM10 crew to do the necessary repairs. Its cosmonauts, the "two Gennadys", Manakov and Strekalov, arrived on Mir on 3 August, and on 9 August as advertised, Solovyov and Balandin routinely ended their mission, making a mockery of the media hype the previous June. The mission lasted 179 days 2 hours 19 minutes.

Milestones

131st manned space flight
68th Soviet manned space flight
61st Soyuz manned mission
8th Soyuz TM manned mission
16th Soviet and 39th flight with EVA operations
Baladin celebrates his 37th birthday in space (30 Jul)

STS-36

Int. Designation	1990-019A
Launched	28 February 1990
Launch Site	Pad 39A, Kennedy Space Center, Florida
Landed	4 March 1990
Landing Site	Runway 23, Edwards Air Force Base, California
Launch Vehicle	OV-104 Atlantis/ET-33/SRB BI-036/SSME #1 2019; #2 2030; #3 2027
Duration	4 days 10 hrs 18 min 22 sec
Callsign	Atlantis
Objective	6th classified DoD shuttle mission

Flight Crew

CREIGHTON, John Oliver, 45, USN, commander, 2nd mission
Previous mission: STS 51-G (1985)
CASPER, John Howard, 46, USAF, pilot
MULLANE, Richard Michael, 45, USAF, mission specialist 1, 3rd mission
Previous missions: STS 41-D (1984); STS-27 (1988)
HILMERS, David Carl, 40, USMC, mission specialist 2, 3rd mission
Previous missions: STS 51-J (1985); STS-26 (1988)
THUOT, Pierre Joseph, 34, civilian, mission specialist 3

Flight Log

This DoD classified military mission by the orbiter Atlantis was always going to be a quiet affair, other than the usual comical revelation of exactly what the classified payload was going to be. In this case, it was a digital imaging and electronic signals intelligence satellite, which was to be deployed on the Shuttle's eighteenth orbit, comparatively late in the proposed four-day mission, by a new system called the Stabilised Payload Deployment System, SPDS. This was fixed to the payload in the payload bay before launch and was to be used to rotate the satellite clear of the Shuttle before release by spring-loaded pistons.

Another innovation for the mission was its high inclination of 62°, ostensibly 5° over the safety limits for launches from the Kennedy Space Center, but on a trajectory which would not quite take it over land. As if to veil this fact as much as possible, Atlantis was first scheduled for a night launch on 16 February, which was eventually moved to 22 February, at 01:00 hrs local time. Before the crew could board the Shuttle, however, the weather caused concern and for the first time since Apollo 9, a US mission was delayed by the illness of the crew. In this case, it was commander John Creighton, who was suffering from an upper respiratory tract infection.

Commander Creighton photographs views out of the overhead windows of Atlantis

His illness delayed the launch until 25 February and the count reached $T - 31$ seconds when a range safety computer went on the blink. By the time it had been fixed, the liquid oxygen was too cold for the SSMEs. The launch was cancelled the following day due to winds at altitude and a low cloud deck and was routinely postponed for 48 hours. On 28 February, the weather looked to win again, but with just seconds of the launch window remaining, Atlantis lit up the night sky at 02:50 hrs local time, heading for its unique launch azimuth. Deployment of the classified satellite took place as planned and the mission ended quietly with a landing at Edwards Air Force Base at $T + 4$ days 10 hours 18 minutes 22 seconds.

It was revealed later that the satellite had apparently broken apart in orbit and two of the six resulting fragments had re-entered soon after. The satellite may have failed to fire its rocket stage to reach operational orbit. All in all, it seems that STS-36 was an expensive waste of a mission.

Milestones

132nd manned space flight
64th US manned space flight
34th Shuttle flight
6th flight of Atlantis

STS-31

Int. Designation	1990-037A
Launched	24 April 1990
Launch Site	Pad 39B, Kennedy Space Center, Florida
Landed	29 April 1990
Landing Site	Runway 22, Edwards AFB, California
Launch Vehicle	OV-103 Discovery/ET-34/SRB BI-037/SSME #1 2011 #2 2031 #3 2107
Duration	5 days 1 hr 16 min 6 sec
Call sign	Discovery
Objective	Deployment of the Hubble Space Telescope facility

Flight Crew

SHRIVER, Loren James, 46, USAF, commander, 2nd mission
Previous mission: STS-51C (1985)
BOLDEN Jr., Charles Frank, 44, USMC, pilot, 2nd mission
Previous mission: STS 61-C (1986)
McCANDLESS II, Bruce, 53, USN, mission specialist 1, 2nd mission
Previous mission: STS 41-B (1984)
HAWLEY, Steven Alan, 39, mission specialist 2, 3rd mission
Previous missions: STS 41-D (1984); STS 61-C (1986)
SULLIVAN, Kathryn Dwyer, 39, mission specialist 3, 2nd mission
Previous mission: STS 41-G (1984)

Flight Log

The launch of the Hubble Space Telescope (HST) deployment mission was originally set for 18 August, but was moved up to 12 April, then 10 April, following the Flight Readiness Review. This was the first time a Shuttle launch had been advanced following the FRR and not put back. However, the 10 April attempt was scrubbed at $T - 4$ minutes due to a faulty valve in APU #1. The battery was replaced and the payload batteries on Hubble were recharged. On 24 April, the count was briefly halted at $T - 31$ seconds when a fuel valve line failed to shut. This was soon traced to a software failure and was overridden by engineers, allowing the count to continue.

Following a nominal ascent, most of the rest of FD 1 was spent preparing for the deployment of the telescope, which included powering up the RMS 2 hours 54 minutes into the flight. The cabin pressure was lowered in order to reduce the time the EVA crew of McCandless and Sullivan would need to pre-breathe pure oxygen should a contingency EVA be required. About 4.5 hours into the mission the umbilical power connection to the telescope was activated. The next day, the two spacewalkers got themselves partially dressed in their coolant garments, to save time should they need

The Hubble Space Telescope, still in the grasp of the RMS, is back-dropped over Cuba and the Bahamas. The solar arrays and high-gain antenna have yet to be deployed. The EVA handrails to support future Shuttle service missions are clearly visible across the main structure of the telescope

to exit the airlock in support of HST deployment. Steve Hawley lifted the telescope out of the payload bay using the RMS. Once the end effector had grasped the starboard grapple fixture of the telescope, the five latches that restrained Hubble in the bay were released.

With the telescope out of the payload bay, its solar arrays were deployed. There was some concern early in the process when they became stuck, and at one point, it looked as though the EVA crew would have to go out and assist in the unfurling of the arrays. Eventually, by disengaging the tension warning system, the arrays unfurled to their full length. Nine hours after lifting the telescope out of the bay, Hawley released it from the grip of the RMS. Discovery then completed two separation burns to move away from the telescope. Until the RMS was stowed, the EVA crew remained in the airlock in case they were required to manually retract the arm for entry and landing.

Following the release of the telescope, the crew focused on their programme of secondary and mid-deck experiments, which included monitoring particles in the payload bay, a protein crystal growth experiment, radiation-monitoring equipment, polymer membrane processing and a student experiment to determine the effects of microgravity on electrical arcs. From their 600 km altitude vantage point, the crew also recorded spectacular images of the Earth. This was the highest apogee in the programme to date, and only Gemini 10 and 11 in 1966 and the nine Apollo lunar missions had ever taken astronauts higher. The IMAX camera was flown to record mission events from outside the crew compartment and a hand-held IMAX captured images from inside the flight and mid-deck. Sequences from STS-31 footage were later used in the IMAX movie presentation *The Blue Planet* in IMAX theatres. On FD 4, HST controllers managed to open the aperture door of the telescope and, with the astronauts no longer required to support the telescope, the crew turned their attention to preparations for landing on FD 6. Over the coming weeks, the telescope was checked out in orbit. Unfortunately, about two months after its deployment, it became apparent that the mirror on the telescope was not focusing as designed due to a production error. It was decided that a set of corrective optics would have to be developed and then installed, during the first scheduled servicing mission in 1993. This, however, was not the fault of the astronauts or the mission of STS-31, which was a complete success.

Milestones

133rd manned space flight
65th US manned space flight
35th Shuttle mission
10th flight of OV-103 Discovery
1st use of carbon brakes at landing
1st launch set earlier than planning following FRR
Highest orbit in Shuttle programme to date (600 km)

SOYUZ TM10

Int. Designation	1990-067A
Launched	1 August 1990
Launch Site	Pad 1, Baikonur cosmodrome, Kazakhstan
Landed	10 December 1990
Landing Site	69 km NW of Arkalyk
Launch Vehicle	R7 (11A511U2); spacecraft serial number 061A
Duration	130 days 20 hrs 35 min 51 sec
Call sign	Vulkan (Volcano)
Objective	Delivery of the seventh resident crew (EO-7) to the Mir space station; rewiring of base block; repair of Kvant EVA hatch

Flight Crew

MANAKOV, Gennady Mikhailovich, 40, Soviet Air Force, commander
STREKALOV, Gennady Mikhailovich, 49, civilian, flight engineer, 4th mission
Previous missions: Soyuz T3 (1980), Soyuz T8 (1983), Soyuz T10-1 (1983); Soyuz T11 (1984)

Flight Log

The two Gennadys comprised the seventh Mir resident crew and were launched with four live Japanese quails for the Inkubator 2 experiment on board Mir. They would be used in "adaptation to weightlessness" experiments. During the two-day flight to Mir, one of the older quails "laid" an egg and this was returned to Earth with the TM9 crew. The TM10 spacecraft docked with the rear port of Kvant 1 on 3 August. Following the period of handover from the TM9 crew, which included a rather extensive review of where everything was stored, the new crew had a relatively quiet residency aboard the station. Their mission had been delayed ten days to allow the Mir-6 crew to complete their commissioning of systems aboard the Kristall module.

During their residency aboard Mir, Manakov and Strekalov had the primary engineering task of rewiring the base block's power supply, as well as attempting to repair the Kvant 2 EVA hatch that had been damaged during the previous mission. They would also continue the wide programme of scientific work aboard the complex. After a long and frustrating wait, Strekalov finally achieved his goal of a long-duration mission, having previously flown to Salyut 6 and 7 on short visiting missions. He was also a hardened veteran space explorer, having been a crew member of the 1983 Soyuz T8 docking abort and the T10-1 launch pad abort. In boarding Mir, he became one of the first cosmonauts to visit three separate space stations. The only EVA of the mission had been planned for 19 October, but Strekalov developed a head cold, delaying it until 30 October. When the two cosmonauts inspected the damaged

The crew of Soyuz TM10: Manakov (left) and Strekalov

hinge plate they were scheduled to replace, it was found to be deformed beyond repair. Instead, they installed a special latch to ensure that the hatch could be closed and used until fully repaired. With the repair task not deemed to be urgent, it was deferred to the next resident crew, who would fit a replacement unit. The EVA lasted 3 hours 45 minutes.

During this mission, the station was supplied by two Progress cargo craft. Progress M4 docked on 17 August, delivering power cables and TV equipment for the upcoming Japanese commercial mission. Before it was undocked, the crew attached a small experiment to the docking assembly, which was activated on 17 September when the ferry undocked. During station keeping, about 100 metres from Mir, artificial plasma was created around the Progress and this was filmed by the cosmonauts. Progress M5, which arrived on 29 September, carried more TV equipment for the Japanese mission. It also featured the first Raduga recoverable capsule system that could return about 150 kg of experiment material, the trade-off being a reduction in the cargo capacity of the vehicle. The Raduga capsule featured a truncated cone that would eject from the descending Orbital Module at about 120 km, just prior to the module's fiery destruction in the atmosphere. At 15,000 metres, air pressure sensors successfully triggered the parachute deployment and Raduga was successfully retrieved. It returned 115 kg of payload.

The 7th expedition completed their mission on 10 December, returning to Earth in the TM10 capsule along with the first Japanese cosmonaut, TV journalist Toyohiro Akiyama, who had arrived with the 8th expedition crew in Soyuz TM11 on 4 December.

Milestones

134th manned space flight
69th Soviet/Russian manned space flight
62nd Soyuz manned space flight
9th Soyuz TM manned space flight
17th Soviet and 40th flight with EVA operations
7th Mir resident crew
1st use of the Raduga return capsule
Strekalov celebrates his 50th birthday in orbit (28 October)

STS-41

Int. Designation	1990-090A
Launched	6 October 1990
Launch Site	Pad 39B, Kennedy Space Center, Florida
Landed	10 October 1990
Landing Site	Runway 22, Edwards AFB, California
Launch Vehicle	OV-103 Discovery/ET-32/SRB BI-040/SSME #1 2011; #2 2031; #3 2107
Duration	4 days 2 hrs 10 min 4 sec
Call sign	Discovery
Objective	Deployment of Ulysses solar polar probe by IUS-17/PAM-S upper stages; secondary payload bay experiments included Shuttle Solar Backscatter Ultraviolet hardware; Intelsat Solar Array Coupon

Flight Crew

RICHARDS, Richard Noel, 44, USN, commander, 2nd mission
Previous mission: STS-28 (1989)
CABANA, Robert Donald, 41, USMC, pilot
MELNICK, Bruce Edward, 40, USCG, mission specialist 1
SHEPHERD, William McMichael, 41, USN, mission specialist 2, 2nd mission
Previous mission: STS-27 (1988)
AKERS, Thomas Dale, 39, USAF, mission specialist 3

Flight Log

Originally intended to be deployed from Challenger by the liquid-fuelled Centaur upper stage during the STS 61-F mission in May 1986, the joint NASA/ESA Ulysses solar polar probe mission was delayed by the loss of Challenger in the STS 51-L accident of January 1986. The decision not to fly Centaur stages on the Shuttle over safety concerns and to use the IUS/PAM upper stages instead meant that Ulysses would miss the 1986 launch window. It soon became clear that the Shuttle would not be ready for the June 1987 window and, to ease the 1989 launch schedule, NASA rescheduled the mission to October 1990. Difficulties with the leaking propulsion systems on Atlantis and Columbia during the summer of 1990 placed added pressure to launch STS-41 on time but, despite three short delays due to ground equipment and the weather problems, STS-41 finally left the ground just 12 minutes into the 2.5 hour window.

The crew successfully deployed the IUS combination carrying Ulysses just 6 hours 1 minute 42 seconds after leaving the launch pad. Following the deployment of their primary payload, the crew of STS-41 concentrated on the variety of mid-deck and

Ulysses atop of the IUS/PAM-S upper stages is back-dropped against the blackness of deep space at the start of its five-year mission to the Sun

payload bay experiments for the remainder of their short mission. Though the flight of STS-41 lasted only just over 4 days and is one of the shortest missions in the programme, the primary payload mission has lasted much longer. After more than 16 years in space, the Ulysses probe continues to function, transmitting important solar and interplanetary data back to Earth. To a degree, therefore, the "mission" of STS-41 continues.

Just over an hour after the deployment, the first stage of the IUS burned for 110 seconds, boosting the spacecraft from 29,237 kph to 36,283 kph. The second stage burned for 106 seconds, further increasing the speed to 41,158 kph, before the PAM-S fired for 88 seconds, resulting in a speed of 54,915 kph. Ten minutes later, the spacecraft was separated from the upper stage to begin its long flight towards the Sun via Jupiter. The probe made its 375 km closest approach to Jupiter on 8 February 1992. Its first southern polar zone pass between 26 June and 6 November reached 80°S (13 September). Its first northern polar pass occurred between 19 June and 30 September 1995 and saw the official completion of its primary mission. Its closest approach at 1.34 AU occurred on 12 March 1995. It took almost five years from launch to the second polar pass, though it took only 8 hours to journey the 382,942 km from Earth to the orbit of the Moon, a trip that took Apollo astronauts three days to complete. Ulysses completed its second pass of both poles in 2001. Its third southern polar pass is planned for 2006/2007 and its third northern polar pass for 2007/2008.

Milestones

135th manned space flight
66th US manned space flight
36th Shuttle flight
11th Discovery flight
3rd Shuttle solar system deployment mission
1st three stage IUS deployment mission
1st solar polar probe
1st US Coast Guard officer (Melnick) to fly in space

STS-38

Int. Designation	1990-097A
Launched	15 November 1990
Launch Site	Pad 39A, Kennedy Space Center, Florida
Landed	20 November 1990
Landing Site	Runway 33, Shuttle Landing Facility, KSC, Florida
Launch Vehicle	OV-104 Atlantis/ET-40/SRB B-039/SSME #1 2019; #2 2022; #3 2017
Duration	4 days 21 hrs 54 min 31 sec
Call sign	Atlantis
Objective	7th dedicated classified DoD mission

Flight Crew

COVEY, Richard Oswalt, 44, USAF, commander, 3rd mission
Previous missions: STS 51-I (1985), STS-26 (1988)
CULBERTSON Jr., Frank Lee, 41, USN, pilot
MEADE, Carl Joseph, 40, USAF, mission specialist 1
SPRINGER, Robert Clyde, 48, USMC, mission specialist 2, 2nd mission
Previous mission: STS-29 (1989)
GEMAR, Charles Donald "Sam", 35, US Army, mission specialist 3

Flight Log

Originally scheduled for launch in July 1990, when a liquid hydrogen leak was found on Columbia (STS-35), three precautionary mini tanking tests on Atlantis also confirmed hydrogen leaks on its ET. These could not be repaired on the pad, and the stack was returned to the processing area for repairs on 9 August. The STS-38 stack was parked outside the VAB overnight to allow STS-35 to be rolled out to its pad. Unfortunately, a hail storm that night caused minor tile damage which also needed repairing. Atlantis was returned to the VAB for mating on 2 October, but during hoisting operations a platform beam that should have been removed from the aft compartment fell off, causing more (but thankfully minor) damage, which was quickly repaired. The stack was returned to the Pad on 12 October and a fourth tanking test went smoothly. The revised launch date was set at 9 November. However, on 31 October, the USAF announced another delay to the launch, this time due to "anomalies discovered during cargo testing." The night-time launch was rescheduled for 15 November and this time occurred without incident despite some concerns with the weather at the Cape. Atlantis lifted off 18 minutes into its four-hour launch window.

As this was a classified DoD mission, the air-to-ground communications and reporting of crew activities and mission events ceased after confirmation that Atlantis

A happy crew indicate a successful conclusion to the mission shortly after exiting Atlantis. L to r Covey, Springer, Gemar, Culbertson and Meade

had safely reached orbit, but this did not stop the speculation as to what the mission of STS-38 was intended to achieve. Media reports indicated that the payload bay was full of sensors, including high-resolution digital cameras that might be used to monitor activities in the Persian Gulf, particularly the Iraqi invasion of Kuwait that led to the First Gulf War. However, a USAF spokesman indicated that Atlantis had launched into an orbital inclination that would take the Shuttle well south of Iraq "for much of the time." This gave rise to comments that the payload might be an electronic eavesdropping satellite called "Magnum", rather than a photoreconnaissance satellite. The deployment of the payload could have been at any time during the orbital phase, but media reports indicated that this operation had been carried out two days into the mission. The payload was later identified as an advanced data relay satellite for use with the Crystal imaging reconnaissance platform.

Whatever it was, its deployment from the orbiter was monitored by amateur astronomers on Earth. Their reports indicated that the deployed payload was behaving very mysteriously in ways never seen before, suggesting that the satellite might have malfunctioned and that the Shuttle crew might have been required to retrieve it. NASA and the Air Force remained silent, which only served to fuel speculation that there was perhaps an unannounced EVA by the crew. Records have shown that the RMS was carried on classified missions STS 51-C, STS-27, and the later unclassified DoD mission STS-39, but not on STS-38. Therefore, if the satellite had to be retrieved or attended to, the lack of RMS meant that the only other option was a contingency EVA. If such an event had occurred – and there is still no evidence that an EVA was accomplished – astronauts Springer (EV1) and Meade (EV2) would have been assigned the task, supported by Culbertson (IV). A year after the mission, an issue of *Space News* dated 18–24 November 1991 included an interview with Don Stager, the Vice President of TRW's military wing. Stager talked about the upcoming deployment of the DSP satellite during STS-44, and indicated that "a couple of [military shuttle] launches ago, there was a situation that was not understood". He indicated that sunlight glinting off the solar arrays had caused a problem, which may have explained the strange movements observed by the amateur astronomers. Exactly what occurred during the deployment sequence will remain classified for many years to come, however.

Although the activities of the crew were classified, at least one voice message from Atlantis was released. Commander Dick Covey requested that a message of support be sent to the men and women of Desert Shield from the crew of Atlantis. The crew wished them peace and a speedy return home. The astronauts were thinking of them and their families as they orbited the Earth.

The landing was intended to be at Edwards AFB, but unacceptable crosswinds and continuing adverse conditions led to a late decision to delay the landing by 24 hours and take Atlantis back to the Cape. The last landing there, in April 1985 (STS 51-D), led to a landing left of the centreline, locked right side landing brakes and a blown tyre. Because of this, landings at KSC were eschewed in favour of Edwards AFB, whose dry lake bed surfaces surrounding the runways offered more flexibility until improvements were completed at the Cape. This time, however, Atlantis came home without incident, landing on Runway 33 and rolling out about 2,750 metres to wheel stop.

Milestones

136th manned space flight
67th US manned space flight
37th Shuttle flight
7th Atlantis flight
7th and final fully classified DoD Shuttle mission
1st KSC landing for Atlantis
Meade celebrates his 40th birthday in space (16 Nov)

STS-35

Int. Designation	1990-106A
Launched	2 December 1990
Launch Site	Pad 39B, Kennedy Space Center, Florida
Landed	10 December 1990
Landing Site	Runway 22, Edwards AFB, California
Launch Vehicle	OV-102 Columbia/ET-35/SRB BI-038/SSME #1 2024; #2 2012; #3 2028
Duration	8 days 23 hrs 5 min 8 sec
Call sign	Columbia
Objective	Round-the-clock Astro-1 payload operations in the celestial sphere in UV and X-Ray astronomy

Flight Crew

BRAND, Vance DeVoe, 59, civilian, commander, 4th mission
Previous missions: ASTP (1975); STS-5 (1982); STS 41-B (1984)
GARDNER Jr., Guy Spence, 42, USAF, pilot, 2nd mission
Previous mission: STS-27 (1988)
HOFFMAN, Jeffrey Alan, 46, civilian, mission specialist 1, 2nd mission
Previous mission: STS 51-D (1985)
LOUNGE, John Michael "Mike", 44, civilian, mission specialist 2, 2nd mission
Previous mission: STS 51-I (1985)
PARKER, Robert Allan Ridley, 53, civilian, mission specialist 3, 2nd mission
Previous mission: STS-9 (1983)
DURRANCE, Samuel Thornton, 47, civilian, payload specialist 1
PARISE, Ronald Anthony, 39, civilian, payload specialist 2

Flight Log

Astro 1 should have flown in March 1986, the next mission following STS 51-L. Its objective was to combine UV studies with observations of Halley's Comet by the Large Format Camera, but the loss of Challenger put paid these plans and delayed the mission into mid-1990. The Large Format Camera was no longer needed and was replaced by the Broad Band X-Ray Telescope. During the summer of 1990, the launch slipped six months due to several problems, particularly those associated with hydrogen leaks, causing NASA serious manifest, safety and public relations problems. The 16 May launch was delayed due to a faulty Freon coolant loop valve, while the 30 May launch was scrubbed due hydrogen leaks during tanking and forced a rollback to the VAB to complete repairs. While in the VAB, the Astro payload was also serviced. Columbia was rolled back to the pad in August, but two days prior to the 1 September launch date, an avionic box malfunctioned on the Broad Band X-Ray Telescope

Ships that pass in the night. Columbia (left) is rolled past Atlantis (right) on its way to Pad 39A on 9 August 1990. Atlantis was parked in front of the VAB following its rollback from Pad A for repairs to its LH lines

(BBXRT). Then the 6 September launch was scrubbed when high concentrations of hydrogen were again detected in the aft compartment. The 18 September launch was postponed when more leaks were discovered and Columbia was transferred to Pad B on 8 October to make room for Atlantis (STS-38 on Pad A). Finally, a tropical storm (Klaus) forced another rollback to the VAB on 9 October before a final tanking test on 14 October cleared the spacecraft for launch.

The night launch on 2 December came a few hours prior to the launch of the Soyuz TM11 craft from Russia and for a time, there were eleven space explorers in orbit aboard three different spacecraft: the seven astronauts aboard STS-35, two cosmonauts on Mir and three cosmonauts (including the first Japanese in space) on Soyuz TM11. Aboard Columbia the primary controller failed almost as soon as observations started, and the back-up system failed shortly afterwards. The crew reverted to a manual procedure (one they had not trained for) to allow observations to continue. Despite losing the faster automatic system, the crew became quite skilled at aiming the Instrument Pointing System (IPS) towards a target and following step-by-step procedures to operate the telescopes. As if this wasn't enough, the waste water dump system gave further problems due to a clogged drain, but the crew overcame this as well using spare containers. It had been planned to observe 230 objects during the mission and despite all their problems the crew, working in two 12-hour shifts, accomplished 70 per cent of the pre-mission objectives, in spite of the mission also being cut short by one day due to impending bad weather at Edwards AFB.

The Hopkins UV Telescope (HUT) made over 100 observations of 75 hot-stars, galactic nuclei and quasars. The Ultraviolet Imaging Telescope (UIT) collected over 900 images of supernova, planetary nebulae, galaxies and clusters of galaxies. The Wisconsin UV Photo-Polarmeter Experiment (WUPPE) gathered data on over 70 objects including galactic clusters and supernova remnants, and the BBXRT collected data on over 75 objects, including active galactic nuclei, quasars and accretion discs. In addition to live classroom lessons from space with students in Alabama and Maryland, hundreds of ham radio contacts were made using the Shuttle Amateur Radio Experiment SAREX. In all, over 390 observations of 135 high-priority objects were obtained. It is clear that if this had been an unmanned satellite then its mission would have failed. The intervention of the crew to recover from system failures saved the mission and provided scientists with new data that would keep them busy for years. Astro 1 was so successful, despite the delays and setbacks, that the mission helped to secure a second flight of the package of instruments as Astro 2 five years later, although the planned Astro 3 mission did not fly.

Milestones

137th manned space flight
68th US manned space flight
38th Shuttle mission
10th flight of OV-102 Columbia
1st dedicated Shuttle astronomy mission
1st Payload Specialists since STS-51L (1986)
1st crew to include four professional astronomers (Durrance, Hoffman, Parise and Parker)

SOYUZ TM11

Int. Designation	1990-107A
Launched	2 December 1990
Launch Site	Pad 1, Baikonur Cosmodrome, Kazakhstan
Landed	26 May 1991 (Akiyama landed 10 December 1990)
Landing Site	61 km southeast of Dzhestkazgan
Launch Vehicle	R7 (11A511U2); spacecraft serial number 61
Duration	175 days 1 hr 51 min 42 sec
	Akiyama duration: 7 days 21 hrs 54 min 40 sec
Call sign	Derbent (Derbent)
Objective	Mir EO-8 resident crew programme; 1st commercial (Japanese) Soviet manned space flight

Flight Crew

AFANASYEV, Viktor Mikhailovich, 41, Soviet Air Force, commander EO-8
MANAROV, Musa Khiromanovich, 39, civilian, flight engineer EO-8, 2nd mission
Previous mission: Soyuz TM-4/Mir EO-3 (1987)
AKIYAMA, Toyohiro, 48, civilian, research cosmonaut

Flight Log

The eighth Mir resident crew should have been Afanasyev and veteran flight engineer Vitali Sevastyanov, who had flown on Soyuz 9 in 1970 and Soyuz 18-B in 1975. They had trained for a 1989 flight which was cancelled following the decision to leave Mir temporarily unmanned. The two men rotated to back-up Soyuz TM10 but in June 1990, Sevastyanov was medically grounded and was replaced by Manarov.

Professional Japanese journalist Akiyama was TBS's chief foreign correspondent. He spent most of the two-day flight to Mir strapped into his couch suffering from Space Adaptation Syndrome and never fully recovered during his week in space. His TV company had paid $12 million for the mission. He took a significant amount of film and video during his stay on Mir, making live broadcasts (of up to 10 minutes) from space whenever the station passed over Japan. He also performed several Japanese medical and biological experiments. Akiyama landed back on Earth, with much relief, with the EO-7 crew aboard Soyuz TM10 on 10 December, leaving the EO-8 crew to continue their occupancy of Mir. His flight caused resentment among the Soviet media, given that a heavy-smoking and very unfit foreigner had been chosen over one of their own to become the first professional journalist in space. After his mission Akiyama returned to work for TBS, but is now a farmer in a very remote part of Japan.

The Soyuz TM11 crew of Akiyama (left), Afanasyev (centre) and Manarov

The work aboard the Mir during this residency focused on space manufacturing using the Krater V, Gallar and Optizon furnaces. Astronomical observations were accomplished using the Roentgen observatory located in Kvant 1 and part of the Earth resources observation programme focused on activities in the Middle East during the First Gulf War. The crew performed four EVAs during their stay on Mir. The first (7 January 1991) lasted 5 hours 18 minutes and included repairs to the Kvant 2 hatch and setting up support structures for the Strela crane to be used in future transfers of solar arrays. The second (23 January), an EVA that lasted 5 hours 33 minutes, set up the 45 kg jib near the multiple docking adapter. When folded, this measured just 2 metres but when deployed, it stretched for 12 metres and could move 750 kg loads between two points on the left of the station, avoiding any solar arrays. EVA 3 (26 January) lasted 6 hours 20 minutes and saw the framework to accommodate Kristall solar arrays erected on Kvant 1.

On 21 March, during the automatic docking of Progress M7, the Kurs automatic guidance system failed, causing the Progress to miss the station by a mere 457 metres. Two days later, the second attempt almost ended in disaster when the Progress again sailed past the station, but this time at a distance of only 4.6 metres and almost hitting solar panels and aerials on the station. The problem was suspected to be in the Kurs system on the Kvant 1 module of Mir and not on the Progress, so Afanasyev and Manarov undocked the Soyuz from the front port on 26 March and attempted a docking at the aft port, only to find the Kurs system there was at fault. They performed a manual docking successfully and two days later, the Progress docked with the front port of the station. The problem was thought to have been caused by one of the cosmonauts inadvertently knocking the Kurs antenna at the rear port during an earlier EVA. This was confirmed during the 3 hour 34 minute fourth EVA (25 April), which revealed that one of the dishes was missing. This would need to be replaced by the next resident crew. For the time being, the Progress re-supply craft would have to use the forward port.

The EO-9 crew arrived at Mir aboard Soyuz TM12 on 20 May, along with British cosmonaut Helen Sharman. Sharman would spend a week on Mir before returning with the EO-8 crew in Soyuz TM11 on 26 May. The landing of TM11 was hard, as the spacecraft was moved laterally across the ground by a gust of wind, causing it to tip over and roll several times. This caused the crew to lose their sense of direction for a few moments, and was a vivid reminder that the expected "soft-landing" was frequently anything but.

Milestones

138th manned space flight
70th Soviet/Russian manned space flight
63rd Soyuz manned space flight
10th Soyuz TM manned space flight
18th Soviet and 41st flight with EVA operations
1st commercial flight with a paying passenger
1st Japanese in space
1st professional journalist in space (Akiyama)
Manarov exceeds Yuri Romanenko's 430 days cumulative time in space record on 6 February
Afanasyev celebrates his 42nd birthday in space (31 Dec)
Manarov celebrate his 40th birthday in space (22 Mar)

7

The Fourth Decade: 1991–2000

STS-37

Int. Designation	1991-027A
Launched	5 April 1991
Launch Site	Pad 39B, Kennedy Space Center, Florida
Landed	11 April 1991
Landing Site	Runway 33, Edwards Air Force Base, California
Launch Vehicle	OV-104 Atlantis/ET-37/SRB BI-042/SSME #1 2019; #2 2031; #3 2107
Duration	5 days 23 hrs 32 min 44 sec
Call sign	Atlantis
Objective	Deployment of the Gamma Ray Observatory (GRO), the second of NASA's four great observatories; EVA Development Flight Experiments

Flight Crew

NAGEL, Steven Ray, 44, USAF, commander, 3rd mission
Previous missions: STS 51-G (1985); STS 61-A (1985)
CAMERON, Kenneth Donald, 41, USMC, pilot
GODWIN, Linda Maxine, 38, civilian, mission specialist 1
ROSS, Jerry Lynn, 43, USAF, mission specialist 2, 3rd mission
Previous missions: STS 61-B (1985); STS-27 (1988)
APT, Jerome "Jay", 41, civilian, mission specialist 3

Flight Log

Atlantis left the pad almost on time with low-level clouds causing the only delay, of four minutes. Atlantis now carried newly upgraded general purpose computers. The primary payload, the Gamma Ray Observatory, was deployed during FD 3 (7 April), but its high-gain antenna failed to deploy on command. Ross (EV1) and Apt (EV2) performed an unscheduled EVA, the first since April 1985 (STS 51-D) to manually deploy the antenna, permitting the observatory to be successfully released into orbit. The two EVA astronauts had been preparing to exit the Shuttle in the event of something going wrong with the deployment, and as the GRO was lifted out of the payload bay by the RMS, they were checking out their suits. The solar panels of the GRO were opened to their full span of 21 m, although the high-gain antenna it unlatched did not deploy its 5-metre boom. As the procedures for the contingency EVA were faxed up to the crew, Ross and Apt donned their suits and prepared to exit the vehicle. Meanwhile, the crew fired the thrusters on the Shuttle to try to shake the boom loose, but without success. During the 4 hour 26 minute EVA, Ross tried to push the boom free. When that did not work, they set up a work platform to proceed with the manual deployment sequence they had practised four times in the Weightless

Still in the grasp of the RMS the Compton Gamma Ray Observatory is held above Jay Apt during the successful 7 April EVA to free its high-gain antenna

Environment Training Facility, or WETF, during training. Finding adequate hand holds was a problem, especially during the night time pass of the orbit. While Apt checked to ensure they were not damaging the boom, Ross removed a locking pin and pulled the antenna boom to its deployment position, then used a wrench to lock it into position. While they were outside, they also took the opportunity to perform some of the planned EVA Development Flight Experiment activities by evaluating hand rails, measuring the forces imparted on the foot restraints during the performance of simple tasks, and performing translation exercises. When the GRO was ready for deployment they returned to the airlock, but did not re-pressurise it in case they were needed again. They watched the deployment from the vantage point of the airlock hatch.

The following day (8 April), both men were back outside for the scheduled EVA (5 hours 47 minutes). This time, the two astronauts assembled a 14.6-metre track down the port side of the payload bay and fixed the Crew and Equipment Translation Aid (CETA) cart to it. This was an evaluation of the type of cart that would be installed on the space station to aid movement over long distances, saving the astronauts' energy. They also evaluated using the RMS out over the aft of the payload bay at varying speeds, using strain gauges to measure the slippage of the arm's brakes. They found that RMS-based tasks took longer to perform than expected and also they found themselves suffering from the cold due to excessive EMU cooling, giving rise to concerns that the same might occur during space station construction EVAs, especially on the night-side passes. New EVA gloves that were tested proved disappointing, despite excellent results obtained on Earth. The crew also recommended that back-to-back EVAs should be avoided due to crew fatigue. During the post-flight debriefing, Apt reported that the right-hand index finger of his EVA glove had sustained an abrasion and inspections revealed that the palm bar had penetrated the glove bladder by about 1 cm. Had the palm bar come out of the glove again during EVA, it was estimated that the leakage rate would not have been sufficient to activate the secondary oxygen pack, but it was clear that more work was needed on the glove design before the more arduous EVAs planned for the space station.

With the GRO deployed and two EVAs accomplished, the crew worked on their mid-deck experiments, including testing components of the Space Station Heat Pipe Advanced Radiator Element, to better understand the fluid transfer process at work in microgravity. They also processed chemicals with the BioServa apparatus, operated the Protein Crystal Growth apparatus, and made contact with several hundred amateur radio operators across the world. Due to unacceptable winds at the primary site at Edwards in California, and bad weather at the Cape, the homecoming of Atlantis was delayed by a day from 10 April, with the crew taking the opportunity to photograph the Earth.

The Gamma Ray Observatory included four instruments that observed the electromagnetic spectrum from 30 keV to 30 GeV. Subsequently renamed after Dr. Arthur Folly Compton, who won the Nobel Prize in physics for his work on the scattering of high-energy photons by electrons, the observatory initially worked well, but after six months problems developed in the onboard tape recorders, with high error rates in the data. This forced NASA to use the TDRS to relay data to Earth in real time instead of storing it and downloading it later. Despite the reduction in the amount of data returned, this problem did not prevent the completion of a planned all-sky survey by November 1992. The Compton Observatory's results have been important, but are less well known than the high-profile images from Hubble. Its studies of solar flares, pulsars, X-ray binary systems, and numerous high-energy emissions have all contributed to a better understanding of gamma ray sources in deep space. The observatory was safely de-orbited and re-entered Earth's atmosphere on 4 June 2000, nine years after its deployment from Atlantis.

Milestones

139th manned space flight
69th US manned space flight
39th Shuttle mission
8th flight of Atlantis (OV-104)
24th US and 42nd flight with EVA operations
Mission completed first decade of STS flight operations
1st US EVA since December 1985

STS-39

Int. Designation	1991-031A
Launched	28 April 1991
Launch Site	Pad 39B, Kennedy Space Center, Florida
Landed	6 May 1991
Landing Site	Runway 15, Shuttle Landing Facility, Kennedy Space Center
Launch Vehicle	OV-103 Discovery/ET-46/SRB BI-043/SSME #1 2026; #2 2030; #3 2029
Duration	8 days 7 hrs 22 min 23 sec
Call sign	Discovery
Objective	Unclassified DoD mission devoted to military scientific experiments focusing on the Shuttle's orbital environment in wavelengths ranging from IR to UV

Flight Crew

COATS, Michael Lloyd, 45, USN, commander, 3rd mission
Previous missions: STS 41-D (1984), STS-29 (1989)
HAMMOND Jr., Blaine Lloyd, 38, USAF, pilot
HARBAUGH, Gregory Jordan, 34, civilian, mission specialist 1
McMONAGLE, Donald Ray, 38, USAF, mission specialist 2
BLUFORD, Guion, 48, USAF, mission specialist 3, 3rd mission
Previous missions: STS-8 (1983); STS 61-A (1985)
VEACH, Charles Lacy, 46, civilian, mission specialist 4
HIEB, Richard James, 35, civilian, mission specialist 5

Flight Log

The launch of this unclassified DoD mission was originally scheduled for 9 March, but significant cracks were discovered on all four hinges on the two ET umbilical door mechanisms during processing at Pad 39A. The stack was rolled back to the VAB on 7 March and the tank sent to the OPF for repairs. After the stack was returned to the pad on 1 April, the launch attempt scheduled for 23 April was postponed due to problems with a high-pressure oxidiser turbo-pump for SSME #3 during pre-launch loading. After replacement and testing, launch was rescheduled again, this time to 28 April.

STS-39 was one of the most complicated Shuttle missions to date. The purpose of the mission was to fly an unclassified DoD programme to enhance US national security by gathering scientific data that was essential to the development of advanced missile detection systems. The crew, working a two-shift system (Red Team: Hammond, Veach, Heib; Blue Team: Harbaugh, McMongagle, Bluford – Coats working

This view of the payload bay of Discovery reveals some of the STS-39 payload, including the top of the STP-1 payload on the Hitchhiker carrier, and the AF-675 package comprising CIRRIS-1A, FAR UV, HUP, QIMMS and the URA

with either as required), also completed a variety of sophisticated experiments, including the deployment of five separate spacecraft. The Shuttle Pallet Satellite II (SPAS-II) supported both an infrared and an imaging telescope that studied the Earth's limb, the aurora, the orbiter's environment, and the stars both during free-flight and while attached to the RMS. Also aboard SPAS-II was the Infrared Background Signature Survey (IBSS), which was used to image and measure the spectral nature of rocket exhaust plumes by observing both firings of Discovery's RCS from different attitudes, and the ejection of three sub-satellites deployed from canisters in the payload bay which released chemical gases. Simultaneous observations of these gas releases were made by Earth-based instruments at Vandenberg AFB in California. The other classified deployable payload was designated the Multi-Purpose Release Canister.

Space Test Payload 1 (AFP-675) comprised five instruments designed to observe the atmosphere, aurora and stars in the infrared, far ultraviolet and X-ray wavelengths. The Cryogenic Infrared Radiance Instrument for Shuttle (CIRRIS) used an infrared detector chilled by super-cold (cryogenic) liquid helium to study airglow and auroral emissions form Earth's upper atmosphere. The coolant was used faster than anticipated, which made this experiment a priority over SPAS II/IBSS and delayed the latter by 24 hours. Mike Coats reported that passing through the auroral displays was "just like flying through a curtain of light". The rescheduled experiment returned

50 per cent more data than planned. STP-1 also included the FAR UV Cameras, the Uniformly Redundant Array, the Horizon UV Program and the Quadruple Ion–Neutral Mass Spectrometer. When two tape recorders failed these instruments were adversely affected, but the crew demonstrated the value of humans in space by performing a complicated bypass repair, rerouting data via an orbiter antenna and via TDRS to the ground, fulfilling the objectives for these experiments.

The crew also took advantage of their orbital inclination to take colour and infrared pictures of important surface features and phenomena on Earth, including Lake Baikal in Russia, oil field fires in Kuwait, and the results of a devastating typhoon in the Indian Ocean and fires in Central America, whose smoke palls had drifted over Texas and as far east as Florida. The crew landed at the SLF in Florida due to unacceptably high winds at Edwards AFB.

Milestones

140th manned space flight
70th US manned space flight
40th Shuttle mission
12th mission of OV-103 Discovery
8th DoD Shuttle mission
1st unclassified DoD Shuttle mission
1st flight crew to comprise 7 NASA astronauts

SOYUZ TM12

Int. Designation	1991-034A
Launched	18 May 1991
Launch Site	Pad 1, Site 5, Baikonur Cosmodrome, Kazakhstan
Landed	10 October 1991 (Artsebarsky on TM12);
	25 March 1992 (Krikalev on TM13);
	26 May 1991 (Sharman on TM11)
Landing Site	67 km southeast of Arkalyk
Launch Vehicle	R7 (11A511U2); spacecraft serial number (7K-M) 062
Duration	144 days 15 hrs 21 min 50 sec (Artsebarsky);
	311 days 20 hrs 1 min 54 sec (Krikalev);
	7 days 21 hrs 14 min 20 sec (Sharman)
Call sign	Ozon (Ozone)
Objective	Delivery of the 9th main crew to Mir and operation of UK Juno programme by Sharman

Flight Crew

ARTSEBARSKY, Anatoly Pavolich, 34, Soviet Air Force, commander
KRIKALEV, Sergei Konstantinovich, 32, civilian, flight engineer, 2nd mission
Previous mission: Soyuz TM7 (1988)
SHARMAN, Helen Patricia, 27, civilian, UK cosmonaut researcher

Flight Log

The agreement to fly a UK citizen to a Soviet space station originated in 1986, but was not signed until 1989. Financial problems dogged the project but in November 1989, four finalists were named. Sharman was named as primary candidate in February 1991, with Major Tim Mace of the British Army as her back-up. During the docking with the Mir complex, erroneous readings were produced by the rendezvous equipment. Artsebarsky had to dock manually, while Sharman operated cameras from the Descent Module and Krikalev observed the docking from the forward window in the Orbital Module. After performing a week of experiments focusing on medical tests, physical and chemical research, as well as contacting nine British schools by radio, Sharman returned to Earth with the Mir EO-8 crew on TM11. The landing on TM11 was very hard, with the capsule rolling several times and resulting in disorientation and bruising to the crew inside.

The primary objective of the ninth main Mir crew (EO-9) involved a construction programme, with up to eight EVAs planned for their five-month tour aboard the station. Their first task, however, was to relocate TM12 from the front to the aft port to permit the arrival of Progress M8. This was followed by the release of a small mini-satellite, MAK-1, from the experiment airlock in the base block. MAK-1 was designed

The Soyuz TM12 prime crew of Artsebarsky (left), Sharman and Krikalev

to study Earth's ionosphere, but a system failure rendered the satellite inoperable. The first of what turned out to be six EVAs by this crew (24 June, 4 hours 53 minutes) involved the removal and replacement of the failed Kurs approach system. EVA 2 (28 June, 3 hours 24 minutes) involved the deployment of the US-developed TREK device for studying cosmic rays. The other four EVAs (15 July, 5 hours 56 minutes; 19 July, 5 hours 28 minutes; 23 July, 5 hours 34 minutes and 27 July, 6 hours 49 minutes) focused on the construction of the Sofora girder structure. The crew deployed the Hammer and Sickle national flag of the USSR atop the girder during the sixth EVA, but Artsebarsky's visor fogged up from his excursions during this EVA, requiring Krikalev to help him back to the hatch. The crew also continued space physics investigations, astrophysical observations, a range of technological experiments and observations of the Earth and its weather phenomena.

This five-month mission was flown at the time of the failed Soviet coup of August 1991 and the demise of the Soviet Union in the following months. Despite media reports suggesting he would be stranded alone in space, Krikalev was never "alone". He was asked to remain on board the station when funding difficulties affected the flights of TM13 and TM14 later in the year. When the original crews of these two missions were merged into one new crew, only Alexander Volkov was qualified to remain on the station. Krikalev was asked to extend his mission until March when the next scheduled Russian resident crew would arrive. He agreed, and completed an unplanned nine-month stay in space as part of the EO-10 crew.

Milestones

141st manned space flight
71st Soviet manned space flight
64th Soyuz manned space flight
11th Soyuz TM flight
12th manned Mir mission
9th main crew
19th Soviet and 43rd flight with EVA operations
1st UK citizen (Sharman) in space
1st woman to visit Mir (Sharman)
Krikalev celebrates his 33rd birthday in space (27 Aug)
Artsebarsky celebrates his 35th birthday in space (9 Sep)

STS-40

Int. Designation	1991-040A
Launched	5 June 1991
Launch Site	Pad 39B, Kennedy Space Center, Florida
Landed	14 June 1991
Landing Site	Runway 22, Edwards AFB, California
Launch Vehicle	OV-102 Columbia/ET-41/SRB BI-044/SSME #1 2015; #2 2022; #3 2027
Duration	9 days 2 hrs 14 min 20 sec
Call sign	Columbia
Objective	Spacelab Life Sciences-1 payload operations (18 experiments)

Flight Crew

O'CONNOR, Bryan Daniel, 44, USMC, commander, 2nd mission
Previous mission: STS 61-B (1985)
GUTIERREZ, Sidney McNeill, 39, USAF, pilot
BAGIAN, James Phillip, 39, civilian, mission specialist 1, 2nd mission
Previous mission: STS-29 (1989)
JERNIGAN, Tamara Elizabeth, 32, civilian, mission specialist 2
SEDDON, Margaret Rhea, 43, civilian, mission specialist 3, 2nd mission
Previous mission: STS 51-D (1985)
GAFFNEY, Francis Andrew "Drew", 44, civilian, payload specialist 1
HUGHES-FULFORD, Millie Elizabeth, 46, civilian, payload specialist 2

Flight Log

The launch, originally set for 22 May, was postponed less than 48 hours beforehand due to the discovery of a leaking LH transducer in the orbiter MPS. This was removed and replaced during a leak test in 1990. Then, one of the five General Purpose Computers (GPC) failed, along with one of the multiplexer–demultiplexers that control orbiter hydraulics ordnance and OMS/RCS functions in the aft compartment. One LH and two LO transducers were replaced in the propellant flow system and three LO transducers were replaced in the manifold area, while three further LH transducers were removed and the opening plugged. The rescheduled launch for 1 June was again postponed, despite several attempts to calibrate IMU 2. After it was replaced and tested, the launch was rescheduled again, this time for 5 June. This launch proceeded without incident.

The crew worked a single-shift system to complete the research programme. The primary objectives of the mission's 18 investigations required a larger crew than normal. In addition to the seven astronauts (and one mannequin), there were also

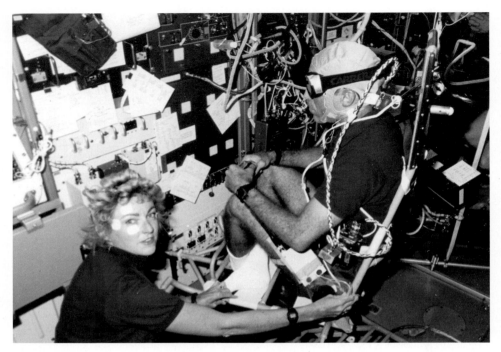

Bagian is in a rotating chair, wearing an accelerometer and electrodes to record head motion and horizontal and vertical eye movements during rotation. This vestibular experiment activity is monitored and assisted by Hughes-Fulford

2,478 jellyfish and 29 lab rats. The humans were involved in ten investigations, with a further seven involving the rodents and one with the jellyfish. In the most detailed and interrelated physiological measurements made on US astronauts since Skylab in 1973/74, the investigations on the crew focused on seven human body systems: the cardiovascular/cardiopulmonary, haematological, muscular, skeletal, vestibular, immune and renal–endocrine systems. The research also included pre- and post-flight medical studies on the crew. In one of these pre-flight investigations, a catheter was inserted into a vein of PS Gaffney before the flight and advanced to a point near his heart. This was designed to monitor blood pressure changes upon his arrival on orbit.

The rats were contained in two groups, one located in the Animal Enclosure Modules (AEM) on the mid-deck of Columbia and the other in a Research Animal Holding Facility (RAHF) in the Spacelab module. The rodents were used for research into muscle, bone and inner-ear functions and for certification of the holding facilities for future use. The jellyfish were encased in flasks and bags filled with artificial seawater. They were filmed to observe their swimming motions for later comparison with a control group on Earth. The crew also evaluated the workstations, the glove box, the medical restraint system and the intravenous pump for future Spacelab and space station use.

Thanks to careful use of their available electrical power, the crew were able to gain an extra flight day in order to continue to collect data. They also continued the programme of photography of the Earth's features, including taking video of the 19 km high yellowish ash plume erupting from Mount Pinatubo in the Philippines during the mission. The twelve Getaway Special (GAS) experiments in the payload bay included research into forming ball bearings, crystal growth and ultra light metals in space; soldering in space; and studying the effects of cosmic particles on computer disks to determine their impact on data storage. Early in the mission, it was thought that a piece of material on the port side payload bay door (used to protect the payload bay from dust contamination on the ground) might interfere with nominal door closing. Bagian and Jernigan would have performed the EVA if required, but ground-based studies concluded there was no hazard and the doors closed properly. After the mission, the astronauts continued to participate in a variety of medical tests. Seddon, Bagian, Gaffney and Hughes-Fulford remained at Edwards for an additional week of medical tests after the rest of the crew returned to Houston.

Milestones

142nd manned space flight
71st US manned space flight
41st flight of Space Shuttle
11th flight of Columbia
1st Spacelab Life Science mission
5th dedicated Spacelab mission
4th flight of Spacelab Long Module configuration
1st flight of payload specialists since STS-51L
Gaffney celebrates his 45th birthday in space (6 Jun)

STS-43

Int. Designation	1991-054A
Launched	2 August 1991
Launch Site	Pad 39A, Kennedy Space Center, Florida
Landed	11 August 1991
Landing Site	Runway 15, Shuttle Landing Facility, Kennedy Space Center, Florida
Launch Vehicle	OV-104 Atlantis/ET-47/SRB BI-045/SSME #1 2024; #2 2012; #3 2028
Duration	8 days 21 hrs 21 min 25 sec
Call sign	Atlantis
Objective	Deployment of the fifth Tracking and Data Relay Satellite (TDRS-E) by IUS-15 upper stage

Flight Crew

BLAHA, John Elmer, 43, USAF, commander, 3rd mission
Previous missions: STS-29 (1989), STS-33 (1989)
BAKER, Michael Allen, 37, USN, pilot
LUCID, Shannon Wells, 48, civilian, mission specialist 1, 3rd mission
Previous missions: STS 51-G (1985), STS-34 (1989)
LOW, George David, 35, civilian, mission specialist 2
ADAMSON, James Craig, 45, US Army, mission specialist 3

Flight Log

The primary objective of STS-43 was achieved barely twelve hours into the mission, when the fifth TDRS satellite was deployed and then placed in its operational circular orbit after two burns of the IUS. Following the mission, the crew returned to Earth with data and results from four payload bay experiments, eight mid-deck payloads, thirteen Detailed Test Objectives (DTO) and nine Detailed Supplementary Objectives (DSO), most of which were linked to extended-duration missions in orbit.

The deployment of a TDRS satellite from a Shuttle rarely attracted headline news and these became some of the "quieter" missions in the programme. However, the work conducted by the crew after the deployment contributed to the development of techniques and procedures that would be significant for missions carried out years later on Mir and ISS. The launch was originally set for 23 July, but was delayed by a day to replace a faulty integrated electronics assembly (which controlled the separation of the ET from the orbiter). Five hours before the second launch attempt, the mission was postponed again due to a faulty main engine controller on SSME #3. The launch was reset for 1 August, but was again delayed, this time due to cabin pressure

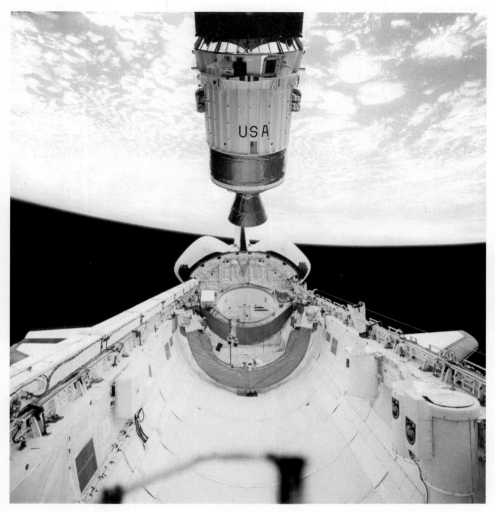

TDRS-E leaves the payload bay of Atlantis atop IUS-15 just six hours after launch from KSC. When it reached its operational station it was renamed TDRS-5, replacing TDRS-3 at 174° west longitude. The GAS canisters are seen to the right of frame along the side of the payload bay wall

vent valve problems. It was postponed a further 24 hours due to infringements of Return-To-Launch-Site (RTLS) weather parameters.

When the mission did finally launch, the crew observed and photographed auroras and lightning discharges, along with four hurricanes in the Pacific Ocean. Measurements of solar UV radiation and UV backscatter radiation from Earth's clouds were also obtained, which would be used to corroborate readings from instruments aboard NIMBUS 7 and the NOAA 9 and 11 satellites, which measured ozone concentration in the upper atmosphere. The In-Flame Preparation in Microgravity experiment was part of an improved fire safety technology investigation connected with safety in future manned spacecraft, while studies of the operation of two large heat pipe radiation elements were linked to the space station. The crew also worked with protein crystallisation, polymer membrane processing and biomedical and fluid science experiments. They also evaluated new Shuttle computers, associated software and improved cursor control devices. In the medical experiments, data was obtained from use of the Lower Body Negative Pressure Device, as well as monitoring cardiovascular performance of the crew in anticipation of future investigations during long space flights. Blaha and Lucid would both participate in their own long-duration flights aboard space station Mir in 1996 and 1997.

Milestones

143rd manned space flight
72nd US manned space flight
42nd Space Shuttle mission
9th mission for Atlantis
5th TDRS Shuttle deployment mission
1st female astronaut to make three space flights (Lucid)
1st planned landing at KSC since January 1986

STS-48

Int. Designation	1991-063A
Launched	12 September 1991
Launch Site	Pad 39A, Kennedy Space Center, Florida
Landed	18 September 1991
Landing Site	Runway 22, Edwards AFB, California
Launch Vehicle	OV-103 Discovery/ET-42/SRB BI-046/SSME #1 2019; #2 2031; #3 2107
Duration	5 days 8 hours 27 min 38 sec
Call sign	Discovery
Objective	Deployment of the Upper Atmosphere Research Satellite (UARS)

Flight Crew

CREIGHTON, John Oliver, 48, USN, commander, 3rd mission
Previous missions: STS 51-G (1985); STS-36 (1990)
REIGHTLER Jr., Kenneth Stanley, 40, USN, pilot
GEMAR, Charles Donald "Sam", 36, US Army, mission specialist 1, 2nd mission
Previous mission: STS-38 (1990)
BUCHLI, James Frederick, 46, USMC, mission specialist, 4th mission
Previous missions: STS 51-C (1985); STS 61-A (1985); STS-29 (1989)
BROWN, Mark Neil, 40, USAF, mission specialist, 2nd mission
Previous mission: STS-28 (1989)

Flight Log

There was only a fourteen-minute delay to the launch of STS-48, caused by a faulty communications link between KSC and MCC in Houston. The landing, however, was scheduled for KSC but was diverted to Edwards due to bad weather around the Cape.

The 6,577 kg UARS was deployed by the RMS on the third day of the flight for a planned 18-month primary mission that would make the most extensive study yet conducted of the upper level of Earth's atmosphere (called the troposphere). Astronauts Gemar and Buchli had trained to open the satellite's solar panels and release its antennas during a contingency EVA, should the automatic systems fail. In the event, this training was not put into practice as the deployment went according to plan. The initial lock-on to TDRS proved difficult, but a back-up system resolved the communication problems with the satellite. Full operations were planned from mid-October 1991, although some instruments began sending data as soon as they were deployed while the rest of the satellite's payload was being checked out.

While the primary objective of the STS-48 mission was the deployment of UARS to study the upper atmosphere, the crew was also busy with investigations aimed at supporting future programmes. Here, Mark Brown (left) and James Buchli work with the Structural Test Article, a scale model of the space station truss designed to test vibration characteristics on the joints of the truss structure in microgravity

In addition to deploying UARS, the crew worked on proton crystal growth experiments; a zero-gravity dynamic experiment studying how fluids and structures react in weightlessness; research into creating polymer membranes to be used as filters for use in industrial refining processes; an experiment that researched the effects of space flight on rodents; and studies of various radiation levels inside the orbiter from gamma rays, cosmic rays and other radiation sources. They also assisted in calibrating USAF optical instruments in Hawaii, and a particle monitor in the payload bay that measured contaminants during launch that might affect the payload being carried, as well as continuing the programme of Earth observations and photography carried on almost all Shuttle missions. On FD 4, Discovery took evasive action to avoid a piece of Soviet space debris that was predicted to pass only 350 metres below the orbiter.

The UARS satellite had a design life of just three years but some ten years after its deployment, six of its ten instruments were still operating. From June 1992 through to 1999, the satellite experienced power and equipment failures but continued to gather data from the surviving instruments. On 20 May 2005, UARS surpassed 5,000 days in orbit with five instruments still working, but by 21 August, the spacecraft had suffered a short circuit in Battery #2, signifying the end of useful operations. One of its three batteries had already been lost in June 1997. UARS was planned for decommissioning during 2006.

Milestones

144th manned space flight
73rd US manned space flight
43rd Shuttle mission
13th mission for Columbia
1st major flight element for NASA's *Mission to Planet Earth* programme
1st time a Shuttle required evasive action to avoid space debris

SOYUZ TM13

Int. Designation	1991-069A
Launched	2 October 1991
Launch Site	Pad 1, Site 5, Baikonur Cosmodrome, Kazakhstan
Landed	25 March 1992 (Volkov A.);
	10 October 1991 (Aubakirov/Viehbock)
Landing Site	85 km southeast of Arkalyk, Kazakhstan (TM13)
Launch Vehicle	R7 (11A511U2), spacecraft serial number (7K-M) 063
Duration	175 days 2 hrs 52 min 43 sec (Volkov A.);
	7 days 22 hrs 12 min 59 sec (Aubakirov/Viehbock)
Call sign	Donbass (Donbass)
Objective	Delivery of E0-10 commander; completion of politically-motivated Kazakh experiment programme; completion of AustroMir experiment programme

Flight Crew

VOLKOV, Alexandr Alexandrovich, 43, commander, 3rd mission
Previous missions: Soyuz T14 (1985); Soyuz TM7 (1988)
AUBAKIROV, Toktar Ongarbayevich, 45, civilian, Kazakhstan cosmonaut researcher
VIEHBOCK, Franz, civilian, 31, Austrian cosmonaut researcher

Flight Log

The docking of Soyuz TM13 to the Mir complex was one of the smoothest yet seen in the Russian programme. Aubakirov's primary objective was photography of Earth resources, especially of his native Kazakhstan and in particular related to Project Aral-91, which monitored dust and aerosol particles blowing off recently exposed parts of the Aral Sea. The programme included two biotechnological experiments and a series of standard medical tests, as well as his participation in several of the on-going Mir experiments. The cosmonaut's photographs helped supplement the ground- and air-based monitoring of the particles. Viehbock's Austrian experiment programme had been supplemented by the delivery of 150 kg of apparatus aboard Progress M9, which had docked earlier. The programme featured fourteen experiments comprising ten biomedical, three materials processing and one Earth observation investigation. Both cosmonaut researchers returned with EO-9 commander Artsebarsky in TM12 on 10 October, leaving Volkov and Krikalev to continue the EO-10 programme.

Krikalev's adaptation to life aboard Mir helped ease the transition from the EO-9 to the EO-10 residency. One major task was to re-qualify the automatic Kurs docking system on the front port of Mir using Soyuz TM13. After several test approaches on 15 October, the system performed a successful automated docking. Just four days

The Soyuz TM13 crew of Aubakirov (left), Volkov (centre) and Viehbock

later, however, Progress M10 aborted its initial approach, only successfully docking with Mir on 21 October. Clearly, Kurs at the front port was still having difficulties.

While the two cosmonauts resumed their research programme of Earth resources, materials processing, astrophysical studies and biomedical experiments, down on Earth, the Soviet Union was in turmoil. The demise of the USSR occurred on 25 December 1991, with the country officially ceasing to exist by the turn of the year to be replaced by the Russian state and the Commonwealth of Independent States (CIS). The once-proud Soviet space programme now became the Russian space programme and was almost immediately dogged by severe budget restrictions, interstate disputes and great uncertainty. Work on Mir continued, but the question was for how long. The ocean-going tracking ships used since the 1960s were phased out in order to save money and resulted in Mir being out of radio contact with mission control for up to nine hours every day. In early 1992, flight controllers went on strike for higher rates of pay during the flight of Progress M11, but did not interfere with the docking.

The crew's only EVA on 20 February was quickly revised when the heat exchanger in Volkov's Orlan DMA pressure suit failed. Restricted to remaining close to the EVA hatch, he could only assist Krikalev, who was now on his seventh EVA in less than a year (and set a new cumulative EVA total of over 36 hours). Volkov assisted in installing equipment near the hatch, but could not operate the Strela boom to move

Krikalev to the Kvant module. Krikalev had to move hand-over-hand across the station to the worksite. Once there, he dismantled the equipment used to build Sofora in 1991, and then cleaned Kvant 1's camera lenses and collected samples of the solar cell added to the third (top) array base block in 1988.

In March 1992, after hosting the German Mir-92 mission operated by cosmonaut Klaus Flade, Volkov and Krikalev returned to Earth along with the German cosmonaut, but to a new homeland. Again, all three cosmonauts who landed had travelled to Mir on different vehicles.

Milestones

145th manned space flight
72nd Soviet (now Russian) manned space flight
65th Soyuz manned space flight
12th Soyuz TM manned flight
20th Soviet/Russian and 44th flight with EVA operations
1st flight of ethnic Kazakh in space
1st Austrian in space

STS-44

Int. Designation	1991-080A
Launched	24 November 1991
Launch Site	Pad 39A, Kennedy Space Center, Florida
Landed	1 December 1991
Landing Site	Runway 05R (Lakebed), Edwards AFB, California
Launch Vehicle	OV-104 Atlantis/ET-53/SRB BI-047/SSME #1 2015; #2 2030; #3 2029
Duration	6 days 22 hrs 50 min 44 sec
Call sign	Atlantis
Objective	Deployment of the Defense Support Program (DSP) Satellite by IUS-14; Terra Scout Experiment; Military Man in Space Experiment

Flight Crew

GREGORY, Frederick Drew, 50, USAF, commander, 3rd mission
Previous missions: STS 51-B (1985); STS-33 (1989)
HENRICKS, Terence Thomas "Tom", 39, USAF, pilot
VOSS, James Shelton, 42, US Army, mission specialist 1
MUSGRAVE, Franklin Story, civilian, mission specialist 2, 4th mission
Previous missions: STS-6 (1983); STS 51-F (1985); STS-33 (1989)
RUNCO Jr., Mario, 39, USN, mission specialist 3
HENNEN, Thomas John, 39, US Army, payload specialist 1

Flight Log

STS-44 deployed one of NORAD's (North American Air Defence Command) Tactical Warning and Attack Assessment System satellites. This series of space-borne detector systems has been used and upgraded since 1970 and provides detection and reports of real time space launches, missile launches and nuclear detonations across the globe using infrared sensors to detect the heat from missile plumes or nuclear explosions. This satellite, codenamed "Liberty", was deployed six hours after a spectacular night launch. This came after the 19 November launch was scrubbed due to the failure of one of five gyroscopes in the Redundant Inertial Measurement Unit of IUS-14. A replacement was fitted and the launch rescheduled for 24 November.

The day after the deployment of "Liberty", the crew of Atlantis were awoken by a special message from actor Patrick Stewart, better known as Jean-Luc Picard, Captain of the Star Ship USS Enterprise in *Star Trek: The Next Generation*. "Picard" reminded the crew of their ten-day mission "to explore new methods of remote sensing and observation of the planet Earth. To seek out new data on radiation in space and

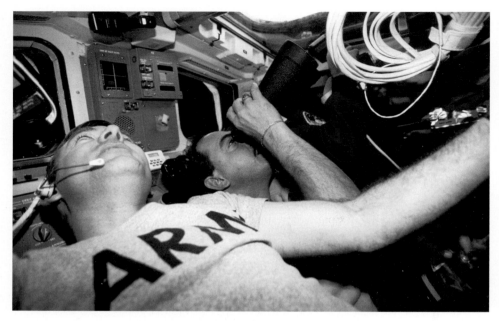

Voss (in foreground) looks at Earth while Hennen continues his Terra Scout observations. In addition to naked eye and binocular observations, a device called the Space-borne Direct View Optic Systems (SPADVOS) was used for selected ground points

a new understanding on the effects of microgravity on the human body. To boldly go where ... 255 men and women have gone before."

Tom Hennen, the US Army PS, operated the Terra Scout package which was sponsored by the US Army Intelligence Center. This suite of experiments was designed to allow a trained imagery analyser to observe targets of military interest from the vantage point of the Shuttle in orbit. For this mission, there would be thirty such targets. The Military Man-in-Space Experiment was designed to evaluate the ability of a space-borne observer to gather important information about ground troops, equipment and facilities. In addition to Hennen's Terra Scout package, there was also a range of monitoring and observation experiments, used to record aspects of the flight of a Shuttle in orbiter. The Shuttle Activation Monitor (SAM) measured the radiation environment on board the orbiter and its effect on gamma ray detectors. The Cosmic Radiation Effects and Activation Monitor (CREAM) gathered data on cosmic rays and radioactivity on board the vehicle, while the third-generation Radiation Monitoring Equipment (RME) measured ionising radiation aboard the orbiter and the crew's exposure to it. In addition, the USAF Maui Optical System used an electrical–optical system located on the Hawaiian island to observe Shuttle jet firings, water dumps and encounters with atomic oxygen. The Interim Operations Contamination monitor in the cargo bay of Atlantis had already proven successful, measuring contamination in the payload bay during launch. Finally, the Ultraviolet Plume

Instrument (UPI) sensor in a US DoD satellite located in geosynchronous orbit attempted to observe Atlantis as a method of fine-tuning the sensor.

In addition to continuing observations of the Earth and weather phenomena, the crew undertook a range of experiments as part of an on-going programme of medical investigations. These studies were connected to studying the effects of weightlessness on crew members and methods of counteracting such effects. The programme, originally planned for ten days, was designed to provide baseline data for the future extended-duration orbiter medical programme on missions lasting between 12 and 17 days from 1992. The treadmill suffered a bearing failure and exercises had to be modified to include squatting actions, using the back muscles rather than those in the legs. For the first time, Dr. Musgrave was able to perform medical experiments in space – on his previous missions, he had fulfilled the role of flight engineer looking after the orbiter. This time, he could call upon his surgical and medical skills as he was not assigned to either orbiter or EVA duties.

On FD 7, one of three Inertial Measurement Units (IMU) failed. Strict flight rules meant that the orbiter would have to land as soon as possible, and at Edwards rather than the planned landing at the Cape to take advantage of the wide runways. Although the Shuttle will fly perfectly well with just two IMUs operating, if another failure had occurred, it would have posed a serious risk to the navigational systems and would have left the orbiter with no back-up unit. With the mission now designated a Minimum Duration Flight (MDF), the crew were bitterly disappointed at having to come home three days early. Despite the early return of STS-44 and some difficulties with on-orbit equipment, the mission still achieved almost 90 per cent of its pre-mission objectives.

Milestones

146th manned space flight
74th US manned space flight
44th Shuttle mission
10th Atlantis mission
9th DoD Shuttle mission
2nd declassified DoD Shuttle mission
1st NCO to fly in space (Hennen)

STS-42

Int. Designation	1992-002A
Launched	22 January 1992
Launch Site	Pad 39A, Kennedy Space Center, Florida
Landed	30 January 1992
Landing Site	Runway 22, Edwards AFB, California
Launch Vehicle	OV-103 Discovery/ET-52/SRB BI-048/SSME #1 2026; #2 2022; # 3 2027
Duration	8 days 1 hr 14 min 44 sec
Call sign	Discovery
Objective	Operation of International Microgravity Laboratory 1 (IML-1) payload; fifty-five experiments devoted to space medicine and manufacturing utilising a Spacelab Long Module

Flight Crew

GRABE, Ronald John, 46, USAF, commander, 3rd mission
Previous missions: STS 51-J (1985); STS-30 (1989)
OSWALD, Stephen Scott, 40, civilian, pilot
THAGARD, Norman Earl, 48, civilian, mission specialist 1, payload commander, 4th mission
Previous missions: STS-7 (1985); STS 51-B (1985); STS-30 (1989)
READDY, William Francis, 39, civilian, mission specialist 2
HILMERS, David Carl, 41, USMC, mission specialist 3, 4th mission
Previous missions: STS 51-J (1985); STS-26 (1988); STS-36 (1990)
BONDAR, Roberta Lynn, 46, civilian, Canadian payload specialist 1
MERBOLD, Ulf Dietrich, 50, civilian, ESA payload specialist 2, 2nd mission
Previous mission: STS-9 (1983)

Flight Log

As Discovery landed at Edwards AFB in California at the end of the IML-1 mission, the crew were told that their mission had provided a preview of both space station operations and the kind of international cooperation that would be part of future space exploration. As a new Russia emerged from the turmoil of the break-up of the Soviet Union, the Freedom Space Station was itself struggling to survive. But there was a glimmer of hope in the potential cooperation of the Russians in the future programme. However, there was still much to be done on Earth before any hardware would fly in space, but the mission of STS-42 and the first flight of the International Microgravity Laboratory had demonstrated that such cooperation was feasible.

The international crew of IML-1 pose for the traditional in space "star-burst" portrait inside the Spacelab module. At top centre is MS Hilmers, and clockwise are commander Grabe, MS Readdy, ESA PS Merbold, payload commander Thagard, pilot Oswald and Canadian PS Bondar. The rotating chair used often in biomedical tests is partially obscured in the centre of frame

Both the launch and landing of Discovery passed without incident, apart from a one-hour delay to the launch to evaluate indications of power surges from one of the fuel cells. With the vehicle cleared for launch, and safely on orbit just over an hour later, Shuttle operations in 1992 opened with one of the most successful Spacelab missions of all. With the crew operating in two shifts for round-the-clock activity (Red – Readdy, Hilmers, Merbold; Blue – Grabe, Oswald, Thagard, Bondar), operations primarily focused on the adaptation of the human nervous system to low gravity and on the effects of microgravity on other life forms. These included shrimp eggs, lentil seedlings, fruit fly eggs and bacteria. There was also a programme of materials processing experiments, including crystal growth from a range of substances such as enzymes, mercury iodide and a virus. The secondary payloads carried included twelve GAS canisters attached to a GAS Bridge Assembly in the payload bay. This contained numerous US and international experiments, ranging from materials processing to investigations into the development of animal life in weightlessness.

The IML experiment programme was a cooperative effort between the space agencies of the United States (NASA), Europe (ESA), Canada (CSA), France

(CNES), Germany (DARA) and Japan (NASDA). The GAS experiments also originated from multiple countries (Australia, China, Federal Republic of Germany, Japan, Sweden and the United States). There were also two student experiments flown, as well as the IMAX large-format camera and a package of on-going small mid-deck experiments. In all, over 200 scientists from sixteen countries participated in the flight and investigation programme.

Though minor problems occurred, they were all overcome with no adjustments to the flight plan, nor loss of science results. On 24 January, the Mir space station passed within 39 nautical miles of Discovery and the crew reported that the sunlight reflecting off the station looked as bright as planet Mercury when seen after sunset from Earth. On board Discovery, Thagard observed the Russian space station that he would live aboard just three years later. Towards the end of the flight, mission managers concluded that the crew had conserved their consumables so well that they would be able to stay an extra day in orbit to continue their science experiment programme.

IML-1 was the first of a series of four or five such flights that were envisaged over a ten-year period (one flight every two years), dedicated to the study of life and materials sciences and providing important data for planning and executing follow-on research on Space Station Freedom. Such was the success of IML-1 that the prospect of international cooperation on Freedom looked assured, even if the programme itself was floundering due to complexity and cost. Ironically, the revised ISS programme would signal the demise of Spacelab missions due to limited resources. The IML series was reduced from a ten-year programme to just two missions.

Milestones

147th manned space flight
75th US manned space flight
45th Shuttle mission
14th flight of Discovery
1st flight of IML configuration
5th flight of Spacelab Long Module
Readdy celebrates his 40th birthday in space (24 Jan)
Hilmers celebrates his 42nd birthday in space (28 Jan)
1st female Canadian in space (Bondar)

SOYUZ TM14

Int. Designation	1992-014A
Launched	17 March 1992
Launch Site	Pad 1, Site 5, Baikonur Cosmodrome, Kazakhstan
Landed	10 August 1992 (Viktorenko/Kaleri/M. Tognini) 24 March 1992 (Flade in TM13)
Landing Site	136 km east of Dzhezkazgan
Launch Vehicle	R7 (11A511U2): spacecraft serial number (7K-M) 64
Duration	145 days 14 hrs 10 min 32 sec (Viktorenko/Kaleri); 7 days 21 hrs 56 min 52 sec (Flade)
Call sign	Vityaz (Viking)
Objective	Delivery of the 11th expedition crew to Mir; German Mir-92 mission

Flight Crew

VIKTORENKO, Alexandr Stepanovich, 44, Russian Air Force, commander, 3rd mission
Previous missions: Soyuz TM3 (1987), Soyuz TM8 (1989)
KALERI, Alexandr Yuriyevich, 35, civilian, flight engineer
FLADE, Klaus-Dietrich, 39, German Air Force, cosmonaut researcher

Flight Log

The fully automated docking of TM14 to Mir was final confirmation that the Kurs rendezvous system had been repaired. After being bumped off the crew for TM13, Kaleri finally made it to Mir, alongside German cosmonaut Klaus Flade who conducted fourteen German experiments during his week aboard the station. His programme included materials processing experiments and Flade would also provide baseline biomedical data in preparation for extended orbital operations on the ESA Columbus laboratory, part of the Freedom Space Station programme. He would return to Earth with Volkov and Krikalev in the TM13 spacecraft, after they had spent the week briefing the new resident crew and packing their equipment for the return to Earth.

At this time, there was a strong possibility that the cash-starved Russian Space Agency might be forced to temporarily abandon Mir until new funds could be secured to support further manned operations. The EO-11 crew were therefore never sure when they might be called back to Earth. This residency was also very "quiet", with the cosmonauts continuing the on-going programme of Earth observations, materials processing, biomedical studies and astrophysical observations, balanced with routine maintenance, housekeeping and unloading the Progress supply vehicles. The docking of Progress M13 was aborted on 2 July due to a fault in the onboard software, but

Formal crew portrait of the TM14 cosmonauts. L to r: German cosmonaut Flade, EO-11 commander Viktorenko and EO-11 FE Kaleri

reprogramming by operators on the ground resolved the problem, allowing a safe docking two days later to deliver some of the experiments for the upcoming French mission.

On 8 July, the crew performed the only EVA of their residency (of 2 hours 3 minutes) to examine the gyrodynes on the outside of Kvant 2. A dozen gyrodynes stabilised the station as it orbited the Earth. Similar to gyroscopes, these spinning devices generated angular momentum to maintain Mir's orientation to the Sun, which was essential for the solar arrays to be able to absorb energy to produce electricity for use on the station. Though the gyrodynes consumed considerable power to start with, once they were spinning, they would run for some time with minimal energy consumption. Five of the six units on Kvant 1 had exceeded their five-year design life but four of the six on Kvant 2 had failed. During this EVA, the cosmonauts wielded large shears to cut through thermal insulation on Kvant 2 to reach the gyrodynes and inspected and photographed the units for engineers back on the ground as part of an evaluation for future EVA operations to remove and replace them. The cosmonauts also evaluated binoculars that were compatible with the Orlan suit's visor to allow inspection of the more remote areas of Mir, where it would be difficult, if not impossible, for a cosmonaut to get to.

This was a quiet tour of duty on the space station. The two cosmonauts completed a programme of agricultural photography and spectral observation before dividing their time between these commitments and their astrophysical observations. As the crew completed these studies, the onboard furnaces were being run in semi-automated mode. Towards the end of their residency the crew received the EO-12 cosmonauts and French cosmonaut researcher Michel Tognini, who would complete his own research programme during a 12-day hand-over period, and return with the EO-11 cosmonauts.

Milestones

148th manned space flight
73rd Russian manned space flight
21st Russian and 45th flight with EVA operations
14th Soyuz flight to Mir
11th main Mir crew
9th visiting crew (Flade)
66th Soyuz manned mission
13th Soyuz TM manned mission
Viktorenko celebrates his 45th birthday in space (29 Mar)
Kaleri celebrates his 36th birthday in space (13 May)

STS-45

Int. Designation	1992-015A
Launched	24 March 1992
Launch Site	Pad 39A, Kennedy Space Center, Florida
Landed	2 April 1992
Landing Site	Runway 33, Kennedy Space Center, Florida
Launch Vehicle	OV-104 Atlantis/ET-44/SRB B-049/SSME #1 2024; #2 2012; #3 2028
Duration	8 days, 22 hrs 9 min 28 sec
Call sign	Atlantis
Objective	Operation of Atmospheric Laboratory for Applications and Science-1 (ATLAS-1), mounted on non-deployable Spacelab pallets in the payload bay

Flight Crew

BOLDEN Jr., Charles Frank, 45, USMC, commander, 3rd mission
Previous missions: STS 61-C (1986); STS-31 (1990)
DUFFY, Brian, 38, USAF, pilot
SULLIVAN, Kathryn Dwyer, 40, civilian, mission specialist 1, payload commander, 3rd mission
Previous missions: STS 41-G (1984); STS-31 (1990)
LEESTMA, David Cornell, 42, USN, mission specialist 2, 3rd mission
Previous missions: STS 41-G (1984); STS-28 (1989)
FOALE, Colin Michael "Mike", 35, civilian, mission specialist 3
FRIMOUT, Dirk "Dick" Dries David Damiaan, ESA payload specialist 1
LICHTENBERG, Byron Kurt, 43, civilian, payload specialist 2

Flight Log

Originally called Earth Observation Missions (EOM) in the mid-1980s, the redesignated Atmospheric Laboratory for Applications and Science (ATLAS) series of missions were part of Phase I of NASA's *Mission to Planet Earth*, a large-scale unified study of planet Earth as a single dynamic system. There were plans for up to ten ATLAS missions through one solar cycle of eleven years, but changes in the launch manifest of the Shuttle in the 1990s due to budget limitations, the space station programme and US participation in Mir docking missions altered these plans. The significant reduction in Shuttle/Spacelab science missions curtailed the series to just three flights and utilised previously flown experiments and Spacelab pallet hardware wherever possible. The research fields studied as part of these missions were in atmosphere science, solar physics, Earth science, space plasma research and astronomy. ATLAS-1 incorporated fourteen investigations using twelve instruments and

Illustrating the busy flight deck area during a Spacelab mission flying pallets only, pilot Duffy floats in the foreground while PS Frimout (left) works with an ESA spectrometer experiment and commander Bolden fetches an IR filter pouch prior to taking pictures

support systems provided by international partners, including the United States, Belgium, France, Germany, Japan, the Netherlands, Switzerland and the United Kingdom, with ESA providing operational support for the European investigations.

The 23 March launch attempt was delayed for one day when higher than allowed concentrations of liquid oxygen were detected in the aft compartment of Atlantis during pre-launch tanking operations. When these leaks could not be reproduced during simulations, it was concluded that the reading had come from plumbing in the main propulsion system that was not thermally conditioned to cryogenic propellants. The launch on 24 March was delayed by thirteen minutes due to low-level clouds at the KSC Shuttle Landing Facility.

Once safely in orbit, the crew powered up the ATLAS systems and split into their two twelve-hour-shift teams. The Blue Team consisted of Duffy, Sullivan and Frimout, and the Red Team of Leestma, Foale and Lichtenberg. Commander Bolden was not assigned to a shift and worked with either as required. They would remain in these teams for the duration of the mission, affording round-the-clock operation of the science package in the payload bay and mid-deck. In addition to working with the ATLAS package, the crew operated the SAREX ham radio system and contacted operators in Wales, Norway, Canada, India and the USA. Subsequent contact was made to schools in Barcelona in Spain, Harrogate in England and Augsburg in Germany and with students in Massachusetts, California and Houston (where commander Duffy spoke with his son who was attending the Ed White Elementary School near the Johnson Space Center). SAREX was used several times to attempt to contact the cosmonauts on Mir and though the cosmonauts could be heard, no confirmation was received from Mir that they had received the signals from Atlantis. As the first Belgian in space, Frimout received a call from Prince Philippe of Belgium during his visit to the US Payload Operations Control Center at the Marshall Space Flight Center in Huntsville, Alabama. Commander Bolden completed experiments that assessed visual performance parameters and collected samples from cabin humidity for post-flight analysis, as part of on-going studies of Shuttle flight operations and environmental conditions. The astronauts continued extensive Earth photography as part of the long-established Shuttle programme objective, and used the Space Shuttle Backscatter Ultraviolet/A (SSBUV/A) spectrometer to gather data that would corroborate satellite data on the Earth's ozone layer. There was also a single GAS canister carrying a crystal growth experiment.

On FD 6, the mission management team determined that there were again sufficient consumables aboard Atlantis to allow a one-day extension to the mission. The initial reports of the scientific results from ATLAS were impressive, recording the most extensive collection of scientific data ever gathered on Earth's atmosphere. In excess of one billion bits of data were downlinked to Earth, sufficient, if written on paper, to form a stack over 3.38 km high. The information from this flight alone was expected to keep scientists and researchers busy for the next eight years.

Milestones

149th manned space flight
76th US manned space flight
46th Shuttle mission
11th Atlantis mission
1st African American mission commander
1st flight of ATLAS payload
3rd Spacelab pallet only mission
1st Belgian citizen in space

STS-49

Int. Designation	1992-026A
Launched	7 May 1992
Launch Site	Pad 39B, Kennedy Space Center, Florida
Landed	16 May 1992
Landing Site	Runway 22, Edwards AFB, California
Launch Vehicle	OV-105 Endeavour/ET-43/SRB BI-050/SSME #1 2030; #2 2015; #3 2017
Duration	8 days 21 hrs 17 min 38 sec
Call sign	Endeavour
Objective	Capture, repair and redeployment of stranded satellite INTELSAT VI (F-3); Assembly of Space Station by EVA Methods demonstration (ASSEM)

Flight Crew

BRANDENSTEIN, Daniel Charles, 49, USN, commander, 4th mission
Previous missions: STS-8 (1983); STS 51-G (1985); STS-32 (1990)
CHILTON, Kevin Patrick, 36, USAF, pilot
HIEB, Richard James, 36, civilian, mission specialist 1, 2nd mission
Previous mission: STS-39 (1991)
MELNICK, Bruce Edward, 42, US Coast Guard, mission specialist 2, 2nd mission
Previous mission: STS-41 (1990)
THUOT, Pierre Joseph, 36, mission specialist 3, 2nd mission
Previous mission: STS-36 (1990)
THORNTON, Kathryn Cordell Ryan, 39, civilian, mission specialist 4, EV3, 2nd mission
Previous mission: STS-33 (1989)
AKERS, Thomas Dale, 40, USAF, mission specialist 5, 2nd mission
Previous mission: STS-41 (1990)

Flight Log

The maiden flight of the Endeavour (the replacement orbiter for Challenger which was lost in the 1986 launch accident) was an impressive mission that clearly demonstrated the value of having astronauts on board to overcome technical problems. Whether there was a need for Endeavour itself was a question long debated, but it was the loss of Challenger that finally secured the construction of the new vehicle from the structural spares that had been factored into the orbiter construction programme several years before. There was no such contingency to replace Columbia seventeen years later.

Three astronauts hold onto the 4.5-ton Intelsat VI satellite after completing a six-handed "capture". L to r are astronauts Hieb, Akers and Thuot, who stands on the end of the RMS. This first three-person EVA was the third attempt at grappling the satellite

Following the 6 April 1992 flight readiness firing of Endeavour's three main engines, the management team decided to replace all three due to irregularities that had arisen in two of the high-pressure oxidiser turbo-pumps. The launch of STS-49 was set for 4 May, but was rescheduled for 7 May, with an early launch window that would offer better lighting conditions for photo-documentation of the behaviour of the new vehicle during its first ascent. The lift-off on 7 May was delayed by 34 minutes due to bad weather at the transoceanic abort landing site, as well as technical problems with one of Endeavour's master event controllers.

The primary objective of the flight was the capture and redeployment of the Intelsat VI satellite that had been launched aboard a Titan rocket on 14 March 1990.

During the launch, the second stage of the Titan had not separated, preventing the satellite's ascent into a geosynchronous orbit. Quick thinking by ground controllers triggered the separation of the satellite from the unused Perigee Kick Motor (PKM) that was still attached to the Titan stage, and careful use of onboard liquid propellant allowed the satellite to reach a thermally stable 299 × 309 nautical mile orbit. Subsequent data analysis suggested it would be more cost-effective to bring a new kick motor up to the stranded satellite on the Shuttle than to return it to the ground and relaunch it. The Hughes Aircraft Company's Space and Communications Group worked with NASA to construct special hardware to support the EVA operations that would be required.

The crew split into two EVA teams. Thuot (EV1) and Hieb (EV2) were termed the Intelsat EVA crew for the satellite retrieval and redeployment, while Thornton (EV3) and Akers (EV4) would work on the planned evaluation of space station construction techniques. A specially designed capture bar would be used to capture the satellite but in the event, this did not work as planned. During the first attempt on 10 May, Thuot had been unable to attach the capture bar, causing the satellite to bounce away and tumble at even greater rates the more he tried. The following day, the rotation of the satellite had slowed sufficiently for Thuot to gently move the bar into place. This time, however, the latches on the bar failed to fire, causing the satellite to drift off once again. It became evident that the planned method of capture would not work and during 11 May, as the crew rested, plans were formulated for another attempt. This would be the last chance, as Endeavour had only enough propellant aboard to support one more rendezvous. The following day, the crew practised getting three pressure-suited astronauts into an airlock that was designed to accommodate just two. Other astronauts simulated the operation in the water tank at JSC and the crew played an important role in the final decision to try the first three-person EVA.

On 13 March, Thuot, Hieb and Akers ventured outside and placed themselves in foot restraints 120° apart around the payload bay. Hieb was stationed on the starboard wall of the bay and Akers stood on a borrowed strut from the ASSEM experiment, while Thuot rode the RMS. Brandenstein and Chilton flew Endeavour and gently closed in on the satellite, allowing all three EVA astronauts to reach up and grasp the satellite by the three electric motors that would deploy the satellite's cylindrical solar panel. Over a difficult 85 minutes, the capture bar was finally attached to the satellite, allowing the RMS to manipulate it over the payload bay and onto the new kick motor. Thuot and Hieb then used hand-operated ratchets to pull the satellite down and latch it into place with four clamps. They then connected two electrical umbilicals. With the astronauts back in the airlock, but with an open hatch in case they were still required, the new PKM was ejected from the payload bay. Thirteen minutes later, a procedural error on the checklist prevented the initial deployment. The EVA set a new duration record, surpassing that of the Apollo 17 astronauts on the Moon in December 1972. The new engine worked perfectly on 14 May, placing the satellite on its way to geosynchronous orbit.

The same day, Thornton and Akers were in the payload bay performing the ASSEM EVA demonstration. Originally scheduled for two EVAs, the Intelsat difficulty had curtailed this to a single excursion. The pair assembled a pyramid out of struts to simulate a station truss section, then attached a triangular pallet to this to simulate the mass of a major component such as a propulsion module. Scheduled for three hours, the astronauts found that it took twice as long as expected and proved very tiring, forcing other activities planned for this EVA to be cancelled. The mission had been extended by two days to complete its primary objectives and this took a toll on the crew. They also had to complete the rest of their experiment programme, including the Commercial Proton Crystal Growth experiment, the UV Plume Instrument experiment and the USAF Maui Optical Site experiment. Post-flight inspections revealed negligible damage and the crew and flight controllers reported only minor problems. The new vehicle had joined the fleet in style.

Milestones

150th manned space flight
77th US manned space flight
47th Shuttle mission
1st flight of Endeavour
25th US and 46th flight with EVA operations
1st Shuttle mission to feature four EVAs
1st use of landing drag parachute
1st three-person EVA
1st astronaut attachment of rocket motor to orbiting satellite

STS-50

Int. Designation	1992-034A
Launched	25 June 1992
Launch Site	Pad 39A, Kennedy Space Center, Florida
Landed	9 July 1992
Landing Site	Runway 33, Kennedy Space Center, Florida
Launch Vehicle	OV-102 Columbia/ET-50/SRB BI-051/SSME #1 2019; #2 2031; #3 2011
Duration	13 days 19 hrs 30 min 4 sec
Call sign	Columbia
Objective	Operation of first US Microgravity Laboratory payload utilising the Spacelab pressurised module

Flight Crew

RICHARDS, Richard Noel, 45, USN, commander, 3rd mission
Previous missions: STS-28 (1989), STS-41 (1990)
BOWERSOX, Kenneth Duane, 36, USN, pilot
DUNBAR, Bonnie Jean, 43, civilian, mission specialist 1, payload commander, 3rd mission
Previous missions: STS 61-A (1985), STS-32 (1990)
BAKER, Ellen Louise, 39, civilian, mission specialist 2, 2nd mission
Previous mission: STS-34 (1989)
MEADE, Carl Joseph, 41, USAF, mission specialist 3, 2nd mission
Previous mission: STS-38 (1990)
DELUCAS, Lawrence James, 41, civilian, payload specialist 1
TRINH, Eugene Huu-Chau, 41, civilian, payload specialist 2

Flight Log

The longest flight to date in the Shuttle programme was made possible after an extensive modification programme for Columbia at the Rockwell facility in California. The upgrades comprised over fifty modifications, including the installation of a drag chute and the first fitting to any orbiter of the Extended-Duration Orbiter hardware (incorporating the EDO cryogenic pallet). The EDO pallet carried additional hydrogen and oxygen supplies in the cargo bay. Other system improvements included upgraded carbon dioxide filters and stowage provision for cabin waste, additional food supplies and equipment.

USML-1 was the first of a planned series of at least four flights of the pressurised Spacelab module, which should have flown every two or three years. It was designed to advance US microgravity research efforts in several disciplines but, like many of these science-orientated Shuttle/Spacelab missions, the USML series was cancelled after

In Orbiter Processing Facility High Bay 3, workers continue to establish the mechanical interfaces between the USML-1 laboratory and Columbia. The first Extended-Duration Orbiter Pallet that allowed a mission duration of 13 days is visible to the left of the science module

only two missions due to the changing priorities in favour of the Shuttle–Mir and ISS programmes. USML-1 featured 31 experiments, ranging from manufacturing crystals for possible semiconductor use, to studies of the behaviour of liquids in microgravity. The flight also featured an experiment in manufacturing polymers as filters for terrestrial industries and another flight of the Shuttle Amateur Radio experiment. In addition, the EDO Medical Project (EDOMP) was a series of medical investigations designed to provide further data and experience in the development of counter-measures against the adverse effects of space flight on the human body. A significant focus of this research was in the re-adaptation process upon return to Earth, looking for potential problems that might hinder a station crew in the event of an emergency escape and recovery situation.

During the two-shift operation (in which Richards, Bowersox, Dunbar and DeLucas formed the Red Team, and Baker, Mead and Trinh the Blue Team), the crew worked with a whole range of equipment for the USML payload, including four experiments in the Crystal Growth Furnace, three experiments in the Drop Physics Module and sixteen experiments using the Glove Box. There was also a surface tension-driven convection experiment, a solid surface combustion experiment, a space acceleration measurement experiment and four biological experiments in the mid-deck.

The landing was delayed by a day due to rain at the primary landing site at Edwards AFB. Mission controllers hoped to land at Edwards, where Columbia would have had more room on the runway (and substantial overshoot capacity on the dry lake beds), given that it was returning with 104,328 kg of payload and flying new landing systems. However, the landing at Kennedy occurred without incident and saved precious processing time. The mission eclipsed all previous US manned space flight durations save for those of the three Skylab missions in 1973–1974. It also set a new US duration record for a spacecraft – as opposed to a space station – mission, surpassing the Gemini 7 record set in 1965.

Milestones

151st manned space flight
78th US manned space flight
48th Shuttle mission
12th flight of Columbia
1st Extended-Duration Orbiter (EDO) mission
1st landing of OV-102 at KSC
1st use of new synthetic tread tyres
1st flight of USML laboratory configuration
6th Spacelab Long Module mission

SOYUZ TM15

Int. Designation	1992-046A
Launched	27 July 1992
Launch Site	Pad 1, Site 5, Baikonur Cosmodrome, Kazakhstan
Landed	1 February 1993 (Solovyov/Avdeyev) 10 August 1992 (Tognini, with Mir EO-11 crew)
Landing Site	100 km northwest of Arkalyk
Launch Vehicle	R7 (11A511U2); spacecraft serial number (7K-M) #65
Duration	188 days 21 hrs 41 min 15 sec
	Tognini: 13 days 18 hrs 56 min 14 sec
Call sign	Rodnik (Spring)
Objective	Delivery of Mir-12 resident crew; French *Antares* mission (Tognini)

Flight Crew

SOLOVYOV, Anatoly Yakovlevich, 44, Russian Air Force, commander, 3rd mission
Previous missions: Soyuz TM5 (1988); Soyuz TM9 (1990)
AVDEYEV, Sergei Vasilyevich, 36, civilian, flight engineer
TOGNINI, Michel Ange Charles, 42, French Air Force, cosmonaut researcher

Flight Log

The Kurs system failed once again during the automated docking approach of Soyuz TM15 to Mir, forcing commander Solovyov to conduct a manual docking. During Tognini's twelve days on Mir, a programme of ten experiments were completed, encompassing medical and technological experiments under the French *Antares* programme. This was the third French flight to a space station, but the first commercial one. The earlier flights had proven so productive that CNES had arranged a series of missions every two years, building up a valuable database of orbital operations experience that could be applied to future programmes, such as the Freedom Space Station or dedicated French Shuttle/Spacelab missions that were under consideration (but which did not materialise). Tognini would return with the EO-11 crew in TM14.

The two Russian cosmonauts continued the rotation of resident crew teams on Mir, operating the onboard instruments during their work shifts, and ensuring that some would continue to operate autonomously while they were busy doing other things or were sleeping. Progress M14 docked on 18 August and its cargo included a 700 kg Vynosnaya Dvigatyelnaya Ustanovka (Outer Engine Unit), which was located in place of the tanker unit on the supply vessel. After it was automatically unloaded by commands from the ground on 2 September, the cosmonauts' task would be to install

The international Soyuz TM15 crew of Tognini (left), Solovyov (centre) and Avdeyev

it on top of the Sofora girder that was mounted on Kvant. This new unit would improve the attitude control capabilities of the complex. The cosmonauts' first excursion (3 September, 3 hours 56 minutes) saw them relocate the VDU unit from Progress to the worksite and prepare the girder for accepting the device. Four days later, the crew were back at the worksite (7 September, 5 hours 8 minutes) and bent the Sofora girder at a hinge point, one-third of the way down its length, to make the area where the unit would be placed more accessible. They fitted a communications cable along the girder and also took the opportunity to remove the USSR flag (or what was left of it after orbital debris and UV deterioration had taken their toll) that had been deployed the previous year. Four days later, the crew went outside for their third EVA (11 September, 5 hours 44 minutes) to attach the VDU atop the Sofora girder, which they then straightened to its full length. The final EVA of this residency (15 September, 3 hours 33 minutes) saw the cosmonauts relocate the Kurs antenna to the Kristall module. This would enable the next crew to arrive (aboard Soyuz TM16) to dock there, as the Progress M15 spacecraft would still be attached at the aft port when the new crew arrived. Solovyov and Avdeyev also took the opportunity to remove solar cells and material samples from the exterior of the station for return to Earth.

Towards the end of their residency, the crew used the base block airlock to eject a 16.5 kg satellite, called MAK-2, which would study the characteristics of the ionosphere. On 8 November, there was a "near miss" when the 55 kg Cosmos 1508 satellite (launched in 1983) passed within 300 metres of Mir. The end of the Progress M14 mission also saw the return of the fifth Raduga (Rainbow) payload recovery system, carrying approximately 150 kg of samples from Mir back to Earth.

Milestones

152nd manned space flight
74th Russian manned space flight
22nd Russian and 47th flight with EVA operations
15th Soyuz flight to Mir
12th main Mir crew
10th visiting crew (Tognini)
67th Soyuz manned mission
14th Soyuz TM manned mission
Avdeyev celebrates his 37th birthday in space (1 Jan)
Solovyov celebrates his 45th birthday in space (16 Jan)

STS-46

Int. Designation	1992-049A
Launched	31 July 1992
Launch Site	Pad 39B, Kennedy Space Center, Florida
Landed	8 August 1992
Landing Site	Runway 33, Kennedy Space Center, Florida
Launch Vehicle	OV-104 Atlantis/ET-48/SRB-BI052/SSME #1 2032; #2 2033; #3 2027
Duration	7 days 23 hrs 15 min 3 sec
Call sign	Atlantis
Objective	Deployment of ESA's European Retrievable Carrier (EURECA) and operation of joint NASA/ISA Tethered Satellite System (TSS)

Flight Crew

SHRIVER, Loren James, 48, USAF, commander, 3rd mission
Previous missions: STS 51-C (1985); STS-31 (1990)
ALLEN, Andrew Michael, 36, USMC, pilot
NICOLLIER, Claude, 47, civilian, ESA mission specialist 1
IVINS, Marsha Sue, 41, civilian, mission specialist 2, 2nd mission
Previous mission: STS-32 (1990)
HOFFMAN, Jeffrey Alan, 47, civilian, mission specialist 3, payload commander, 3rd mission
Previous missions: STS 51-D (1985); STS-35 (1990)
CHANG-DIAZ, Franklin Raymond de Los Angeles, 42, civilian, mission specialist 4, 3rd mission
Previous missions: STS 61-C (1986); STS-34 (1989)
MALERBA, Franco, 46, civilian, Italian Space Agency payload specialist

Flight Log

The launch of STS-46 was delayed just 45 seconds at $T-5$ minutes, to verify that the APUs were ready to start. The deployment of the European Space Agency's European Retrievable Carrier (EURECA) was delayed by one day due to a problem with its data-handling system. Following deployment from Atlantis using the RMS, EURECA's thrusters were fired to boost the platform to its planned operating altitude of about 500 km. The firing was planned to last 24 minutes, but lasted only six minutes due to unexpected altitude data from EURECA. The problem was resolved and the engines were restarted to place the payload in its operational orbit during the sixth day of the mission. EURECA was subsequently retrieved and returned to Earth during the STS-57 mission in 1993.

The EURECA satellite is hoisted above Atlantis's payload bay by the RMS prior to deployment. The 16-mm lens gives this 35-mm frame a "fish eye" effect. The Tethered Satellite System in centre frame is stowed in the payload bay prior to its planned operations later in the mission

The delay to the EURECA deployment also delayed the Tethered Satellite System experiment for a day. The objective of TSS was to demonstrate the technology of long-tethered systems in space and to demonstrate that such systems were useful for research. The investigations planned for the system on this mission included a variety of space plasma physics and electrodynamics investigations. TSS could operate in the upper reaches of the atmosphere at an altitude higher than the operating range of balloons but below that of orbiting satellites, providing prolonged data gathering far beyond that of sounding rockets. The experiment, if successful, would probably lead to follow-on research into the use of tether systems for generating electrical power, spacecraft propulsion, broadcasting from space, studying the atmosphere, using the atmosphere as a wind tunnel and controlled microgravity experiments.

The 518 kg satellite featured a 1.6-meter sphere mounted on both a pallet in the cargo bay and on the Spacelab Mission Peculiar Equipment Support Structure (MPESS) that supports TSS orbiter-based science instruments. The sphere had an electrically conductive surface and carried its science instruments mounted on extendable booms. The extended boom satellite support structure measured twelve metres when fully extended above the payload bay and the motorised reel used to deploy the satellite could hold up to 108 km of tether (on STS-46, this was limited to 20 km). A data acquisition system would acquire data from the satellite and control it when deployed. The programme envisaged 30 hours of deployed activity, with twelve experiments gathering data on the satellite, the support structure and the environment in which it was flying.

During this mission, the system suffered several failures. The No. 2 umbilical failed to retract from the tethered satellite and the satellite itself failed to deploy on the first "flyaway" attempt. The deployment was also punctuated by an unplanned stop at 179 metres, a second at 256 metres, and the inability to either deploy or retrieve the satellite at 224 metres. During STS-46, the satellite reached a maximum distance of 256 metres, instead of the planned 20 kilometres on the initial deployment, due to a jammed tether line. Despite numerous attempts over several days to free the tether, TSS operations were curtailed and the satellite successfully stowed for return to Earth. Post-flight investigations revealed that a protruding $\frac{1}{4}$-inch bolt had hampered deployment operations. Slack tether during the deployment operations was also likely to have resulted in the cable snagging in the Upper Tether Control Mechanism.

Frustrated by their setbacks with TSS, the crew nevertheless completed a range of secondary experiments and payloads, working on a two-shift system. Allen, Nicollier and Malerba formed the Blue Team, while Ivins, Hoffman and Chang-Diaz were the Red Team. Mission commander Shriver worked with either team. There were six NASA experiments located in the payload bay. These were designed to study the effects of the space environment on materials and equipment that were planned for future use on Space Station Freedom. The 70 mm IMAX Cargo Bay Camera was also in the payload bay and was remotely controlled by the crew from the aft flight deck to film scenes from the mission for use in future IMAX films. There were also three secondary payloads located in the mid-deck area, which the crew worked on during their flight.

The mission was extended by one day in order to complete science activities. This would be the last flight of Atlantis prior to a scheduled inspection and modification period. This was later extended to include additional modifications that would allow Atlantis to dock with the Mir space station. Atlantis was shipped to Rockwell in October 1992. Its next mission would be STS-66 in 1994.

Milestones

153rd manned space flight
79th US manned space flight
49th Shuttle mission
12th flight of Atlantis
6th flight of Shuttle pallet mission
1st European mission specialist (Nicollier)
1st European RMS operator (Nicollier)
1st Italian in space (Malerba)
TSS-1 was the longest structure ever flown in space (256 metres)
Allen celebrates his 37th birthday in space (4 Aug)

STS-47

Int. Designation	1992-061A
Launched	12 September 1992
Launch Site	Pad 39B, Kennedy Space Center, Florida
Landed	20 September 1992
Landing Site	Runway 33, Kennedy Space Center, Florida
Launch Vehicle	OV-105 Endeavour/ET-45/SRB BI-053/SSME #1 2026; #2 2022; #3 2029
Duration	7 days 22 hrs 30 min 23 sec
Call sign	Endeavour
Objective	Spacelab J (SL-J) research objectives utilising the pressurised Spacelab module

Flight Crew

GIBSON, Robert Lee "Hoot", 45, USN, commander, 4th mission
Previous missions: STS 41-B (1984); STS 61-C (1986); Cdr STS-27 (1988)
BROWN Jr., Curtis Lee, 35, USAF, pilot
LEE, Mark, 40, USAF, mission specialist 1, payload commander, 2nd mission
Previous mission: STS-30 (1989)
APT, Jerome "Jay", 43, civilian, mission specialist 2, 2nd mission
Previous mission: STS-37 (1991)
DAVIS, (Nancy) Jan, 37, civilian, mission specialist 3
JEMISON, Mae Carol, 34, civilian, science mission specialist
MOHRI, Mamoru Mark, 43, civilian, Japanese payload specialist

Flight Log

The 50th flight of the Space Shuttle series featured a joint venture in materials science and life science experiments between NASA and the Japanese NASDA agency. The mission began with the first on-time launch since November 1985, carrying an experiment programme of 24 materials and 20 life science experiments into orbit. Of the 44 investigations, 35 were sponsored by NASDA, seven were from NASA and two were joint efforts. As with previous missions, many of the experiments were designed to help prepare the astronauts for future work on space station and long-duration missions.

As one of the partners in the space station programme, Japan was eager to gain some experience in a dedicated Spacelab mission (termed Fuwatto 92, or First Materials Processing Test – FMPT) for its first Shuttle flight, rather that having a PS fly on a Shuttle mission with a smaller, more generic experiment programme. Aside from the main Spacelab experiment payload, the crew worked in two shifts (Red – Brown, Lee, Mohri; Blue – Apt, Davis, Jemison; Gibson worked with either shift as

The prime and alternative crew members inside the Spacelab J laboratory installed in the cargo bay of Endeavour during STS-47 launch-processing in July 1992. Kneeling from left Chiaki Mukai alternative Japanese PS; Davis; Takao Doi alternative Japanese PS. Standing from left are Brown, Lee, Apt, Mohri, Gibson, Jemison and Stanley Koszelak, an alternative PS who served as back-up to Jemison

required) on five mid-deck secondary payloads. Endeavour also carried twelve GAS canisters, ten of which were experiments and the other two acting as ballast mass. Of the five mid-deck secondary payloads, one was an Israeli experiment to study the ability of oriental hornets to orientate their combs in microgravity. In fact, the hornets were part of a very comprehensive "crew". In addition to the seven astronauts, STS-47 carried four female frogs, thirty chicken eggs, 180 oriental hornets, about 400 adult fruit flies, 7,200 fly larvae and two Japanese carp. The menagerie became known as "Hoot's Ark" after commander Robert "Hoot" Gibson (whose nickname came from a famous cowboy film star of the silent and early sound western movies).

Media interest focused on Jemison, both as the first African American woman in space and as a science mission specialist. She was the first (and, to date, only) career astronaut to be officially assigned to perform the tasks of a payload specialist. Jemison had been frustrated with her experiences at NASA for some time prior to this flight and declined the chance of a second mission, resigning from the astronaut office in March 1993 to pursue other interests. Media attention also focused on the first married couple in space (Lee and Davis), with the inevitable question arising of "will they or won't they become the first members of the 300 km high club." Rumours about this persisted, despite claims that no "marital experiments" were planned and the fact that Lee and Davis worked on different shifts throughout the mission.

Like several others, this mission was extended a day to gather more science from the experiment package. Offshore reports of rain around the Cape area were a factor in delaying the landing of Endeavour when it was finally due to return. Mission controllers passed up the first opportunity to land at KSC, but further analysis indicated that the cloud would not encroach over the Cape, allowing the Shuttle to land at the second opportunity.

Milestones

154th manned space flight
80th US manned space flight
50th Shuttle mission
2nd flight of Endeavour
7th Spacelab Long Module mission
1st on-time Shuttle launch since November 1985
1st married couple on same mission (Lee and Davis)
1st African American female in space (Jemison)
1st flight of a science mission specialist (Jemison)
1st Japanese to fly on the Shuttle (Mohri)

STS-52

Int. Designation	1992-071A
Launched	22 October 1992
Launch Site	Pad 39B, Kennedy Space Center, Florida
Landed	1 November 1992
Landing Site	Runway 33, Shuttle Landing Facility, Kennedy Space Center, Florida
Launch Vehicle	OV-102 Columbia/ET-55/SRB BI-054/SSME #1 2030; #2 2015; #3 2034
Duration	9 days 20 hrs 56 min 13 sec
Call sign	Columbia
Objective	Deployment of Laser Geodynamic Satellite II (LAGEOS II) and operation of US Microgravity Payload 1 (USMP-1)

Flight Crew

WETHERBEE, James D., 39, USN, commander, 2nd mission
Previous mission: STS-32 (1990)
BAKER, Michael A., 38, USN, pilot, 2nd mission
Previous mission: STS-43 (1991)
VEACH, Lacy, 49, civilian, mission specialist 1, 2nd mission
Previous mission: STS-39 (1991)
SHEPHERD, William Michael, 43, USN, mission specialist 2, 3rd mission
Previous missions: STS-27 (1988); STS-41 (1990)
JERNIGAN, Tamara E., 32, civilian, mission specialist 3, 2nd mission
Previous mission: STS-40 (1991)
MACLEAN, Steven Glenwood, 37, civilian, Canadian payload specialist 1

Flight Log

The original mid-October launch date for STS-52 slipped when it was decided to exchange No. 3 SSME over concerns about possible cracks in the LH coolant manifold on the engine nozzle. The revised launch on 22 October was delayed by two hours due to crosswinds at the Shuttle Landing Facility, violating the Return-To-Launch-Site criteria. There were also heavy clouds at the Banjul trans-oceanic abort landing site. Despite concerns about the weather, the decision was made to proceed with the launch, in spite of higher than permitted wind speeds at launch. This caused some controversy at the time, but NASA stated that they felt the launch was safe and was performed within the intent of the rule.

Once in orbit, the crew activated the USMP-1 payload, which contained three experiments mounted on two MPESS structures in the payload bay. The Lambda Point Experiment studied the properties of liquid helium in microgravity, while the

The Space Vision System (SVS) experiment is seen in the grasp of the RMS above the payload bay. Target spots placed on the Canadian Target Assembly (CTA) satellite were photographed and monitored as the arm moved around the payload bay holding the satellite. Computers measured the changing position of the dot pattern and provided real-time TV display of location and orientation of the CTA. This was an evaluation to aide future RMS operations in guiding the RMS more precisely during berthing and deployment activities

French Space Agency (CNES) and French Atomic Energy Commission (CEA)-sponsored Material pour L'Etude des Phénomènes Intéressant la Solidification sur Terre et en Orbite (MEPHISTO) included crystal growth experiments. The Space Acceleration Measurement System (SAMS) had flown on previous Shuttle missions to measure and record accelerations which could affect onboard experiments. Located in the payload bay of the orbiter and operated by ground-based science teams independently of the flight crew, these experiments were a "dress rehearsal" for telescience operations on space station and other free-flying satellites.

The LAGEOS II satellite was successfully deployed at the end of FD 1 by spin stabilisation. Two subsequent firings of the solid rocket stages placed the geodynamics satellite in its 5,900 km orbit inclined at 52° to the equator. The previous satellite, LAGEOS I, which had been launched on a Delta expendable launch vehicle in 1976, was located at 110° inclination. The 426 laser reflectors on LAGEOS II provided accurate mapping of the Earth's surface by using ground-based laser ranging systems and ground-based tracking stations worldwide. The possible applications for data collected by LAGEOS included calculations of the shifting of crustal plates, as well as rotation rates, tides and polar motion of the Earth. This data was also beneficial for global monitoring of regional fault movements in earthquake-prone areas of the Earth. By having two satellites in orbit, the data could be cross-referenced for confirmation and greater accuracy. The satellite was a joint project between NASA and the Agenzia Spaziale Italiana (ASI), the Italian Space Agency. The upper stage used for deployment of the satellite was the Italian Research Interim Stage (IRIS), also built by ASI and being evaluated on this flight for the first time for potential use on future Shuttle missions as an operational upper stage.

Canadian PS MacLean was responsible for the Canadian Experiments-2 (CANEX-2) programme of seven experiments, located both in the cargo bay and on the mid-deck of Columbia. His programme continued and expanded upon the work begun by Canadian astronauts aboard STS 41-G (Garneau) and STS-42 (Bondar), as part of Canadian involvement in Shuttle and Space Station operations. The primary experiment in the CANEX-2 package was the Space Vision System (SVS), in which a computerised "eye" would assist an astronaut in operating the RMS in situations where light or field of vision was restricted. The remaining CANEX-2 experiments included research into materials exposure (sample plates attached to the Canadian-built RMS), liquid-metal diffusion, phase partitioning in liquids, measurements of the Sun, photo-spectrometer in the atmosphere, the orbiter glow phenomena and space adaptation tests and observations.

The crew also worked with a number of mid-deck and payload bay secondary experiments, including the ESA-supplied ASP, a three-independent-sensor package that was designed to determine spacecraft orientation. There was also an experiment to control pressure in cryogenic fuel tanks in low gravity (which would have application to Space Station and long-duration space systems operations), protein crystal growth experiments, fluid mixing in microgravity, heat pipe performance in space, an experiment to study the proprietary protein molecule on twelve rodents, and an investigation into Shuttle RCS plume burn contamination.

One of the more challenging aspects of this mission regarding the payload was whether the capability of the Shuttle was being fully utilised on this flight. Commander Wetherbee commented from space that the experiment package his crew were dealing with was imposing time and power constraints on the mission, and that the crew were having a tough time staying out of each other's way while performing the mid-deck experiments. That the mission was so successful was made possible by very good pre-flight planning of both crew and experiment time.

Milestones

155th manned space flight
81st US manned space flight
51st Shuttle mission
13th flight of Columbia
1st flight of USMP payload
1st flight of Italian IRIS upper stage on Shuttle
Baker celebrates his 39th birthday in space (27 Oct)

STS-53

Int. Designation	1992-086A
Launched	2 December 1992
Launch Site	Pad 39A, Kennedy Space Center, Florida
Landed	9 December 1992
Landing Site	Runway 22, Edwards AFB, California
Launch Vehicle	OV-103 Discovery/ET-49/SRB BI-055/SSME #1 2024; #2 2012; #3 2017
Duration	7 days 7 hrs 19 min 47 sec
Call sign	Discovery
Objective	Deployment of classified DoD payload (DOD-1); operation of two secondary and nine mid-deck experiments

Flight Crew

WALKER, David Mathiesan, USN, commander, 3rd mission
Previous missions: STS 51-A (1984); STS-30 (1989)
CABANA, Robert Donald, USMC, pilot, 2nd mission
Previous mission: STS-41 (1990)
BLUFORD Jr., Guion Stewart, USAF, mission specialist 1, 4th mission
Previous missions: STS-8 (1983); STS 61-A (1985); STS-39 (1991)
VOSS, James Shelton, US Army, mission specialist 2, 2nd mission
Previous mission: STS-44 (1991)
CLIFFORD, Michael Richard Uram, US Army, mission specialist 3

Flight Log

STS-53 was Discovery's 15th mission, its first since STS-42 the previous year. During the intervening 23 months, 78 major modifications had been made to the orbiter while still at KSC. These included the addition of a drag parachute for landing and the capability of redundant nose wheel steering. The launch was delayed by one hour 25 minutes to allow the sunlight to melt ice on the ET that had accumulated thanks to overnight temperatures of −4°C.

The initial activity after reaching orbit was the deployment of a military satellite on FD 1. The satellite remains classified, although the payload was later identified as the third Advanced Satellite Data Systems Intelligence Relay Satellite. Once that deployment had been completed, the remainder of the mission became declassified. The crew continued with their experiment programme of two cargo bay and nine mid-deck experiments, most of which were instigated by the Defense Department Space Test Program Office, headquartered at Los Angeles AFB in California.

The experiment payload on Discovery included the Shuttle Glow Experiment/ Cryogenic Heat Pipe Experiment, which measured and recorded electrically charged

The end of one phase of Shuttle operations as Discovery lands on Runway 22 at Edwards AFB, signalling the final flight of dedicated DoD Shuttle missions. Almost eight years earlier in January 1985, the Discovery orbiter completed the first dedicated DoD mission STS 51-C landing at Kennedy

particles as they struck the tail of the orbiter. The second part of this experiment provided research into the use of super-cold LO pipelines for spacecraft cooling. Also in the payload bay was the NASA Orbital Debris Radar Calibration Spheres (ODERACS) experiment, designed to improve the accuracy of ground-based radars in detection, identification and tracking of orbital space debris. In the mid-deck, the Microcapsule in Space and Space Tissue Loss experiments were devoted to medical research, while the Vision Function Test measured the changes in astronauts' vision that might occur in the microgravity environment. The Cosmic Radiation Effects and Activation Monitor (CREAM) recorded levels of radiation inside the mid-deck, as did the Radiation Monitoring Experiment. There was a joint USN, US Army and NASA experiment for the crew to locate 25 preselected ground sites with a one nautical mile accuracy. This was an evaluation of detecting laser beams from space and the use of such beams in ground-to-spacecraft communications. Other experiments included a photographic assessment of cloud fields for DoD systems, while the Fluid Acquisition and Resupply Experiment studied the motion of liquids in microgravity during simulated refuelling of propellant tanks with distilled water. There were also seven medical tests, including a re-flight of the rowing machine rather than the treadmill for physical exercise.

This crew dubbed themselves "the Dogs of War Crew", as they represented all four branches of the US armed forces. Their training team had been called "Bad Dog" and these combined to have the STS-53 crew become known as the "Dog Crew" (and they often quipped that they were "working like dogs" throughout their mission). Walker was known as "Red Dog", Cabana was known as "Mighty Dog" and Clifford, being the rookie, was known as "Puppy Dog". Bluford became "Dog Gone" and Voss became "Dog Face". The crew mascot was known as "Duty Dog" and a stowaway that looked over the crew during the mission (a rubber dog mask hung over an orange launch and entry suit) was known as "Dog Breath".

The landing was originally scheduled for KSC but was diverted to Edwards due to clouds in the vicinity of the SLF. Following the landing, a small leak was detected in the forward thrusters, delaying the egress of the crew until fans and winds dissipated the leaking gas.

Milestones

156th manned space flight
82nd US manned space flight
52nd Shuttle mission
15th flight of Discovery
10th and final dedicated DoD Shuttle mission

STS-54

Int. Designation	1993-003A
Launched	13 January 1993
Launch Site	Pad 39B, Kennedy Space Center, Florida
Landed	19 January 1993
Landing Site	Runway 33, Kennedy Space Center, Florida
Launch Vehicle	OV-105 Endeavour/ET-51/SRB BI-056/SSME #1 2019; #2 2033; #3 2018
Duration	5 days 23 hrs 38 min 19 sec
Call sign	Endeavour
Objective	Deployment of TDRS-F by IUS-13; EVA operations, procedures and training exercise

Flight Crew

CASPER, John Howard, 48, USAF, commander, 2nd mission
Previous mission: STS-36 (1990)
McMONAGLE, Donald Ray, 38, USAF, pilot, 2nd mission
Previous mission: STS-39 (1991)
RUNCO Jr., Mario, 39, USN, mission specialist 1, 2nd mission
Previous mission: STS-44 (1991)
HARBAUGH, Gregory Jordan, 35, civilian, mission specialist 2, 2nd mission
Previous mission: STS-39 (1991)
HELMS, Susan Jane, 33, USAF, mission specialist 3

Flight Log

The launch of STS-54 was delayed by just seven minutes over concerns about upper-atmospheric wind levels. The primary objective of the mission was the deployment of the sixth Tracking and Data Relay Satellite by IUS, which was accomplished on FD 1.This was the fifth successful deployment, with TDRS-B having been lost in the January 1986 Challenger accident.

Also located in the payload bay was the "hitchhiker experiment", the Diffuse X-ray Spectrometer (DXS), which was designed to collect X-ray radiation data from diffuse sources in deep space. Despite some difficulties in operating this experiment, it did return good science data. Additional experiments included one to record microgravity acceleration, and a solid surface combustion experiment that recorded the rate of flame spread and the temperature of burning filter paper. A UV Plume Imager, mounted on a free-flying satellite, observed the orbiter and obtained data on the vehicle during controlled conditions. The crew also continued the long programme of photography of the Earth surface begun in the 1960s. This was catalogued at JSC as the Earth Observation Project, and liaised with various global scientific research

Harbaugh (back to camera) carries Runco along the starboard side of Endeavour's payload bay during the 17 January EVA. He is evaluating the ability of astronauts to move a "bulky" object about in space by hand

efforts, allowing Shuttle crews to document specific features and phenomena of interest.

The FD 5 EVA (17 Jan, 4 hours 28 minutes) was the first of a series of EVAs planned for the next three years, leading up to start of space station construction (planned at that time to be 1996). With only two missions having included EVA operations since the return to flight in 1988, there was an urgent need to develop experience, techniques and procedures on extensive EVA operations prior to constructing the space station. On STS-54, the two EVA astronauts (Harbaugh EV1 and Runco EV2) tested their ability to freely move around the payload bay, to climb into the foot restraints without using their hands, and to simulate carrying large objects in the microgravity environment. This was also useful in evaluating the possibility of rescuing an incapacitated crew member during an EVA. They also evaluated the differences and similarities in these tasks between training simulations and actual orbital operations. After the EVA, they answered detailed questions on their experiences and, following the mission's return to Earth (which was delayed by one orbit due to ground fog at the Cape), went back into the WETF to evaluate the tasks again, for comparison with the actual EVA and to help improve future training procedures.

The crew also participated in the Physics of Toys experiment during a 45-minute live TV transmission. Staff at the Museum of Natural Science helped to develop the course, which was designed to generate interest in science at schools. The toys were selected by a team of physicists. The Physics of Toys experiment was a follow-on to the Toys in Space package flown on STS 51-D in 1985.

Milestones

157th manned space flight
83rd US manned space flight
26th US and 48th flight with EVA operations
53rd Shuttle mission
3rd flight of Endeavour
6th TDRS deployment mission

SOYUZ TM16

Int. Designation	1993-005A
Launched	24 January 1993
Launch Site	Pad 1, Site 5, Baikonur Cosmodrome, Kazakhstan
Landed	22 July 1993 (with French cosmonaut J.P. Haigneré)
Landing Site	140 km East of Dzhezkazgan
Launch Vehicle	R7 (11A511U2); spacecraft serial No. (7K-M) 101
Duration	179 days 0 hrs 43 min 46 sec
Call sign	Vulkan (Volcano)
Objective	Delivery of the 13th resident crew to Mir

Flight Crew

MANAKOV, Gennady Mikhailovich, 42, Russian Air Force, commander, 2nd mission
Previous mission: Soyuz TM10/Mir EO-7 (1990)
POLESHCHUK, Alexandr Fedorovich, 39, civilian, flight engineer

Flight Log

The flight of the Soyuz TM16 spacecraft had a different background to most Soyuz ferry missions. Originally, it was intended to use the Soyuz as a one-man "rescue vehicle", on standby for the first manned flight of the Buran Space Shuttle. If required to bring the two Buran pilots back to Earth, the Soyuz would dock with Buran's Androgynous Peripheral Docking Assembly (APAS-89) to allow internal transfer between the two spacecraft. However, the first manned flight of Buran slipped into 1990 and the profile changed so that the Soyuz TM would dock directly with Buran and then go on to dock with Mir. Three spacecraft were ordered for rescue missions with Buran (serial numbers 101, 102 and 103). Spacecraft 101 eventually flew as TM16 when it became clear that Buran would never fly with a crew aboard. The Russian space shuttle programme was finally abandoned in 1992. TM16 would also utilise the outermost docking port of the Kristall module on Mir, which was intended for use with Buran but would now be tested by TM16 in preparation for the forthcoming US Shuttle–Mir docking missions. This programme had been agreed in June 1992 between the US and Russia, as part of the Phase 1 joint space station (Freedom) programme starting in 1995.

The EO-13 resident crew was the first two-man crew since August 1990, and only the fifth two-man TM launch in 15 missions. The planned Israeli commercial flight was cancelled and no third-seat replacement was scheduled. The docking occurred without a hitch, qualifying both the system and the port for docking vehicles off the longitudinal axis of the target spacecraft. After testing the structural integrity of the seven-spacecraft combination, the crew got down to the hand-over period with the

The Progress M18 is shown undocked from the Mir complex, with the Progress M17 (right at the aft Kvant port) and Soyuz TM16 (bottom at the Kristall docking port) attached to the station. This photo was taken from the approaching TM17 spacecraft (ESA image)

previous resident crew. The EO-12 crew eventually departed in TM15 on 1 February 1993.

During their busy residency, in addition to the continuing Mir science programme and normal housekeeping and maintenance chores, the Vulkan crew would work with four Progress spacecraft (Progress M15 through M18). When Progress M15 undocked, the attached Znamya (New Light, or Banner) 40 kg triangular reflector was dragged out of the rolling spacecraft by centrifugal force, unfurling to 20 m. It reflected sunlight to generate a 4,000-m-wide spot of light on Earth, demonstrating the feasibility of using space mirrors to illuminate polar regions during periods of extended darkness. The Progress later re-rendezvoused with Mir at a distance of 200 m and, using a monitor and recently fitted control columns in the Mir base block, Manakov was able to put the Progress through a series of manoeuvres to demonstrate that it was possible for a cosmonaut on Mir to automatically manoeuvre a vehicle nearby and, eventually, to dock remotely if the automatic system failed. The next Progress (M16) was undocked and re-docked by the cosmonaut using the new TORU manual docking system. M15 could not have done this, as it carried the older Znamya package.

During the first of three planned EVAs (19 Apr, 5 hours 25 minutes), the cosmonauts installed electrical drives on the side of Kvant, ready for deployable solar arrays that would be installed later. Using the Strela boom, the cosmonauts transferred one of the containers to the support framework that had been installed on Kvant during 1991. They noted that one of the Strela handles had become detached and had floated away, requiring a replacement to be shipped to Mir on the next Progress (M18). As a result, the second EVA planned for 23 April was cancelled and the third EVA became the mission's second excursion on 18 June (4 hours 23 minutes). During this EVA, the crew repaired the boom by replacing the handle and also installed the second electrical drive. They also completed a TV documentation of the exterior of the station. The crew landed in TM16 on 22 July, along with French cosmonaut Jean-Pierre Haigneré, who had arrived aboard Soyuz TM17 with the EO-14 crew for a three-week mission during the hand-over period.

Milestones

158th manned space flight
75th Russian manned space flight
23rd Russian and 49th flight with EVA operations
16th Soyuz flight to Mir
13th Mir resident crew
68th Soyuz flight
15th Soyuz TM flight
1st seven-docked-spacecraft configuration
Heaviest mass for the Mir complex to date (90 tons)
Manakov celebrates his 43rd birthday in space (1 Jun)

STS-56

Int. Designation	1993-023A
Launched	7 April 1993
Launch Site	Pad 39B, Kennedy Space Center, Florida
Landed	17 April 1993
Landing Site	Runway 33, Kennedy Space Center, Florida
Launch Vehicle	OV-103 Discovery/ET-54/SRB BI-058/SSME #1 2024; #2 2033; #3 2018
Duration	9 days 6 hrs 8 min 24 sec
Call sign	Endeavour
Objective	Operation of ATLAS-2 science package located in the payload bay; SPARTAN-201 free flying astronomy platform

Flight Crew

CAMERON, Kenneth Donald, 43, USMC, commander, 2nd mission
Previous mission: STS-37 (1991)
OSWALD, Stephen Scott, 41, USNR, pilot, 2nd mission
Previous mission: STS-42 (1992)
FOALE, Colin Michael, 36, civilian, mission specialist 1, payload commander, 2nd mission
Previous mission: STS-45 (1992)
COCKRELL, Kenneth Dale, 42, USNR, mission specialist 2
OCHOA, Ellen Lauri, 34, civilian, mission specialist 3

Flight Log

The 6 April launch attempt for STS-56 was halted at $T - 11$ seconds by the orbiter's computers. Incorrectly configured instrumentation on the LH high-point bleed valve in the main propulsion system indicated it was in an "off" position instead of the "on" position. The launch was set for 8 April, after a 48-hour scrub was implemented.

Once again the crew worked in two shifts throughout the mission. Foale and Cockrell operated the Red Shift, while Cameron, Oswald and Ochoa worked the Blue Shift. All seven ATLAS instruments had flown on ATLAS-1 in 1992 and were scheduled to fly a third time on ATLAS 3 in 1994. The ATLAS payload continued the collection of data regarding the relationship between energy from the Sun and the middle atmosphere of Earth, and how such a relationship affects the ozone layer. The other primary payload was deployed by the RMS on 11 April. The Shuttle Pointing Autonomous Research Tool for Astronomy-201 (SPARTAN-201) was a free-flying science platform whose configuration was designed to study the velocity and acceleration of solar wind particles and to make observations of the Sun's

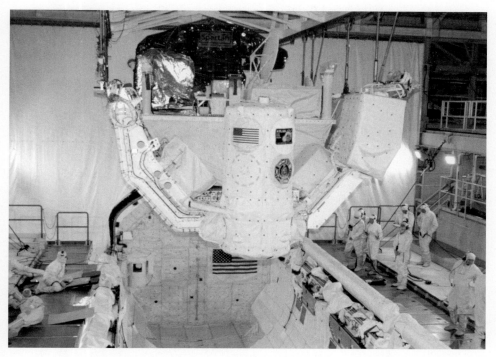

Mounted on a Spacelab pallet at centre is the primary payload of STS-56, ATLAS-2. The SPARTAN-201, with a protective covering over its instruments, is mounted directly behind ATLAS-2. This photo was taken in Bay 3 of the Orbiter Processing Facility at KSC during cargo-processing in February 1993

corona. The data collected during its independent flight was stored onboard tape recorders for play-back after its return to Earth. The RMS retrieved the SPARTAN-201 on 13 April.

In addition to the primary payload, the flight included a suite of mid-deck payloads, focusing on biomedical and life science research, Earth location targeting equipment and optical calibration tests. In addition, SAREX was flown again, with the crew making numerous radio contracts with schools across the globe. They also reported brief contact with the Russian Mir space station, the first such confirmed contact between the Shuttle and space station using amateur equipment.

The landing of STS-56 was originally planned for 16 April, but was delayed due to bad weather conditions. The payload commander on this mission, Mike Foale, indicated that he did not wish to fly the third ATLAS mission having flown the first two, as he did not wish to become known as "Mr. Atlas" for the rest of his astronaut career. He suggested Ellen Ochoa as the next PC. This format of re-flying crew members on the same series of missions freed up training time and utilised actual flight experience to provide an on-going flow of CB contact with series payloads from one flight to the next.

Milestones

159th manned space flight
84th US manned space flight
54th Shuttle mission
16th flight of Discovery
4th Spacelab pallet-only mission
1st amateur radio contact between Shuttle and Mir
Cockrell celebrates his 43rd birthday in space (9 Apr)

STS-55

Int. Designation	1993-027A
Launched	26 April 1993
Launch Site	Pad 39A, Kennedy Space Center, Florida
Landed	6th May 1993
Landing Site	Runway 22, Edwards AFB, California
Launch Vehicle	OV-102 Columbia/ET-56/SRB BI-057/SSME #1 3031; #2 2109; #3 2029
Duration	9 days 23 hrs 39 min 59 sec
Call sign	Columbia
Objective	Operation of the Spacelab D2 research programme located in the Long Module configuration

Flight Crew

NAGEL, Steven Ray, 47, USAF, commander, 4th mission
Previous missions: STS 51-G (1985), STS 61-A (1985); STS-37 (1991)
HENRICKS, Terence Thomas "Tom", 41, USAF, pilot, 2nd mission
Previous mission: STS-44 (1991)
ROSS, Jerry Lynn, 45, USAF, mission specialist 1, payload commander, 4th mission
Previous missions: STS 61-B (1985); STS-27 (1988); STS-37 (1991)
PRECOURT, Charles Joseph, 37, USAF, mission specialist 2
HARRIS Jr., Bernard Anthony, 36, civilian, mission specialist 3
WALTER, Ulrich, 38, civilian, German payload specialist 1
SCHLEGEL, Hans William, 41, civilian, German payload specialist

Flight Log

Getting STS-55 off the ground proved to be one of the more frustrating tasks of the Shuttle programme. The launch was originally set for late February 1993 but slipped back after problems arose with the turbine blade-tip seal retainers in the high-pressure oxidiser turbo pumps of the SSMEs. The option chosen was to replace the turbo-pumps at the pad, pushing the launch back to 14 March. This new launch date slipped when a hydraulic flex hose burst during a Flight Readiness Test. All twelve lines were removed and three of them had to be replaced before they could all be reinstalled. The revised 21 March launch was delayed by 24 hours due to a one-day delay in the launch of a preceding Delta II mission. Then, at $T - 3$ seconds on 22 March, the launch was aborted again by orbiter computers, this time because the #3 engine had failed to ignite. The third pad abort in the programme (the others being STS 41-D in 1984 and STS 51-F in 1985) was later traced to contamination during manufacture that had caused overpressure and precluded full engine ignition. All three engines were

German PS Walter works at the fluid physics experiment in the Spacelab D-2 science module aboard Columbia

replaced with spare units. The next attempt, on 24 April, was scrubbed when one of three IMUs gave possibly faulty readings and a 48-hour delay was scheduled to allow the removal and replacement of the IMU. Finally, on 26 April, the launch proceeded without incident. Following the launch, Pad A was scheduled for a period of refurbishment and modification which would last until February 1994.

The second German Spacelab mission featured 88 experiments in materials and life sciences, technology applications, Earth observations, astronomy and atmospheric physics. The crew would work in two shifts. The Red Shift comprised Precourt, Harris and Schlegel, while Nagel, Ross, Henricks and Walter worked the Blue Shift. After all the dramas of getting the mission off the ground, the crew encountered further problems in orbit. An overheating orbiter refrigerator/freezer unit in the mid-deck necessitated the use of a back-up to store samples, while a leaking nitrogen link in the waste water systems had to be fixed by the crew. The mission also suffered from a loss of communications for about 90 minutes due to an errant command from Mission Control in Houston (MCC-H). Columbia flew in a gravity gradient mode for most of the flight, which meant that onboard consumables were used at a reduced level. The mission management team determined that there was sufficient electrical power available to extend the mission by a day, which also meant that at landing Columbia had logged sufficient duration to bring the cumulative total flight time across the fleet (Columbia, Challenger, Discovery, Atlantis and Endeavour) to 365 days 23 hours 48 minutes – just over one year in space on 55 missions. The landing of STS-55 was originally set for KSC, but was moved to Edwards due to cloud cover over the Shuttle Landing Facility area at the Cape.

Most of the experiments were provided by the German Space Agency and ESA, with a number being supplied by Japan and three by NASA. The French Space Agency, CNES, was also involved in the mission. This was the final "national" Spacelab mission from Germany, due in part to NASA's reluctance to reduce the costs of flying such a large payload. Germany (and many other nations) decided that in future they would fly their experiments as part of International Spacelab missions. Despite this, valuable experience and information was gathered from the mission that would have relevance to the Columbus module that was being designed by ESA for the US space station (Freedom) programme.

Milestones

160th manned space flight
85th US manned space flight
55th Shuttle mission
14th flight of Columbia
8th Spacelab Long Module mission
2nd dedicated German Spacelab mission
Accumulated flight time for all Shuttles exceeds 1 year

STS-57

Int. Designation	1993-057A
Launched	21 June 1993
Launch Site	Pad 39B, Kennedy Space Center, Florida
Landed	2 July 1993
Landing Site	Runway 33, Shuttle Landing Facility, Kennedy Space Center, Florida
Launch Vehicle	OV-105 Endavour/ET-58/SRB BI-059/SSME #1 2019; #2 2034; #3 2017
Duration	9 days 23 hrs 44 min 54 sec
Call sign	Endeavour
Objective	Retrieval of the ESA EURECA reusable platform; inaugural flight of the SpaceHab laboratory module; 2nd EVA operations and training exercise

Flight Crew

GRABE, Ronald John, 47, USAF, commander, 4th mission
Previous missions: STS 51-J (1985); STS-30 (1989); STS-42 (1992)
DUFFY, Brian, 39, USAF, pilot, 2nd mission
Previous mission: STS-45 (1992)
LOW, George David, 37, mission specialist 1, payload commander, 3rd mission
Previous missions: STS-32 (1990); STS-43 (1991)
SHERLOCK, Nancy Jane, 34, US Army, mission specialist 2
WISOFF, Peter Jeffrey Karl, 34, civilian, mission specialist 3
VOSS, Janice Elaine, 36, civilian, mission specialist 4

Flight Log

The original mid-May launch of STS-57 was rescheduled to June to allow both launch and landing to occur in daylight. The 3 June launch date slipped when a decision was made to replace the high-pressure oxidiser turbo-pump on SSME #2 after concerns over a misplaced inspection stamp on a spring in the pump. This also allowed time to investigate an unexplained loud noise heard after Endeavour arrived at the pad, which was eventually traced to a ball strut tie rod assembly inside the 43 cm LH line. The 20 June attempt was scrubbed at $T - 5$ minutes due to low clouds and rain at the SLF, as well as weather concerns at all three TAL landing sites.

SpaceHab was a commercially developed pressurised laboratory designed to double the workspace available for crew-tended experiments. Originally developed to meet the need for commercial scientific payloads that could be accommodated in the mid-deck, such experiments and payloads did not require the use of the Spacelab pressurised modules and could be flown on non-Spacelab missions. However, the

High over Baja California, astronauts Low (nearest camera) and Wisoff attach a mobile foot restraint to the end of the RMS. Also in frame are the new SpaceHab augmentation laboratory module (foreground) and the captured EURECA (at rear of payload bay)

mid-deck locker space available on a Shuttle flight was limited and additional room was required, hence the development of SpaceHab. Though the commercial aspect of Shuttle operations did not materialise as originally foreseen, SpaceHab nevertheless evolved into an effective logistics carrier for later Shuttle–Mir and ISS operations. The inaugural flight contained 22 experiments (half supplied by NASA) that covered materials and life sciences, as well as a waste water recycling experiment that was planned for use on the space station. In addition, there was a hitchhiker experiment in the payload bay designed to collect data on X-ray radiation from diffuse sources in deep space. As with most Shuttle missions, several secondary experiments were carried in the mid-deck in addition to those located in the SpaceHab module. These additional experiments supported the materials and life sciences experiments flown in the module.

During FD 4, the crew retrieved the ESA EURECA carrier using the RMS. During attempts to stow the EURECA in the payload bay, ground controllers were unable to command the two antennas to fold, so the astronauts spent the start of their EVA period on 25 June manually folding both antennas to enable the carrier to be returned to Earth. This vehicle had been deployed from STS-46 in 1992 on the first of five planned flights of the carrier, each lasting six months over a ten-year period. A second flight opportunity in 1995 was passed up by ESA due to lack of funds and the 1997 launch opportunity was also not taken up. The EURECA reusable carrier remained unused after its first mission.

In the second of the series of generic EVAs planned for 1993, Low (EV1) and Wisoff (EV2) spent 5 hours 50 minutes outside Endeavour on FD 5. In addition to manually folding the EURECA antennas, the two astronauts became the first crew to use the Shuttle airlock as part of the payload bay tunnel extension. All previous Shuttle-based EVAs had been directly out of the airlock in the mid-deck. The astronauts evaluated moving a large mass (another astronaut) while attached to the foot restraints on the RMS, or located on the side of the payload bay. They also evaluated the movement of safety tethers while on the RMS and worked with different tools to gauge the stability of the restraints while tightening or loosening a bolt. These tests were important evaluations for both the space station and for the forthcoming Hubble Telescope service planned for mission STS-61 later in the year. While working in the night side of the orbit, the EVA crew experienced low temperatures which caused shivers, numbness and painful hands.

Landing attempts on 29 and 30 June were waived off due to unacceptable cloud cover and rain at KSC, the first time since STS 61-C in January 1986 that a mission had received two waive-offs. Following their eventual landing on 2 July, the crew talked with the STS-51 crew, who were conducting a training exercise in Discovery on Pad 39B. This was the first orbiter-to-orbiter conversation since the orbiting STS 51-D crew talked with the STS 51-B crew at KSC in 1985. Following STS-57, Endeavour was scheduled for an extended inspection period.

Milestones

161st manned space flight
86th US manned space flight
56th Shuttle mission
4th flight of Endeavour
27th US and 50th flight with EVA operations
1st flight of SpaceHab augmentation module

SOYUZ TM17

Int. Designation	1993-043A
Launched	1 July 1993
Launch Site	Pad 1, Site 5, Baikonur Cosmodrome, Kazakhstan
Landed	14 January 1994
	22 July 1993 (Haigneré with EO-13 crew in TM16)
Landing Site	215 km west of Karaganda
Launch Vehicle	R7 (11A511U2); spacecraft serial number (7K-M) 66
Duration	196 days 17 hrs 45 min 22 sec (Tsibliyev and Serebrov);
	20 days 16 hrs 8 min 52 sec (Haigneré)
Call sign	Sirius (Sirius)
Objective	Mir resident crew 14; French *Altair* mission (Haigneré)

Flight Crew

TSIBLIYEV, Vasily Vasilyevich, 39, Russian Air Force, commander
SEREBROV, Alexandr Alexandrovich, civilian, flight engineer, 4th mission
Previous missions: Soyuz T7 (1982); Soyuz T8 (1983); Soyuz TM8 (1989)
HAIGNERÉ, Jean-Pierre, French Air Force, cosmonaut researcher

Flight Log

Two days after launch, Soyuz TM17 was station-keeping some 200 metres from Mir. With all docking ports occupied, the Soyuz could not immediately dock, so the crew filmed the departure of Progress M18 from the front port of Mir. TM17 took its place a short time later. Haigneré's previous training as back-up to the last French visiting mission enabled his "*Altair*" mission to be advanced by six months, but this also meant that there was insufficient time to develop a whole new science programme. Haigneré would therefore use some of the experiments left on board by Tognini almost a year before, together with about 100 kg of new science hardware that had been delivered aboard Progress M18. Haigneré also completed a programme of medical investigations during his time aboard Mir before returning to Earth with the EO-13 crew aboard TM16 on 22 July.

Settling down to their stay on Mir, the EO-14 crew prepared themselves for some of the EVA tasks that had been reassigned to them from the previous residency. Before they could begin their EVA programme, however, the station had to endure an unusually intense Perseid meteor shower, which peaked on the night of 12–13 August. The cosmonauts remained at work, although plans were available for their emergency recovery should the need arise. They also constantly monitored the hits recorded on the station around the clock, including ten window impacts that resulted in craters ranging between 1.5 and 4 mm in diameter. Impact sensors on the station recorded a particle flux of around 2,000 times greater than normal annual showers. The crew also

The Soyuz TM17 crew of Serebrov (left), Tsibliyev (centre) and French cosmonaut Haigneré

recorded about 240 micrometeoroids burning up in the atmosphere. NASA had delayed the STS-51 mission so that it would not fly through the shower and incur damage.

The EVA series began with two EVAs to construct the Rapana girder on top of Kvant (16 Sep for 4 hours 18 minutes and 20 Sep for 3 hours 13 minutes). This was a girder technology experiment to analyse new materials that could be incorporated into similar structures for the proposed Mir 2 programme. On their third EVA (28 Sep for 1 hour 52 minutes), the cosmonauts conducted an observation programme on the outside of Mir, finding a 5 mm hole through one of the solar arrays, surrounded by a number of panels that had been cracked several cm across. It was not possible to determine whether this damage had been caused during the recent Perseid shower. This inspection work was completed on their fourth and fifth EVAs (22 Oct for 38 minutes and 29 Oct for 4 hours 12 minutes) and in addition, the crew deployed and retrieved sample cassettes from the exterior of the station.

In October, the crew was asked to extend their mission into January, as the Energomash factory in Samara which produced the upgraded engines used on the Soyuz U vehicle would not release any more engines until it was paid, and there was no money available. This stop–start nature of the Russian programme would become a regular occurrence as the once state-driven economy and space programme gave way to the vagaries of the corporate machine. Progress M19 and M20 included Raduga sample return capsules (as had M18), the latter having launched on a former military R7, the only one available at Baikonur for its mission to keep the station supplied. In December, it was announced that Russia and the US, together with their international partners, had agreed to cooperate in the construction of what was now being called the

International Space Station. This evolution of the abandoned design studies and budget reviews of the US-led Freedom space station would include Mir as a key part of its preparations.

On 14 January, after the arrival of the next resident crew and the brief hand-over period, Tsibliyev and Serebrov undocked Soyuz TM17 for their trip home. Instead of the normal reverse away from the port, however, it was decided to fly TM17 in proximity to Kristall to take close-up images that would benefit the Shuttle–Mir programme. These images were supposed to enable NASA Shuttle pilots to become familiar with the aerials and other appendages on the module. Unfortunately, Tsibliyev failed to realise that his translation control was in stand-by mode and was unable to prevent TM17 from striking Kristall a glancing blow approximately 1 m from the spherical docking system station. Aboard Mir, the new resident crew felt no impact and a later inspection by TM18 when it was being relocated to another docking port did not reveal any serious damage.

Milestones

162nd manned space flight
76th Russian manned space flight
24th Russian and 51st flight with EVA operations
17th manned Soyuz flight to Mir
14th Mir resident crew
69th manned Soyuz flight
16th manned Soyuz TM flight

STS-51

Int. Designation	1993-058A
Launched	12 September 1993
Launch Site	Pad 39B, Kennedy Space Center, Florida
Landed	22 September 1993
Landing Site	Runway 15, Shuttle Landing Facility, Kennedy Space Center, Florida
Launch Vehicle	OV-103 Discovery/ET-59/SRB BI-060/SSME #1 2031; #2 2034; #3 2032
Duration	9 days 20 hrs 11 min 11 sec
Call sign	Discovery
Objective	Deployment of the Advanced Communication Technology Satellite (ACTS); deployment and retrieval of the Orbiting and Retrieval Far and Extreme UV Spectrograph-Shuttle Pallet Satellite (ORFEUS-SPAS); EVA procedures and demonstration test

Flight Crew

CULBERTSON Jr., Frank Lee, 44, USN, commander, 2nd mission
Previous mission: STS-38 (1990)
READDY, William Francis, civilian, pilot, 2nd mission
Previous mission: STS-42 (1992)
NEWMAN, James Hansen, 36, civilian, mission specialist 1
BURSCH, Daniel Wheeler, 36, USN, mission specialist 2
WALZ, Carl Erwin, 38, USAF, mission specialist 3

Flight Log

The original launch attempt on 17 July was scrubbed during the $T - 20$ minute hold due to premature and unexplained charging of pyrotechnic initiator controllers on the LH vent arm umbilical and the SRB hold-down bolts. The problem was traced to a faulty control card on the Mobile Launch Platform (MLP). After an abbreviated countdown which commenced on 23 July, the 24 July launch attempt was halted at $T - 19$ seconds due to problems with the APU turbine assembly in one of the two hydraulic power units on the right SRB. The APUs were replaced on the pad. The launch was rescheduled for 4 August and again for 12 August because of concerns over the Perseid meteor shower which was due to peak on 11 August. The 12 August attempt was aborted at the $T - 3$ second mark due to a faulty sensor which was monitoring fuel flow on SSME #2. This resulted in the fourth pad abort in the programme and the second of 1993. All three engines were subsequently changed out on the pad. The launch was then scheduled for 10 September, but following loss of

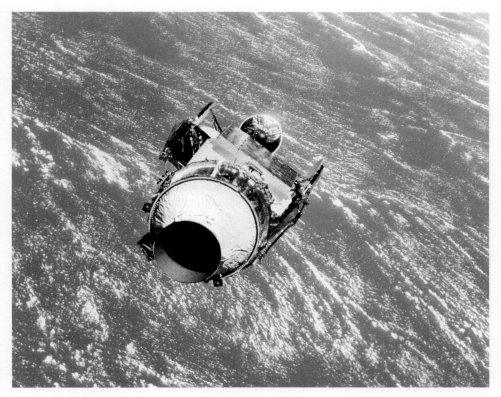

The ACTS satellite, with its attached TOS upper stage, is shown following release from Discovery, completing the first major objective of the mission

communications with the Mars Observer spacecraft and the NOAA-123 satellite, NASA slipped the launch 48 hours in order to review the design, production and testing of ACTS prior to committing it to launch and deployment.

The ACTS was successfully deployed during FD 1 and, some 45 minutes after leaving the payload bay of Discovery, the TOS was fired to take the satellite towards its geostationary operational orbit. The second deployment of the mission came on FD 2, when ORFEUS-SPAS was released by RMS for six days of independent data collection, the first of a planned series of ASTRO-SPAS missions. The SPAS also carried the IMAX camera, which recorded spectacular images of Discovery flying in orbit both during release and recapture of the pallet satellite.

As ACTS was being checked out in its geosynchronous orbit and ORFEUS was away gathering data, the crew of Discovery settled down to their own science programme of payload bay and mid-deck experiments. These included the exposure to space of selected materials for a short duration, protein crystal growth, chromosome and plant cell division, high-resolution Shuttle glow spectrograph photography of the aurora, an investigation into polymer membranes processing, further calibration of the Air Force Maui Optical Site and radiation monitoring inside the crew compartment. The crew also continued the programme of Earth resources and phenomena observations.

On FD 5 (16 Sep), Walz (EV1) and Newman (EV2) performed the third and final generic EVA (7 hours 5 minutes) to evaluate tools, tethers, foot restraints and mobility. Part of this programme was connected to the forthcoming Hubble Service Mission manifested for STS-61. This time the RMS was not used to support the EVA, because it was needed later in the flight to retrieve ORFEUS-SPAS. This also provided a "minimum equipment scenario", with the crew making optimum use of materials already aboard for other purposes. This could prove essential for a mission requiring EVA where no RMS was available.

The landing of STS-51 on 21 September was called off due to the possibility of rain showers within 48 km of the SLF. Discovery eventually came home to the first end-of-mission night landing at the Cape.

Milestones

163rd manned space flight
87th US manned space flight
57th Shuttle mission
17th flight of Discovery
28th US and 52nd flight with EVA operations
1st use of Transfer Orbit Stage
1st end of mission night landing at KSC
3rd and final test and demonstration EVA

STS-58

Int. Designation	1993-065A
Launched	18 October 1993
Launch Site	Pad 39B, Kennedy Space Center, Florida
Landed	1 November 1993
Landing Site	Runway 22, Edwards AFB, California
Launch Vehicle	OV-102 Columbia/ET-57/SRB BI-061/SSME #1 2024; #2 2109; #3 2018
Duration	14 days 0 hrs 12 min 32 sec
Call sign	Columbia
Objective	Operation of the second dedicated Spacelab Life Sciences payload using the Spacelab Long Module configuration

Flight Crew

BLAHA, John Elmer, 51, USAF, commander. 4th mission
Previous missions: STS-29 (1989); STS-33 (1989); STS-43 (1991)
SEARFOSS, Richard Alan, 37, USAF, pilot
SEDDON, Margaret Rhea, 45, civilian, mission specialist 1, payload commander, 3rd mission
Previous missions: STS 51-D (1985); STS-40 (1991)
McARTHUR Jr., William Surles, 42, US Army, mission specialist 2
WOLF, David Alexander, 37, civilian, mission specialist 3
LUCID, Shannon Wells, 50, civilian, mission specialist 4, 4th mission
Previous missions: STS 51-G (1985); STS-34 (1989); STS-43 (1991)
FETTMAN, Martin Joseph, 36, civilian, payload specialist 1

Flight Log

The first attempt at launching STS-58 on 14 October was scrubbed at the $T-31$ second mark, as a result of a failed range safety computer. The next attempt on 15 October was scrubbed at $T-9$ minutes due to a failed S-band transponder aboard Columbia. The launch on 18 October was also delayed, but only by a few seconds due to an aircraft straying into the launch exclusion zone.

Over the next 14 days, the crew, working a single-shift system, conducted the SLS-2 research programme and other research objectives, including the Orbiter Acceleration Research Experiments, SAREX, and Pilot In-flight Landing Operations Trainer (PILOT), a portable laptop computer simulator that allowed the commander and the pilot to maintain their proficiency for approach and landing on longer missions.

The SLS payload included 14 experiments focusing on four areas: regular physiology, cardiovascular/cardiopulmonary, musculoskeletal and neuro-science experiments. The Rotating Dome Experiment was used in conjunction with the first

Rhea Seddon spins the rotating chair as PS Fettman serves as a test subject during the SLS-2 mission

flight prototype of the Astronaut Science Advisor (ASA), a laptop computer program designed to assist the crew member in conducting experiments to increase the efficiency of activities. This was also termed the "principle investigator in a box". Of the fourteen experiments, eight focused on the astronauts, while the other six were conducted on the 48 rodents aboard. Six of the rodents were killed and dissected during the mission, yielding the first tissue samples collected during a space mission which were not altered by re-exposure to the Earth's gravity.

During the mission, the crew collected over 650 different samples from the rodents and themselves. This greatly increased the database of life science research and this work continued, at least for the "payload crew" (Seddon, Fettman, Lucid and Wolf) after landing. For the first week after the end of the mission, these four astronauts gave regular blood and urine samples to reveal how the body readjusted to gravity after two weeks in space. The blood samples were collected over a period of 45 days after landing. The combined data from SLS-1 and SLS-2 helped to build a more comprehensive picture of how animals and humans adapted to space flight and readapted to life back on Earth, an important milestone in developing protocols and research programmes for the space station. There were plans to fly a third SLS, which could have become a dedicated French Spacelab mission, but this was not pursued due to budget restrictions, launch manifest constraints and the introduction of the Shuttle–Mir programme, which drew resources away from Spacelab missions.

Milestones

164th manned space flight
88th US manned space flight
58th Shuttle mission
15th flight of Columbia
2nd flight of SLS series
9th Spacelab Long Module mission
Longest Shuttle mission to date
4th longest US spaceflight (after three Skylab missions)
1st veterinarian to fly in space (Fettman)
1st tissue samples collected during a space flight (rodents)

STS-61

Int. Designation	1993-075A
Launched	2 December 1993
Launch Site	Pad B, Launch Complex 39, Kennedy Space Center, Florida
Landed	12 December 1993
Landing Site	Runway 33, Shuttle Landing Facility, KSC, Florida
Launch Vehicle	OV-105 Endeavour/ET-60/SRB BI-063/SSME #1 2019; #2 2033; #3 2017
Duration	10 days 19 hrs 58 min 37 sec
Call sign	Endeavour
Objective	First Hubble Service Mission (SM-1)

Flight Crew

COVEY, Richard Oswalt, 47, USAF, commander, 4th mission
Previous missions: STS 51-I (1985); STS-26 (1988); STS-38 (1990)
BOWERSOX, Kenneth Duane, 37, USN, pilot, 2nd mission
Previous mission: STS-50 (1992)
THORNTON, Kathryn Cordell Ryan, 41, civilian, mission specialist 1, 3rd mission
Previous missions: STS-33 (1989); STS-49 (1992)
NICOLLIER, Claude, 49, civilian, mission specialist 2, 2nd mission
Previous mission: STS-46 (1992)
HOFFMAN, Jeffery Alan, 49, civilian, mission specialist 3, 4th mission
Previous missions: STS 51-D (1985); STS-35 (1990); STS-46 (1992)
MUSGRAVE, Franklin Story, 58, civilian, mission specialist 4, payload commander, 5th mission
Previous missions: STS-6 (1983); STS 51-F (1985); STS-33 (1989); STS-44 (1991)
AKERS, Thomas Dale, 42, USAF, mission specialist 5, 3rd mission
Previous missions: STS-41 (1990); STS-49 (1992)

Flight Log

Described as one of the most challenging manned missions ever attempted, the crew of STS-61 completed a record-breaking five back-to-back EVAs during the first on-orbit service of the Hubble Space Telescope. Many of their tasks were completed sooner than expected, allowing the few contingencies that did occur to be dealt with smoothly. The original launch was to have occurred from Pad 39A at KSC, but following the rollout of the stack to the pad, contamination was discovered in the payload change-out room. As a result, the STS-61 launch was moved to Pad B.

At an altitude of 522 km above the Earth, Musgrave (top) and Hoffman are seen riding on the RMS during the fifth and final EVA of the mission to service the Hubble Telescope, one of the most successful space missions to date. The west coast of Australia forms the backdrop to the scene

The move occurred without incident on 15 November but the first launch attempt on 1 December was scrubbed due to adverse weather conditions at the SLF.

A series of service missions had always been part of the HST programme. At regular intervals, a Shuttle would be sent to repair, replace or upgrade onboard instruments, equipment or systems prolonging the operational life of the facility and improving the quality and quantity of scientific discoveries over the planned fifteen-year life of the telescope. With the focusing difficulties encountered shortly after deployment from STS-31 in 1990, some media reports incorrectly labelled this flight as rescue mission, specially organised to save the telescope. The mission did restore the telescope to full working order, but the corrective optics were incorporated into a far more extensive, and already planned, servicing operation. Rendezvous with Hubble was achieved on FD 3, with the RMS grapple and berthing in the payload bay completed the same day. The telescope was berthed upright in the payload bay of the Shuttle, but remained under the command of the Space Telescope Operations Control Center (STOCC) located at the Goddard Space Flight Center. Following each servicing task, the STOCC controllers verified the interfaces between the new or serviced hardware and the telescope, ensuring at each stage that the telescope would be capable of independent operations once released from the payload bay.

Over a five-day period (4–8 December), the EVA team of four astronauts worked in pairs to complete the complex and demanding programme to restore the telescope to full working order. During the first EVA, four gyros that were situated in pairs in two Rate Sensing Units were replaced, along with two Electronic Control Units that directed the RSUs and eight electrical fuse plugs. The first EVA (7 hours 54 minutes on 4 Dec, conducted by Hoffman (EV1) and Musgrave (EV2)) was the second longest in the US programme to date and the only problem encountered was difficulty in closing the compartment doors after replacing the RSUs. During the next EVA (6 hours 36 minutes on 5 Dec, conducted by Thornton (EV3) and Akers (EV4)), one of the primary objectives of the servicing mission was completed, that of installing new solar arrays. The old arrays were scheduled to be returned to Earth for examination after over three years in space, but one of them refused to fully retract due to a kink in the framework and had to be jettisoned. The other was stowed in the payload bay without difficulty.

The third EVA (6 hours 47 minutes on 6 Dec, the second for Hoffman and Musgrave) was designed to replace the Wide Field/Planetary Camera (WF/PC), one of the five scientific instruments on the telescope, in a four-hour operation. In fact, the astronauts accomplished the exchange with the improved WF/PCII (an upgraded spare modified to compensate for the flawed mirror) in just forty minutes. Two magnetometers were also installed in the top of the telescope. EVA 4 (6 hours 50 minutes on 7 Dec, the second for Thornton and Akers) included the replacement of another primary instrument, the High-Speed Photometer, with the Corrective Optics for Telescope Axial Replacement (COSTAR) unit – often dubbed Hubble's "spectacles" – which redirected light to three of the four remaining instruments, thus compensating for the flaw in the primary mirror. The astronauts also installed a co-processor that improved the memory and speed of the onboard computer. During this EVA, Tom Akers achieved a new cumulative record for an American astronaut on EVA (29 hours 39 minutes), surpassing the 20-year-old record set by Gene Cernan on Apollo 17 at 24 hours 14 minutes. Kathy Thornton became the record-holder for female EVA astronauts at 21 hours 10 minutes. Both had performed EVAs on STS-49 in 1992. During the final EVA (7 hours 21 minutes on 9 Dec, the third by Hoffman and Musgrave), the astronauts replaced the Solar Array Drive Electronics (SADE) unit, as well as installing the Goddard High Resolution Spectrograph Redundancy (GHRS) equipment and placing two protective covers over the original magnetometer.

During FD 8, prior to the final EVA, the Shuttle's orbit was boosted to 595 km. At this height, the telescope would be released on FD 9, after deployment of the twin boom antennas, unfurling of solar arrays and checking of onboard systems. The redeployment was delayed several hours when ground controllers had to troubleshoot erratic telemetric data from the telescope's systems monitor. This had occurred before and was not connected to the recent servicing by the astronauts. The mission ended one orbit earlier than planned to allow the crew two landing opportunities at KSC.

Milestones

165th manned space flight
89th US manned space flight
59th Shuttle mission
5th flight of Endeavour
1st Hubble servicing mission
29th US and 53rd flight with EVA operations
1st flight of ESA astronaut as MS2
US astronaut cumulative EVA record – Akers
World female cumulative EVA record – Thornton

SOYUZ TM18

Int. Designation	1994-001A
Launched	8 January 1994
Launch Site	Pad 1, Site 5, Baikonur Cosmodrome, Kazakhstan
Landed	9 July 1994 (Afanasyev and Usachev)
	22 Mar 1995 (Polyakov, with the EO-17 crew aboard TM20)
Landing Site	112 km north of Arkalyk
Launch Vehicle	R7 (11A511U2); spacecraft serial number (7K-M) 67
Duration	182 days 0 hrs 27 min 2 sec (Afanasyev and Usachev); 437 days 17 hrs 58 min 31 sec (Polyakov)
Call sign	Derbent (Derbent)
Objective	Fifteenth main Mir crew and the beginning of a record-breaking extended-duration orbital flight of 14 months for Dr. Polyakov

Flight Crew

AFANASYEV, Viktor Mikhailovich, 45, Russian Air Force, 2nd mission
Previous mission: Soyuz TM11/Mir EO-8 (1990)
USACHEV, Yuri Vladimirovich, 36, civilian NPO Energiya, flight engineer
POLYAKOV, Valery Vladimirovich, 51, civilian, cosmonaut researcher, 2nd mission
Previous mission: Soyuz TM6/Mir EO-3/4 (1988)

Flight Log

The launch of TM18 had been delayed due to the unavailability of the more powerful Soyuz U launch vehicle to lift the three-man crew. The flight of Polyakov was a logical step in the Russian quest for long-duration space flight experience and medical data. When Polyakov devised the programme for a second space flight, he aimed for an 18-month duration, but delays forced him to curtail the duration to 14 months, as he could not remain aboard Mir when the first NASA astronaut arrived on the station, which at the time was planned for early 1995. Though he carried out his own research programme, Polyakov still participated in other tasks, working with three different resident crews until his return in March 1995.

One of the first tasks on this mission was to relocate the Soyuz TM18 ferry from the aft port to the front port of the base block, which occurred on 24 January. During the short flight, the crew flew past the Kristall module and reported only minor scratches on its hull from where TM17 had struck it. Work for this resident crew was again limited by plans for the upcoming American docking missions, particularly the integration of Soyuz and Progress launches and amendments to subsequent

Record-breaker Polyakov (left) with the rest of the Soyuz TM18 crew, Afanasyev (centre) and Usachev

resident crews to accommodate the joint programme with the Americans. At the time of Afanasyev and Usachev's flight, this national and international coordination was difficult to implement, mainly due to the lack of funds, and soon the launch of the next resident crew had slipped from April to July. However, the EO-15 crew continued their research along the lines of previous resident crews, but also conducted medical and technical experiments sponsored by German institutes. Polyakov had also supplemented the science payload with smaller items brought up in his personal baggage on TM18.

In February, Sergei Krikalev was launched on the American Shuttle mission STS-60, the first time that cosmonauts had been in space at the same time on different missions in spacecraft belonging to different nations. The following month, Progress M22 was also delayed for three days from 19 March when heavy snowfall at the launch site resulted in snow drifts covering the rail network to a depth of up to seven metres, making it impossible to move the spacecraft and its booster from the assembly building to the launch pad.

At the end of March, the cosmonauts on Mir participated in an experiment with a Swedish satellite called Freja, designed to study space plasma and magnetosphere physics in Earth's magnetosphere and ionosphere. Launched in 1992 by the Chinese, Freja was located 1,770 km above the Alaskan coast when the crew of Mir, situated 383 km above the Pacific south of Alaska, fired an electron beam gun at it. At the time of the experiment, a Canadian ground station monitored the operation. Despite its scientific aim of determining how charged particle beams were scattered in the atmosphere, the media still reported the experiment as a test of Russian "Star Wars" weapons.

The difficulties that the post-Soviet Russia was undergoing were brought home to the cosmonauts aboard Mir in May, during unloading of the Progress M23 re-supply craft. They found that some of the food containers intended for the orbiting crew had been tampered with by ground staff, and items that should have been there were missing.

Milestones

166th manned space flight
77th Russian manned space flight
18th manned Mir mission
15th Mir resident crew
70th manned Soyuz mission
17th manned Soyuz TM mission
1st Mir resident mission without scheduled EVAs
Polyakov sets world endurance record for one flight of 437 days 17 hrs, and a career record of 678 days 16 hrs on two flights
Polyakov celebrates his 52nd birthday in space (27 Apr)

STS-60

Int. Designation	1994-006A
Launched	3 February 1994
Launch Site	Pad 39A, Kennedy Space Center, Florida
Landed	11 February 1994
Landing Site	Runway 15, Shuttle Landing Facility, Kennedy Space Center, Florida
Launch Vehicle	OV-103 Discovery/ET-61/SRB BI-062/SSME #1 2012; #2 2034; #3 2032
Duration	8 days 7 hrs 9 min 22 sec
Call sign	Discovery
Objective	Wake Shield Facility 1 operations; SpaceHab 2; first flight of a Russian cosmonaut on a US mission

Flight Crew

BOLDEN Jr., Charles Frank, 47, USMC, commander, 4th mission
Previous missions: STS 61-C (1986); STS-31 (1990); STS-45 (1992)
REIGHTLER Jr., Kenneth Stanley, 42, USN, pilot, 2nd mission
Previous mission: STS-48 (1991)
DAVIS, Nancy Jan, 40, civilian, mission specialist 1, 2nd mission
Previous mission: STS-47 (1991)
SEGA, Ronald Michael, 41, civilian, mission specialist 2
CHANG-DIAZ, Franklin Ramon, 43, civilian, mission specialist 3, 4th mission
Previous missions: STS 61-C (1986); STS-34 (1989); STS-46 (1992)
KRIKALEV, Sergei Konstaninovich, 35, civilian, Russian mission specialist 4, 3rd mission
Previous missions: Soyuz TM7 (1988); Soyuz TM12 (1991)

Flight Log

As part of the agreement between the US and Russia on manned space flight operations, the first Russian cosmonauts (Mir veterans Sergei Krikalev and Vladimir Titov) arrived at NASA JSC in Houston in late October 1992, to train for STS-60 and STS-63. Their training would be an abbreviated form of NASA mission specialist (not candidate) training, which would include RMS operations and EVA training using American EMU hardware. The training flow for the cosmonauts took into account their vast experience in the Russian programme. In February 1993, Krikalev was assigned the prime position on STS-60 with Titov as his back-up, reversing the roles for STS-63. These flights were the precursor missions that would include up to ten (later reduced to seven) long-duration American residencies on Mir, Shuttle dockings and further Russian cosmonauts incorporated into crews for Shuttle–Mir

Russia's first cosmonaut to fly on the American Shuttle, Sergei Krikalev is seen on the aft flight deck of Discovery during the STS-60 mission. He uses the SAREX equipment to talk with American students in Maine and holds a camcorder for recording in-flight activities

missions. This all fell under Phase 1 of the International Space Station programme, the sixteen-nation cooperative programme that essentially replaced both the US Space Station Freedom and Russian Mir 2 programmes.

The mission of STS-60 had been postponed from November 1993 and rescheduled for January 1994, but a leaking aft RCS thruster in the orbiter forced a further delay while it was investigated. On the third attempt, the launch occurred without incident and the SpaceHab experiments were activated shortly after reaching orbit. The twelve experiments in the programme included four in materials sciences, seven in life sciences and a space dust collection experiment. On the third flight day, the crew attempted to deploy the Wake Shield Facility, but radio interference and problems in reading status information from the facility meant that the attempt had to be abandoned. The following day, the deployment was again cancelled when the facility's attitude control system developed problems. For two days, the facility remained on the end of the robot arm, conducting abbreviated experiments and being berthed during FD 6. The Wake Shield Facility was planned as a deployable and retrievable experiment platform that would leave a vacuum wake in low-Earth orbit, which was calculated as being 10,000 times greater than that created on Earth in the laboratory. Within this "ultra-vacuum" environment, it was planned to grow defect-free thin film layers of semi-conductor materials (such as gallium arsenide).

Two experiments deployable from GAS canisters (six orbital debris calibration spheres – ranging from 5 to 15 cm in diameter, and a German satellite that measured acceleration forces) were carried on the flight, as well as three other GAS canister experiments. There was also a capillary pump experiment and an auroral photography experiment included in the science package. Krikalev had been given crew roles in photography and TV tasks, as well as maintenance activities during the flight. He supported SpaceHab systems and RMS operations and participated in the Earth observation programme. He was also assigned roles in some of the secondary experiments on the mission. His primary role was participation in the joint in-flight medical and radiological investigations. Krikalev also used the SAREX ham radio equipment to talk with operators in Moscow. After a one-orbit waive-off due to unfavourable weather at KSC, the mission ended with a perfect landing, a successful opening of the series of cooperative missions that would lead to the ISS programme.

The most significant achievement was that a cosmonaut could be trained to work with an American crew on an American mission in such a short time. This was a strong reflection of the capabilities of Krikalev, who took to both the English language and the American training system more easily than Titov. It was also a compliment to the vast experience both men brought from their own programme. Krikalev had more than 400 days experience from two missions and Titov had logged a year in space on one mission. The Russians seemed to integrate well with the American programme, but the question now was whether the Americans could do the same when they had to spend far longer in Russia training for a residency on Mir.

Milestones

167th manned space flight
90th US manned space flight
60th Shuttle mission
18th flight of Discovery
1st Russian cosmonaut on Shuttle
1st cosmonaut not to be launched from Baikonur or land in Russian territory
1st flight of Wake Shield Facility
100th Get Away Special (GAS) payload

STS-62

Int. Designation	1994-015A
Launched	4 March 1994
Launch Site	Pad 39B, Kennedy Space Center, Florida
Landed	18 March 1994
Landing Site	Runway 33, Shuttle Landing Facility, Kennedy Space Center, Florida
Launch Vehicle	OV-103 Discovery/ET-62/SRB BI-064/SSME #1 2031; #2 2109; #3 2029
Duration	13 days 23 hrs 16 min 41 sec
Call sign	Discovery
Objective	United States Microgravity Payload (USMP)-2; Office of Aeronautics and Space Technology-2 payload

Flight Crew

CASPER, John Howard, 50, USAF, commander, 3rd mission
Previous missions: STS-36 (1990); STS-54 (1993)
ALLEN, Andrew Michael, 38, USMC, pilot, 2nd mission
Previous mission: STS-46 (1992)
THUOT, Pierre Joseph, 38, USN, mission specialist 1, 3rd mission
Previous missions: STS-36 (1990); STS-49 (1992)
GEMAR, Charles Donald ("Sam"), 38, US Army, mission specialist 2, 3rd mission
Previous missions: STS-38 (1990); STS-48 (1991)
IVINS, Marsha Sue, 42, civilian, mission specialist 3, 3rd mission
Previous missions: STS-32 (1990); STS-46 (1992)

Flight Log

The 11 March launch attempt was postponed at the $T-11$ hours mark when the forecast indicated that the weather would not clear in time for the launch. The launch itself proceeded without a problem, but the retrieval of the SRBs and their parachutes was delayed by two days as the recovery ships could not be deployed due to high seas.

This was the second flight under the USMP programme and the payload also featured the OAST-2 package. The OAST-1 package flew on STS-2 in 1981 and included Earth observation experiments, but this time the six experiments focused on space-related technology with potential application for satellites, circuits, sensors, processors and the International Space Station. The USMP-2 payload comprised five experiments that focused on the effects of the microgravity environment on materials and fundamental sciences. The experiments included the Advanced Automated Directional Solidification Furnace, the Critical Fluid Light Scattering Experiment,

Located in the payload bay are elements of USMP-2 and OAST-2 experiments. Also in frame is the RMS used during activities featuring the Dexterous End Effector, a series of operations and observations of the RMS in one-hour sessions to develop improvements to RMS operating techniques

the Isothermal Dendritic Growth Experiment, Materials for the Study of Interesting Phenomena of Solidification on Earth and in Orbit and the Space Acceleration Measurement System. On orbit, the crew would activate both the USMP-2 experiments and the OAST-2 package, but they would be controlled by investigators and controllers on the ground at the Marshall Spacelab Mission Operations Control Center. USMP-2 was the main focus of the early part of the STS-62 mission, before the orbit of Columbia was lowered by about 20 nautical miles to favour the OAST package more. By flying the orbiter with an EDO pallet and in a gravity gradient mode, this flight was another step towards future space station research operations.

While the payload bay experiments were being manipulated via the ground, the crew focused on the mid-deck and other payload bay investigations. The Shuttle

Solar Backscatter UV/A and Limited Duration Space Environment Candidate Material Exposure experiments were in the payload bay, while on the mid-deck, further research was conducted in protein crystal growth, generic bioprocessing, zero-gravity dynamics, auroral photography and Earth observations, keeping the crew busy. In addition, the crew were occupied with evaluating the Dexterous End Effector, a new magnetic grapple fixture that was being evaluated in space for possible use on future RMS operations. The astronauts also completed a programme of biomedical activities, linked to the Extended-Duration Orbiter Medical Project, aimed at providing a better understanding of, and baseline data for, counteracting the effects of prolonged orbital space flight.

Milestones

168th manned space flight
91st US manned space flight
61st Shuttle mission
16th flight of Columbia
2nd flight of USMP payload
3rd Extended-Duration Orbiter (EDO) mission

STS-59

Int. Designation	1994-020A
Launched	9 April 1994
Launch Site	Pad 39A, Kennedy Space Center, Florida
Landed	20 April 1994
Landing Site	Runway 22, Edwards AFB, California
Launch Vehicle	OV-105 Endeavour/ET-63/SRB BI-063/SSME #1 2028; #2 2033; #3 2018
Duration	11 days 5 hrs 49 min 30 sec
Call sign	Endeavour
Objective	Operation of the Space Radar Laboratory (SRL)-1

Flight Crew

GUTIERREZ, Sidney McNeill, 42, USAF, commander, 2nd mission
Previous mission: STS-40 (1990)
CHILTON, Kevin Patrick, 39, USAF, pilot, 2nd mission
Previous mission: STS-49 (1992)
APT, Jerome "Jay", 44, civilian, mission specialist 1, 3rd mission
Previous missions: STS-37 (1991); STS-47 (1992)
CLIFFORD, Michael Richard Uram, 41, USAF, mission specialist 2, 2nd mission
Previous mission: STS-53 (1992)
GODWIN, Linda Maxine, 41, civilian, mission specialist 3, payload commander, 2nd mission
Previous mission: STS-37 (1991)
JONES, Thomas David, 39, civilian, mission specialist 4

Flight Log

In 1991, NASA expanded its many-faceted programme of Earth studies into a global examination of how Earth's systems (air, water, land and life) interact with each other and affect or influence changes in the global climate. The new programme was designated *Mission to Planet Earth* and was divided into phases. The initial phase began in 1991, using satellites such as UARS (deployed from STS-48) and dedicated Shuttle missions (such as the ATLAS series) supported by airborne and ground-based studies. Part of this first phase was the Space Radar Laboratory series. Originally a programme of three missions, STS-59 was the first of the eventual two that actually flew.

The primary advantage of radar imaging is the ability to gather data over virtually any region of the Earth regardless of weather conditions. A similar programme was undertaken by the Magellan probe at the shrouded planet Venus. Observational

Mission specialist Tom Jones monitors a number of cameras on the aft flight deck which are fixed on targets of opportunity in support of the SRL instruments in the payload bay

imaging radar had been carried previously on STS-2 in 1981 (SIR-A) and STS 41-G in 1984 (SIR-B), and it was the latest variety (SIR-C) that was carried aboard STS-59. The system comprised C-Band (6 cm wavelength) and L-band (23 cm wavelength) radars – four in total – with separate horizontally (H) and vertically (V) polarised units for both bands, and steered electronically. In addition, the mission carried X-SAR, a mechanically-pointed single radar unit (X-band, 3 cm wavelength). There was also an experiment that analysed ocean radar data supplied by SIR-C, as well as MAOS, a carbon monoxide monitoring, sensing and Earth photography package that had flown previously on STS-2 and STS 41-G.

The 7 April launch of STS-59 was postponed for a day at the $T - 27$ hour mark in order to facilitate an inspection of the metal vanes inside the SSME high-pressure oxidiser pre-burner pumps. The 8 April launch was scrubbed due to bad weather, with low clouds and high crosswinds at the SLF and cloud around the launch pad, but Endeavour launched without incident the following day. Once in orbit, the spacecraft was configured for orbital operations and the crew split into their two shifts, with the Red Shift (Gutierrez, Chilton and Godwin) starting their sleep period and the Blue Shift (Apt, Clifford and Jones) commencing the first series of data gathering. They worked from the aft flight deck as the mission carried no Spacelab or SpaceHab module. After some initial set-up problems, the information came streaming in and was stored on VCR data cassettes. There were 180 such cassettes aboard the Shuttle, enough to support the planned 50 hours of data collection while covering an estimated 50 million square km of the Earth.

This was an international mission, with 49 science investigators and over 100 scientists from 13 nations making up the international science team. The mission focused on the "dry season", with SRL-2 (STS-68) planned to cover the "wet season" later in the year. This allowed the scientists to compare data from the same sites under different global climate conditions. STS-59 obtained over 133 hours of data (32 terabits, or 32 trillion bits). With SIR-C/X-SAR eventually examining approximately 70 million km of the Earth – representing 12 per cent of the Earth's total surface and 25 per cent of its land masses – there was enough data to fill 20,000 encyclopaedic volumes. The data-gathering operations were the equivalent of 45 TV stations operating at the same time. Even with advances in digital processing, it would still take five months to process a complete set of images and another nine months of detailed processing after that. At the close of the mission the crew had imaged over 400 sites, including 19 primary observation sites (called super-sites) in Brazil, Michigan, North Carolina and Central Europe.

The crew also found time to work on a variety of secondary and mid-deck payloads and to use the SAREX equipment to talk with both the Russian cosmonauts on Mir and with US astronauts Norman Thagard and Bonnie Dunbar, who were training at TsPK in Moscow. STS-59 also carried three GAS candidate experiments, sponsored by researchers in France, Japan and New Mexico. Perhaps most importantly, the mission also carried the Toughened Uni-piece Fibrous Insulation (TUFI), an improved Thermal Protection System (TPS) tile. Scheduled for a six-flight evaluation on all four orbiters, if successful, it was hoped that this fibrous insulation would prove more resilient to impacts in specific areas of the orbiter, such as between the engines, near the landing gear doors and around the orbital manoeuvring thrusters. Several of these new tiles were placed on the base heat shield of Endeavour between the three main engines, for evaluation during the flight and primarily during entry. Post-flight examination revealed no damage on the six tiles installed for the test. The tests ultimately proved successful and TUFI was added to the TPS on the Shuttle from 1996.

Two landing opportunities at Kennedy were cancelled on 19 April due to bad weather. A third chance of bringing Endeavour home to the Cape on 20 April was also waived off, in favour of landing at Edwards.

Milestones

169th manned space flight
92nd US manned space flight
62nd Shuttle mission
6th flight of OV-105
1st flight of SRL payload configuration
1st flight of TUFI – improved thermal protection tile samples

SOYUZ TM19

Int. Designation	1994-036A
Launched	1 July 1994
Launch Site	Pad 1, Site 5, Baikonur Cosmodrome, Kazakhstan
Landed	4 November 1994
Landing Site	88 km northeast of Arkalyk
Launch Vehicle	R7 (11A511U2); spacecraft serial number (7K-M) 68
Duration	125 days 22 hrs 53 min 36 sec
Call sign	Agat (Agate)
Objective	Delivery of the 16th main Mir crew; continuation of Polyakov's record residency and medical experiment programme

Flight Crew

MALENCHENKO, Yuri Ivanovich, 32, Russian Air Force, commander
MUSABAYEV, Talgat Amangeldyevich, 43, Russian Air Force, flight engineer

Flight Log

The mission was delayed when the launch shroud designed to protect the spacecraft during its ascent through the atmosphere could not be delivered on time. By the end of June, the shroud had arrived and been installed on the vehicle in preparation for the launch. This was the first all-rookie crew since the Soyuz 25 mission in 1977, as mission planners finally began to have confidence both in the Kurs docking system and in the ability of the cosmonauts to take over manual control of the spacecraft if necessary to complete a docking approach. To align Russian operations with the proposed first American astronaut launch to Mir in March 1995, the EO-16 residency would be only for four months. Most of their first few weeks in space were spent in Earth observation photography, in particular of Kazakhstan around the Aral Sea region.

Confidence in the cosmonauts docking to Mir may have increased, but there were still problems in getting the Progress re-supply craft to link up successfully. On 27 August, the first automated docking of Progress M24 failed. Three days later, the craft bumped into Mir's forward port before drifting away, hitting a solar array as it went by. With onboard supplies running low, the cosmonauts faced the prospect of abandoning Mir in late September and mothballing it for up to four months if they could not get a Progress to dock to the station. If this had been the case, it was hoped that another crew, and fresh supplies, would have been able to reach the station before its stabilisation propellant ran out. However, all this became academic when Malenchenko successfully used the TORU docking system aboard Mir to skilfully dock Progress M24 by remote control. The docking was critical to a number of events planned for the next Progress supply vehicle and if the station had had to be

Malenchenko (left) and Musabayev launched to Mir aboard Soyuz TM19 to operate the sixteenth residency aboard the station

abandoned, Polyakov's record space flight would have been curtailed. It was also fortunate that Malenchenko had docked M24, because it contained over 275 kg of ESA hardware in preparation for the Euro Mir 94 programme that Ulf Merbold would be running alongside the next resident crew. Without this docking, Merbold's entire mission would have been in doubt.

The EO-16 crew completed two EVAs during their short stay on Mir. The first (9 Sep, 5 hours 6 minutes) focused on an inspection of the docking port hit by Progress M24 and the tear in the thermal blanket caused by the Soyuz TM17 incident. During the EVA, Polyakov monitored his two colleagues from within the space station. After replacing cassettes exposed to space on the outside of Kvant 2, the cosmonauts found that the damage caused by TM17, near to where Kristall joined the base block, was very light. They repaired the 30 cm × 40 cm gap in the thermal insulation blanket and subsequently found that Progress M24 had caused no serious damage to the transfer compartment of Mir. The second EVA (14 Sep, 6 hours 1 minute) was another inspection, this time of the movable arrays on Kristall, which were designed to be relocated on Kvant over a series of EVAs. They also looked at the mounting brackets and solar array drives on Kvant, which would house the arrays. Space exposure cassettes were removed from Rapana and the Sofora was inspected before a new amateur radio antenna was erected. Dr. Polyakov once again monitored EVA operations from inside the Mir, as he was not trained for EVA himself.

Milestones

170th manned space flight
78th Russian manned space flight
25th Russian and 54th flight with EVA operations
19th manned Mir mission
16th Mir resident crew
71st manned Soyuz mission
18th manned Soyuz TM mission

STS-65

Int. Designation	1994-039A
Launched	8 July 1994
Launch Site	Pad 39A, Kennedy Space Center, Florida
Landed	23 July 1994
Landing Site	Runway 33, Shuttle Landing Facility, Kennedy Space Center, Florida
Launch Vehicle	OV-102 Columbia/ET-64/SRB BI-066/SSME #1 2019; #2 2030; #3 2017
Duration	14 days 17 hrs 55 min 00 sec
Call sign	Columbia
Objective	Second flight of the International Microgravity Laboratory using a Spacelab Long Module configuration

Flight Crew

CABANA, Robert Donald, 45, USMC, commander, 3rd mission
Previous missions: STS-41 (1990); STS-53 (1992)
HALSELL Jr., James Donald, 37, pilot
HIEB, Richard James, 38, civilian, mission specialist 1, payload commander, 3rd mission
Previous missions: STS-39 (1991); STS-49 (1992)
WALZ, Carl Erwin, 38, USAF, mission specialist 2, 2nd mission
Previous mission: STS-51 (1993)
CHIAO, Leroy, 33, civilian, mission specialist 3
THOMAS, Donald Alan, 39, civilian, mission specialist 4
MUKAI, Chiaki, 41, civilian, Japanese payload specialist 1

Flight Log

Following a smooth countdown, the mission of STS-65 carrying the IML-2 science payload got off to a perfect start. Once in orbit, the crew divided into the two teams (Red Shift – Cabana, Halsell, Hieb and NASDA PS Mukai; Blue Shift – Walz, Chiao and Thomas), working around the clock to operate not only the IML-2 science programme, but also a range of secondary and mid-deck experiments. This flight carried more than twice the experiments flown on IML-1 two years before and was supported by an international team of 210 scientists representing six space research organisations (ESA, CSA, CNES, DARA, NASDA and NASA).

The life sciences programme consisted of fifty experiments, divided into bio-processing, space biology, human physiology and radiation biology. Part of these investigations required the European Biorack facility, which was making its third trip into space. The Biorack housed 19 experiments, featuring chemicals and biological

The first Japanese woman to fly in space, Chiaki Mukai, is shown entering the IML-2 Spacelab module from the connecting tunnel from the mid-deck of Columbia during the 15-day mission

samples that included bacteria, mammalian and human cells, isolated tissues and eggs, sea urchin larvae, fruit flies and plant seedlings. Thirty materials-processing experiments were also conducted, using nine facilities. In the Protein Crystallisation Facility (flying for the second time), approximately 5,000 video images were taken of crystals grown during the mission. This mission also advanced the concept of remote telescience, with researchers on the ground able to monitor their experiments in real time as they were operated aboard the orbiter. At the end of the mission, the Spacelab Mission Operations Control Center at Huntsville in Alabama reported that over 25,000 payload commands had been issued, a new record.

In addition to the IML investigations the mission also flew the Orbital Acceleration Research Experiment (OARE), the Commercial Protein Crystal Growth (CPCG) and the Military Application of Ship Track (MAST) payloads, as well as the SAREX amateur radio equipment. The Air Force Maui Optical Site (AMOS), which did not require equipment, was also part of the research programme of this flight. On top of all this, there were also more than a dozen Detailed Test Objectives and more than fifteen Detailed Supplementary Objectives assigned to the mission, as well as the ongoing programme of biomedical studies as part of the EDO Medical Project (EDOMP), and the Earth photography and observation programme.

The crew also set up a video to record the experience of riding in the crew cabin during launch and entry for the first time. On 20 July, the crew honoured the 25th anniversary of the Apollo 11 Moon landing, noting that the historic mission also featured a spacecraft called Columbia (the Command and Service Module). The

22 July landing was waived off due to the possibility of rain showers around the Cape but the next day the conditions were good to support a return to Earth. This was the final flight of Columbia prior to its scheduled modification and refurbishment period at Rockwell's facility in California. OV-102 left the Cape in October 1994 and returned in April 1995 to begin preparations for its next mission on STS-73.

Milestones

171st manned space flight
93rd US manned space flight
63rd Shuttle mission
17th flight of Columbia
4th EDO mission
2nd flight of IML configuration
1st Japanese woman to fly in space (Mukai)
Longest single flight to date by a female (Mukai)
1st use of video-tape to record lift-off and re-entry from inside flight deck

STS-64

Int. Designation	1994-059A
Launched	9 September 1994
Launch Site	Pad 39B, Kennedy Space Center, Florida
Landed	20 September 1994
Landing Site	Runway 4, Edwards AFB, California
Launch Vehicle	OV-103 Discovery/ET-66/SRB BI-068/SSME #1 2031; #2 2109; #3 2029
Duration	10 days 22 hrs 49 min 57 sec
Call sign	Discovery
Objective	LITE laser pulse studies of Earth atmosphere experiment; SPARTAN-201 astronomy free-flyer

Flight Crew

RICHARDS, Richard Noel, 48, USN, commander, 4th mission
Previous missions: STS-28 (1989); STS-41 (1990); STS-50 (1992)
HAMMOND Jr., Blaine, 42, USAF, pilot, 2nd mission
Previous mission: STS-39 (1991)
LINENGER, Jerry Michael, 39, USN, mission specialist 1
HELMS, Susan Jane, 36, USAF, mission specialist 2, 2nd mission
Previous mission: STS-54 (1993)
MEADE, Carl Joseph, 43, USAF, mission specialist 3, 3rd mission
Previous missions: STS-38 (1990); STS-50 (1992)
LEE, Mark Charles, 42, USAF, mission specialist 4, 3rd mission
Previous missions: STS-30 (1989); STS-47 (1992)

Flight Log

Weather conditions delayed the launch of STS-64 by almost two hours into a two-and-a-half-hour window, but otherwise the launch was untroubled. Once on orbit, the Lidar-in-space Technology Experiment (LITE), mounted on a Spacelab pallet in the payload bay, was activated on FD 1 and became operational the next day. It operated for almost a week of activities, resulting in what official reports called a "highly successful technology test." The Lidar (light detection and radar) method of optical radar used laser pulses instead of radio waves to study the atmosphere of Earth, as part of the NASA *Mission to Planet Earth* programme. Sixty-five groups of researchers from twenty countries took part in the experiment, which also employed simultaneous airborne and ground-based measurements to verify the data collected by the LITE payload. The experiments operated for 53 hours, of which 43 hours were of high-rate data quality. Atmosphere "sites" located high above northern Europe, Indonesia and the South Pacific area, Russia and Africa were targeted and from

Meade tests the new SAFER system 130 nautical miles above the Earth. Hardware supporting the Lidar-in-space Technology Experiment (LITE) is at the lower right. The photo was taken from the RMS shoulder joint camera. The robot arm is also captured in the scene upper right

the data collected, new information on the structure of clouds, storm systems, dust clouds and pollutants in the atmosphere was obtained. Furthermore, the data was used to understand the effects of forest fires and how reflective the surface of the Earth was at different points and changing times of the day, in varying "seasonal" conditions.

On FD 5, the Shuttle Pointed Autonomous Research Tool for Astronomy-201 (SPARTAN-201) was deployed by the RMS. This was the unit's second mission and was designed to investigate the acceleration and velocity of solar wind, as well as taking measurements of the Sun's corona. The collected data was stored on board for downloading once back on Earth and the vehicle was retrieved on FD 7.

During FD 8, Lee (EV1) and Meade (EV2) performed the only EVA of the mission, but one which was a milestone in the preparations for expanded EVA operations at ISS. During the EVA, the two astronauts evaluated the Simplified

Aid For EVA Rescue (SAFER). The RMS remained active and on hand in case of problems. The SAFER unit was designed to provide a usable back-up if an astronaut became untethered during EVA. In some circumstances, the Shuttle would be capable of manoeuvring to "scoop up" a stranded astronaut (though this has not yet been necessary), but the ISS is far less manoeuvrable, so an alternative personal safety system would be required. This unit was a scaled-down version of the MMU flown during 1984 and was designed for emergency situations only (but with built-in back up systems). Propulsion came from 24 fixed-position thrusters. The 1.36 kg nitrogen supply, which could be recharged from the orbiter nitrogen system, could provide about a 3 m/sec change in velocity until the gas was expelled. The unit also had an attitude control system and a 28-volt battery pack, which could be charged in orbit. During the EVA, both astronauts flew several short translation and rotation sequences, with data recorded in the SAFER unit for analysis after the mission. The unit was an outstanding success, as the astronauts soon learned that it used less nitrogen than predicted. They also evaluated the SAFER attitude hold system by manually tumbling each other. Despite Meade rolling Lee faster than planned, the attitude control system in Lee's unit worked perfectly to correct his rotation. Both astronauts replenished their SAFERs about seven times during the EVA and the only problems during the excursion were Meade reporting that his feet had gone cold, and that evaluation of the Electronic Cuff Check (ECC) list, which was designed to replace the paper cuff checklists that had been used since Apollo 12, proved disappointing.

Aside from the LITE payload, STS-64 also carried the Shuttle Plume Impingement Flight Experiment, a 10 m RMS extension that was designed to collect data on the RCS thrusters, which would help in understanding their effects in close proximity to large space structures such as Mir or ISS. As with all Shuttle flights, a suite of mid-deck experiments was carried on this mission, many of which had flown before. The mission was extended by a day to maximise the collection of data and was increased by a further 24 hours on 19 September due to storms at the Cape. The following day, two attempts at landing at the Cape were also abandoned due to the weather, so the mission was diverted to Edwards for the third landing window of the day.

Milestones

172nd manned space flight
94th US manned space flight
64th Shuttle mission
19th flight of Discovery
30th US and 55th flight with EVA operations
1st flight of LITE
1st untethered US EVA for 10 years
1st tests of SAFER

STS-68

Int. Designation	1994-062A
Launched	30 September 1994
Launch Site	Pad 39A, Kennedy Space Center, Florida
Landed	11 October 1994
Landing Site	Runway 22, Edwards AFB, California
Launch Vehicle	OV-105 Endeavour/ET-65/SRB BI-067; SSME #1 2028; #2 2033; #3 2026
Duration	11 days 5 hrs 46 min 8 sec
Call sign	Endeavour
Objective	Operation of the Space Radar Laboratory (SRL)-2 in the payload bay

Flight Crew

BAKER, Michael Allen, 40, USN, commander, 3rd mission
Previous missions: STS-43 (1991); STS-52 (1992)
WILCUTT, Terrence Wade, 44, USMC, pilot
SMITH, Steven Lee, 35, civilian, mission specialist 1
BURSCH, Daniel Wheeler, 37, USN, mission specialist 2, 2nd mission
Previous mission: STS-51 (1993)
WISOFF, Peter Jeffrey Karl, 36, civilian, mission specialist 3, 2nd mission
Previous mission: STS-57 (1993)
JONES, Thomas David, 39, civilian, mission specialist 4, payload commander, 2nd mission
Previous mission: STS-59 (1994)

Flight Log

After the success of SRL-1, it was hoped that the second mission in the series would be equally rewarding. In order to ensure a smooth (and shorter) flow between the missions, it was decided to place the payload in the same orbiter (Endeavour) and to assign MS Tom Jones from the STS-59 crew as payload commander for STS-68. Endeavour returned from Edwards on 2 May and the SRL payload was removed from the vehicle in the Orbiter Processing Facility on 8 May for inspection, cleaning and maintenance. It was returned to the payload bay on 21 June and by 27 July, Endeavour was back on the pad, ready to support a planned 18 August launch. That attempt was scrubbed at $T - 1.9$ seconds, when the orbiter's computers shut down the three SSMEs after they had detected unacceptably high discharge temperatures in the high-pressure oxidiser turbine for SSME #3. This required a return to the VAB to replace all three engines. The launch was rescheduled for 30 September.

The Space Radar Laboratory-2 payload in the cargo bay of Endeavour during the 11-day mission

Following the format of SRL-1, the crew of STS-68, working a single-shift system, soon settled down to operate their main payload and host of other mid-deck and secondary experiments once on orbit. The same 400 sites and 19 "super-sites" were targeted as on SRL-1 to provide comparison data during a different season. Unfortunately, there would be no SRL-3 mission to provide a third set of data from a different time of the year (December or January). In addition to the programme's scheduled activities, the crew took the opportunity to record other images and impressions of Earth's weather and environmental conditions, including the eruption of the Kliuchevskoi volcano in Kamchatka. They also studied fires in British Columbia, Canada (which had been set for forest management purposes) and used the MAPS equipment to take readings to better understand the carbon monoxide emissions from burning fires.

As well as flying over the same places as on the STS-59 mission (at one point flying the Endeavour just nine metres from where it had flown the previous April), by making small changes to their orbit, the STS-68 crew could take images of areas they had flown over just 24 or 48 hours previously. These interferometric passes were made over central North America, the Amazonian rain forests of Brazil in South America, and the volcanoes in the Kamchatka peninsula in Russia. Images like these taken over long periods of time could provide important data on the movements of the Earth's surface – of even a few centimetres – that could be invaluable in detecting the pre-eruptive changes in volcanoes or movements in major fault lines prior to earthquakes.

One of the radar observations was of a man-made phenomenon, a deliberate and controlled spillage of oil in the North Sea. This was designed to see if the radar could determine the difference between oil spills and naturally produced fish and plankton oils. Four hundred litres of diesel oil and 100 litres of algae-produced natural oil were dumped into the water for comparison. After the data was collected, it took just two hours for the stand-by recovery vessels to clean up the spillages.

After a one-day extension to the mission, STS-68 was diverted to Edwards from KSC because of bad weather in the vicinity of the Cape. Post-flight evaluation of the mission revealed that there had been 923 attempted data sweeps, of which 910 were successful (98.59%). Of the 292 "super-site" data-gathering attempts, the crew achieved 289 takes (98.97%) and from the 724 X-SAR attempts, 719 were acquired (99.31%). The volume of data collected equated to approximately 25,000 encyclopaedic volumes worth. STS-68 had gathered data from 83 million km^2, or about nine per cent of the total Earth surface. Though SRL-3 was not flown, the data from the first two missions would help in planning the SRTM mission flown in 2000.

Milestones

173rd manned space flight
95th US manned space flight
65th Shuttle mission
7th flight of Endeavour
2nd flight of SRL payload combination

SOYUZ TM20

Int. Designation	1994-063A
Launched	4 October 1994
Launch Site	Pad 1, Site 5, Baikonur Cosmodrome, Kazakhstan
Landed	22 March 1995
Landing Site	54 km northeast of Arkalyk
Launch Vehicle	R7 (11A511U2); spacecraft serial number (7K-M) 69
Duration	169 days 5 hrs 21 min 35 sec (Viktorenko/Kondakova); 31 days 12 hrs 35 min 56 sec (Merbold returned with EO-16 aboard TM19)
Call sign	Vityaz (Viking)
Objective	Delivery of the 17th Mir resident crew; ESA EuroMir94 mission (Merbold)

Flight Crew

VIKTORENKO, Alexandr Stepanovich, 47, Russian Air Force, commander, 4th mission
Previous missions: Soyuz TM3 (1987); Soyuz TM8 (1989); Soyuz TM14 (1992)
KONDAKOVA, Yelena Vladimirovna, 37, civilian, flight engineer
MERBOLD, Ulf Dietrich, 53, civilian, ESA research engineer, 3rd mission
Previous missions: STS-9 (1983); STS-42 (1992)

Flight Log

As Soyuz TM20 approached Mir's docking port on automatic, it yawed to the side, forcing Viktorenko to take manual control to complete the docking without further incident. Aboard the Soyuz was Shuttle veteran and ESA astronaut Ulf Merbold, who would be staying aboard the station for a month to operate the EuroMir94 experiment programme. Hastily assembled (and underfunded by ESA), the experiment programme was heavily dependent upon equipment left aboard Mir during earlier international visits by Austrian, French and German cosmonauts, and on the facilities of the station itself. The experiment programme featured 30 experiments: 23 in life sciences, 4 in materials sciences and 3 in technology. Merbold's mission was a precursor to the planned 135-day EuroMir95 mission scheduled for the following year, and for operations on the ESA Columbus module that was planned for ISS. A fault on the Kristallisator furnace prevented Merbold from performing his materials experiments. The replacement would not arrive at the station until after Merbold had returned to Earth.

With six cosmonauts on Mir for a month, the drain on resources was beginning to tell. On 1 October, during the recharging of Soyuz TM20's batteries, TV equipment could not be recharged at the same time. A short circuit on the computer that oriented

The crew of Soyuz TM20 included ESA astronaut Merbold (right) and Russian cosmonaut Viktorenko (left). In the centre is Kondakova, only the third Russian female to fly in space since the manned programme began in 1961.

the solar arrays to face the Sun meant that the station was unable to replenish its power and the batteries had drained. This necessitated the use of reaction control propellant on the Soyuz TM in order to realign the station, point the arrays at the Sun and restore power via a back-up computer. These interruptions affected Merbold's science programme, which had to be adjusted around the periods of lost power. On 3 November, the EO-16 crew and Merbold boarded Soyuz TM19 to test the Kurs docking system. TM19 undocked and backed away from the station to 190 metres, then successfully re-docked automatically and the crew re-entered the station for 24 hours. Had the docking failed then the crew would have completed an emergency return to Earth. In the event, a nominal landing was achieved the following day. Merbold returned with 16 kg of life science samples he had collected during his month on the station. The collection included 125 saliva, 85 urine and 34 blood samples.

When Progress M25 arrived on 13 November, it delivered spares for the furnace, which allowed the experiments planned for Merbold's visit to be completed.

By 18 November, the Mir base block had completed 50,000 orbits of Earth since its launch in February 1986. For a while, it looked as though Viktorenko and Polyakov would have to perform EVAs in late November, to move the Kristall arrays in preparation for the arrival of the Spektr module in December. However, that launch soon slipped (in part due to the Americans' late shipping of equipment and Russian customs bureaucracy) and the EVAs were cancelled.

On 9 January 1995, Polyakov set a new single-mission endurance record of one year and one day, and with over 600 days to his credit from two missions, he was already by far the most experienced space traveller with another three months to go in the flight. Given the uncertainties over the safety of flying such a long mission, and his potential exposure to ambient radiation, Polyakov slept in the Kristall module shielded by the batteries, in order to avoid putting his colleagues at any undue risk. Viktorenko and Kondakova occupied the two individual cabins in the base block. On 11 January, in order to review repairs to the Kurs system, the cosmonauts conducted another test, undocking TM20 and pulling back 160 meters before completing a successful automatic re-docking. A month later, a new spacecraft arrived at Mir, but this one would not be docking. One of the objectives of US Shuttle Discovery's STS-63 mission was to test launch and rendezvous windows for the later docking missions. For a few hours on 6 February, Discovery flew in close proximity to Mir, approaching to approximately 10 meters in a simulated docking approach. This was the closest that a manned Russian and American spacecraft had been to each other in space since ASTP in July 1975. After a photographic fly-around exercise, Discovery departed to continue its own mission, leaving Mir's cosmonauts to resume preparations for their return to Earth.

After Norman Thagard arrived with the next resident crew on Soyuz TM21 on 16 March 1995 and the period of hand-over operations were completed, the EO-17 crew of Viktorenko and Kondakova returned to Earth on 22 March, along with Polyakov. The latter had set a single-mission record of 438 days, which is unlikely to be surpassed for decades. Indeed, there are no plans to try to surpass it for the foreseeable future. Despite such a long flight, the doctor–cosmonaut insisted in walking unaided to the medical tent once he was helped from the Descent Module. Polyakov had performed over 1,000 tests in a programme of 50 medical experiments, losing 15 per cent of his bone density. This took a few months to regain, and even then the recovery was not total. This was despite a strict regime of two hours exercise per day which he observed strictly every day in space. The loss of oxygen-bearing red blood cells in the early weeks of his flight was countered by adjusting Mir's environmental control system, but the soles of his feet had softened as he had not "walked" for several months. He had proved that a flight of 14 months was possible (a duration which could support a manned round trip mission to Mars), but at a cost. Polyakov's bravery and determination to complete both the exercise programme and the flight stand out as one of the milestones in manned space flight history, one whose legacy will only really be seen when the first crews are dispatched towards Mars.

Milestones

174th manned space flight
79th Russian manned space flight
20th Manned Mir mission
17th Mir resident crew
72nd manned Soyuz flight
19th manned Soyuz TM flight

1st female cosmonaut assigned to a long-duration mission
Longest single space flight by a female (Kondakova)
Polyakov sets a new world record for one flight of 437 days 17 hrs and a career record in two space flights of 678 days 16 hrs

STS-66

Int. Designation	1994-073A
Launched	3 November 1994
Launch Site	Pad 39B, Kennedy Space Center, Florida
Landed	14 November 1994
Landing Site	Runway 22, Edwards AFB, California
Launch Vehicle	OV-104 Atlantis/ET-67/SRB BI-069/SSME #1 2030; #2 2034; #3 2017
Duration	10 days 22 hrs 34 min 2 sec
Call sign	Atlantis
Objective	ATLAS-3; CRISTA-SPAS free-flying pallet satellite

Flight Crew

McMONAGLE, Donald Ray, 42, USAF, commander, 3rd mission
Previous missions: STS-39 (1991); STS-54 (1993)
BROWN Jr., Curtis Lee, 38, USAF, pilot, 2nd mission
Previous mission: STS-47 (1992)
OCHOA, Ellen Lauri, 36, civilian, mission specialist 1, payload commander, 2nd mission
Previous mission: MS STS-56 (1993)
TANNER, Joseph Richard, 44, civilian, mission specialist 2
CLERVOY, Jean-François André, 35, civilian, ESA mission specialist 3
PARAZYNSKI, Scott Edward, 33, civilian, mission specialist 4

Flight Log

This was the third flight of the same seven-instrument ATLAS payload that had previously flown on STS-45 in 1992 and STS-56 in 1993, making this one of the most comprehensive efforts to gather data about the energy output of our Sun and the chemical make-up of Earth's atmosphere. The six astronauts would operate a two-shift system. McMonagle would lead the Red Shift, with Ochoa and Tanner, while Brown, Clervoy and Parazynski worked the Blue Shift.

The only slight delay to the launch of STS-66 was caused by weather conditions at the transoceanic abort sites. Atlantis was on its first mission since returning from the Rockwell Palmdale facility, where it had received new nose wheel steering capability, improved internal plumbing and electrical connections that would allow the EDO pallet kit to be fitted when required. Additional electrical wiring was also installed in preparation for fitting the Orbiter Docking System for the first Shuttle–Mir docking missions, which Atlantis was manifested to fly in the summer of 1995.

ATLAS-3's Atmospheric Trace Molecule Spectrometer (ATMOS) collected more data on trace gases in our atmosphere on this flight than on its previous flights

Joe Tanner works among several lockers on the mid-deck of Atlantis during the Atlas 3 mission. While the payload in the cargo bay was being operated, the crew members worked on secondary experiments. Here, Tanner works with protein crystal growth support equipment that represents continued research into the structures of proteins and other macro-molecules such as viruses. Such work never usually made the headlines during these missions but it was as important as the major payload, helping to develop understanding and experience of operating small scientific experiments and hardware in space as a prelude to ISS

combined. The Shuttle Backscatter UV Spectrometer recorded ozone measurements which were used to calibrate the ozone monitor on the ageing NOAA-9 satellite. The Active Cavity Radiometer Irradiance Monitor (ACRIM) that obtained precise measurements of the Sun's total radiation for 30 orbits (about 1,350 minutes) was also used to calibrate another spacecraft, the UARS satellite. Other instruments recorded measurements of the Sun in the various radiation categories. Before a malfunction shut down the Millimeter Wave Atmospheric Sounder (MAS), it collected nine hours of data on water vapour, chlorine, carbon monoxide and ozone in Earth's atmosphere at altitudes of 20–100 km.

The Cryogenic Infrared Spectrometers and Telescopes for the Atmosphere–Shuttle Pallet Satellite (CRISTA-SPAS) was a second primary payload and was classed as a joined mission with ATLAS-3 with a single set of scientific objectives. Released by RMS on FD 2, it flew behind Atlantis at a distance of about 40–70 km, collecting data for over eight days on the medium- and small-scale distribution of trace gases in the middle atmosphere. The instruments on CRISTA-SPAS also recorded the amounts of hydroxyl and nitric acid that destroy the ozone in the middle atmosphere and lower thermosphere from 40 to 120 km. This represented the first complete global mapping of hydroxyl in our atmosphere, and helped to define a more detailed model and understanding of how energy is balanced throughout the layers.

When the satellite was retrieved and stowed in the payload bay prior to entry, the astronauts adopted the R-Bar rendezvous approach, which was the same as the one that would be used in the upcoming Shuttle–Mir missions, saving propellant and reducing the risk of contamination from thrusters' jets. For this approach, the active spacecraft (in this case, Atlantis) approaches its passive target (CRISTA-SPAS) by flying along an imaginary line (bar) aligned with the radius (R) of the Earth. Approaching from "above" is called a "negative R-Bar" but the approach from below (used on STS-66 and for Shuttle–Mir), known as "positive R-Bar", depends upon the differential gravity from vertical separation to act as a brake and slow down the rate of closure. At the end of the STS-66 mission, Atlantis was diverted to a landing at Edwards because of high winds, rain and cloud cover at the Cape caused by tropical storm Gordon.

Milestones

175th manned space flight
96th US manned space flight
66th Shuttle mission
13th flight of OV-1094 Atlantis
3rd and final flight of ATLAS payload

STS-63

Int. Designation	1995-004A
Launched	3 February 1995
Launch Site	Pad 39B, Kennedy Space Center, Florida
Landed	11 February 1995
Landing Site	Runway 15, Shuttle Landing Facility, KSC, Florida
Launch Vehicle	OV-103 Discovery/ET-68/SRB BI-070/SSME #1 2035; #2 2109; #3 2029
Duration	8 days 6 hrs 28 min 15 sec
Call sign	Discovery
Objective	Mir Rendezvous (near-Mir) mission; SpaceHab 3; EVA Development Flight Test

Flight Crew

WETHERBEE, James Donald, 42, USN, commander, 3rd mission
Previous missions: STS-32 (1990); STS-52 (1992)
COLLINS, Eileen Marie, 38, USAF, pilot
HARRIS Jr., Bernard Anthony, 38, civilian, mission specialist 1, 2nd mission
Previous mission: STS-55 (1993)
FOALE, Colin Michael, 38, civilian, mission specialist 2, 3rd mission
Previous missions: STS-45 (1992); STS-56 (1993)
VOSS, Janice Elaine, 38, civilian, mission specialist, 2nd mission
Previous mission: STS-57 (1993)
TITOV, Vladimir Georgievich, 48, Russian Air Force, Russian mission specialist 4, 3rd mission
Previous missions: Soyuz T8 (1983); Soyuz T10 launch pad abort (1983); Soyuz TM4 (1987)

Flight Log

As originally planned, Mir should have been visited by the Soviet Space Shuttle Buran but when the first Shuttle finally reached the space complex, it was an American, not a Russian one. The STS-63 mission achieved a number of milestones in space history for both the US and the Russian space programmes, and was another significant step towards cooperative efforts for the forthcoming ISS. With only a five-minute window to rendezvous with Mir, Discovery's countdown was refined to include additional holding time at the $T - 6$ hour and $T - 9$ minute points. The 2 February launch was postponed at $T - 1$ day when one of the three IMUs on Discovery failed.

Starting on FD 1, a series of thruster burns brought Discovery to a rendezvous with Mir on FD 4. The approach was expected to be as close as 10 metres, but with three of the Shuttle's 44 RCS thrusters used for small manoeuvres leaking prior to

Astronaut Foale (on the RMS) attempts to grab the SPARTAN 204 as Harris looks on during the first EDFT EVA. The roof of the SpaceHab module is in foreground

rendezvous, the Russians voiced some concerns and it took some considerable discussions and exchange of technical information to convince them that it was safe to proceed. Wetherbee brought the Shuttle to a station-keeping distance of 122 metres, then closed to about 11.2 metres with the crews excitedly talking to each other. Cosmonaut Titov, who was aboard Discovery, had spent over a year on Mir in the late 1980s and talked extensively with his colleagues on the station. This was the first time American and Russian spacecraft had been this close for almost 20 years and the next step in the schedule was the planned docking of STS-71 in June. For now, the close approach was a useful demonstration of a skill that American astronauts had not used in conjunction with another manned spacecraft since the Apollo era – proximity operations. Discovery was eventually withdrawn back to 122 metres and Wetherbee executed a one-and-a-quarter orbit loop around Mir as the astronauts conducted a detailed photographic survey of the station. On board Mir, the EO-17 crew reported no vibrations or movement of the station's solar array panels during the manoeuvres. This was an excellent start to Shuttle–Mir operations and is often termed the near-Mir mission.

In addition to the Mir rendezvous, STS-63 featured the usual complement of mid-deck and payload bay secondary experiments, plus the third flight of SpaceHab. This flight of the commercially-developed augmentation module included 20 experiments, with 11 biotechnology experiments, three advanced materials development experiments, four demonstrations of technology and a pair of hardware experiments supporting acceleration technology. In past flights, crew time was taken up with caretaking the experiments but on this flight, developments in remote monitoring and data transfer reduced direct crew involvement and allowed principle investigators to monitor and control their own experiments. A new robotic device to change samples, called Charlotte, was also flown as an evaluation of automated systems that would allow the crew to focus their efforts on other areas of the flight plan. SPARTAN-204 was lifted out of the cargo bay on FD 2 by the RMS and would study the orbiter glow phenomena and firings of the jet thrusters on the Shuttle. It was later released for a 40-hour free-flight, during which time its Far UV Imaging Spectrograph studied a range of celestial targets in interstellar space.

The SPARTAN was also planned to be used during the EVA towards the end of the mission. The EVA (9 Feb, 4 hours 39 minutes) was the first in a series of EVA Development Flight Test objectives designed to prepare NASA for ISS assembly activities. Harris (EV1) and Foale (EV2) were meant to handle the 1,134 kg SPARTAN payload to rehearse ISS assembly techniques for translating large masses, but both astronauts reported feeling cold while at the end of the RMS, despite modifications in their suits to keep them warm when away from the somewhat protected environment of the payload bay. One of the final objectives of STS-63 was to test the revised landing surface at the Shuttle Landing Facility. This was expected to decrease wear on the tyres and give the orbiters a better chance of landing in crosswinds, thus offering a greater range of landing opportunities at the Cape to help maintain processing schedules and to meet the launch windows of a tight manifest. Upon landing, the crew received congratulations from the cosmonauts on Mir. The once-independent US and Russian manned space programmes were beginning to merge into one international programme for ISS and this mission was an important step towards that goal.

Milestones

176th manned space flight
97th US manned space flight
67th Shuttle mission
20th flight of Discovery

- 31st US and 56th flight with EVA operations
- 1st orbiter to complete 20 missions
- 1st approach/fly-around with Mir by US Shuttle
- 1st female Shuttle pilot
- 2nd Russian cosmonaut on Shuttle
- 1st EDFT excursion
- 3rd SpaceHab mission
- 1st African American to perform EVA (Harris)

STS-67

Int. Designation	1995-007A
Launched	2 March 1995
Launch Site	Pad 39A, Kennedy Space Center, Florida
Landed	18 March 1995
Landing Site	Runway 22, Edwards AFB, California
Launch Vehicle	OV-105 Endeavour/ET-69/SRB BI-071/SSME #1 2012; #2 2033; #3 2031
Duration	16 days 15 hrs 8 min 48 sec
Call sign	Endeavour
Objective	Astro-2 payload; EDO mission

Flight Crew

OSWALD, Stephen Scott, 43, civilian, commander, 3rd mission
Previous missions: STS-42 (1992); STS-56 (1993)
GREGORY, William George, 37, USAF, pilot
GRUNSFELD, John Mace, 36, civilian mission specialist 1
LAWRENCE, Wendy Barrien, USN, 35, mission specialist 2
JERNIGAN, Tamara Elizabeth, 35, civilian, mission specialist 3, 3rd mission
Previous missions: STS-40 (1991); STS-52 (1992)
DURRANCE, Samuel Thornton, 51, civilian, payload specialist 1, 2nd mission
Previous mission: STS-35 (1990)
PARISE, Ronald Anthony, 43, civilian, payload specialist 2, 2nd mission
Previous mission: STS-35 (1990)

Flight Log

STS-67 was the long-awaited re-flight of the Astro payload, flying the same three telescopes as those on the original Astro mission. It was also the second flight of the two payload specialists who had accompanied the package in 1990. Improvements to the Hopkins UV Telescope (HUT) since Astro-1 had made the instrument three times more sensitive and the mission's 16-day duration was double that of the first mission. This was the latest in the series of EDO missions, maximising the scientific return from the vehicle and payload in preparation for Shuttle–Mir and ISS missions.

After a launch delay of only a minute due to concerns with a heater system in the Flash Evaporator System, the countdown proceeded to launch using a back-up heater. Once on orbit, a leaky RCS thruster briefly delayed activation of the payload, but this, too, was soon resolved. The crew settled down to operating Astro in the familiar two-shift system (Red Shift of Oswald, Gregory, Grunsfeld and Parise; Blue Shift of Lawrence, Jernigan and Durrance), essentially providing 32 days worth of operations in 16 flight days. The crew activated and operated a range of mid-deck and

Back-dropped against the desert of Namibia in Africa the Astro-2 payload is deployed from the payload bay of Endeavour during the two-week astronomy mission. Two GAS canisters are in the lower left foreground and the Igloo, which supports the package of experiments, is in centre foreground

payload bay secondary experiments, as well as the primary Astro payload package. The crew used the discoveries from Astro 1 for their observation programme, as well as targeting new areas. In addition to the HUT instrument, the Wisconsin UV Photo-Polarimeter Experiment (WUPPE) measured UV radiation photometry and polarisation from a range of astronomical objects, while the UV Imaging Telescope (UIT) took wide-field photography of objects in UVB light. Observations were planned daily as the mission proceeded and divided into three-hour blocks covering two orbits, with one of the three telescopes assigned higher priority each time. The Astro-2 programme encompassed 23 different science objectives and achieved all of them.

HUT returned over 200 separate observations of more than 100 celestial objects. It recorded intergalactic helium, and in conjunction with the Hubble Space Telescope, took UV measurements of the aurora of the planet Jupiter, and studied the Jovian moon Io as well as the atmospheres of Venus and Mars. Despite the loss of one of its two cameras, 80 per cent of the scientific objectives of the UIT were also obtained. UIT imaged around two dozen spiral galaxies and obtained the first complete UV images of the Moon. The instrument looked at stars over 100 times hotter than our own Sun, at some of the faintest stars known, and at elliptical galaxies. The WUPPE gathered additional data on the dust clouds in the Milky Way and the nearby Large Magellanic Cloud galaxy. Several types of stars were investigated and the instrument was also used to study three recently exploded novae.

STS-67 also became the first mission to be connected to the internet, with its own Mission Home Page. There were reportedly 2.4 million requests recorded on the site at Marshal Space Flight Center in Huntsville, Alabama, from 200,000 computers across 59 countries. The landing was delayed a day from 17 March due to bad weather at the Cape and when things did not improve the following day, the landing was diverted to Edwards.

Milestones

177th manned space flight
98th US manned space flight
68th Shuttle mission
8th mission for Endeavour
2nd flight of Astro payload
1st internet link-up
5th EDO mission

536 The Fourth Decade: 1991–2000

SOYUZ TM21

Int. Designation	1995-010A
Launched	14 March 1995
Launch Site	Pad 1, Site 5, Baikonur Cosmodrome, Kazakhstan
Landed	7 July 1995 (aboard STS-71/Atlantis)
Landing Site	Runway 15, Shuttle Landing Facility, KSC, Florida
Launch Vehicle	R7 (11A511U2); spacecraft serial number: (7K-M) 70
Duration	115 days 8 hrs 43 min 2 sec
Call sign	Uragan (Hurricane)
Objective	Mir 18 resident crew; first NASA resident astronaut, part of Shuttle–Mir Phase 1 programme

Flight Crew

DEZHUROV, Vladimir Nikolayevich, 32, Russian Air Force, commander
STREKALOV, Gennady Mikhailovich, 54, civilian, flight engineer, 5th mission
Previous missions: Soyuz T3 (1980); Soyuz T8 (1983); Soyuz T10-1 pad abort (1983); Soyuz T11 (1984); Soyuz TM10 (1990)
THAGARD, Norman Earl, 51, civilian, NASA cosmonaut researcher, 5th mission
Previous missions: STS-7 (1983); STS 51-B (1985); STS-30 (1989); STS-42 (1992)

Flight Log

Veteran Shuttle astronaut Norman Thagard became the first American to fly into space on a foreign launch vehicle and spacecraft as crew member of Soyuz TM21. He was also a member of the 18th resident crew on Mir and would participate in a programme of 28 experiments, most of which were focused on biomedical research. When the joint programme was planned, it was expected that most of the American research equipment would be delivered with the Spektr research module, but this was seriously delayed (and in fact did not dock with Mir until half-way through Thagard's residence), so most of Thagard's equipment had to be delivered by Progress re-supply craft. Scheduling problems affected the planning of this mission due to delays in preparing Spektr for launch, which in turn affected the scheduling of STS-71, the first Shuttle–Mir docking mission. Most of the working time of the two Russian cosmonauts was consumed by maintenance and repair tasks, mostly on the environmental and thermal regulation systems of the station. Reports indicated that at least 40 per cent of their working time was taken up with such tasks, as the station was now beginning to show signs of its age.

In preparation for receiving the new module at the station, Progress M23 was undocked from the front port of Mir. Then the two Russian cosmonauts had to complete a series of EVAs to retract the solar arrays on Kristall and reposition them on Kvant to facilitate the rearrangement of the module using the Ljappa arm. The first

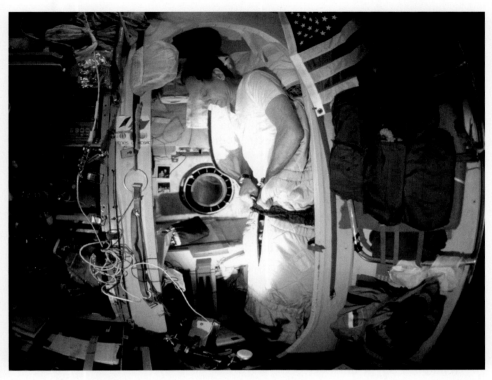

US astronaut Norm Thagard in his sleep restraint aboard Mir, the first of seven NASA astronauts to live and work aboard the Russian station as resident crew members

three EVAs were in support of retracting the solar arrays on Kristall. This module had been located initially where Spektr was due to be docked at the −Y port (negative Y – these locations refer to the facing of the docking ports on the station along the X, Y and Z axes) so it had to be moved. Kristall was moved on 26 May from the −Y port to the −X port, and then to the −Z port on 29 May after the cosmonauts had performed an internal EVA (IVA) to relocate the docking systems. Spektr docked at the −X port on 1 June and the next day was moved to the −Y port, again after an IVA to relocate docking equipment. Kristall was rotated to the −X port on 10 June and Atlantis docked with it on 29 June during the STS-71 mission. After the departure of the Shuttle, Kristall was relocated from the −X port back to the −Z port on 17 July.

The five EVAs, totalling 18 hours 43 minutes were conducted on 12 May (6 hours 14 minutes), 17 May (6 hours 42 minutes), 22 May (5 hours 14 minutes), 29 May (21 minutes IVA) and 2 June (22 minutes IVA). On 5 June, the two arrays of Spektr were deployed but one of them failed to fully open and an unscheduled EVA was planned for 16 June to manually unfurl the array in order to provide sufficient power during the docked phase of the Atlantis STS-71 mission. However, Strekalov argued that the EVA was unnecessary and very hazardous as they had not trained for it.

He refused to conduct the EVA despite pressure from Dezhurov. The electrical power was eventually estimated to be sufficient for the docking, but the cosmonauts were fined 15 per cent of their fee for the mission for failing to conduct their sixth EVA.

Thagard had to contend with several frustrating problems during his stay on the station. A freezer broke down, causing many of the samples he had taken to be lost. He also felt isolated, as communication with Earth was far less frequent on Mir than on the Shuttle, but kept himself informed of what was happening on Earth thanks to contact with a ham radio operator in California. Thagard also lost weight and became bored, although the arrival of the Spektr module helped keep him busy. It quickly became apparent to the Americans that flying with the Russians on a long-duration Mir mission would be very different to flying short missions on the Shuttle. Thagard's flight became the longest by an American astronaut to date, finally surpassing the 84-day record held by the Skylab 4 crew since 1974. Just like the pioneering Skylab long-duration missions, Thagard's 115 days in space proved to be another valuable, if difficult, learning curve for NASA. It was also clear, however, that there would be many more hurdles to overcome during the Phase 1 programme.

Milestones

178th manned space flight
80th Russian manned space flight
18th Mir resident crew
73rd manned Soyuz mission
20th manned Soyuz TM mission
26th Russian and 57th flight with EVA operations
1st US astronaut launched on Russian spacecraft
1st US astronaut crew member on a main Russian crew
1st US resident astronaut on Mir
Thagard celebrated his 52nd birthday in space (3 Jul)

STS-71

Int. Designation	1995-030A
Launched	27 June 1995
Launch Site	Pad 39A, Kennedy Space Center, Florida
Landed	7 July 1995
Landing Site	Runway 15, Shuttle Landing Facility, KSC, Florida
Launch Vehicle	OV-104 Atlantis/ET-70/SRB BI-072/SSME #1 2028; #2 2034; #3 2032
Duration	9 days 19 hrs 22 min 17 sec
Call sign	Atlantis
Objective	1st Mir docking mission; delivery of EO-19 crew; return of EO-18 crew including 1st US NASA Mir resident astronaut (Thagard)

Flight Crew

GIBSON, Robert Lee, 48, USN, commander, 5th mission
Previous missions: STS 41-B (1984); STS 61-C (1986); STS-27 (1988); STS-47 (1992)
PRECOURT, Charles Joseph, 39, USAF, pilot, 2nd mission
Previous mission: STS-55 (1993)
BAKER, Ellen Louise, 42, civilian, mission specialist 1, 3rd mission
Previous missions: STS-34 (1989); STS-50 (1992)
HARBAUGH, Gregory Jordan, 39, civilian, mission specialist 2, 3rd mission
Previous missions: STS-39 (1991); STS-54 (1993)
DUNBAR, Bonnie Jean, 46, civilian, mission specialist 4, 4th mission
Previous missions: STS 61-A (1985); STS-32 (1990); STS-50 (1992)

Mir EO-19 crew up only:

SOLOVYOV, Anatoly Yakovlevich, 47, Russian Air Force, Russian cosmonaut 1, commander EO-19, 4th mission
Previous missions: Soyuz TM5 (1988); Soyuz TM9 (1990); Soyuz TM15 (1992)
BUDARIN, Nikolai Mikhailovich, 42, civilian, Russian cosmonaut 2, flight engineer EO-19

Mir EO-18 crew down only:

DEZHUROV, Vladimir Nikolayevich, 32, Russian Air Force, Russian cosmonaut 1, commander EO-18
STREKALOV, Gennady Mikhailovich, 54, civilian, Russian cosmonaut 2, flight engineer EO-18, 5th mission
Previous missions: Soyuz T3 (1980); Soyuz T8 (1983); Soyuz T10-1 pad abort (1983); Soyuz T11 (1984); Soyuz TM10 (1990)

540 The Fourth Decade: 1991–2000

THAGARD, Norman Earl, 51, civilian, NASA-1 cosmonaut researcher
Mir-18, mission specialist 5, 5th mission
Previous missions: STS-7 (1983); STS 51-B (1985); STS-30 (1989); STS-42 (1992)

Flight Log

The launch of STS-71 was originally scheduled for late May, but was delayed due to the late launch of Spektr and the series of Mir EVAs in support of relocating the new module. Launch attempts on 23 and 24 June were scrubbed due to weather concerns. The docking with Mir occurred on 29 June, using the Earth radius vector approach (R-Bar) in which the orbiter approached Mir from "below". The Orbiter Docking System and an Androgynous Peripheral Docking System acted as the connection

A historic handshake on 29 June 1995 between American astronaut Robert Gibson (STS-71 commander) and Russian cosmonaut Vladimir Dezhurov, Mir-18 commander in the first link-up of a Shuttle and the Mir space complex. This event took place two-and-a-half weeks prior to the 20th anniversary of a similar space greeting, during ASTP in July 1975

point between the Shuttle and the Kristall module. The same day, the EO-19 crew took over from the EO-18 crew as residents on the station.

Over the next five days (or 100 hours), a programme of joint biomedical investigations and logistics transfer operations was conducted, the first such activity between US and Russian spacecraft and crews. In the Spacelab module, more than 15 biomedical and scientific investigations were conducted in seven different fields (cardiovascular and pulmonary functions; human metabolism; neuroscience; sanitation and radiation; hygiene; behavioural performance and biology; fundamental biology and microgravity research). Experiment samples from Mir were gathered and returned to Earth aboard Atlantis, including over 100 urine and saliva samples, 30 blood samples, 20 surface samples and 12 air samples. The returning EO-18 crew members also conducted a programme of exercise and prevention measures to help prepare them for the return to Earth after their three-month mission.

Transferred to Mir were 454 kg of water generated by the orbiter system, which would be used on the station for waste system flushing. Specially designed EVA tools were also transferred, as was a supply of oxygen and nitrogen from the Shuttle ECS to raise air pressure inside Mir and conserve the station's own consumables. A broken Salyut 5-type computer was also returned to Earth aboard Atlantis. Undocking occurred on 3 July, shortly after the EO-19 crew had undocked their Soyuz TM and positioned the small spacecraft to photograph the departure of Atlantis. The Shuttle crew then recorded the re-docking of the Soyuz TM before departing for the return to Earth. The EO-18 crew lay supine in custom-made Russian seat liners to ease their readaptation to gravity due to returning on the Shuttle instead of a Soyuz Descent Module. The complement of eight crew members aboard Atlantis as it came home equalled the largest Shuttle crew in history, STS 61-A in October 1985. The runway was changed from No. 33 to No. 15 just twenty minutes before touchdown, due to concerns over clouds in the area obscuring landing aids.

Milestones

179th manned space flight
99th US manned space flight
69th Shuttle mission
14th flight by OV-104 Atlantis
1st Shuttle–Mir docking
1st space station crew exchange by US Shuttle
100th US human space launch from the Cape in Florida
11th Spacelab Long Module mission
1st and only Spacelab to be part of a Shuttle payload docked to a space station
Largest spacecraft ever in orbit (225 tons)
1st on-orbit change of Shuttle crew
Precourt celebrated his 40th birthday in space (29 Jun)

MIR EO-19

Int. Designation	N/A (launched on STS-71 landed via Soyuz TM-21)
Launched	27 Jun 1995 (see STS-71)
Launch Site	Pad 39A, Kennedy Space Center, Florida
Landed	11 September 1995 (in Soyuz TM-21)
Landing Site	108 km north of Arkalyk
Launch Vehicle	STS-71
Duration	75 days 11 hrs 20 min 21 sec
Call sign	Rodnik (Spring)
Objective	Mir 19 resident crew, launched on Shuttle but landed on Soyuz TM, part of Shuttle–Mir Phase 1 programme and domestic Mir programme

Flight Crew

SOLOVYOV, Anatoly Yakovlevich, 47, Russian Air Force, commander, 4th mission
Previous missions: Soyuz TM5 (1988); Soyuz TM9 (1990); Soyuz TM15 (1992)
BUDARIN, Nikolai Mikhailovich, 42, civilian, flight engineer

Flight Log

The Mir EO-19 crew arrived at the space station via the American Shuttle during the STS-71 mission. The cosmonauts had received basic training on ascent operations (and emergency escape procedures) at NASA for the ascent to orbit, but it was planned for them to return to Earth in Soyuz TM21, so there was no need for them to conduct extensive Shuttle systems training. The EO-19 crew and their back-ups (Onufriyenko and Usachev) each received on average 70 hours training on Shuttle launch, entry and orbital operations, crew and Shuttle systems, and procedures. The two EO-18 cosmonauts, on the other hand, who would only complete re-entry aboard the Shuttle, received about 28.5 hours each.

After completing the hand-over procedures from the EO-18 cosmonauts, the EO-19 crew transferred their Soyuz seat liners into the Soyuz DM and officially became the resident crew members of the station. On 2 July, the EO-19 cosmonauts undocked their Soyuz TM21 spacecraft from the Kvant module to dock at the forward port, freeing the rear port to receive further Progress re-supply craft. This event also allowed them to photograph the undocking of Atlantis from the station with the returning EO-18 crew. Then the astronauts on Atlantis photographed the re-docking of the Soyuz at the front port of the station before departing to begin the return to Earth.

Safely back on Mir, the two cosmonauts began their short residency, which again was mainly focused on maintenance and repair although they managed to complete

The 19th Mir resident crew consisted of the veteran Solovyov (left) and the rookie Budarin

some materials-processing operations. Three EVAs were conducted. The first on 14 July (5 hours 34 minutes) was used to inspect the −Z port where Kristall was to be relocated on 17 July. They found nothing to prevent the relocation, and also used the EVA to unfurl the Spektr arrays using a NASA-provided tool. Solovyov and Budarin also inspected an antenna and a malfunctioning solar array drive motor on the Kvant 2 module. The next EVA on 19 July (3 hours 8 minutes) included the retrieval of a US-provided detector and preparations for installing a joint Belgian/French/Russian infrared spectrometer. A failed cooling system in Solovyov's Orlan suit curtailed activities, forcing the cosmonaut to remain near the Kvant 2 hatch and use umbilical cooling supplied from the module instead of the integral backpack. After suit repair work inside Mir, the third EVA on 21 July (5 hours 50 minutes) was used to complete the tasks scheduled for the cancelled second excursion. It was during this EVA that Solovyov became the record-holder for total career EVA time, at 41 hours 49 minutes, surpassing that of Krikalev set in 1992.

Milestones

1st Russian space station crew launched by US Shuttle
1st primary crew flying as passengers on ascent

STS-70

Int. Designation	1995-035A
Launched	13 July 1995
Launch Site	Pad 39B, Kennedy Space Center, Florida
Landed	22 July 1995
Landing Site	Runway 33, Shuttle Landing Facility, KSC, Florida
Launch Vehicle	OV-103 Discovery/ET-71/SRB BI-073; SSME #1 2036; #2 2019; #3 2017
Duration	8 days 22 hrs 20 min 5 sec
Call sign	Discovery Objective Deployment of TDRS-G by IUS-26

Flight Crew

HENRICKS, Terence Thomas "Tom", 43, USAF, commander, 3rd mission
Previous missions: STS-44 (1991); STS-55 (1993)
KREGEL, Kevin Richard, 38, civilian, pilot
THOMAS, Donald Alan, 40, civilian, mission specialist 1, 2nd mission
Previous mission: STS-65 (1994)
CURRIE, Nancy Jane, 36, US Army, mission specialist 2, 2nd mission
Previous mission: STS-57 (1993)
WEBER, Mary Ellen, 32, civilian, mission specialist 3

Flight Log

STS-70 should have launched prior to STS-71. However, on 31 May, after northern flicker woodpeckers at Pad 39B had poked approximately 200 holes in the foam insulation of the ET, attempts to repair the damage at the pad were unsuccessful. The resulting rollback to the VAB for repairs forced the mission to be rescheduled after STS-71. The media coverage of the woodpecker activities (from the wildlife nature reserve around the Cape), prompted two JSC employees to design a comic STS-70 mission emblem, adding a smiling *Woody Woodpecker* cartoon character – a tongue-in-cheek joke that the flight crew enjoyed. They were also amused by the use of the *Woody Woodpecker* cartoon's theme tune as a wake-up call on FD 2. The countdown to launch on 13 July proceeded relatively smoothly, with only a short, 55-second hold to verify range safety system signals from the destruct system on the ET. The lift-off marked the shortest time between the landing of one mission and the launch of the next (just 6 days) in the programme. However, post-flight inspection of the right-hand SRM nozzle revealed a gas path in internal joint number 3 that extended from the motor chamber up to, but not beyond, the primary O-ring. A similar gas path had been noted on STS-71 and had been revealed on other missions, but the incidents on STS-71 and STS-70 were the first to show a slight heat effect on the primary O-ring.

Tom Henricks on the aft flight deck of Discovery aims towards a site on Earth with the TV camera and other hardware for the HERCULES-B system. For this third-generation space-based geolocation system, a Xybion multispectral camera was integrated with the Hercules geolocation hardware. Previously, a NASA electronic still camera was used on HERCULES-A, flown on STS-53 and STS-56

The problem would require investigation and repair on future SRB/SRMs scheduled for flight (see STS-69).

The launch of STS-70 included the first flight of a new Block I SSME (#2036) that featured improvements to increase the reliability and safety margins of the engines. The first use of three Block I improved engines was scheduled for STS-73. Some six hours into the mission, the crew deployed the final TDRS satellite, which would act as an on-orbit operational spare. With the completion of the primary task, the crew spent the remainder of their mission working on a range of mid-deck experiments.

These focused on plant growth and development, the hormone system of insects, the performance of a bioreactor in microgravity for the growth of individual cells, and a range of experiments aimed at studying the effects of space flight on mammalian development. The crew also conducted commercial protein crystal growth experiments, research into space tissue loss, tests of hand-held space-based geolocation systems, the production of pharmaceuticals in weightlessness, examinations of the effects of ships on the marine environment, radiation monitoring and experiments to understand the chemistry and dynamics of thruster emissions, outgassing and other debris on the Shuttle's exterior hardware and surfaces.

The mission was also to be run from the new upgraded Mission Control Room in Building 30 at the JSC facility near Houston, Texas. Following the deployment of TDRS, the controllers on the next shift operated from the new MCC (called the White Flight Control Room – FCR, pronounced "flicker"). The old room had been used since Gemini 4 in June 1965 and became famous as the Apollo mission control room during the lunar landing missions. It would be turned into a national monument and tourist attraction at JSC. Orbital operations continued in the new room, but landing operations were handled from the old room. Until early 1996, all Shuttle launch and landing phases would still be controlled from the old FCR, but would gradually be moved into the new room.

The first landing opportunity for STS-70, on 21 July at KSC, was waived off due to fog and low visibility, as was the first attempt on 22 July. Following the landing, Discovery was prepared for shipment to California for a period of refurbishment and modifications. It was due to return to duty in the summer of 1996, to prepare for the second Hubble service mission planned for early 1997 (STS-82).

Milestones

180th manned space flight
100th US manned space flight
70th Shuttle mission
21st flight of Discovery
7th and final TDRS deployment mission
1st use of new MCC in Houston, JSC
Quickest turnaround from landing (STS-71) to launch (STS-70) between missions – 6 days

SOYUZ TM22

Int. Designation	1995-047A
Launched	3 September 1995
Launch Site	Pad 1, Site 5, Baikonur Cosmodrome, Kazakhstan
Landed	29 February 1996
Landing Site	108 km northeast of Arkalyk
Launch Vehicle	R7 (11A511U2); spacecraft serial number (7K-M) 71
Duration	179 days 1 hr 41 min 46 sec
Call sign	Uran (Uranus)
Objective	20th Mir resident crew; EuroMir95 mission; host to STS-74 docking mission

Flight Crew

GIDZENKO, Yuri Pavlovich, 33, Russian Air Force, commander
AVDEYEV, Sergei Vasilyevich, 39, civilian, flight engineer, 2nd mission
Previous mission: Soyuz TM15 (1992)
REITER, Thomas, 37, German Air Force, cosmonaut researcher

Flight Log

The next resident crew arrived at Mir on 5 September. Thomas Reiter, a German ESA astronaut, was flying the EuroMir95 mission, and had received complete training for an extended mission. He was also qualified for assignment as a Soyuz flight engineer and for EVAs using the Orlan DMA pressure garment. The EuroMir95 mission included an EVA and a programme of 41 experiments that incorporated 18 biomedical, 10 technical and 8 materials-processing experiments. On 17 October, it was announced that the mission had been extended by another three weeks and that a second EVA was planned for Reiter. This agreement was also in part due to financial difficulties in the Russian programme, in particular problems with paying workers to complete the fabrication of launchers and spacecraft. The resulting delays meant that the launch of the next resident crew had to be put back, so Reiter's eventual 44-day mission extension gave him a 179-day stay aboard Mir.

For the first EVA (20 Oct 1995, 5 hours 16 minutes), Reiter accompanied Avdeyev to install the European Science Exposure Facility (ESEF) and to exchange experiment cassettes on a Russian experiment designed to trap ambient particle debris and to record the velocity, mass and trajectory of each impact particle. The second EVA (8 Dec 1995, 29 minute IVA) was conducted by both Russian cosmonauts from inside the base block docking node, to transfer the docking drogue from the −Z port to the +Z port in order to receive Priroda, the final Mir module, the following year. During this EVA, Reiter remained inside the Soyuz TM22 Descent Module for safety reasons. For the final excursion (8 Feb 1996, 3 hours 6 minutes), Reiter was accompanied by

The three Mir EO-20 crew members in the Docking Module delivered to Mir by STS-74. In the centre is ESA cosmonaut researcher Thomas Reiter, holding the camera is Sergey Avdeyev, and on the left is commander Yuri Gidzenko

Gidzenko. The pair had been trained for this excursion by radio instructions from the ground, as a third EVA was not part of the original flight plan. The first task was to move the SDPK MMU (used in early 1990) outside the Kvant −Z hatch in order to make more room inside the airlock area. They also removed dust collectors, but were unable to remove a faulty antenna as they did not have the required tools.

Once again, maintenance and repair tasks took priority over science, one of the problems being the sheer amount of equipment and unwanted hardware being built up on the station. Indeed, Reiter finally found his centrifuge after mislaying it for two months. This residency was visited by STS-74, the second Shuttle docking mission, in November. The Shuttle also delivered the Russian-supplied Docking Module, enabling the docking facility to be moved further away from the main modules for better vision during future Shuttle dockings. On 20 February 1996, the base block of the Mir station reached the milestone of a decade in orbit, and it looked as though (funds permitting) the station would be maintained for some years, perhaps past the turn of the century, by which point the ISS would be in orbit and operational. During the flight, Reiter compiled an educational video called *Riding High*, which documented his life aboard the station, as well as the visit of Atlantis, his EVAs and the science research programme. ESA was pleased with the results from the two EuroMir missions and planned a 45-day EuroMir97 with Reiter's back-up Christer Fuglesang. However, support and funding from ESA member states was not forthcoming and the mission was not flown, partly due to the desire to progress beyond Mir to ISS, where ESA would have its own science module – Columbus.

Milestones

181st manned space flight
81st Russian manned space flight
74th Soyuz manned mission
21st Soyuz TM manned mission
20th main Mir crew
27th Russian and 58th flight with EVA operations
1st EVA by two flight engineers
1st EVA by German cosmonaut
Final manned launch of Soyuz U2 launch vehicle
Avdeyev celebrates his 40th birthday in space (1 Jan)

STS-69

Int. Designation	1995-048A
Launched	7 September, 1995
Launch Site	Pad 39A, Kennedy Space Center, Florida
Landed	18 September 1995
Landing Site	Runway 33, Shuttle Landing Facility, KSC, Florida
Launch Vehicle	OV-105 Endeavour/ET-72/SRB BI-074/SSME #1 2035; #2 2109; #3 2029
Duration	10 days 20 hrs 28 min 56 sec
Call sign	Endeavour
Objective	Wake Shield Facility-2 operations; SPARTAN 201-03 operations; EVA Development Flight Test

Flight Crew

WALKER, David Mathiesen, 51, USN, commander, 4th mission
Previous missions: STS 51-A (1984); STS-30 (1989); STS-53 (1992)
COCKRELL, Kenneth Dale, 45, civilian, pilot, 2nd mission
Previous mission: STS-56 (1993)
VOSS, James Shelton, US Army, mission specialist 1, payload commander, 3rd mission
Previous missions: STS-44 (1991); STS-53 (1992)
NEWMAN, James Hansen, 38, civilian, mission specialist 3, 2nd mission
Previous mission: STS-51 (1993)
GERNHARDT, Michael Landen, 39, civilian, mission specialist 4

Flight Log

Set originally for 5 August, the launch of STS-69 was postponed indefinitely pending a full review of the SRM nozzle damage found on the two previous missions. This problem was found to be caused by the Room Temperature Vulcanising (RTV) process when insulation material was applied; small air pockets in the material could later form pathways for the hot gases during motor operation. The joint design was deemed to be sound and non-destructive inspections were developed to examine the insulation material on specific joints. Development, testing and application of the repair procedures was conducted, and these could be performed on STS-69 while it was still on the pad, alleviating the need for a rollback. The nozzles on the motors assigned to STS-73 and 74 were repaired at KSC but these did not impact upon processing schedules. The planned 31 August launch of STS-69 was scrubbed about 5.5 hours prior to lift-off when Fuel Cell 2 indicated higher than allowed temperatures during activation. It was removed and replaced, and the launch was rescheduled for 7 September.

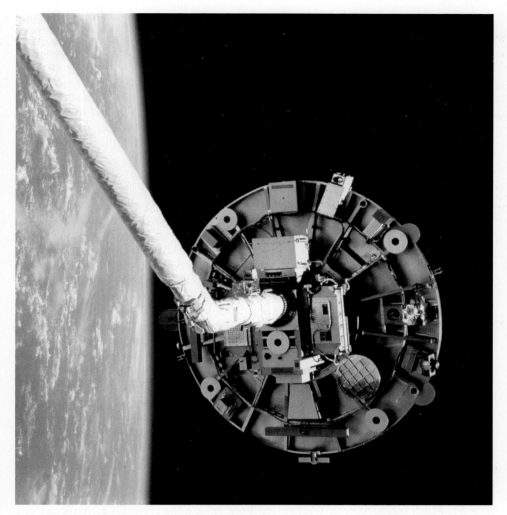

The Wake Shield Facility is attached to the RMS during its second of four planned flights on the Shuttle

The deployment of the SPARTAN-201 occurred during FD 2 and was the third of a series of four flights of the facility. SPARTAN-201 was designed to study the outer atmosphere of our Sun and its transition into the solar wind that constantly flows past the Earth. At the same time as SPARTAN was being operated, the Ulysses space craft (deployed from STS-41 in October 1990) was flying over the Sun's northern polar regions, obtaining data that would expand our understanding of solar wind from different regions of the Sun. Concerns about the performance of the two onboard instruments were raised after two days of gathering data, but subsequent analysis revealed that both instruments had operated normally and had gathered the expected amounts of data. The question of why the SPARTAN was rotating and in a different attitude for retrieval than expected would require further analyses.

The second flight of the Wake Shield Facility featured another first. Released on FD 5, the WSF manoeuvred away from Endeavour instead of the other way around, as on previous deployments. Seven thin film growths for the next generation of advanced electronics were planned during the mission, but after only three successful growths, the facility placed itself in safe mode. It was decided to extend the facility's free-flight by 24 hours to allow a 20-hour break in operations while still achieving seven growths of material. However, when controllers commanded the resumption of operations, they could not generate the flow of the thin material, so the facility was shut down again for a six-hour cool-down period. This time it started up correctly and a fourth film was grown, before the facility was finally retrieved on FD 8. The WSF-2 was deployed on the RMS and placed over the side of the cargo bay for an Air Force experiment designed to collect data on the build-up of electrical fields around orbiting spacecraft.

In addition to the range of medical and secondary payloads in the payload bay, the crew also completed the second in the series of EVA development tests aimed at providing experience and developing procedures applicable to ISS. On FD 10, Voss (EV1) and Gerhardt (EV2) completed their 6 hour 46 minute EVA, where they evaluated improvements in the thermal protection of the EMUs (as a result of the crew's experiences during the STS-63 EVA earlier in the year) and reported improved comfort levels. Improvements included finger tip heaters in the gloves, which were powered by 3.7 volt lithium batteries. Sensors on their EMU boots and on the PLSS backpacks also recorded exterior temperatures. In between activities at a tool board, each astronaut was "cold-soaked" 9 metres above the payload bay on the RMS for 45 minutes while they completed repetitive tool-handling exercises. The two astronauts also evaluated a range of new EMU helmet lights, tools restraints and techniques that were being developed for possible use in construction missions at ISS. They took turns in removing micrometeoroid/debris shields from a work board, tested power tools, and manipulated items such as an antenna, an orbital replacement unit box and electrical conduits, both while restrained and "free-floating".

This mission was also the second flight of a "dog crew", a tradition begun on STS-53, which included both Walker and Voss among the crew. Dog crew call signs originated from Walker who, due to his red hair, was called "Red Dog". His old station wagon was also known as the "dog mobile" and was the impetus for "dog tag" names for his crew members. Dog Crew II thus became: Red Dog (Walker), Cujo (Cockrell); Dog Face (Voss); Pluto (Newman) and Under Dog (Gerhardt).

Milestones

182nd manned space flight
101st US manned space flight
71st Shuttle mission

9th flight of Endeavour
32nd US and 59th flight with EVA operations
2nd flight of Wake Shield Facility
3rd flight of SPARTAN 201 facility
1st Shuttle mission featuring deployment and retrieval of two different payloads on the same mission
2nd EDFT exercise

STS-73

Int. Designation	1995-056A
Launched	20 October 1995
Launch Site	Pad 39B, Kennedy Space Center, Florida
Landed	5 November 1995
Landing Site	Runway 33, Shuttle Landing Facility, KSC, Florida
Launch Vehicle	OV-102 Columbia/ET-73/SRB BI-075; SSME: #1 2037; #2 2031; #3 2038
Duration	15 days 21 hrs 52 min 28 sec
Call sign	Columbia
Objective	USML-2 operations using Spacelab Long Module

Flight Crew

BOWERSOX, Kenneth Duane, 38, USN, commander, 3rd mission
Previous missions: STS-50 (1992); STS-61 (1993)
ROMINGER, Kent Vernon, 39, USN, pilot
COLEMAN, Catherine Grace "Cady", 34, USAF, mission specialist 1
LOPEZ-ALEGRIA, Michael Eladio, 37, USN, mission specialist 2
THORNTON, Kathryn Cordell Ryan, 43, civilian, mission specialist 3, payload commander, 4th mission
Previous missions: STS-33 (1989); STS-49 (1992); STS-61 (1993)
LESLIE, Fred Weldon, 43, civilian, payload specialist 1
SACCO Jr., Albert, 46, civilian, payload specialist 2

Flight Log

Getting STS-73 back on the ground after a highly successful USML-2 mission was achieved at the first attempt. Getting the mission off the ground in the first place, however, had proven a little more frustrating, equalling STS 61-C (12–18 Jan 1986) for the most launch scrubs. Shortly after tanking began for the 25 September launch, a leak in the SSME #1 main fuel valve caused the first scrub. The reset 5 October launch was postponed a day due to the effects of Hurricane Opal and the 6 October attempt was scrubbed when hydraulic fluid was inadvertently drained from hydraulic system 1 after the SSME #1 fuel valve replacement. Reset again to 7 October, the launch was scrubbed once more, this time due to a faulty Main Events Controller that required replacement. The 14 October launch was rescheduled for the following day to allow additional time to inspect the SSME oxidiser ducts on Columbia, as a result of cracks found in a test engine at Stennis. A faulty GPC was also replaced. The 15 October attempt was scrubbed at $T - 5$ minutes because of low clouds and rain and the launch was reset for 19 October, pending a successful Atlas launch on 18 October. However, the Atlas launch was delayed, forcing the STS-73 launch to be rescheduled once more,

Payload commander Thornton and commander Bowersox observe liquid drop activity at the Drop Physics Module in the Spacelab science laboratory aboard Columbia, as part of the USML-2 mission

for 20 October. Even that countdown was delayed 3 minutes, due to a range computer glitch.

When Columbia finally made it to orbit, the crew built upon the success of USML-1 (on which Bowersox was a crew member) with many of the experiments flying a second time. The astronauts worked in two shifts. The Red Shift consisted of Bowersox, Thornton, Rominger and Sacco. The Blue Shift comprised Lopez-Alegria, Coleman and Leslie. There were five areas of research conducted on the mission: fluid physics; materials science; biotechnology; combustion science; and commercial space processing. Generally the experiments performed well, and many re-confirmed findings recorded during USML-1, while the flight also recorded new and unique findings confirming the decision to fly a second USML mission. Flying for the first time was a droplet combustion experiment in which 25 droplets of a variety of fuels were ignited. The studies revealed that larger droplets could be observed in microgravity than on Earth and they burned ten times longer in space. Results from Astroculture experiments indicated that edible food could be grown in a plant growth facility, which has enormous potential for future applications on larger space facilities, deep space missions or future bases on the Moon or Mars. A record number of protein crystal growth samples (approximately 1,500) were flown on this mission. Apart from their Spacelab work the crew taped a ceremonial first pitch for game five of the Baseball World Series, which marked the first time in baseball history that the thrower was not actually in the ballpark for the pitch.

Milestones

183rd manned space flight
102nd US manned space flight
72nd Shuttle mission
18th flight of Columbia
2nd flight of USML series
12th flight of Spacelab Long Module
6th EDO mission

STS-74

Int. Designation	1995-061A
Launched	12 November 1995
Launch Site	Pad 39A, Kennedy Space Center, Florida
Landed	20 November 1995
Landing Site	Runway 33, Shuttle Landing Facility, KSC, Florida
Launch Vehicle	OV-104 Atlantis/ET-74/SRB BI-076/SSME #1 2012; #2 2026; #3 2032
Duration	8 days 4 hrs 30 min 44 sec
Call sign	Atlantis
Objective	Mir docking mission; delivery of Russian-built Docking Module

Flight Crew

CAMERON, Kenneth Donald, 45, USMC, commander, 3rd mission
Previous mission: STS-37 (1991); STS-56 (1993)
HALSELL Jr., James Donald, 39, USAF, pilot, 2nd mission
Previous mission: STS-65 (1994)
HADFIELD, Chris Austin, 36, Canadian Air Force, mission specialist 1
ROSS, Jerry Lynn, 47, USAF, mission specialist 2, 5th mission
Previous missions: STS 61-B (1985); STS-27 (1988); STS-37 (1991); STS-55 (1993)
McARTHUR Jr., William Surles, 44, US Army, mission specialist 3

Flight Log

Because of the planned rendezvous and docking with Mir, this Shuttle mission had only a very small seven-minute window in which to launch. The 11 November launch was scrubbed as a result of bad weather at the TAL sites and the launch was rescheduled to the following day. The original plan had been to have a crew exchange on this mission. Thagard's back-up on the first NASA residency (Bonnie Dunbar) was originally scheduled to remain on Mir after STS-71 departed, but this option was not followed, so STS-74 was the only docking mission on which no US astronaut was exchanged or returned. Instead, the mission focused on the delivery of hardware and logistics. It did feature an international flavour, however, reflecting the plans for ISS in the coming years. Canadian astronaut Chris Hadfield was part of the Shuttle crew, and the Canadian-developed RMS was carried once more. The payload bay of Atlantis carried the Russian-built Docking Module and solar array, along with the US-built Orbiter Docking System and a joint US/Russian-built solar array. And of course, on board Mir were the two Russian and one German cosmonaut, together with a range of Russian and European equipment and experiments.

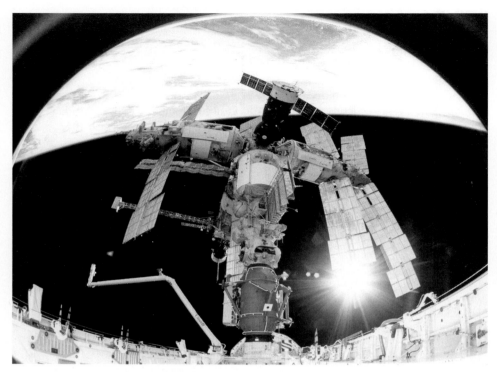

Atlantis is seen docked with Mir high above central Canada in this IMAX camera image, which provides this 65-mm fish-eye perspective. The recently delivered Docking Module is shown connecting the Shuttle to Mir and affording better clearance for Shuttle dockings

The Russian Docking Module (which, when permanently attached to the Kristall module would give better clearance for further Shuttle dockings) was lifted out of the payload bay by Hadfield, who was operating the RMS. It was positioned just above the Orbiter Docking System, carried on all docking flights at the front of the payload bay to permit physical connection between the Shuttle and the space station. Cameron then fired the downward-facing jets on the Shuttle to move the vehicle "up" to dock, with the Docking Module held on the RMS. The docking between the Docking Module on Atlantis and Mir occurred on FD 4 and for the next three days, the crews of Atlantis and Mir completed a joint programme of activities. This included the transfer of the control of the DM to the main Mir crew. There was also 453.6 kg of water transferred across to the station, along with gifts such as Canadian maple leaf candies and the second guitar to be delivered to the station. New lithium hydroxide canisters were also delivered, which would be used in the event of a further failure of the ECS, requiring further "scrubbing" of the air inside the station. Experiment samples were transferred to Atlantis for the return to Earth and on 18 November, Atlantis separated from the DM to begin its fly around of the station and the journey home, leaving the Mir crew to continue their six-month mission.

In January 1996, NASA pronounced itself happy with the success of Shuttle–Mir missions. Continued discussions with the Russians had resulted in expansion of the programme and two further dockings were included in the Phase 1 programme, bringing the total dockings to nine. Two further long-duration visits by American astronauts were also likely, bringing the total US residencies on the station to seven prior to the commencement of ISS construction.

Milestones

184th manned space flight
103rd US manned space flight
73rd Shuttle mission
15th flight of Atlantis
2nd Shuttle–Mir docking mission

STS-72

Int. Designation	1996-001A
Launched	11 January 1996
Launch Site	Pad 39B, Kennedy Space Center, Florida
Landed	20 January 1996
Landing Site	Runway 15, Shuttle Landing Facility, KSC, Florida
Launch Vehicle	OV-105 Endeavour/ET-75/SRB BI-077/SSME #1 2028; #2 2039; #3 2036
Duration	8 days 22 hrs 1 min 47 sec
Call sign	Endeavour
Objective	Retrieval of Japanese Space Flyer Unit; deployment and retrieval of OAST-Flyer; EDFT-03

Flight Crew

DUFFY, Brian, 42, USAF, commander, 3rd mission
Previous missions: STS-45 (1992); STS-57 (1993)
JETT Jr., Brent Ward, 37, USN, pilot
CHIAO, Leroy, 35, civilian, mission specialist 1, 2nd mission
Previous mission: STS-65 (1994)
SCOTT, Winston Elliott, 45, USN, mission specialist 2
WAKATA, Koichi, 32, civilian, Japanese mission specialist 3
BARRY, Daniel Thomas, 42, civilian, mission specialist 4

Flight Log

The launch of STS-72 was delayed for 23 minutes due both to problems with ground sites and the need to avoid a potential collision with an item of space debris. On FD 3, Japanese MS Wakata used the RMS to grasp the Japanese Space Flyer Unit (SFU), which had originally been launched in March 1995 aboard an H-2 rocket from the Tanegashima Space Centre in Japan. Over a ten-month period, more than a dozen onboard instruments and experiments had been operating in a research programme that encompassed materials and biological science. Prior to grappling the unit with the RMS, the twin solar arrays had to be jettisoned after it was found that they were not correctly retracted.

The next day, the Office of Aeronautics and Space Technology Flyer (OAST-Flyer) was deployed, again by Wakata using the RMS, on an independent two-day flight that extended to approximately 72 km from Endeavour. Attached to the SPARTAN platform were four experiments that investigated spacecraft contamination, global positioning technology, laser ordnance devices and an amateur radio package. The flyer was retrieved on FD 6. In addition to the deployment and retrieval operations, the crew had a programme of payload bay and mid-deck secondary

The Japanese Space Flyer Unit (SFU) is retrieved using the Shuttle's RMS. The yet-to-be-deployed OAST Flyer satellite is seen in the payload bay at bottom centre

experiments to conduct, which mainly consisted of studies in ozone concentrations in the atmosphere, a laser to accurately measure the distance between the Earth's surface and the orbiter, and a range of biological and biomedical experiments.

The crew also completed two EVAs as part of the EDFT programme of preparation for extensive EVA activities during ISS construction. During the first EVA (15 Jan, 6 hours 9 minutes), astronauts Chiao (EV1) and Barry (EV2) evaluated a new portable work platform and the Rigid Umbilical Structure, which was being developed as a possible retention device for ISS fluid and electrical lines. During the second EVA (17 Jan, 6 hours 54 minutes), this time conducted by Chiao and Scott (EV3), the portable work platform was again evaluated and the astronauts also tested the design of a utility box, another item under development for ISS, which would hold avionics and fluid line connections. During the EVA, Scott tested his suit in severe cold temperatures of up to −75°C, to find out whether the revised design would keep him warm during the test. In fact, the 35-minute test resulted in temperatures of −122°C being recorded, providing a tough test of the suit's extremities (fingers and feet) and coolant loop bypass system. Scott reported that he was aware of the low temperatures but remained comfortable and though had he been working rather than staying still, he determined that he would have felt warmer in either situation.

Milestones

185th manned space flight
104th US manned space flight
74th Shuttle mission
10th flight of Endeavour
33rd US and 60th flight with EVA operations
3rd EDFT exercise

SOYUZ TM23

Int. Designation	1996-011A
Launched	21 February 1996
Launch Site	Pad 1, Site 5, Baikonur Cosmodrome, Kazakhstan
Landed	2 September 1996
Landing Site	107 km southwest of Arkalyk
Launch Vehicle	R7 (11A511U – # 651); spacecraft serial number (7K-M) #72
Duration	193 days 19 hrs 7 min 35 sec
Call sign	Skif (Scythian)
Objective	21st Mir resident crew programme; worked with NASA-2 astronaut (Lucid)

Flight Crew

ONUFRIYENKO, Yuri Ivanovich, 35, Russian Air Force, commander
USACHEV, Yuri Vladimirovich, 38, civilian, flight engineer, 2nd mission
Previous mission: Soyuz TM18 (1994)

Flight Log

After taking over from the EO-20 crew, the next resident crew for Mir had a busy programme of activities planned for their mission. They were to perform six EVAs, support the arrival of the final Mir science module (Priroda), and work with the second American to live on Mir, Shannon Lucid. In addition there was a change of cosmonauts in the subsequent resident crew and an extension both to Lucid's mission and to the cosmonauts' own stay on Mir.

The "two Yuris", as the press called them, were expecting to conduct a four-and-a-half month residency, but by July, continued problems over the construction of Soyuz rockets at the Samara factory meant that the next mission would have to slip and the cosmonauts were told they would have to fly a six-month mission. Their first EVA (15 May, 5 hours 52 minutes) featured the installation of a second Strela boom to allow easier movement to the repositioned Kristall module. Other EVAs would be performed after the arrival of STS-76 and the transfer of Shannon Lucid to the station.

Shannon Lucid had arrived at Mir on 24 March for what was planned as a 140-day mission but, again in July, following technical problems with the Shuttle SRBs and the decision to replace those assigned to STS-79 (the mission that would bring her home), her own residency was extended, eventually reaching 188 days, and setting a new US endurance record both for a single mission and for a career total. During the other EVAs conducted by the two Russian cosmonauts, Lucid would remain inside the station, monitoring the activities of her two colleagues.

"The two Yuris", Usachev (left) and Onufriyenko with the second US resident astronaut Shannon Lucid

The science module Priroda arrived on 26 April, and the following day was transferred to the left radial (+Z) port, completing the Mir complex. The module contained an array of remote-sensing instruments for ecological and environmental investigations. It also included over 700 kg of NASA equipment and experiments for the American resident crew members to use during their science programmes. Part of the task of the EO-21 resident crew was to begin changing the inside of the new module from its launch configuration to allow it to be used on orbit over the next few years.

The two Yuri's conducted their other five EVAs from mid-May to mid-June (21 May for 5 hours 20 minutes; 24 May for 5 hours 43 minutes; 30 May for 4 hours 20 minutes; 6 Jun for 3 hours 34 minutes; and 14 Jun for 5 hours 42 minutes). Their activities included the installation of the Mir Cooperative Solar Array, the installation of a multi-spectral scanner, retrieval and installation of sample cassettes, the assembly of the Stombus (Ferma-3) 5.9-metre truss, and the deployment of a Symmetric Aperture Radar antenna. They also constructed (over two of the EVAs) a 1.2-m replica Pepsi can from aluminium struts and decorated nylon sheets. They filmed each other near the replica and then disassembled it for return to Earth. The video footage was to be used in a PepsiCo commercial campaign.

Towards the end of their mission, the two cosmonauts handed over to the 22nd resident crew, who had arrived on 19 August. Also arriving with the new crew was

French cosmonaut Claudie André-Deshays, who would return to Earth with the EO-21 cosmonauts two weeks later. For a while, six cosmonauts were on board the station, and Usachev noted the increase in noise levels aboard the station, though it still seemed roomy enough for now.

Milestones

186th manned space flight
82nd Russian manned space flight
75th manned Soyuz mission
22nd manned Soyuz TM mission
28th Russian and 61st flight with EVA operations
21st Mir resident crew

STS-75

Int. Designation	1996-012A
Launched	22 February 1996
Launch Site	Pad 39B, Kennedy Space Center, Florida
Landed	2 September 1996
Landing Site	Runway 33, Shuttle Landing Facility, KSC, Florida
Launch Vehicle	OV-102 Columbia/ET-76/SRB BI-078/SSME #1 2029; #2 2034; #3 2017
Duration	15 days 17 hrs 40 min 21 sec
Call sign	Columbia
Objective	Tethered Satellite System Re-flight 1; US Microgravity Payload #3

Flight Crew

ALLEN, Andrew Michael, 40, USMC, commander, 3rd mission
Previous missions: STS-46 (1992); STS-62 (1994)
HOROWITZ, Scott Jay, USAF, pilot
HOFFMAN, Jeffrey Alan, civilian, mission specialist 1, 5th mission
Previous missions: STS 51-D (1985); STS-35 (1990); STS-46 (1992); STS-61 (1993)
CHELI, Maurizio, Italian Air Force, ESA mission specialist 2
NICOLLIER, Claude, Swiss Air Force, ESA, mission specialist 3, 3rd mission
Previous missions: STS-46 (1992); STS-61 (1993)
CHANG-DIAZ, Franklin Ramon, 45, civilian, mission specialist 4, payload commander, 4th mission
Previous missions: STS 61-C (1986); STS-34 (1989); STS-46 (1992)
GUIDONI, Umberto, 41, civilian, Italian payload specialist 1

Flight Log

This mission was the re-flight of the US/Italian Tethered Satellite System that was previously flown on STS-46 in 1992. The crew worked a two-shift system. Allen, Horowitz, Cheli and Guidoni worked the Red Shift while Hoffman, Nicollier and Chang-Diaz formed the Blue Shift. Allen and Hoffman were also designated the White Shift, allowing them to work between shifts as necessary. The deployment of TSS-1R was delayed on STS-75 for 24 hours so that the flight crew could troubleshoot problems with the system's computers. On FD 3, the tether was extended almost to its maximum 20.5 km length and the satellite was sending back excellent data when the tether suddenly snapped (at 19.7 km) without prior warning. Orbital parameters at the time meant that the satellite separated from Columbia and the crew were placed in no danger. The day after the loss of the satellite, the remaining tether was reeled in and

The frayed end of the TSS is seen at the end of the support boom. The TSS broke free during operations on 25 February (FD 3) just short of its 20.5-km full deployment range

the deployer retracted. Tether Optical Phenomenon operations continued through to FD 14 and there was some consideration given to a rendezvous and possible retrieval of the free-flying satellite. However, there was insufficient propellant aboard and projections precluded any detailed plans for an attempt from being pursued.

Despite this loss, the satellite had recorded useful data for five hours up to the point of separation. Analysis of the data revealed that voltages as high as 3,500 volts developed across the tether, and this achieved current levels of 480 milliamps. In addition, scientists received important data on how satellite thruster gas interacted with Earth's ionosphere, measurements of ionised shock waves around a satellite for the first time, and data on plasma wakes created by the movement of a body through the electrically charged ionosphere. Some experiments were also conducted using the free-flying satellite and its attached tether prior to its re-entry and destruction in the upper atmosphere. Despite the loss, the data gathered during the TSS-1R operation resulted in a number of space physics and plasma theories being revised or overturned.

The other major payload of this mission was the USMP-3 package, which included re-flights of US and international experiments. Most of the operation of USMP experiments was via telescience, with principle investigators located at the Marshall Space Flight Center Spacelab Mission Operations Control Center in Huntsville, Alabama. There were five major experiments in materials science and a glove box on the mid-deck was used to perform a series of combustion experiments to better understand combustion processes and improve fire safety for the ISS. Protein crystal growth experiments included processing nine proteins extracted from the tropical rain forests of Costa Rica into crystals to further the understanding of their molecular structures. This was a joint US, Costa Rican and Chilean experiment, with application for the pharmaceutical treatment of Chagas Disease, an incurable ailment that affects over 15 million people in Latin America. The landing of Columbia on 9 March occurred after a 24-hour waive-off due to unfavourable weather conditions. The first attempt on 9 March was also abandoned due to bad weather, but it cleared to allow a landing at the Cape at the second opportunity. On FD 7, Hoffman surpassed Kathy Thornton's Shuttle space flight record of 978 hours 18 minutes, and by the end of the mission both he and Chang-Diaz had achieved over 1,000 hours cumulative flight time aboard the Shuttle.

In June 1996, a joint NASA and ASI (the Italian Space Agency) investigation board report was released, which determined that the tether failed as a result of "arcing and burning of the tether, which led to a tensile failure after a significant portion of the tether had burned away." The board concluded that external penetration (but not space debris or micrometeoroids) or a defect in the tether caused a breach in the layer of insulation surrounding the conductor. This would have allowed a current to jump or arc from the copper wire in the tether to a nearby electrical ground. By examining the data and the frayed end of the tether, the board concluded that there was nothing to preclude any follow-on mission, after sufficient improvements had been made to ensure that the problem would not recur. The data received from the abbreviated experiments indicated that, on the whole, the major objectives had been met, the programme was mostly successful, and it was viable for future development.

Milestones

187th manned space flight
105th US manned space flight
75th Shuttle mission
19th flight of Columbia
2nd flight of TSS
7th EDO mission

STS-76

Int. Designation	1996-018A
Launched	22 March 1996
Launch Site	Pad 39B, Kennedy Space Center, Florida
Landed	31 March 1996
Landing Site	Runway 22, Edwards AFB, California
Launch Vehicle	OV-104 Atlantis/ET-77/SRB BI-079/SSME #1 2035; #2 2109; #3 2019
Duration	9 days 5 hrs 15 min 53 sec Lucid 188 days 4 hrs 0 min 11 sec (landed on STS-79)
Call sign	Atlantis
Objective	Third Shuttle–Mir docking mission; transfer of NASA-2 astronaut (Lucid) to Mir EO-21 resident crew

Flight Crew

CHILTON, Kevin Patrick, 41, USAF, commander, 3rd mission
Previous missions: STS-49 (1992); STS-59 (1994)
SEARFOSS, Richard A., 39, USAF, pilot, 2nd mission
Previous mission: STS-58 (1993)
SEGA, Ronald Michael, 43, civilian, mission specialist 1, 2nd mission
Previous mission: STS-60 (1994)
CLIFFORD, Michael Richard Uram, 43, US Army, mission specialist 2, 3rd mission
Previous missions: STS-53 (1992); STS-59 (1994)
GODWIN, Linda Maxine, 43, civilian, mission specialist 3, 3rd mission
Previous missions: STS-37 (1991); STS-59 (1994)

NASA-2 up only:

LUCID, Shannon Wells, 53, civilian, mission specialist 4, EO-21 research cosmonaut, NASA-2 board engineer, 5th mission
Previous missions: STS 51-G (1985); STS-34 (1989); STS-43 (1991); STS-58 (1993)

Flight Log

Lessons were learned from Thagard's stay on Mir. Missing his family was one of the main problems he identified, so NASA organised regular contact for subsequent NASA astronauts who stayed on Mir. Thagard was thankful that he did not have to endure an extended mission of six months, but Shannon Lucid was forced to do exactly that. The O-rings from the retrieved SRBs on STS-75 showed a different problem from that seen on STS-71 and 70. This time, the gas path went through the

Clifford is seen during the 27 March EVA at the restraint bar on the Docking Module of Mir. Godwin is out of frame. During the EVA, both astronauts were careful not to venture beyond the 4.6 DM onto Kristall, due to Russian fears of accidental damage. NASA would have similar concerns if cosmonauts inadvertently ventured from Mir across to an unfamiliar Shuttle during a future EVA

adhesive but not past it. This had been seen before, but never with two different gas paths on both motors. A review deemed that the STS-76 units would be safe to fly, but the 21 March attempt was scrubbed due to concerns over high winds. The revised launch occurred without further difficulty, but APU 3 shut down prior to entering orbit. Mission management concluded that the system would remain stable, however, and would still support a full-duration mission.

The docking with Mir occurred on FD 3, and for the next five days the astronauts worked with the EO-21 cosmonauts to transfer 680 kg of water to the station in fifteen Contingency Water Containers (CWC). Two tons of equipment were also transferred into Mir, while experiment samples and unwanted equipment were taken back into Atlantis for the return to Earth. Lucid transferred to the Mir resident crew two hours after the hatches were opened following docking. She relocated her personal Soyuz seat support into the Descent Module of Soyuz TM23. Also transferred over to the space station were Mir Glove Box stowage experiments, which were located in the station's Docking Module. While docked, the Shuttle crew activated or worked on a number of secondary experiments located in the Shuttle's middeck, SpaceHab module and payload bay.

On FD 6, Godwin (EV1) and Clifford (EV2) performed the first US EVA at a space station since the final Skylab EVA in February 1973. This was also the first time that activities were performed around the docking interfaces of two different spacecraft since Apollo 9 in March 1969. Godwin and Clifford attached four Mir Environmental Effects Payload (MEEP) experiments to the station's Docking Module. Over an 18-month period, these experiments were designed to examine the environment around the station. Both astronauts wore a SAFER unit (first tested on STS-64) during an EVA which lasted 6 hours 2 minutes.

Undocking occurred on FD 8, leaving Lucid aboard Mir at the start of a planned two-year continuous American presence on the station by successive astronauts. The landing of STS-76 was planned for 31 March, but was brought forward due to anticipated rain and clouds around the Cape area. In the event, both the 30 and 31 March attempts were waived off due to the bad weather and Atlantis was eventually diverted to Edwards. Following the problem with APU 3 during ascent, extra precautions were instigated for the landing, particularly in requiring more conservative weather criteria. After the 30 March waive-off, the re-opening of the payload bay doors to expose the orbiter's radiators was interrupted by indications that latches 9–12 on both sides had failed to open properly. A visual inspection from SpaceHab confirmed that they had operated correctly, however, and it was determined that faulty micro-switches had given erroneous indications of problems. The doors closed nominally for landing. Despite the loss of three of the primary RCS thrusters out of the set of 38, the re-entry of STS-76 was not adversely affected.

Milestones

188th manned space flight
106th US manned space flight
76th Shuttle mission
16th flight of Atlantis
3rd Shuttle–Mir docking mission
34th US and 62nd flight with EVA operations
4th SpaceHab mission (4th single module)
1st SpaceHab space station mission
1st US EVA at a space station for 22 years

STS-77

Int. Designation	1996-032A
Launched	19 May 1996
Launch Site	Pad 39B, Kennedy Space Center, Florida
Landed	29 May 1996
Landing Site	Runway 33, Shuttle Landing Facility, KSC, Florida
Launch Vehicle	OV-105 Endeavour/ET-78/SRB BI-080/SSME #1 2037; #2 2040; #3 2038
Duration	10 days 0 hrs 39 min 18 sec
Call sign	Endeavour
Objective	SpaceHab 4, single-module configuration; SPARTAN/IAE free-flyer

Flight Crew

CASPER, John Howard, 52, USAF, commander, 4th mission
Previous missions: STS-36 (1990); STS-54 (1993); STS-62 (1994)
BROWN Jr., Curtis Lee, 40, USAF, pilot, 3rd mission
Previous missions: STS-47 (1992); STS-66 (1994)
THOMAS, Andrew Sydney Withiel, 44, civilian, mission specialist 1
BURSCH, Daniel Wheeler, 38, USN, mission specialist 2, 3rd mission
Previous missions: STS-51 (1993); STS-68 (1994)
RUNCO Jr., Mario, 44, civilian, mission specialist 3, 3rd mission
Previous missions: STS-44 (1991); STS-54 (1993)
GARNEAU, Marc Joseph Jean-Pierre, 47, civilian, Canadian mission specialist 4, 2nd mission
Previous mission: STS 41-G (1984)

Flight Log

After rescheduling the original 16 May launch date due to other programmes at the Eastern Test Range of the Cape, STS-77 was launched on time three days later. This mission featured four rendezvous activities with two different payloads. The largest payload was the SpaceHab 4 module, with a mass of about 1,300 kg of experiments and support equipment. Its research programme encompassed twelve commercial space product development packages in biotechnology, polymers and agriculture, and electronic materials. Over 90 per cent of the payloads on the mission were sponsored by NASA's Office of Space Access and Technology (OSAT). In addition to SpaceHab 4, Endeavour's cargo included the SPARTAN-207 free-flyer carrier, from which the Inflatable Antenna Experiment would be deployed, and a suite of four technology experiments, designated TEAMS, in the payload bay. One of these was a cooperative experiment between the US, Canada and Germany, in which

The Inflatable Antenna Experiment (IAE), part of the SPARTAN 207 payload, nears completion of its inflation process. The IAE experiments provided further groundwork for future technology development in inflatable space structures, which will be launched and then inflated like a balloon on orbit

various samples of electronic and semi-conductor material were heated using float zone techniques, designed to produced large, ultra-pure crystals of semi-conductor materials such as gallium arsenide.

On FD 2, SPARTAN-207 was deployed using the RMS. Attached to the SPARTAN was the 60 kg IAE structure, mounted on three struts, which was inflated to its full size of about 15 metres in diameter. After 90 minutes, the structure was ejected, allowing the SPARTAN to be retrieved later in the mission. The objective of the inflation experiment was to gather data about its on-orbit behaviour by recording the deployment and inflation sequence on film. This would provide designers with important information for the design of future deployable and inflatable structures in space. The potential benefits of such structures include lower development costs, lower mass and volume inside launch vehicles, and the possibility of utilising a smaller launch vehicle to get them into orbit.

Other deployment and rendezvous activities on the fourth day of the mission included the release of the Passively Aerodynamically-stabilised Magnetically-damped Satellite (PAMS), one of four Technology Experiments for Advancing Missions in Space (or TEAMS) that were mounted on a hitchhiker carrier in the

574   The Fourth Decade: 1991–2000

payload bay of the orbiter. The other TEAMS experiments included a GPS attitude and navigation experiment, a Vented Tank Re-supply Experiment and a Liquid Mass Thermal Experiment. There was also a range of secondary payloads in technology and biology, and further GAS canisters on the flight. The Aquatic Research Facility was a joint NASA/Canadian experiment for investigating a wide range of small aquatic species. For this mission, it included starfish, muscles and sea urchins, and it was hoped that the facility would provide scientists with the opportunity to investigate the process of fertilisation, formation of embryos and development of calcified tissue, as well as the feeding behaviour of small aquatic organisms while in the microgravity environment.

Commander Casper took the opportunity to talk with Shannon Lucid on her 65th day on Mir during the mission. In addition to a smooth, on-time launch, Endeavour showed no significant on-orbit problems and completed a first opportunity landing at Edwards, the smoothest, most trouble-free flight for some time.

Milestones

189th manned space flight
107th US manned space flight
77th Shuttle mission
11th flight of Endeavour
5th flight of SpaceHab (5th single module)
1st flight of the Aquatic Research Facility (ARF)

STS-78

Int. Designation	1996-036A
Launched	20 June 1996
Launch Site	Pad 39B, Kennedy Space Center, Florida
Landed	7 July 1996
Landing Site	Runway 33, Shuttle Landing Facility, KSC, Florida
Launch Vehicle	OV-102 Columbia/ET-79/SRB BI-081/SSME #1 2041; #2 2039; #3 2036
Duration	16 days 21 hrs 47 min 45 sec
Call sign	Columbia
Objective	Life and Microgravity Spacelab programme

Flight Crew

HENRICKS, Terence Thomas "Tom", 43, USAF, commander, 4th mission
Previous missions: STS-44 (1991); STS-55 (1993); STS-70 (1995)
KREGEL, Kevin Richard, 39, civilian, pilot, 2nd mission
Previous missions: STS-70 (1995)
LINNEHAN, Richard Michael, 38, civilian, mission specialist 1
HELMS, Susan Jane, 38, USAF, mission specialist 2, 3rd mission
Previous missions: STS-54 (1993); STS-64 (1994)
BRADY Jr., Charles Eldon, 44, USN, mission specialist 3
FAVIER, Jean-Jacques, 47, civilian, French payload specialist 1
THIRSK, Robert Brent, 42, civilian, Canadian payload specialist 2

Flight Log

The launch of STS-78 not only occurred on time, it also featured the first use of a video TV camera transmitting images from the flight deck. Filming began with the ingress of the crew into their seats and finished at MECO. The video link was also used during the descent on 7 July. During post-launch assessment of the SRBs, it was found that a hot-gas path had penetrated the motor field joints up to, but not past, the O-ring capture feature. This was the first time that a Redesigned SRM (RSRM) had shown penetration into the J-joint, although flight safety was not compromised and all performance data indicated that the design specifications were met. The problem was attributed to new, environmentally friendly adhesive and cleaning fluid used in the area. The rather quick turnaround of the commander/pilot pairing from STS-70 (less than one year) was done to evaluate the effects of such a short gap between flights on mission training and preparation time. NASA had often reviewed proposals to re-fly an "orbiter" crew (commander, pilot and MS2/flight engineer), or to re-fly an MS on a mission with a similar science payload to reduce training time, but this was yet to be

PS Favier prepares a sample for the Advanced Gradient Heating Facility while wearing instruments that measure upper-body movement. In a typical science scene aboard the Spacelab Long Module, several experiments are being performed at the same time. MS Helms and commander Henricks work in the background, while in the foreground MS Linnehan tests his muscle response with the Handgrip Dynamometer

fully implemented given the frequent, real-time changes to the Shuttle manifest. On this flight, the crew worked a single shift.

During the mission, the longest flight of the Shuttle to date, five space agencies and research scientists from more than ten countries participated in the 40 experiments flown on LMS. The experiments were grouped into life sciences and materials sciences research. The life sciences experiments included research into human physiology and space biology, while the materials sciences experiments encompassed basic fluid physics, advanced semi-conductor and metal alloy materials processing, and medical research into protein crystal growth. The mission also expanded the use of telescience, to the point where four locations in Europe and four remote locations in the US were utilised by investigators involved in the mission. This was a demonstration of the way science activities were being planned for ISS operations. Video-imaging was also valuable to assist the crew in completing some of the in-flight maintenance procedures required during the flight.

Whereas previous life sciences investigations had focused on the changes in the microgravity environment on the human body, those on STS-78 examined why such changes occurred. There were extensive studies of sleep cycles, circadian rhythms and task performance in microgravity, as well as studies into bone and muscle loss in space. Biopsy tissue samples were taken both before and after flight to record changes from one-G to microgravity and then back again.

Columbia's RCS engines were pulsed as a test to try boosting the vehicle's altitude without disturbing the delicate instruments in the Spacelab module. This was in preparation for the next Hubble servicing mission (STS-82), in which the space telescope's orbit would need to be raised without damaging its fragile solar arrays.

Milestones

190th manned space flight
108th US manned space flight
78th Shuttle mission
20th flight of Columbia
13th flight of Spacelab Long Module configuration
8th EDO flight
New Shuttle flight duration record
1st live downlink during orbiter ascent and descent
Henricks celebrates his 44th birthday in space (5 Jul)

SOYUZ TM24

Int. Designation	1996-047A
Launched	17 August 1996
Launch Site	Pad 1, Site 5, Baikonur Cosmodrome, Kazakhstan
Landed	2 March 1997
Landing Site	128 km east of Dzhezkazgan
Launch Vehicle	R7 (11A511U); spacecraft serial number (7K-M) #073
Duration	196 days 17 hrs 26 min 13 sec
	Deshays 15 days 18 hrs 23 min 37 sec
Call sign	Freget (Frigate)
Objective	Mir 22 resident crew programme; French *Cassiopée* mission; worked with NASA 2, 3 and 4 crew members

Flight Crew

KORZUN, Valery Grigoryevich, 46, Russian Air Force, commander
KALERI, Alexandr Yuriyevich, 40, civilian, flight engineer, 2nd mission
Previous mission: Soyuz TM14 (1992)
ANDRÉ-DESHAYS, Claudie, 39, civilian, cosmonaut researcher

Flight Log

The original crew for this mission were supposed to have been Expedition 22 crew members Gennady Manakov (commander) and Pavel Vinogradov (flight engineer), along with French cosmonaut Claudie André-Deshays. However, just one week before launch, Manakov failed a regular medical check and the back-up crew of Korzun and Kaleri took the place of the Russian prime crew. Since she was not part of the main crew, André-Deshays was able to remain on the mission. The change did not have much effect on Shannon Lucid, who was nearing the end of her stay on Mir, but the next US resident astronaut, John Blaha, had not trained with either Korzun or Kaleri. This would prove a challenging hurdle for the American to overcome after he replaced Lucid on Mir in September.

The French *Cassiopée* programme included cardiovascular and neurosensory investigations. There was also a technology experiment that recorded vibrations aboard the station while there were more than the normal resident crew aboard, as well as a materials-processing experiment. After two weeks aboard Mir, André-Deshays returned with the EO-21 crew, leaving Lucid to complete her residency with the new cosmonauts prior to the arrival of Blaha on STS-79. André-Deshays commented that two weeks was nowhere near sufficient time to adjust to life on Mir and to complete all her experiment programme, but with two more French long-duration visits already booked, this was not expected to be a problem for future French cosmonauts.

The Soyuz TM24 crew of Kaleri (left), Korzun and André-Deshays

On 7 September, Lucid surpassed Kondakova's record for the longest female space flight. Ten days later, with her return Shuttle flight already in space, she surpassed Reiter's 179-day record for a visiting cosmonaut. When Blaha took over, he continued the programme of experiments started by Lucid, but also brought some new ones. During his stay on Mir, it appeared he was enjoying his residency, and it was only after he came home that he revealed it had been a difficult mission. He was replaced on Mir in January 1997 by Jerry Linenger, who would remain with the EO-22 crew for a month before the next resident crew took over.

The EO-22 cosmonauts completed two EVAs during their stay on Mir; the first on 2 December lasted 5 hours 58 minutes, and the second, a week later, lasted 6 hours 38 minutes. Their work outside included the completion of MCSA cable installation. They also relocated the Rapana girder to the top of the new Stombus girder on the underside of Kvant and on their second EVA, they had to reinstall antennae they had dislodged during their first excursion. Back inside the station, in addition to the regular maintenance and housekeeping chores, the cosmonauts continued the research programme and assisted their various visitors with their research objectives.

One of the more challenging events at the end of their residency was the fire on 24 February 1997, which was the result of a split chemical burner in the Vika unit in Kvant 1. The use of "candles" to supplement the oxygen output of the Elektron regeneration system was typical when more than three cosmonauts were aboard Mir. But this time, the unit had split and released oxygen into the electronics, causing a jet of flame to shoot across the area. For an hour or so, the crew (by this point including Linenger) wore oxygen masks until the station's filtration system finally began dissipating the thick black smoke. The crew reported some eye irritation, but no lingering damage was found from smoke inhalation, although filtration masks were still worn for a few days afterwards as a precaution. The faulty canister was stored in the Soyuz TM24 Descent Module for return to Earth and post-flight examination. Both EO-22 cosmonauts made the return journey with German cosmonaut Ewald.

Milestones

191st manned space flight
83rd Russian manned space flight
76th manned Soyuz mission
23rd manned Soyuz TM mission
29th Russian and 63rd flight with EVA operations
4th French long-duration mission (16 days)

STS-79

Int. Designation	1996-057A
Launched	16 September 1996
Launch Site	Pad 39A, Kennedy Space Center, Florida
Landed	26 September 1996
Landing Site	Runway 15, Shuttle Landing Facility, KSC, Florida
Launch Vehicle	OV-104 Atlantis/ET-82/SRB BI-083/SSME #1 2012; #2 2031; #3 2033
Duration	10 days 3 hrs 18 min 26 sec
	Blaha 128 days 5 hrs 27 min 55 sec (landed on STS-81)
Call sign	Atlantis
Objective	4th Shuttle–Mir docking mission; delivery of the NASA-3 Mir EO-22 crew member (Blaha) and return of NASA-2 Mir EO-21 crew member (Lucid)

Flight Crew

READDY, William Francis, 44, civilian, commander, 3rd mission
Previous missions: STS-42 (1992); STS-51 (1993)
WILCUTT, Terrence Wade, 46, USMC, pilot, 2nd mission
Previous mission: STS-68 (1994)
APT, Jerome "Jay", 47, civilian, mission specialist 1, 4th mission
Previous missions: STS-37 (1991); STS-47 (1992); STS-59 (1994)
AKERS, Thomas Dale, 45, USAF, mission specialist 2, 4th mission
Previous missions: STS-41 (1990); STS-49 (1992); STS-61 (1993)
WALZ, Carl Erwin, 41, USAF, mission specialist 3, 3rd mission
Previous missions: STS-51 (1993); STS-65 (1994)

NASA-3 Mir crew member up only:

BLAHA, John Elmer, 54, USAF, mission specialist 4, EO-22 cosmonaut researcher, NASA board engineer 3, 5th mission
Previous missions: STS-29 (1989); STS-33 (1989); STS-43 (1991); STS-58 (1993)

NASA-2 Mir crew member down only:

LUCID, Shannon Wells, 53, civilian PhD, mission specialist 4, EO-21 cosmonaut researcher, NASA board engineer 2, 5th mission
Previous missions: STS 51-G (1985); STS-34 (1989); STS-43 (1991); STS-58 (1993)

Flight Log

During her six-month stay on Mir, Shannon Lucid had conducted research in advanced technology, Earth sciences, fundamental biology, human life sciences,

Carl Walz totes a bag carrying an Orlan DMA Unit #18 spacesuit brought back to Earth for analysis. Other stowage bags and sample return are shown in the frame

microgravity research and space sciences. She also won admiration for keeping cheerful despite delays in getting her home, firstly for the technical problems with the SRBs, and then for weather problems with two hurricanes. She kept track of the time on Mir by wearing pink socks each Sunday, and relieved any boredom by reading several books. Her daughter had given her one novel, but she had not included the sequel, which was a little frustrating to the orbiting astronaut.

The original 31 July launch of STS-79 was delayed when the two SRBs were swapped as the adhesives used on them were the same as that on STS-78, where a hot-gas path into the J-joints on the motor field joints was discovered. The SRB set intended for STS-80, which used the older type of adhesive, was fitted to STS-79 while Atlantis was back in the VAB due to the weather threat from Hurricane Bertha. The new launch, set for 12 September, was further delayed to 16 September when Atlantis was rolled back to the VAB a second time due to the threat of Hurricane

Fran. The third launch attempt occurred on time, and though APU 2 powered down prematurely 13 minutes into the flight, mission management deemed it safe to continue with a nominal full-term mission.

The docking of STS-79 with Mir occurred on FD 3 (18 September) and a few hours after opening the hatches, Lucid and Blaha exchanged places, with Blaha becoming a member of the Mir resident crew (EO-22) and Lucid replacing him as MS4 on the Shuttle crew. During the five days of joint operations, over 1,600 kg of supplies were transferred to the space station, including food, water and three new experiments. About 900 kg of material was transferred back to the Shuttle, including experiment samples and unwanted equipment.

The crew also operated three experiments that remained in the Shuttle during their stay at Mir. One was an extreme translation furnace that allowed space-based processing up to 1,600°C. The second was a commercial protein crystal growth experiment, and the third was the Mechanics of Granular Materials experiment designed to study the behaviour of cohesionless granular material, which was particularly applicable to understanding how the surface of the Earth responds during earthquakes and landslides. The vernier jets of Atlantis were used near the end of the mission to lower the Shuttle's orbit slightly. This was another test of operations planned for the upcoming Hubble service mission planned for 1997, in which the jets could refine and raise the orbit of the telescope while it was still in the payload bay.

When Lucid was approached by the team assigned to carry her off the Shuttle after her six-month mission, she dismissed them, determining to walk of the vehicle herself. This she managed to do, walking (with some assistance) to the Crew Transfer Vehicle.

Milestones

192nd manned space flight
109th US manned space flight
79th Shuttle mission
17th flight of Atlantis
4th Shuttle–Mir docking
6th SpaceHab mission (1st double module)
1st US resident crew exchange
New female world space flight endurance record (Lucid)
1st double rollback to VAB

STS-80

Int. Designation	1996-065A
Launched	19 November 1996
Launch Site	Pad 39B, Kennedy Space Center, Florida
Landed	7 December 1996
Landing Site	Runway 33, Shuttle Landing Facility, KSC, Florida
Launch Vehicle	OV-102 Columbia/ET-80/SRB BI-084/SSME #1 2032; #2 2026; #3 2029
Duration	17 days 15 hrs 53 min 18 sec
Call sign	Columbia
Objective	Wake Shield Facility 3/ORFEUS-SPAS II

Flight Crew

COCKRELL, Kenneth Dale, 46, civilian, commander, 3rd mission
Previous missions: STS-56 (1993); STS-69 (1995)
ROMINGER, Kent Vernon, 40, USN, pilot, 2nd mission
Previous mission: STS-73 (1995)
JERNIGAN, Tamara Elizabeth, 37, civilian, mission specialist 1, 4th mission
Previous missions: STS-40 (1991); STS-52 (1992); STS-67 (1995)
JONES, Thomas David, 41, civilian, mission specialist 2, 3rd mission
Previous missions: STS-59 (1994); STS-68 (1994)
MUSGRAVE, Story Franklin, 61, civilian, mission specialist 3, 6th mission
Previous missions: STS-6 (1983); STS 51-F (1985); STS-33 (1989); STS-44 (1991); STS-61 (1993)

Flight Log

The original launch date for this mission of 31 October was slipped to 8 November due both to the removal of its SRBs to the STS-79 mission, and as a precaution due to concerns over Hurricane Fran. Engineers wanted more time to analyse the booster nozzles from STS-79 and with a 13 November Atlas launch planned, STS-80 was rescheduled for 15 November. Although the Atlas launch was scrubbed, STS-80 actually slipped further, to 19 November, in order to clear the bad weather that was predicted to lie around the Cape for several days. A three-minute delay to the launch was caused by concerns over hydrogen conditions in the aft engine compartment and when the SRBs were examined post-retrieval, they indicated some erosion – although far less than on STS-79.

The mission was a successful demonstration of the deployment and retrieval of two separate free-flying research spacecraft. The ORFEUS-SPAS II was deployed on FD 1 and became a two-week independent mission before retrieval on FD 15. The mission of the satellite was devoted to astronomical observations at very short

MS Tom Jones uses the controls at the aft flight deck of Columbia to conduct tests with the captured WSF, seen though the window on the end of the RMS at frame centre

wavelengths, using three primary scientific instruments and a secondary payload. The primary instruments were the ORFEUS 2.4 m focal length telescope, the Far UV spectrograph and the Extreme UV Spectrograph. The secondary payload was the Interstellar Medium Absorption Profile Spectrograph. The objective was to investigate the nature of hot stellar atmospheres and the cooling mechanisms of white dwarf stars, to determine the nature of accretion disks around collapsed stars, to investigate supernova remnants and interstellar media, and to examine potential star-forming regions. During the two-week mission, no significant problems were reported and all mission goals were achieved. Some 422 observations of about 150 astronomical objects were conducted, including the Moon, nearby stars, distant stars in the Milky Way, stars in other galaxies, active galaxies and quasar 3C273. With more sensitive instruments giving better quality data, almost twice as much information and data was obtained than on the first ORFEUS flight.

On FD 4 the WSF-3 was deployed. This time it was highly successful and achieved the maximum seven growths of thin film semi-conductor material, with the satellite performing almost flawlessly. It was retrieved during FD 7 for return to Earth. As with most Shuttle missions, the flight also carried a range of mid-deck and payload bay experiments that formed part of the secondary payloads and objectives.

Two planned six-hour EVAs by Jernigan (EV1) and Jones (EV2), designed to gather further knowledge and experience in preparation for the ISS programme, had to be abandoned when a stuck EVA hatch thwarted attempts to leave the airlock during EVA 1. Despite crew attempts during the mission to discover the cause, it was not until Columbia was back on the ground that engineers discovered that a small screw had become loose in the internal assembly and had lodged in an actuator, the gearbox-type device that operated the linkage to secure the hatch. When a new actuator was installed, the hatch worked perfectly. Though unable to complete the EVAs, the crew did still manage to evaluate a new pistol grip tool – resembling a hand-held drill – in the mid-deck during the mission.

The two-day waive-off for landing due to weather conditions in Florida resulted in this mission becoming the longest in Shuttle history and gave the crew an extra opportunity to look at the view out of the window. For Musgrave, this was particularly poignant, as this would be his last flight. He was fully appreciative of the chance to take a leisurely view of the Earth from orbit, knowing he would not be returning for a seventh mission. A condition of his being able to fly this mission was that he would retire from the active flight list when it was over. Musgrave set a record of six space Shuttle flights (equalling John Young's career space flight record of Gemini, Apollo and Shuttle flights) and became the oldest person in space at the age of 61. He also became the only astronaut to fly on each of the five Shuttles capable of orbital flight, including twice on Challenger.

Milestones

193rd manned space flight
110th US manned space flight
80th Shuttle mission
21st flight of Columbia
New Shuttle mission duration record
3rd flight of Wake Shield Facility
Musgrave becomes the oldest person in space, aged 61

STS-81

Int. Designation	1997-001A
Launched	12 January 1997
Launch Site	Pad 39B, Kennedy Space Center, Florida
Landed	22 January 1997
Landing Site	Runway 33, Shuttle Landing Facility, KSC, Florida
Launch Vehicle	OV-104 Atlantis/ET-83/SRB BI-082/SSME #1 2041; #2 2034; #3 2042
Duration	10 days 04 hrs 55 min 21 sec
	Linenger 132 days 4 hrs 0 min 21 sec (landing on STS-84)
Call sign	Atlantis
Objective	5th Shuttle–Mir docking; delivery of NASA 4 (Linenger) Mir EO-23 crew member; return of NASA 3 (Blaha) Mir EO-22 crew member

Flight Crew

BAKER, Michael Allen, 43, USN, commander, 4th mission
Previous missions: STS-43 (1991); STS-52 (1992); STS-68 (1994)
JETT Jr., Brent Ward, 38, USN, pilot, 2nd mission
Previous mission: STS-72 (1996)
WISOFF, Peter Jeffrey Karl, 38, civilian, mission specialist 1, 3rd mission
Previous missions: STS-57 (1993); STS 68 (1994)
GRUNSFELD, John Mace, 38, civilian, mission specialist 2, 2nd mission
Previous mission: STS-67 (1995)
IVINS, Marsha Sue, 45, civilian, mission specialist 3, 4th mission
Previous mission: STS-32 (1990); STS-46 (1992); STS-62 (1994)

NASA 4 Mir crew member up only:

LINENGER, Jerry Michael, 40, USN, mission specialist 4, Mir EO-23 cosmonaut researcher, NASA board engineer 4, 2nd mission
Previous mission: STS-64 (1994)

NASA 3 Mir crew member down only:

BLAHA, John Elmer, 54, USAF, mission specialist 4, Mir EO-22 cosmonaut researcher, NASA board engineer 3, 5th mission
Previous missions: STS-29 (1989); STS-33 (1989); STS-43 (1991); STS-58 (1993)

Flight Log

John Blaha became the only pilot–astronaut to complete a long-duration residency mission aboard Mir. Before he left Earth, he knew his stay on the station would be

Valeri Korzun (left) works with Mike Baker and Brent Jett to unstow a gyrodyne device for attitude control, and then transfer it to Mir. They are pictured in the SpaceHab double module which is packed with logistics to transfer to the space station

tough. The crew he had trained with (Manakov and Vinogradov) had been replaced by their back-ups (Korzun and Kaleri) shortly before launch, and this new pairing were strangers to the American. Blaha's first month was a difficult one, with bouts of depression, but he overcame this by talking to NASA ground controllers in Moscow who read up NFL scores during the season. He was unable to vote in the US Presidential election, however, as legal complications in his Houston voting district prevented him from securing a computer electronic ballot in time. Blaha also became the first American to spend Christmas and New Year in orbit since the crew of Skylab 4 in 1973. During his stay on Mir, Blaha operated a range of experiments that had been used by Lucid, together with a new tissue growth experiment, a protein crystal growth experiment, a study of alloy crystallisation and a number of technology experiments, some of which were linked to body motion during his time aboard the station. Similar experiments were completed during the Skylab missions.

Atlantis docked with Mir on 14 January and shortly after transferring to the station, Linenger exchanged his Soyuz seat liner with that of Blaha, marking the point that Linenger took over as the Mir resident. During five days of docked operations, the joint crews transferred over 2,700 kg of logistics to Mir, including about 725 kg of water, 516 kg of US science equipment and 1,000 kg of Russian logistics and equipment. Over 1,088 kg of material was transferred to Atlantis for return to Earth, including the first plants to complete a lifecycle in space – a crop of wheat grown from seed to seed.

The crew also evaluated the Treadmill Vibration Isolation and Stabilisation System (TVIS), which was located on the Shuttle but was intended for use on the Russian segment of ISS. Other ISS-related investigations included the firing of the vernier jets of Atlantis to record the stability of docked spacecraft and gather further engineering data on the behaviour of large masses docked in space. Atlantis undocked, with Blaha on board, on 19 January and conducted what was becoming a traditional fly-around of the space complex before heading for landing. The orbiter touched down during the second landing opportunity three days later.

Milestones

194th manned space flight
111th US manned space flight
81st Shuttle mission
18th flight of Atlantis
5th Shuttle–Mir docking
7th SpaceHab flight (2nd double module)
Linenger celebrates his 41st birthday in space (16 Jan)

SOYUZ TM25

Int. Designation	1997-003A
Launched	10 February 1997
Launch Site	Pad 1, Site 5, Baikonur cosmodrome, Kazakhstan
Landed	14 August 1997
Landing Site	128 km east of Dzhezkazgan
Launch Vehicle	R7 (11A511U); spacecraft serial number (7K-M) 74
Duration	184 days, 22 hrs, 7 min 40 sec
Call sign	Sirius (Sirius)
Objective	Mir EO-23 resident crew; German Mir 97 programme; they also worked with NASA 4 (Linenger) and 5 (Foale) astronauts

Flight Crew

TSIBLIYEV, Vasily Vasilyevich, 42, Russian Air Force, commander, 2nd mission
Previous mission: Soyuz TM17 (1993)
LAZUTKIN, Alexander Ivanovich, 39, civilian, flight engineer
EWALD, Reinhold, 40, German cosmonaut researcher

Flight Log

Following the docking of TM25 to Mir on 12 February and the transfer of the cosmonauts to the station, there were six crew members aboard once again. The Mir EO-22 crew of Korzun and Kaleri now included NASA astronaut Linenger, who had arrived the month before. The Mir EO-23 crew of Tsibliyev and Lazutkin would continue their mission with Linenger after Ewald had completed his 18-day research programme and returned to Earth with the EO-22 cosmonauts. The EO-23 crew would also work with Mike Foale, who would replace Linenger in May 1997. Theirs would be an eventful and challenging residency, with the 23 February fire aboard the station and the 25 June collision of Progress M34 to contend with.

Ewald's research centred on medical experiments to study the effects of microgravity on human performance, the function of both hormones and the cardiovascular system, and related psychological effects. In addition, he had a programme of technological and materials processing experiments. Following the fire on board the station, Ewald completed his experiments and prepared for the return to Earth with the EO-22 crew, leaving Mir on 2 March aboard Soyuz TM24.

The EO-23 crew now settled down to their research programme, but the science schedule was frequently interrupted by essential maintenance work on the Mir's aging systems. One of these repairs was to deal with reported coolant system leaks, which sprayed ethylene glycol into the station's atmosphere, reaching dangerous levels at

Mir EO-23 commander Tsibliyev operates at the end of the station's Strela boom during the 29 April EVA with Jerry Linenger. At lower left is the Kvant 1 module and above it is the Sofora tower

times. Again, delays in supplying a new Soyuz R7 launch vehicle meant that the Russians were told they would have to extend their mission by six weeks. On 29 April, Jerry Linenger became the first American to perform an EVA from a space station (for 4 hours 49 minutes) since the Skylab crews some 23 years previously. Linenger also became the first US astronaut to use a Russian EVA suit. This was a new variant called Orlan-M, which was the latest update to the basic Orlan suit that had been used since 1978. Orlan-M would also be used on early Russian EVA operations at ISS.

During the EVA, the crew installed the Optical Properties Monitor on Kristall and removed some US experiments that would be returned to Earth on STS-84. The EO-23 crew were scheduled to make two further EVAs in June, but other events precluded these from taking place.

Following the next exchange of NASA crew members in May 1997, Mike Foale became the fifth astronaut to work on the Mir station. He had a busy science programme planned that would complement that of the Russian cosmonauts, though most of their time was still being taken up with housekeeping and maintenance. On 25 June, Tsibliyev attempted to re-dock Progress M34. An attempt to use the TORU remote control system had failed during Progress M33 operations in March and in order for the system to be re-qualified for use, Tsibliyev found himself having to try to perform an operation that he had had little training for. When the Progress approached, the cosmonaut realised that it was not responding to commands as it should have been and tried to guide it past the station for another attempt. However, it collided with the Spektr module, puncturing the pressurised compartment and damaging the solar arrays. As alarms sounded, the air inside the station rapidly began leaking into the vacuum of space. Internal power and instrument connections leading from the core module to Spektr were severed and the crew were able to seal the hatch, shutting off the module (and with it, most of Foale's equipment and personal possessions) and the leak. In the days that followed, Mir suffered other problems, including an inadvertent crew error that caused the station to drift, losing solar lock. It took a Herculean effort that tested all three crew members to finally bring the station back under control and avoid the need for an emergency evacuation. It had been hoped that Tsibliyev and Lazutkin could perform EVAs to effect repairs, but concerns over the health of the commander put these plans on hold until the arrival of the next crew. After much discussion between Russian and American space officials and the astronaut himself, it was decided that Foale would remain on the station with the replacement crew, as planned.

Even the end of the EO-23 mission was not trouble-free. The re-entry burn occurred on time and the parachute deployed, but the soft-landing rockets that should have fired 1.5 metres off the ground failed, resulting in a 7.5 m/sec landing impact. The seat shock absorbers saved the crew from serious injury. Post-flight investigation revealed that the soft-landing rockets had fired, but at 5.8 km altitude. Initially blamed for the collision with Progress, the crew were eventually cleared of all responsibility, although neither ever flew again. It was later determined that the mass of trash inside the Progress had been miscalculated by a ton. This meant that the manoeuvring and braking burns were not as effective as they should have been, thus making Progress arrive quicker and close in more rapidly than estimated.

Milestones

195th manned space flight
84th Russian manned space flight
77th manned Soyuz mission
24th manned Soyuz TM mission
23rd Mir resident crew
30th Russian and 64th flight with EVA operations
1st US EVA in a Russian spacesuit
Tsibliyev celebrates his 43rd birthday in space (20 Feb)

STS-82

Int. Designation	1997-004A
Launched	11 February 1997
Launch Site	Pad 39A, Kennedy Space Center, Florida
Landed	21 February 1997
Landing Site	Runway 15, Shuttle Landing Facility, KSC, Florida
Launch Vehicle	OV-103 Discovery/ET-81/SRB BI-085/SSME #1 2037; #2 2040; #3 2038
Duration	9 days 23 hrs 37 min 9 sec
Call sign	Discovery
Objective	2nd Hubble Servicing Mission

Flight Crew

BOWERSOX, Kenneth Duane, 40, USN, commander, 4th mission
Previous missions: STS-50 (1992); STS-61 (1993); STS-73 (1995)
HOROWITZ, Scott Jay, 39, USAF, pilot, 2nd mission
Previous mission: STS-75 (1996)
TANNER, Joseph Richard, 47, civilian, mission specialist 1, 2nd mission
Previous mission: STS-66 (1994)
HAWLEY, Steven Alan, 45, civilian, mission specialist 2, 4th mission
Previous missions: STS 41-D (1984); STS 61-C (1986); STS-31 (1990)
HARBAUGH, Gregory Jordan, 39, civilian, mission specialist 3, 4th mission
Previous missions: STS-39 (1991); STS-54 (1993); STS-71 (1995)
LEE, Mark Charles, 44, USAF, mission specialist 4, payload commander, 4th mission
Previous missions: STS-30 (1989); STS-47 (1992); STS-64 (1994)
SMITH, Steven Lee, 38, civilian, mission specialist 5, 2nd mission
Previous mission: STS-68 (1994)

Flight Log

This was the first flight of Discovery after returning from its maintenance down period. The launch had been scheduled for 13 February but was moved up two days to give more flexibility. This mission was the second servicing mission to the Hubble Space Telescope, this time to upgrade and maintain the facility for further orbital use. It would also demonstrate the unique capability of the Shuttle to serve as a satellite-servicing vehicle, and the importance of having humans aboard to respond to unplanned activities. Four EVAs were scheduled, and a fifth was added to repair insulation material on the telescope.

Hubble was recaptured by the RMS and placed in Discovery's payload bay on 13 February. Lee (EV1) and Smith (EV2) participated in EVAs 1, 3 and 5, while

A wide-angle view of the HST in Discovery's payload bay high over Australia during the fifth and final EVA of the STS-82 mission. Steve Smith (centre) and Mark Lee (on RMS) are conducting a survey of handrails on the telescope. In the foreground is the hatch that provides access to the airlock and crew compartment of the Shuttle

Harbaugh (EV3) and Tanner (EV4) conducted EVAs 2 and 4. When one EVA crew was outside, the other provided IV support and EVA choreography, as well as resting and preparing their own EVA equipment for their next excursion. There were over 150 tools and crew aids available to the EVA astronauts on this flight.

During the first EVA, the astronauts replaced the Goddard High Resolution Spectrograph (GHRS) and Faint Object Spectrograph (FOS) with the new Space Telescope Imaging Spectrograph (STIS) and Near-Infrared Camera and Multi-Object Spectrometer (NICMOS). The second EVA saw the replacement of the Far Guidance System (FGS) and out-of-date recorders. The astronauts also installed the Optical Control Electronics Enhancement Kit (OCE-EK). It was on this EVA that cracking and wear to the telescope's insulation material on the Sun-facing side in the direction of orbital travel was noted. EVA 3 was used to replace the older reel-to-reel Engineering and Science Data Recorders (ESDR) with new solid state data recorders. The Data Interface Unit (DIU) was also replaced, as was one of the four Reaction Wheel Assembly Units used to generate spin momentum both to move the telescope and to keep it stable. At the end of this EVA, mission managers decided to add a fifth EVA to repair the thermal insulation damage that had been discovered earlier.

During EVA 4, the Solar Array Drive Electronics (SADE) were replaced and new covers were placed over the magnetometers. The astronauts also installed thermal blankets of multi-layered material over two areas where the insulation had degraded. This was around the light shield section of the instrument near the top of the telescope. While Harbaugh and Tanner were completing this EVA, Horowitz and Lee worked inside Discovery to fabricate new insulation blankets for the telescope from spare material carried on the mid-deck. The fifth and final EVA saw the attachment of several thermal blankets to three equipment compartments at the top of the Support System Module, which contained key data-processing, electronics and scientific instrument and telemetry packages. At the close of this final excursion outside, the astronauts had logged 33 hours 11 minutes of total EVA time.

During the time the telescope was attached to the payload bay, Discovery's manoeuvring engines were fired several times to raise the orbit by 8 nautical miles. The telescope was released on 19 February into its highest orbit to date, of 599 km × 620 km. The landing of Discovery was on the second attempt for 21 February, after the initial opportunity was waived off due to low clouds. The next planned Hubble service missions were manifested for 1999 and 2002.

Milestones

196th manned space flight
112th US manned space flight
82nd Shuttle mission
22nd flight of Discovery
2nd HST Service Mission
35th US and 65th flight with EVA operations
Bowersox exceeds 1,000 hours in space

STS-83

Int. Designation	1997-013A
Launched	4 April 1997
Launch Site	Pad 39A, Kennedy Space Center, Florida
Landed	8 April 1997
Landing Site	Runway 33, Shuttle Landing Facility, KSC, Florida
Launch Vehicle	OV-102 Columbia/ET-84/SRB BI-086/SSME #1 2012; #2 2109; #3 2019
Duration	3 days 23 hrs 12 min 39 sec
Call sign	Columbia
Objective	Material Science Laboratory 1 payload operations

Flight Crew

HALSELL Jr., James Donald, 40, USAF, commander, 3rd mission
Previous missions: STS-65 (1994); STS-74 (1995)
STILL, Susan Leigh, 35, USN, pilot
VOSS, Janice Elaine, 40, civilian, mission specialist 1, payload commander, 3rd mission
Previous missions: STS-57 (1993); STS-63 (1995)
GERNHARDT, Michael Landen, 40, civilian, mission specialist 2, 2nd mission
Previous mission: STS-69 (1995)
THOMAS, Donald Alan, 41, civilian, mission specialist 3, 3rd mission
Previous missions: STS-65 (1994); STS-70 (1995)
CROUCH, Roger Keith, 57, civilian PhD, payload specialist 1
LINTERIS, Gregory Thomas, 39, civilian, payload specialist 2

Flight Log

The original launch on 3 April was delayed by 24 hours on 1 April after it became necessary to add extra thermal insulation to a water coolant line in Columbia's payload bay. There was concern that there was insufficient insulation and that the line might freeze while in orbit. A further 20.5 minute delay on launch day was caused by the need to replace the orbiter access hatch seal.

The mission was planned for sixteen days, supported by an EDO kit. However, when a sudden upward voltage trend was noted in Fuel Cell 2 shortly after reaching orbit, mission rules were implemented and meant an early termination of the flight. Though the vehicle could fly safely on two fuel cells, mission rules state that all three fuel cells need to be operating well to ensure crew safety and provide sufficient back-up capacity during re-entry and landing. Similar problems had been noted with this fuel cell during launch check-ups, but tests cleared the unit for flight. Measures to address

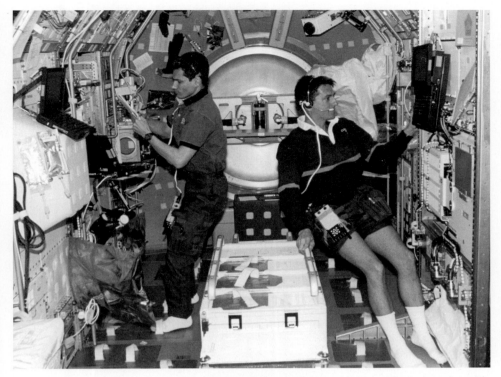

Greg Linteris (left) is seen at the Mid-deck Glove Box (MGBX) while Don Thomas works at the Expedite Processing of Experiments to Space Station (EXPRESS) rack. Despite the shortened mission the crew were able to achieve some science results

the problem on orbit were to no avail and on 6 April, the mission management team opted to terminate the mission at the earliest point.

The crew had been able to conduct some science in the Spacelab module despite the early return. Some of the materials processing experiments and fire-related experiments were conducted, but most of the experiments on board the science laboratory had not been fully activated when the call came to shorten the mission. Shortly after landing, the mission management team indicated that a re-flight of the mission was possible despite an extremely tight manifest for the rest of the year. Halsell commented that his crew had just completed the best training session possible in order to fly – they trained in space!

NASA began to evaluate manifesting STS-83R (Re-flight) to fly after the next Shuttle–Mir docking mission (STS-84), which was scheduled for May. By 24 April, the mission had been re-designated STS-94 (the next available flight number in the manifest) and the remaining 1997 missions were adjusted to accommodate the extra flight.

This would be one of the quickest turnarounds in Shuttle history and the first time a complete crew would re-fly intact and return to orbit to complete an abbreviated mission. By using the same orbiter, configured the same way, and flying the same crew, considerable time would be saved in processing the launch.

Post-flight tests indicated that an undetermined and isolated incident had caused a slight change in the voltage in about 25 per cent of the 96 cells that comprised the fuel cell generation unit, rather than a complete cell failure as at first suspected. More monitoring would be introduced on future missions, as it was determined that Columbia could have flown its full mission without problems. In light of the Challenger accident, the question of safety was raised given the quick turnaround plan, but an independent aerospace safety advisory panel recommended that NASA was capable of quickly flying Columbia again without placing undue risks on the crew or the vehicle.

Milestones

197th manned space flight
113th US manned space flight
83rd Shuttle mission
22nd flight of Columbia
14th flight of Spacelab Long Module
3rd shortened Shuttle mission
10th EDO mission (planned)

STS-84

Int. Designation	1997-023A
Launched	15 May 1997
Launch Site	Pad 39A, Kennedy Space Center, Florida
Landed	24 May 1997
Landing Site	Runway 33, Shuttle Landing Facility, KSC, Florida
Launch Vehicle	OV-104 Atlantis/ET-85/SRB BI-087/SSME #1 2032; #2 2031; #3 2029
Duration	9 days 5 hrs 19 min 56 sec Foale 144 days 13 hrs 47 min 21 sec (landing on STS-86)
Call sign	Atlantis
Objective	6th Shuttle–Mir docking; delivery of NASA 5 Mir EO-24 crew member; return of NASA 4 Mir EO-23 crew member

Flight Crew

PRECOURT, Charles Joseph, 41, USAF, commander, 3rd mission
Previous missions: STS-55 (1993); STS-71 (1995)
COLLINS, Eileen Marie, 40, USAF, 2nd mission
Previous mission: STS-63 (1995)
CLERVOY, Jean-François André, 38, civilian, mission specialist 1, 2nd mission
Previous mission: STS-66 (1994)
NORIEGA, Carlos Ismael, 37, USMC, mission specialist 2
LU, Edward Tsang, 33, civilian, mission specialist 3
KONDAKOVA, Yelena Vladimirovna, 40, civilian, Russian mission specialist 4, 2nd mission
Previous mission: Soyuz TM20 (1994)

NASA 5 Mir resident crewmember up only:

FOALE, Colin Michael, 40, civilian, mission specialist 5, Mir EO-24 cosmonaut researcher, NASA board engineer 5, 4th mission
Previous missions: STS-45 (1992); STS-56 (1993); STS-63 (1995)

NASA 4 Mir resident crewmember down only:

LINENGER, Jerry Michael, 41, USN, mission specialist 5, Mir EO-23 cosmonaut researcher, NASA board engineer 4, 2nd mission
Previous mission: STS-64 (1994)

Flight Log

Atlantis docked with Mir on 16 May, bringing the next American Mir resident crew member (Mike Foale) to begin his residency. The formal hand-over between Foale

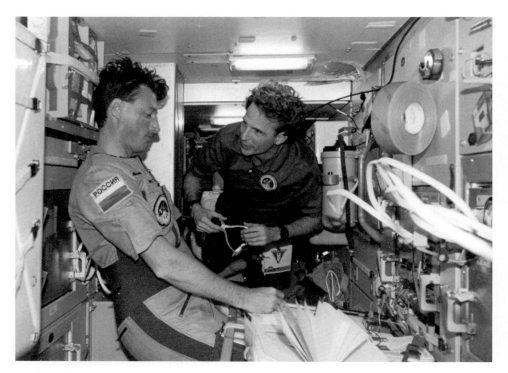

Change of shift on Mir. Jerry Linenger (right) briefs Mike Foale in preparation for the latter's stay on Mir. The photo was taken after Foale moved to the Mir resident crew and Linenger became part of the STS-84 crew, as evidenced by the uniforms they wear. Part of the briefing would have been details on the fire Linenger experienced in February. What no one could have foreseen was the events that Foale was to endure over the next few months

and Linenger occurred the following day. Linenger had spent 123 days on board the station and by the end of his mission, he had become the second most experienced American astronaut, behind Shannon Lucid. During his stay on board the station, Linenger sent regular emails to his family that were posted on the NASA website and later became the focus of the book *Letters from Mir*. He also wrote of his experiences in the book *Off the Planet*.

During the docked phase of the STS-84 mission, the crew transferred approximately 3,400 kg of logistics and supplies to the station, of which about 450 kg was water. During his stay on Mir, Foale had a research programme of 36 investigations (33 on Mir, two on STS-84 and another which included pre- and post-flight participation). These were shared among six disciplines: advanced technology, Earth observations and remote sensing, fundamental biology, human life sciences, space station risk mitigation, and microgravity sciences. Of these experiments, 28 had been performed during earlier missions, and would be continued, repeated or completed by Foale. Seven new experiments were concerned with materials processing, biology, and crystal growth studies. While Atlantis was still docked to Mir, the crew utilised the Biorack facility located in the SpaceHab double module. In addition, they took environmental air samples, continued to monitor radiation levels and photo-documented the exterior of the station through the windows.

On 21 May, Atlantis undocked from Mir with Jerry Linenger on board. There was no fly-around of Mir on this flight, but the orbiter was halted three times as it backed away from the station, allowing a European sensing device to be evaluated. The data would help in the design of rendezvous systems for the proposed Automated Transfer Vehicle (ATV), the unmanned re-supply craft being developed by ESA for the ISS programme. The first landing opportunity for STS-84 on 24 May was waived off due to low clouds around the SLF, but the weather cleared sufficiently to allow a landing on the second opportunity.

Milestones

198th manned space flight
114th US manned space flight
84th Shuttle mission
19th flight of Atlantis
6th Shuttle–Mir docking
8th SpaceHab mission (3rd double module)

STS-94

Int. Designation	1997-032A
Launched	1 July 1997
Launch Site	Pad 39A, Kennedy Space Center, Florida
Landed	17 July 1997
Landing Site	Runway 33, Shuttle Landing Facility, KSC, Florida
Launch Vehicle	OV-102 Columbia/ET-86/SRB BI-088/SSME #1 2037; #2 2034; #3 2033
Duration	15 days 16 hrs 34 min 4 sec
Call sign	Columbia Objective Material Science Laboratory 1 (Re-flight)

Flight Crew

HALSELL Jr., James Donald, 40, USAF, commander, 4th mission
Previous missions: STS-65 (1994); STS-74 (1995); STS-83 (1997)
STILL, Susan Leigh, 35, USN, pilot, 2nd mission
Previous mission: STS-83 (1997)
VOSS, Janice Elaine, 40, civilian, mission specialist 1, payload commander, 4th mission
Previous missions: STS-57 (1993); STS-63 (1995); STS-83 (1997)
GERNHARDT, Michael Landen, 40, civilian, mission specialist 2, 3rd mission
Previous missions: STS-69 (1995); STS-83 (1997)
THOMAS, Donald Alan, 41, civilian, mission specialist 3, 4th mission
Previous missions: STS-65 (1994); STS-70 (1995); STS-83 (1997)
CROUCH, Roger Keith, 57, civilian, payload specialist 1, 2nd mission
Previous mission: STS-83 (1997)
LINTERIS, Gregory Thomas, 39, civilian, payload specialist 2, 2nd mission
Previous mission: STS-83 (1997)

Flight Log

The 84-day turnaround from the landing of STS-83 to the launch of the re-flight mission, designated STS-94, was a new record and an impressive demonstration of the ability and skills of the processing team. The quick turnaround was in part facilitated by servicing the MSL payloads while still in the payload bay of Columbia. The original STS-94 mission was manifested as a "flight opportunity" by Discovery in October 1998, but as no payload had been assigned to that flight, it was the first available flight number to assign the MSL administration and planning documents to. As the same vehicle, crew and payload would be flying, it was in effect a "paper change" to the flight designation, although a new ET, SRBs and different SSMEs were

Susan Still (left) and Janice Voss review In-flight Maintenance (IFM) procedures during one of the daily planning sessions in the Spacelab Science Module in support of the MSL mission. Meanwhile Greg Linteris works at a laptop computer in the background

assigned to support the new mission. There was a delay to the launch due to unacceptable weather around the SLF.

With the crew operating the familiar Red and Blue two-shift system, 33 investigations were completed in the fields of combustion, biotechnology and materials processing. There were 25 primary investigations, four glove box investigations and four accelerometer studies on MSL-1. Some of this work involved evaluating hardware, facilities and procedures in preparation for similar hardware and research programmes that were due to be carried out on ISS. Within the combustion investigations, 144 experiments were planned, and over 200 were actually completed. The TEMPUS electromagnetic containerless processing facility completed over 120 melting cycles of zirconium at temperatures ranging between 340 and 2,000°C. In the "ignition of large fuel droplets" experiment, conducted in the glove box, only 52 test runs were planned, but the crew managed to complete 125 by the end of the mission. There were in excess of 700 crystals of various proteins grown during the 16-day mission and a record number of commands (over 35,000) were sent from the Spacelab Mission Operations Control Center at Marshall Space Flight Center to the MSL-1 experiments.

On 2 July, Don Thomas reported sighting the Mir space station as it passed within 100 km of Columbia. Two days later, as America celebrated Independence Day, the crew sent messages of congratulations to the JPL Pathfinder team in California on the successful landing of the Mars Pathfinder spacecraft on the Red Planet. Three days later, the crew were informed of the successful docking of a Russian Progress (M35) re-supply vehicle with Mir and the following day, Halsell, Gernhardt and Voss used the SAREX equipment aboard Columbia to talk with Mike Foale aboard the Mir space station. On 14 July, the crew reported a minute debris impact with one of the overhead windows, a familiar occurrence which again was no cause for concern over safety. Shuttle windows are often hit by small pieces of space debris during orbital flight. Being multi-layer panels, such small impacts are highly unlikely to jeopardise the integrity of the window or the safety of the crew and vehicle.

Milestones

199th manned space flight
115th US manned space flight
85th Shuttle mission
23rd flight of Columbia
1st re-flight of same vehicle, payload and crew
15th flight of Spacelab Long Module
11th EDO mission

SOYUZ TM26

Int. Designation	1997-083A
Launched	5 August 1997
Launch Site	Pad 1, Site 5, Baikonur Cosmodrome, Kazakhstan
Landed	19 February 1998
Landing Site	31 km southeast of Arkalyk
Launch Vehicle	R7 (11A511U); spacecraft serial number (7K-M) #75
Duration	197 days 17 hrs 34 min 36 sec
Call sign	Rodnik (Spring)
Objective	Mir EO-24 programme; worked with NASA 5 (Foale), 6 (Wolf) and 7 (Thomas) astronauts

Flight Crew

SOLOVYOV, Anatoli Yakovlevich, 49, Russian Air Force, commander, 5th mission
Previous missions: Soyuz TM5 (1988); Soyuz TM9 (1990); Soyuz TM15 (1992); EO-19 (1995)
VINOGRADOV, Pavel Vladimirovich, 43, civilian, flight engineer

Flight Log

The EO-24 cosmonauts arrived at Mir on 7 August to take over from the tired EO-23 crew. This time, there was an abbreviated hand-over period to avoid putting too much strain on the station's systems or departing EO-23 crew. The day after the TM25 spacecraft left Mir, Solovyov, Vinogradov and Foale relocated the TM26 spacecraft from the aft to the front port of Mir, taking close-up pictures of the damaged Spektr module on the way.

Most of the new resident crew's work would concentrate on restoring Mir to operational status again, but they also had to cope with two changes of American crew members. In late September, STS-86 docked with Mir, bringing much-needed supplies and astronaut Dave Wolf to take over from Mike Foale. In January 1998, STS-89 arrived with Andy Thomas aboard, the seventh and final NASA astronaut to work aboard the Mir complex with a resident crew. A couple of days after Wolf returned home, the replacement Mir EO-25 resident crew arrived aboard Soyuz TM27 to continue Mir residency with Thomas. Also aboard TM27 was French cosmonaut researcher Leopold Eyharts, who would return to Earth with Solovyov and Vinogradov in TM26 after completing his research programme.

The seven EVAs conducted by the EO-24 crew made up one of the most extensive spacewalk schedules of the whole Mir programme. Solovyov was outside for all seven of the EVAs, with three different colleagues (one Russian and two American). The first EVA (22 Aug, 3 hours 16 minutes) saw the two cosmonauts re-enter the Spektr

The Soyuz TM26 crew of Solovyov (left) and Vinogradov faced a long schedule of repairs to Mir during their residency

module (while Foale remained in the Soyuz). Upon entering the dark, cold module, the cosmonauts found white crystals floating around and surfaces covered in a layer of frost. The crew reconnected power cables to a new, modified hatch plate to allow the use of the undamaged solar arrays on the Spektr module. They also retrieved several items for Foale from the module before it was permanently sealed. Partial electrical power was restored, but the system would not allow the solar arrays to track the Sun, preventing maximum power output. On the next EVA (6 Sep, 6 hours 0 minutes), Solovyov and Foale conducted an external inspection of Spektr, videoing the exterior for analysis on the ground. The next three EVAs (20 Oct, 6 hours 38 minutes; 3 Nov, 6 hours 4 minutes; and 6 Nov, 6 hours 17 minutes) were completed by the two Russians and included reconnection of cables in an IVA, and relocation of solar arrays from the Kvant module. On the second of these three EVAs, the outer hatch of Kvant 2 failed to hermetically seal. The "C" clamp used since 1990 had finally deteriorated and required replacement. This was initially planned for this crew, but was subsequently postponed for the following resident crew instead. The sixth EVA (9 Jan 1998, 4 hours 4 minutes) saw the two Russians complete a photo-documentation of the Mir exterior, including the Kvant 2 hatch area, as well as retrieving several exterior experiments. The last EVA of the expedition (14 Jan, 6 hours 38 minutes) was completed by Solovyov and Wolf and included the use of a spectro-reflectometer to examine the physical condition of the station's exterior surfaces.

Having launched in the warm August temperatures, the EO-24 crew came home to snow. A helicopter remained close by the grounded Soyuz with its rotors spinning to prevent ice from building up before the crew could be extracted. The three cosmonauts (including Eyharts) were immediately carried away on stretchers and airlifted away from the harsh conditions of the landing zone.

Milestones

200th manned space flight
85th Russian manned space flight
78th manned Soyuz mission
25th manned Soyuz TM mission
24th Mir resident crew
31st Russian and 66th flight with EVA operations

STS-85

Int. Designation	1997-039A
Launched	7 August 1997
Launch Site	Pad 39A, Kennedy Space Center, Florida
Landed	19 August 1997
Landing Site	Runway 33, Shuttle Landing Facility, KSC, Florida
Launch Vehicle	OV-103 Discovery/ET-87/SRB BI-089/SSME #1 2041; #2 2039; #3 2042
Duration	11 days 20 hrs 26 min 59 sec
Call sign	Discovery
Objective	CRISTA-SPAS-02 operations

Flight Crew

BROWN Jr., Curtis Lee, 41, USAF, commander, 4th mission
Previous missions: STS-47 (1992); STS-66 (1994); STS-77 (1996)
ROMINGER, Kent Vernon, 41, USN, pilot, 3rd mission
Previous missions: STS-73 (1995); STS-80 (1996)
DAVIS, Jan, 43, civilian, mission specialist 1, payload commander, 3rd mission
Previous missions: STS-47 (1992); STS-60 (1994)
CURBEAM Jr., Robert Lee, 35, USN, mission specialist 2
ROBINSON, Stephen Kern, 41, civilian, mission specialist 3
TRYGGVASON, Bjarni Vladimir, 52, civilian, Canadian payload specialist 1

Flight Log

Continuing the *Mission to Planet Earth* programme, as well as preparations for the construction of ISS, the STS-85 mission featured a complex payload from Germany, Japan and the US, as well as an international crew.

The primary payload on this flight was the Cryogenic IR Spectrometer and Telescope for the Atmosphere Shuttle Pallet Satellite-2 (CRISTA-SPAS-2), making its second flight on the Shuttle as part of the fourth cooperative mission between the German space agency DARA and NASA. There were three telescopes and four spectrometers on the satellite. Deployed on FD 1, it operated for over 200 hours and was retrieved on FD 10. After completing its primary objective, CRISTA-SPAS was used in a simulation exercise to prepare for the first ISS assembly flight, STS-88, with the payload being manipulated as if it were the Russian Functional Cargo Block (FGB-Zarya) that was to be attached to the US Node 1 (Unity).

The Technology Applications and Science 01 (TAS-01) payload included seven separate experiments to gather data on the topography of the Earth and its atmosphere, to study the energy from the Sun and to evaluate new thermal control devices. International Extreme UV Hitchhiker 02 was a set of four experiments that studied

Canadian astronaut Bjarni Tryggvason inputs data into a computer for the Microgravity Vibration Isolation Mount (MIM) experiment located on the mid-deck of Discovery. Behind him, the use of mid-deck lockers for stowage both inside and outside is evident

UV radiation from stars, the Sun and other solar system sources. The Japanese Manipulator Flight Demonstrator (MFD) consisted of three separate experiments located on a support structure in the payload bay and were designed to test a mechanical arm that was being evaluated for possible inclusion on the Japanese Experiment Module (JEM) planned for ISS. Despite some glitches, a series of exercises was performed by the crew in space and a team of operators on the ground.

There was also a range of in-cabin payloads, including a UV imaging system to observe Comet Hale-Bopp, crystal growth and materials-processing experiments, the Orbiter Space Vision System (to be used during ISS assembly for determining precise alignment and pointing capability). Canadian astronaut Bjarni Tryggvason, principle investigator of the Microgravity Vibration Isolation Mount (MIM), had a major role on the flight in evaluating his own equipment, performing fluid physics experiments to determine sensitivity to spacecraft vibrations when using MIM, and its application to ISS and future research facilities. The MIM had been in operation aboard Mir since April 1996 and was first operated by Shannon Lucid, where it supported a number of Canadian and US experiments in materials science and fluid physics.

The 18 August landing opportunities were waived off due to ground fog in the local area, allowing the crew an extra day on orbit.

Milestones

201st manned space flight
116th US manned space flight
86th Shuttle mission
23rd flight of Discovery

STS-86

Int. Designation	1997-055A
Launched	25 September 1997
Launch Site	Pad 39A, Kennedy Space Center, Florida
Landed	6 October 1997
Landing Site	Runway 15, Shuttle Landing Facility, KSC, Florida
Launch Vehicle	OV-104 Atlantis/ET-88/SRB BI-090/SSME #1 2012; #2 2040; #3 2019
Duration	10 days 19 hrs 20 min 50 sec
	Wolf 127 days 20 hrs 0 min 50 sec (landing on STS-89)
Call sign	Atlantis
Objective	7th Shuttle–Mir docking; delivery of NASA 6 (Wolf) crew member; return of NASA 5 (Foale) crew member

Flight Crew

WETHERBEE, James Donald, 44, USN, commander, 4th mission
Previous missions: STS-32 (1990); STS-52 (1992); STS-63 (1995)
BLOOMFIELD, Michael John, 38, USAF, pilot
TITOV, Vladimir Georgievich, 50, Russian Air Force, mission specialist 1, 4th mission
Previous missions: Soyuz T8 (1983); Soyuz T10 abort (1983); Soyuz TM4 (1987); STS-63 (1995)
PARAZYNSKI, Scott Edward, 36, civilian, mission specialist 2, 2nd mission
Previous mission: STS-66 (1994)
CHRETIEN, Jean-Loup Jacques Marie, 59, French Air Force, mission specialist 3, 3rd mission
Previous missions: Soyuz T6 (1982); Soyuz TM7 (1988)
LAWRENCE, Wendy Barrien, 38, USN, mission specialist 4, 2nd mission
Previous mission: STS-67 (1995)

NASA 6 Mir crew member up only:

WOLF, David Alan, 41, civilian, mission specialist 5, Mir EO-24 cosmonaut researcher, NASA board engineer 6, 2nd mission
Previous mission: STS-58 (1993)

NASA 5 Mir crew member down only:

FOALE, Colin Michael, 40, civilian, mission specialist 5, Mir EO-23 cosmonaut researcher, NASA board engineer 5, 4th mission
Previous missions: STS-45 (1992); STS-56 (1993); STS-63 (1995)

This image of Mir taken by the crew of STS-86 clearly shows the damaged Spektr module and arrays following the collision with a Progress re-supply vessel

Flight Log

Both Scott Parazynski and Wendy Lawrence were originally in line for long flights on the Mir space station. Parazynski had been removed from long-duration training due to the fact that he was too tall to fit in the Soyuz contour seat if he needed to use one for emergency landing (he would have been launched to and from the Mir on the Shuttle under normal circumstances). Lawrence would have followed Foale on Mir, but was deemed too short to fit into an Orlan suit, a requirement introduced after the Progress collision in order to allow American astronauts to support EVA operations to repair the station should the need arise. Lawrence had never completed Orlan EVA training, as it was not part of her original programme to perform an EVA. However, she still remained part of the STS-86 Shuttle crew to visit Mir. In addition, by way of compensation for losing the duration flight she had trained so long for, she was also guaranteed a flight on the STS-89 mission that would exchange Wolf with the final US astronaut, Andy Thomas. For some time, the three astronauts were known as Scott "Too Tall" Parazynski, Wendy "Too Short" Lawrence and Dave "Just Right" Wolf.

Regular reviews of Shuttle–Mir operations occurred prior to each docking mission, but after a fire and a collision in the space of four months, an independent and internal safety assessment was completed before NASA Administrator Dan Goldin would authorise the flight and exchange of NASA crew members. His authorisation came only an hour before the launch of STS-86. The events at Mir had seriously affected Foale's science programme, as most of his equipment had been left in the sealed-off Spektr module. But his contribution to the recovery of the station both during and immediately after the collision had earned him high praise from Russian space officials.

Atlantis docked to Mir for the seventh (and the orbiter's final) time on 27 September, with the exchange between Foale and Wolf accomplished the following day. During the six days of docked operations, the crew moved over four tons of material from SpaceHab/Atlantis to the space station, including over 770 kg of water, plus specimens and hardware for ISS risk mitigation experiments that would monitor the health and safety of the resident crew. A gyrodyne, batteries, three air pressurisation units, an attitude control computer and a range of other logistical items were also transferred to Mir. Coming the other way for the return to Earth were experiment samples and hardware and an old Elektron oxygen generator.

On 1 October, Parazynski (EV1) and Titov (EV2) completed a joint US/Russian EVA, a forerunner to those planned for ISS operations. During the EVA, they attached a 55-kg Solar Array Cap to the Docking Module for a future Russian EVA crew to seal off a suspected leak in Spektr's hull. They also retrieved four Mir Environmental Effects Payloads and continued testing the SAFER units.

After undocking on 3 October, Atlantis completed a fly-around to conduct a visual inspection of the station. This included allowing air into the Spektr module to see if the Atlantis crew could detect seepage or debris particles that would help to locate the breach in the module's hull. Particles were seen but they could not conclusively be deemed to have originated from Spektr. Two landing opportunities were waived on 5 October due to low clouds. This was the last flight of Atlantis before a planned maintenance down period, after which the vehicle would participate in the early construction flights of ISS.

Milestones

202nd manned space flight
117th US manned space flight
87th Shuttle mission
20th flight of Atlantis
7th Shuttle–Mir docking
38th US and 67th flight with EVA operations
9th SpaceHab mission (4th double module)

STS-87

Int. Designation	1997-073A
Launched	19 November 1997
Launch Site	Pad 39B, Kennedy Space Center, Florida
Landed	5 December 1997
Landing Site	Runway 33, Shuttle Landing Facility, KSC, Florida
Launch Vehicle	OV-102 Columbia/ET-89/SRB BI-092/SSME #1 2031; #2 2039; #3 2037
Duration	15 days 16 hrs 34 min 4 sec
Call sign	Columbia
Objective	USMP-4; SPARTAN 201-04

Flight Crew

KREGEL, Kevin Richard, 41, civilian, commander, 3rd mission
Previous missions: STS-70 (1995); STS-78 (1996)
LINDSEY, Steven Wayne, 37, USAF, pilot
CHAWLA, Kalpana, 34, civilian, mission specialist 1
SCOTT, Winston Elliott, 47, USN, mission specialist 2, 2nd mission
Previous mission: STS-72 (1996)
DOI, Takao, 43, civilian, mission specialist 3
KADENYUK, Leonid Konstantinovich, 46, Ukraine Air Force, payload specialist 1

Flight Log

Completing a sixth on-time launch for the year and ending the second year in which eight flights had been completed (the first being 1992), this was a flight of mixed fortunes. The USMP-4 payload performed well, with experiments focusing on materials science, combustion science and fundamental physics. There were other secondary and mid-deck experiments flown as well, including the Collaborative Ukrainian Experiment, which featured ten planet biology experiments.

SPARTAN 201 was on its fourth mission and this time, its experiment programme was geared towards investigating the physical conditions and processes of the hot outer layers of the sun's atmosphere – the Solar Corona. The SPARTAN was also to gather information on the solar wind. Originally, SPARTAN was to be deployed on FD 2, but a companion spacecraft, the Solar and Hemispheric Observatory (SOHO), had a temporary power problem and so the deployment was delayed by 24 hours. On FD 3, the RMS was used to lift the SPARTAN out of the bay, but the spacecraft failed to initiate a pirouette manoeuvre. This indicated a problem with the attitude control system, which would be required for finer pointing towards solar targets. During an attempted recapture, the RMS did not secure a firm grip and when

Winston Scott releases a prototype free-flying experiment, the Autonomous EVA Robotic Camera (AEROCam) Sprint. The EVA was also the first by a Japanese astronaut (Doi – out of frame) and included the capture of the Spartan satellite seen to the right of Scott

it was retracted, it imparted a small rotational spin on the satellite of about 2 degrees per second. The crew tried to match this rotation by firing the orbiter's thrusters for a second grapple attempt, but this was called off by the flight controllers. Instead, a plan was devised for the EVA crew to capture the satellite by hand allowing it to be stowed back into the payload bay.

The original plan for the EVA was amended to include the SPAS capture, which was achieved on 24 November. Scott (EV1) and Doi (EV2) manually grappled the satellite, allowing Chawla to use the RMS to grab the satellite and gently lower it into the payload bay. A review of further operations with SPARTAN would be conducted by mission management prior to trying to release it a second time. After the satellite was secured, the EVA crew continued with their planned programme of activities, designed to support forthcoming ISS assembly missions. This included working with a crane which was installed on the port side of the payload bay. The EVA lasted 7 hours 3 minutes.

After completing most of their experiment programme, the crew received the news that a second EVA would be added to the flight, but the SPARTAN would not be released again. The risk of being unable to retrieve the unit again was too great and the orbiter's fuel reserves were insufficient to support all contingencies. SPARTAN 201-04 therefore would not free-fly again on this mission, though it was later raised on the end of the RMS to test the video and laser sensors of the Automated Rendezvous and Capture System. The EVA crew also deployed the AEROCam Sprint, a prototype free-flying TV camera that could be utilised for remote inspections of the exterior of ISS and for visual inspections of hazardous locations which would be difficult for a suited EVA astronaut to safely reach. This second EVA, on 3 December, lasted 4 hours 59 minutes.

Milestones

203rd manned space flight
118th US manned space flight
88th Shuttle mission
24th flight of Columbia
39th US and 68th flight with EVA operations
1st Japanese to perform EVA (Doi)
4th flight of USMP payload
12th EDO mission
1st EVAs from Columbia

STS-89

Int. Designation	1998-003A
Launched	22 January 1998
Launch Site	Pad 39A, Kennedy Space Center, Florida
Landed	31 January 1998
Landing Site	Runway 15, Shuttle Landing Facility, KSC, Florida
Launch Vehicle	OV-105 Endeavour/ET-90/SRB BI-093/SSME #1 2043; #2 2044; #3 2045
Duration	8 days 19 hrs 46 min 54 sec
	Thomas 140 days 15 hrs 12 min 6 sec (landing on STS-91)
Call sign	Endeavour
Objective	8th Shuttle–Mir docking mission; delivery of NASA 7 (Thomas) Mir EO-25 crew member; return of NASA 6 (Wolf) Mir EO-24 crew member

Flight Crew

WILCUTT, Terrence Wade, 48, USAF, commander, 3rd mission
Previous missions: STS-68 (1994); STS-79 (1996)
EDWARDS Jr., Joe Frank, 39, USN, pilot
REILLY II, James Francis, 43, civilian, mission specialist 1
ANDERSON, Michael Phillip, 38, USAF, mission specialist 2
DUNBAR, Bonnie Jean, 48, civilian, mission specialist 3, payload commander, 5th mission
Previous missions: STS 61-A (1985); STS-32 (1990); STS-50 (1992); STS-71 (1995)
SHARIPOV, Salizhan Shakirovich, Russian Air Force, mission specialist 4

NASA 7 Mir resident crew member up only:

THOMAS, Andrew Sydney Withiel, 46, civilian, mission specialist 5, Mir EO-25 cosmonaut researcher, NASA board engineer 7, 2nd mission
Previous mission: STS-77 (1996)

NASA 6 Mir resident crew member down only:

WOLF, David Alan, 41, civilian, NASA mission specialist 5, Mir EO-24 cosmonaut researcher, NASA board engineer 6, 2nd mission
Previous mission: STS-58 (1993)

Flight Log

Dave Wolf's 119-day residency aboard Mir during his 128-day mission attracted much less attention in the media than the tours of either Linenger or Foale. Wolf was

Salizan Sharipov (centre) signs the long-lived Mir roster on the base block of the space station. Some of the other Mir and Shuttle crew members look on: from left Thomas (back to camera), Solovyov, Wolf, Vinogradov, Edwards (partially obscured) and Dunbar

able to perform more science on board the station and his programme involved six areas of research – advanced technology, Earth science, fundamental biology, human life sciences, microgravity research and risk mitigation of ISS issues. In all there were 35 scientific studies and technology demonstrations comprising the NASA 6 science programme, some of which were continuations of experiments conducted by previous resident NASA astronauts. Wolf also completed a 6 hour 38 minute EVA on 14 January with veteran spacewalker Anatoly Solovyov.

Endeavour was chosen to fly the STS-89 mission instead of Discovery. Because of a schedule of work that needed to be completed on Mir, the Russians requested a postponement of the mission launch. It was initially moved from 15 January to 20 January and finally to 22 January. The docking with Mir occurred on 24 January and the exchange of American resident crew members was made the next day. For a while, it looked like Thomas might not be able to remain aboard Mir. Thomas's Sokol pressure suit for use in the Soyuz would not fit properly and the crew exchange was allowed only after Wolf adjusted his suit to fit Thomas, as Wolf no longer needed it for his return on the Shuttle. Later, Thomas was able to make suitable adjustments to his own suit.

During the four days of joint operations, a total of 3,629 kg of scientific equipment, logistics and other hardware was transferred to Mir. Included in this transfer was over 730 kg of water. During the docked operations Bonnie Dunbar, on her second visit to Mir and who could have conducted the second residence mission after Norman Thagard had the schedule been worked out early enough, acknowledged the upcoming 25th anniversary of the launch of Skylab, America's only national space station to reach orbit. With ISS on the horizon, the Skylab programme, together with Mir and Salyut, had helped to develop techniques and procedures for endurance space flights which were still being referred to in preparation for the new station. Interestingly, Dunbar had been a flight controller during the de-orbiting of Skylab in 1979. Sharipov had only six months training for his Shuttle flight and had relatively few crew responsibilities during the mission. His primary responsibility was in Russian language liaison and in the transfer of logistics across to Mir. STS-89 undocked on 28 January and three days later, Soyuz TM25 docked at Mir to deliver a new resident crew, just hours before Endeavour touched down in Florida.

Milestones

204th manned space flight
119th US manned space flight
89th Shuttle mission
12th flight of Endeavour
8th Shuttle–Mir docking mission
40th US and 69th flight with EVA operations
10th SpaceHab mission (5th double module)
1st and only Endeavour–Mir docking

SOYUZ TM27

Int. Designation	1998-004A
Launched	29 January 1998
Launch Site	Pad 1, Site 5, Baikonur Cosmodrome, Kazakhstan
Landed	25 August 1998
Landing Site	40 km north of Arkalyk
Launch Vehicle	R7 (11A5111U); spacecraft serial number (7K-M) #76
Duration	207 days 12 hrs 51 min 2 sec (Musabayev/Budarin); 20 days 16 hrs 36 min 48 sec (Eyharts, in TM26 with EO-24 crew)
Call sign	Kristall (Crystal)
Objective	Mir EO-25 programme; French *Pegasus* mission

Flight Crew

MUSABAYEV, Talgat Amangeldyevich, 47, Russian Air Force, commander, 2nd mission
Previous mission: Soyuz TM19 (1994)
BUDARIN, Nikolai Mikhailovich, 44, civilian, flight engineer, 2nd mission
Previous mission: Mir EO-19 (STS-71/Soyuz TM21) (1995)
EYHARTS, Leopold, 40, French Air Force, cosmonaut researcher

Flight Log

The original *Pegasus* mission had been scheduled for August 1997, but the collision of Progress with Mir and the subsequent on-orbit difficulties meant that the flight was delayed. There was also a medical issue, with Jean-Pierre Haigneré having injured his leg during a badminton match in July 1997. He was replaced by Eyharts, but as a result, the French mission was moved back to the next Soyuz TM flight. The *Pegasus* science programme was a repeat of that completed by Claudie André-Deshays in 1996 under the *Cassiopée* programme. Eyharts was kept fully occupied during his three weeks on the station, returning with the EO-24 crew on 19 February.

The EO-25 crew now settled down to work with Thomas, the final American resident on Mir, as well as completing the routine maintenance and housekeeping chores all Mir resident crews had to address. The EO-25 crew also completed a programme of six EVAs. The first, on 3 March (1 hour 15 minutes), had to be abandoned when a hatch wrench broke, preventing them from opening the exit. With replacement parts delivered by Progress M38, the two cosmonauts resumed their EVA work in April, delaying their science programme by completing five excursions in the same month (1 Apr for 6 hours 26 minutes; 6 Apr for 4 hours 23 minutes; 11 Apr for 6 hours 25 minutes; 17 Apr for 6 hours 33 minutes; and 22 Apr for 6 hours 21 minutes). Their work included bracing the solar array that had been damaged by the Progress

French cosmonaut Eyharts (left) was launched to Mir aboard Soyuz TM27, along with the next resident crew of Musabayev (centre) and Budarin

collision. As it was still generating some electricity, the Russians wanted to try to repair the solar array rather than disabling it. They also replaced the VDU engine block and stowed the Rapana Truss next to the Sofora girder (for possible future use). During the EVAs, Thomas assisted the EVA crew from inside the space station.

The cosmonauts resumed their science programme after completing their month of EVAs. NASA was pushing for the demise of Mir to allow full concentration on the ISS programme, but Russia was reluctant to do so. With existing contracts to fly one more long French mission and a Slovak visiting mission, there were also reports of selling seats to fare-paying passengers for short missions, generating much needed funds for the programme. This idea of "Soyuz seats for sale" generated interest from wealthy individuals across the world, and led to the prospect of turning Mir into a commercially funded station while ISS was under construction. This did not go down well with the Americans, who basically wanted Mir out of the way so that everyone's full attention could be devoted to the more complex work on ISS. A shortage of hardware and funds was already a familiar and concerning problem in operations at Mir, and the Americans did not want to see a drain on resources from ISS because the Russians were trying to run two space station programmes. For the Russians, however, it was also a question of national pride. They wanted to keep their Mir station in orbit as long as they could.

In June, STS-91 arrived to bring home American astronaut Andy Thomas and so end two years of continuous US occupation of Mir. Among the Shuttle crew was veteran Russian cosmonaut Valery Ryumin, now chief of the Russian side of the Shuttle–Mir programme. His primary role on the mission was to make a thorough inspection of the Mir complex to assess its potential for further use. He concluded that it would take some time to stow everything properly and that a crew of two or three cosmonauts would struggle to keep on top of the tasks. What Mir needed was a crew of six or seven to fully utilise the station. This was not something NASA wanted to hear.

After STS-91 had departed, Musabayev commented that the station was far roomier now there were only two on board. The cosmonauts resumed their science work with materials-smelting experiments, Earth observations and remote sensing, and continued the biological and medical experiment programmes. In August 1998 a new crew arrived at Mir, just three months before the start of ISS construction.

Milestones

205th manned space flight
86th Russian manned space flight
79th manned Soyuz mission
26th manned Soyuz TM mission
25th Mir resident mission
32nd Russian and 70th flight with EVA operations
5th French long-duration mission (21 days)

STS-90

Int. Designation	1998-022A
Launched	17 April 1998
Launch Site	Pad 39B, Kennedy Space Center, Florida
Landed	3 May 1998
Landing Site	Runway 33, Shuttle Landing Facility, KSC, Florida
Launch Vehicle	OV-102 Columbia/ET-91/SRB BI-091/SSME #1 2041; #2 2032; #3 2012
Duration	15 days 21 hrs 49 min 59 sec
Call sign	Columbia
Objective	Neurolab

Flight Crew

SEARFOSS, Richard Alan, 41, USAF, commander, 3rd mission
Previous missions: STS-58 (1993); STS-76 (1996)
ALTMAN, Scott Douglas, 38, USN, pilot
LINNEHAN, Richard Michael, 41, civilian, mission specialist 1, payload commander, 2nd mission
Previous mission: STS-78 (1996)
HIRE, Kathryn Patricia, 38, USN, mission specialist 2
WILLIAMS, Dafydd Rhys, civilian, Canadian, mission specialist 3
BUCKLEY Jr., Jay Clark, 41, civilian, payload specialist 1
PAWELCZYK, James Anthony, 37, civilian, payload specialist 2

Flight Log

The launch of Neurolab was delayed by 24 hours from 16 April due to problems with one of the two network processors aboard Columbia. These format data and voice communications between the Shuttle and the ground and the unit had to be replaced. The payload for this mission was the final flight of the Spacelab Long Module, which had first flown in 1983. The science programme it contained was designated Neurolab, and included 26 experiments grouped together to form one of the most extensive investigations into the most complex and least understood part of the human body – the nervous system. The primary objective of this research was to expand our understanding of how the nervous system develops and functions in microgravity and for such a comprehensive research programme, the test subjects included more than just the seven astronauts. Making the journey with the human crew were rats, mice, crickets, snails and two species of fish.

This mission had its origins in 1991, when NASA proposed a mission to contribute to the "Decade of the Brain". A total of 132 experiment proposals were reduced to 32, with 26 flying on STS-90 and the remaining six reassigned to later

As American residency on Mir draws to a close, so too does another aspect of the Shuttle programme – Spacelab. On 12 February 1998, the Neurolab payload in the Spacelab Long Module is lowered into the cargo bay of Columbia in OPF Bay 1 at KSC. This was the final flight of the European-built Spacelab module system, first flown as Spacelab 1 in November 1983. There had been 16 Spacelab Long Module missions between 1983 and 1998

missions. The research programme covered eight areas. Adult neuronal plasticity studied the ability of neurons to react to different conditions (in this case microgravity) to make new connections in new ways, allowing the neurosystem to compensate for the new environment. This programme used rats as the test subjects. Mammalian development research utilised the rats and mice to answer key questions such as "Can walking be learned without gravity?" Aquatic experiments studied the effects of microgravity on otoliths and statolith development and adaptation in oyster toadfish, swordtail fish and freshwater snails. Neurology research on crickets was used to help understand how much normal development is pre-programmed in the genes and how much requires clues from the environment. The remaining studies, which were carried out by the human crew, included investigations into the autonomic nervous system, sleep, vestibular experiments and sensory, motor and performance studies.

Neurolab was activated 1 hour 45 minutes into the mission. The crew followed a single-shift system, with the science crew participating in or activating most of the experiments while the orbiter crew looked after the spacecraft and its systems. The orbiter crew (commander, pilot and MS2/flight engineer) assisted the science crew as required throughout the mission. This was an international, mission with cooperation between NASA and the space agencies of Canada (CSA), France (CNES) and Germany (DARA) as well as the European Space agency (ESA) and the National Development Agency of Japan (NASDA). Most of the research was conducted as planned, except for the mammalian development studies, which were prioritised due to the unexpectedly high mortality rate of the neo-natal rats aboard (55 of the 96 nine-day-old rodents died).

A week into the mission, the crew worked with engineers on the ground to overcome a problem with a system valve in the Regenerative Carbon Dioxide Removal System, which threatened to cut short the flight. A decision to extend the mission was considered, but when the science community indicated that this was not necessary, and with weather conditions expected to deteriorate after the scheduled 3 May landing, the mission was ended as planned after 16 days. It was the end of a highly successful mission, but also of the Spacelab Long Module series.

Milestones

206th manned space flight
120th US manned space flight
90th Shuttle mission
25th flight of Columbia
16th and final flight of Spacelab Long Module
13th and final EDO mission

STS-91

Int. Designation	1998-034A
Launched	2 June 1998
Launch Site	Pad 39A, Kennedy Space Center, Florida
Landed	12 June 1998
Landing Site	Runway 15, Shuttle Landing Facility, KSC, Florida
Launch Vehicle	OV-103 Discovery/ET-96/SRB BI-091/SSME #1 2047; #2 2040; #3 2042
Duration	9 days 19 hrs 53 min 54 sec
Call sign	Discovery
Objective	9th and final Shuttle–Mir docking; return of NASA 7 (Thomas) Mir EO-25 crew member

Flight Crew

PRECOURT, Charles Joseph, 43, USAF, commander, 4th mission
Previous missions: STS-55 (1993); STS-71 (1995); STS-84 (1997)
GORIE, Dominic Lee Pudwill, 41, USN, pilot
CHANG-DIAZ, Franklin Ramon de Los Angeles, 48, civilian, mission specialist 1, 6th mission
Previous missions: STS 61-C (1986); STS-34 (1989); STS-46 (1992); STS-60 (1994); STS-75 (1996)
LAWRENCE, Wendy Barrien, 38, mission specialist 2, 3rd mission
Previous missions: STS-67 (1995); STS-86 (1997)
KAVANDI, Janet Lynn, mission specialist 3
RYUMIN, Valery Viktorovich, 58, civilian, Russian, mission specialist 4, 4th mission
Previous missions: Soyuz 25 (1977); Soyuz 32 (1979); Soyuz 35 (1980)

NASA 7 Mir EO-25 resident down only:

THOMAS, Andrew Sydney Withiel, 46, civilian, mission specialist 5, Mir EO-25 cosmonaut researcher, NASA board engineer 7, 2nd mission
Previous mission: STS-77 (1996)

Flight Log

Apart from a slight delay in tanking operations, the launch of the final Shuttle mission to Mir proceeded nominally. This was the first docking mission for Discovery, which successfully joined the space station on 4 June. The hatches were opened on the same day and Thomas transferred to the Shuttle crew, ending 130 days of residency on Mir. Prior to his residency, Thomas had been criticised by the Russians for his limited ability to speak the language, but with Russian being the only language spoken on

End of an era. The STS-91 crew and the Mir EO-25 crew pose for the final traditional in-flight NASA Shuttle–Mir crew photo in the core module of the station. L to r Ryumin, Lawrence, Precourt, Thomas, Musabayev, Kavandi, Gorie, Budarin, Chang-Diaz

Mir, he soon became well versed. With the transfer of Thomas to the Discovery crew, a total of 907 days had been logged by the seven resident astronauts aboard Mir. In addition, there had been a US presence in space for 812 consecutive days and on Mir for 802 consecutive days.

During the four days of joint operations, the crews transferred 500 kg of water and a further 2,130 kg of experiments and supplies. US long-term experiments were also moved back into Discovery or the SpaceHab module from the station. The Shuttle crew completed a range of secondary experiments during docked activities. Although he was part of the Shuttle crew, cosmonaut Valery Ryumin, who had spent about a year aboard the Salyut 6 station on two six-month missions in 1979 and 1980, was aboard to evaluate the state of the station and, according to NASA, to confirm the station's condition for mothballing and decommissioning. Instead, he indicated that it was still viable for future operations. One of the less publicised "experiments" was the transfer of a stowed American EMU through the smaller opening of the Orbiter Docking System (ODS), a process which would become a regular operation on ISS. The EMU are normally bundled in a Lower Torso Assembly Restraint Bag (LTARB) for ease of handling, but during ground tests, it was found that it took less time to simply stow the gear and clear a path through loose equipment around the connecting hatches. With no high-fidelity mock-ups on Earth, it was useful to try this method aboard an actual station in space. The information gained would be valuable in planning such transfers on ISS.

After the joint programme had been completed the hatches were sealed and the spacecraft separated on 8 June, marking the final Shuttle docking mission and the conclusion of Phase 1 of the ISS programme. It was now time to move toward the assembly missions for ISS later in the year.

Over the preceding six years, considerable hurdles had been overcome, differences ironed out and a strong partnership formed. This resulted in ten missions and nine dockings to the Mir complex, the residence of seven astronauts on Mir and training for five cosmonauts to fly on the Shuttle. The difficulties, and at times dangers, of long-duration flight were quickly learned by the Americans – something the Russians had been aware of for years. For the Russians, their learning curve was in accepting an international cooperative partner beyond the former Soviet Bloc countries and friendship agreements. The Shuttle (and American money) gave the aging Mir station a prolonged life and Mir gave NASA the experience in space station operations it badly needed before committing to ISS operations. Lessons had been learned the hard way at times, but they were essential lessons. Without Shuttle–Mir, the ISS programme would have been much harder to initiate.

Milestones

207th manned space flight
121st US manned space flight
91st Shuttle mission
24th flight of Discovery
9th and final Shuttle–Mir docking mission
11th SpaceHab mission (6th single module)
1st docking mission for Discovery
1st user of super-lightweight ET
Completion of ISS Phase 1 programme

SOYUZ TM28

Int. Designation	1998-047A
Launched	13 August 1998
Launch Site	Pad 1, Site 5, Baikonur Cosmodrome, Kazakhstan
Landed	8 February 1999
Landing Site	58 km north of Arkalyk
Launch Vehicle	R7 (11A511U); spacecraft serial number (7K-M) 77
Duration	198 days 16 hrs 31 min 20 sec (Padalka returned in TM28); 379 days 14 hrs 51 min 10 sec (Avdeyev landed with EO-27 in TM29); 11 days 19 hrs 41 min 33 sec (Baturin returned with EO-25 in TM27)
Call sign	Altair (Altair)
Objective	Mir EO-26 programme

Flight Crew

PADALKA, Gennady Ivanovich, 40, Russian Air Force, commander
AVDEYEV, Sergei Vasilyevich, 42, civilian, flight engineer, 3rd mission
Previous missions: Soyuz TM15 (1992); Soyuz TM22 (1996)
BATURIN, Yuri Mikhailovich, 49, Russian Air Force, cosmonaut researcher

Flight Log

Before becoming a cosmonaut, Yuri Baturin had been a space physicist. He then became the national security advisor for President Boris Yeltsin and then the Defence Council Secretary. He was also an advisor on space matters and was attached to the 1997 Air Force cosmonaut selection (with the rank of colonel). In April 1998, he completed a 12-day mission to Mir, returning with the EO-25 crew. After his mission he stated that, in his opinion, the Mir complex should be kept operating for at least two years beyond the planned 1999 decommission date.

When Baturin and the E0-25 crew departed Mir, the two remaining cosmonauts pursued their EO-26 programme. This included an internal EVA inside the forward node to reset electrical connectors inside the Spektr module. Their subsequent 10 November EVA (5 hours 54 minutes) featured the deployment of Japanese and French experiments. They then continued their programme of experiments, many of which had been left on board the station by international visitors, enhancing the return from the investment in those experiments.

Mir was almost forgotten in the wake of the launch of the first element of ISS – the Zarya module – in November. This was followed by the first Shuttle mission to add other elements – the US node Unity – the following month. With ISS in orbit, Russia indicated that it was looking for private funds to keep Mir aloft into the new

The Soyuz TM28 crew included Baturin (left), a former advisor to President Boris Yeltsin, along with Padalka (centre) and Avdeyev

millennium, as governmental support would end in 1999 when its commitment to ISS increased. When news came that the Service Module of the new station (Zvezda) would be delayed into 2000 and with it the capability of supporting a resident crew, the call to maintain Mir operations beyond 1999 intensified. There was also discussion about further commercial ventures for the station, including filming part of a movie aboard Mir with actors making a short visiting mission to film scenes in orbit. With news that a new investor might support further use of Mir, there remained the question of fitting in the two missions already planned as the EO-26 residence drew to a close and Mir funding from the government ended. The two options were to either fly a Russian commander with the French flight engineer to see out the planned programme, or to leave Avdeyev aboard Mir to join the Russian and French crew members, making a three-person crew for the remainder of the government-funded occupation of the station. The subsequent Slovak mission could then be launched with the new crew and return with Padalka on TM28.

In the new year, the cosmonauts continued their science programme as news came in that the life of the station was to be extended three years (to 2002), if sufficient non-budgetary (government) funds could be found. This allowed Energiya to tentatively plan a programme through to 2001. Just weeks later, the news came that the "private investor" had pulled out. The EO-27 crew launched in February 1999 were expecting to be the last crew to man the station.

Milestones

208th manned space flight
87th Russian manned space flight
80th manned Soyuz mission
27th manned Soyuz TM mission
26th Mir resident crew
33rd Russian and 71st flight with EVA operations
Avdeyev celebrates his 43rd birthday (1 Jan) on Mir – the third birthday he has spent in space (previously 1993 and 1996)

STS-95

Int. Designation	1998-064A
Launched	29 October 1998
Launch Site	Pad 39B, Kennedy Space Center, Florida
Landed	7 November 1998
Landing Site	Runway 33, Shuttle Landing Facility, KSC, Florida
Launch Vehicle	OV-103 Discovery/ET-98/SRB BI-096/SSME #1 2048; #2 2043; #3 2045
Duration	8 days 21 hrs 43 min 56 sec
Call sign	Discovery
Objective	SpaceHab module; John Glenn's return to space; SPARTAN free-flyer

Flight Crew

BROWN Jr., Curtis Lee, 46, USAF, commander, 5th mission
Previous missions: STS-47 (1992); STS-66 (1994); STS-77 (1996); STS-85 (1997)
LINDSEY, Steven Wayne, 38, USAF, pilot, 2nd mission
Previous mission: STS-87 (1997)
ROBINSON, Stephen Kern, 43, civilian, mission specialist 1, 2nd mission
Previous mission: STS-85 (1997)
PARAZYNSKI, Scott Edward, 37, civilian, mission specialist 2, 3rd mission
Previous missions: STS-66 (1994); STS-86 (1997)
DUQUE, Pedro Francisco, 35, civilian, mission specialist 3
MUKAI, Chiaki, 46, civilian, payload specialist 1, 2nd mission
Previous mission: STS-65 (1994)
GLENN Jr., John Herschel, 77, US Senator, payload specialist 2, 2nd mission
Previous mission: Mercury 6 (1962)

Flight Log

John Glenn, the first American astronaut to orbit the Earth in 1962, had always wanted to return to space. But he had not expected that it would take over 36 years for him to do so. While pursing a business and political career, Glenn had always maintained a close relationship with NASA and had convinced NASA administrators that a set of medical experiments looking at the effects of microgravity on an older person would be of benefit to the near- and long-term goals of the agency. It would also provide an interesting set of comparison data with that taken during his first flight in 1962. Glenn first approached NASA with the idea in 1996, but it took two years to develop the experiment programme and obtain formal authorisation. The flight was also a public relations triumph for the space agency and a swansong of Glenn's career in public life.

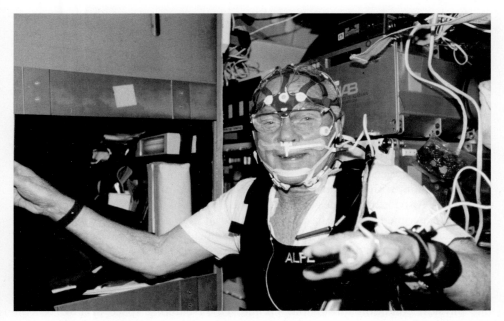

US Senator and former Mercury astronaut John H. Glenn Jr., equipped with sleep-monitoring equipment, floats near his sleep station on the mid-deck of Discovery

The launch progressed smoothly, with only minimal delays caused by a master alarm in the cabin and an aircraft infringement in the restricted airspace around KSC. After the ignition of the three main engines and prior to SRB ignition, the drag chute compartment door fell off, but this never posed a problem for the mission. It was decided that the drag chute would not be deployed during landing rollout.

The primary objective of the flight was a suite of over 80 experiments in the SpaceHab module. These focused on medical and materials research and a series of life sciences investigations that were sponsored by NASA and the space agencies of Canada, Europe and Japan (hence the inclusion of Duque and Mukai on the crew). The latter included cardiovascular studies, sleep studies and blood research. The investigations conducted by and on John Glenn provided useful data that would help to understand the process of aging in humans. The aging process and long space flights have similar common physiological effects and it was felt that information from Glenn and the other astronauts would help not only with long-duration space flight countermeasures, but also to identify early signs of aging and deterioration, which would assist in understanding the process and help in the development of counter-measures for the aging process on Earth. Like all former NASA astronauts, Glenn had been undertaking regular annual physicals at JSC since leaving the programme. These examinations have built up into an impressive database of biomedical studies of space explorers to see what changes, if any, occur as a result of space flight. Glenn's flight at the age of 77, some 36 years after his first flight, was a unique opportunity to expand this data base. His experiments focused on how the absence of gravity affects balance and perception, the immune system response, bone and muscle density, metabolism and blood flow, and sleep.

The flight also included a range of studies on fish and plant specimens and a programme of microgravity materials studies in agriculture, medicine and manufacturing. During the mission, the crew also released a Petite Amateur Naval Satellite (PANSAT), which tested innovative technologies to capture and transmit radio signals normally lost because the original signal was too weak. The Hubble Space Telescope Orbital System Test provided an on-orbit test bed for hardware that would be used during the third Hubble service mission in 1999. The SPARTAN 201 free-flyer was released between FD 4 and 6, carrying a set of re-flight experiments from the 1997 STS-87 mission, with instruments to study the solar corona and gather data on the solar wind. Also located in the payload bay was a Hitchhiker Support Structure (HSS) with six experiments which had solar, terrestrial and astronomical objectives, to obtain data on extreme UV radiation on the sun and atmosphere. A UV spectrograph telescope was included to obtain information on extended plasma sources, such as hot stars and the planet Jupiter.

There were concerns over Glenn's health on both his missions, the first because he was venturing into the unknown and the second over his age. But he had maintained a fine physical condition throughout his life. Upon entering space in 1962, Glenn commented: "Zero-G and I feel fine," and 36 years later he said that he still felt the same way. When he returned to Earth at the end of STS-95, his comment was: "One-G and I feel fine."

Milestones

209th manned space flight
122nd US manned space flight
92nd Shuttle flight
25th flight of Discovery
12th SpaceHab mission (7th single module)
1st flight of SSME Block II engines
Glenn becomes oldest person to fly in space, aged 77

STS-88

Int. Designation	1998-069A
Launched	4 December 1998
Launch Site	Pad 39A, Kennedy Space Center, Florida
Landed	15 December 1998
Landing Site	Runway 15, Shuttle Landing Facility, KSC, Florida
Launch Vehicle	OV-105 Endeavour/ET-097/SRB BI-095; SSME #1 2043; #2 2044; #3 2045
Duration	11 days 19 hrs 17 min 57 sec
Call sign	Endeavour
Objective	ISS assembly flight 2A; mating of Unity docking node to Zarya control module

Flight Crew

CABANA, Robert Donald, 49, USMC, commander, 4th mission
Previous missions: STS-41 (1990); STS-53 (1992); STS-65 (1994)
STURCKOW, Frederick Wilford, 37, USMC, pilot ROSS, Jerry Lynn, 50, USAF, mission specialist 1, 6th mission
Previous missions: STS 61-C (1985); STS-27 (1988); STS-37 (1997); STS-55 (1993); STS-74 (1995)
CURRIE, Nancy Jane, 39, US Army, mission specialist 2, 3rd mission
Previous missions: STS-57 (1993); STS-70 (1995)
NEWMAN, James Hanson, 42, civilian, mission specialist 3, 3rd mission
Previous missions: STS-51 (1993); STS-69 (1995)
KRIKALEV, Sergei Konstantinovich, 40, civilian, Russian, mission specialist 5, 4th mission
Previous missions: Soyuz TM7 (1988); Soyuz TM12 (1991); STS-60 (1994)

Flight Log

This mission initiated the construction of the International Space Station (ISS), a project which had long been proposed but which so often looked as though it would never become reality. In 1984, President Ronald Reagan had challenged NASA to build a space station within a decade. An international team assembled to accomplish the feat, but an over-complicated and expensive design, coupled with the loss of Challenger and doubts over the reliability of the Shuttle had added years to the project. By 1993, the idea was still only on the drawing board and in mock-ups. After several redesigns, a new partnership with Russia helped put the programme back on track. The series of Shuttle–Mir dockings proved that the Shuttle was perfectly capable of doing what it was originally envisioned for back in 1969 – servicing and supplying a space station. A simplified station design helped focus

Shortly after release from Endeavour's cargo bay, the connected Unity and Zarya modules are photographed during a fly-around survey, documenting the completion of a major milestone in the ISS programme with the connection of the first two elements in orbit

the ISS project to the point where the first element of the station was launched on 20 November 1998. This was not an American element, however, but the Russian FGB *Zarya* ("Dawn"), designed to provide electrical power, attitude control and computer command and later serve as a fuel depot and storage facility. The next element of the station would be the link between the US and the Russian elements. Known as Node 1 ("Unity"), it featured six docking ports that would enable the facility to be further expanded.

Unity was the primary payload of STS-88, the first American ISS Shuttle mission, which would use the RMS to attach the module to the forward docking port of Zarya. The launch of STS-88 was postponed by 24 hours on 3 December due to problems with hydraulic system number 1. By the time the problem was cleared, it was too late in the launch process to initiate the final countdown, so the first American element had to wait until the following day to lift off without further incident.

During the approach to Zarya, the crew used their time to prepare Unity by testing the RMS. On 5 December, they attached the end effector to the node, lifting it out of the rear of the payload bay and relocating it in the front of the payload bay along with the Shuttle docking system. This would later allow the crew access through internal hatches from the crew compartment of Endeavour into Unity and on into Zarya. The attachment of Unity to Zarya occurred on 6 December, using the RMS to grasp a grapple feature on Zarya and using the Shuttle's engines to gently nudge the

Unity docking system on to that of Zarya. The embryonic ISS configuration was created. After powering up Unity and checking the integrity of the docking seals and internal atmospheres, the hatches were opened, allowing Cabana and Krikalev to symbolically float into ISS together for the first time.

During the three EVAs (7 Dec for 7 hours 21 minutes; 9 Dec for 7 hours 2 minutes; and 12 Dec for 6 hours 59 minutes), Ross (EV1) and Newman (EV2) removed launch restraint pins on the four hatches on Unity that would be used in future operations, nudged two stuck antennas on Zarya into position, installed sunshades over Unity's data relay boxes, disconnected the umbilicals that were used to mate the units, and installed a handrail, a tool bag and an S-Band communication system. They also tested the SAFER units.

Inside Zarya, Krikalev and Currie replaced a faulty unit, inspected the inside of the module and removed some launch bolts and restraints. The undocking from ISS took place on 13 December. After a fly-around photographic inspection, the crew prepared for landing, having completed one of the most important and critical Shuttle flights. One of the largest international construction projects in history – and certainly the largest off the Earth – had begun.

The STS crew called themselves "Dog Crew 3", since two of them had flown on previous "Dog Crews". Thus, the crew were known as "Mighty Dog" (Cabana), "Devil Dog" (Sturckow), "Hooch" (Ross), "Laika" (Currie), "Pluto" (Newman) and "Spotnik" (Krikalev).

Milestones

210th manned space flight
123rd US manned space flight
93rd Shuttle mission
41st US and 72nd flight with EVA operations
13th flight of Endeavour
1st Shuttle ISS mission
1st Endeavour ISS mission

SOYUZ TM29

Int. Designation	1999-007A
Launched	20 February 1999
Launch Site	Pad 1, Site 5, Baikonur Cosmodrome, Kazakhstan
Landed	28 August 1999
Landing Site	76 km north-northeast of Arkalyk
Launch Vehicle	R7 (11A511U); spacecraft serial number (7K-M) #78
Duration	188 days 20 hrs 16 min 19 sec (Afanasyev/Haigneré); 7 days 21 hrs 56 min 29 sec (Bella – landed with Padalka on TM28)
Call sign	Derbent (Derbent)
Objective	Mir EO-27 programme; French *Perseus* programme; Slovakian *Stefanik* programme

Flight Crew

AFANASYEV, Viktor Mikhailovich, 50, Russian Air Force, 3rd mission
Previous missions: Soyuz TM11 (1990); Soyuz TM18 (1994)
HAIGNERÉ, Jean-Pierre, 50, French Air Force, flight engineer, 2nd mission
Previous mission: Soyuz TM17
BELLA, Ivan, 34, Slovak Armed (Air) Forces, cosmonaut researcher

Flight Log

This was to prove the final in-orbit hand-over of a Mir crew on a station that had been continually manned since September 1989. The short Slovak *Stefanik* scientific mission was reportedly paid for by the Russians writing off a Soviet era debt to Slovakia of US$20 million. France reportedly paid US$20.6 million for the *Perseus* programme, which should have been completed in June but was extended at no extra cost until August.

The Slovak programme encompassed medical experiments, measurements of radiation and observations of the development of quail eggs. Haigneré's *Perseus* programme included the use of equipment brought up in previous French missions as well as four new experiments. The programme focused on life sciences, physics and space technology. Two other experiments were provided by ESA and there were several experiments provided by French high schools working in cooperation with CNES. On 16 April, Haigneré and Afanasyev competed a 6 hour 19 minute EVA in which they were to test a new sealant tool for repairing small holes in the hull. A simulated hole in Kvant was to be used in the test, and the sealant was also to have been used for Spektr, but the hole in the module was never pinpointed. In the test, the valve failed to open and the simulation at Kvant had to be cancelled. The EVA crew did retrieve experiment samples from the exterior of the station, but the deployment of new detectors had to be abandoned as they fell behind schedule.

Slovakia's first cosmonaut Bella (left) was launched aboard Soyuz TM29 with Russian Afanasyev (centre) and Frenchman Haigneré

The three men continued their programmes of biomedical studies, astrophysical and technical experiments and Earth photography, as well as astronomical and solar observations, filling the weeks as they orbited in Mir. In June, Avdeyev surpassed the career record of 681 days accumulated time in space (previously held by Dr. Valery Polyakov). According to some reports, not all the time spent on Mir was harmonious, with Afanasyev not enjoying his third mission to Mir and at times being at odds with Haigneré. Two EVAs by Afanasyev and Avdeyev were completed in July (23 Jul for 6 hours 7 minutes and 27 Jul for 5 hours 22 minutes) to deploy an elliptical 6.4 × 5.2 m reflector antenna that was 1.1 m high. This was a test of a new prototype design for a telecommunications antenna planned for future generations of satellites. It initially refused to deploy and remained furled despite the crew kicking it. During the second EVA, they were able to complete the deployment operation. Over the course of the two EVAs, they also deployed and returned experiments and sample cassettes on the exterior of the station, and during the second EVA they detached the antenna from the Sofora girder, manually pushing it away from the station.

On 25 July, Haigneré spoke over the radio to fellow French astronaut Michel Tognini, who was on Columbia during the STS-93 mission. Towards the end of July and in Early August, the crew's scientific work began to come to an end and for several days the three cosmonauts began winding up their experiments and mothballing the station. Later, they witnessed the effects of the 11 August 1999 total solar eclipse as the shadow passed over southern England, and over the Indian sub-continent one orbit later.

On 27 August, the crew undocked from Mir to complete a landing a few hours later. Afanasyev said that his crew were "abandoning a piece of Russia [with] grief in our souls." According to Russian press releases, there had been over 22,000 scientific experiments in 20 research programmes, utilising over 240 pieces of scientific equipment. A total of 14 tons of scientific hardware had been used on Mir by the 27 main crews and numerous visiting crew members. For now, there did not seem to be any further missions on the horizon, although Mir was kept in autonomous flight while all options were examined. The Russians seemed to have committed themselves to ISS and the end of Mir was approaching.

Milestones

211th manned space flight
88th Russian manned space flight
81st manned Soyuz mission
28th manned Soyuz TM mission
29th Mir resident crew
34th Russian and 73rd flight with EVA operations
6th French long-duration mission (189 days)
Haigneré celebrates his 51st birthday on Mir (19 May)
New duration record of 748 days in space set by Avdeyev over three missions
New duration record of 209 days in space for a non-Russian (Haigneré)

STS-96

Int. Designation	1999-030A
Launched	27 May 1999
Launch Site	Pad 39B, Kennedy Space Center, Florida
Landed	6 June 1999
Landing Site	Runway 15, Shuttle Landing Facility, KSC, Florida
Launch Vehicle	OV-103 Discovery/ET-100/SRB BI-100/SSME #1 2047; #2 0251; #3 2049
Duration	9 days 19 hrs 13 min 57 sec
Call sign	Discovery
Objective	ISS assembly flight 2A.1; logistics mission

Flight Crew

ROMINGER, Kent Vernon, 42, USN, commander, 4th mission
Previous missions: STS-73 (1995); STS-80 (1996); STS-85 (1997)
HUSBAND, Rick Douglas, 41, USAF, pilot
JERNIGAN, Tamara Elizabeth, 40, civilian, mission specialist 1, 5th mission
Previous missions: STS-40 (1991); STS-52 (1992); STS-67 (1995); STS-80 (1996)
OCHOA, Ellen Lauri, 41, civilian, mission specialist 2, 3rd mission
Previous missions: STS-56 (1993); STS-66 (1994)
BARRY, Daniel Thomas, 45, civilian, mission specialist 3, 2nd mission
Previous mission: STS-72 (1996)
PAYETTE, Julie, 35, civilian, Canadian, mission specialist 4
TOKAREV, Valery Ivanovich, 46, Russian Air Force, mission specialist 5

Flight Log

This mission was the first logistics flight to the station in preparation for the arrival of the Russian Service Module *Zvezda* ("Star") scheduled for later in 1999. Due to weight limitations on the previous STS-88 mission, not all the logistics could be taken to the station in one go. STS-96 was originally planned for later in the year, after STS-93 had deployed the Chandra X-ray telescope, but early in 1999 there were problems in the circuitry boards on Chandra which needed to be replaced, forcing the launch to be delayed. In early May, weather damage to the ET intended for STS-96 resulted in further delays for repairs. With the Russian Service Module also being delayed, further ISS Shuttle missions and the arrival of the first resident crew were put back until 2000. This meant there would be a long gap in ISS-related missions between STS-88 and support missions for the first resident crew in 2000. This gap was filled only with the STS-96 logistics mission.

After the STS-96 stack was returned to the VAB for ET tank repairs, during which 460 critical divots out of a total of 650 divots in the ET outer foam were

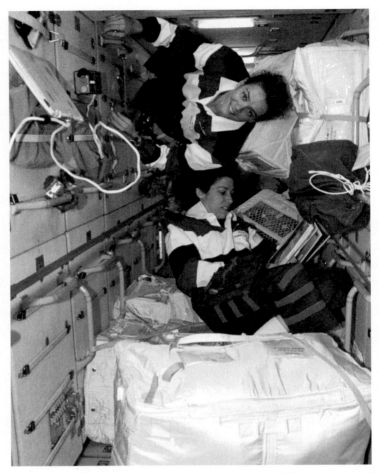

On board the Zarya module, astronauts Julie Payette (top) and Ellen Ochoa handle supplies being moved over from the docked Shuttle Discovery

repaired, the only other concern prior to launch was when a sail-boarder ventured into the SRB recovery zone. Once that was removed, the launch proceeded smoothly. Two days after launch, Discovery completed the first docking with ISS. The Shuttle remained docked to ISS for 138 hours, during which members of the crew spent over 79 hours inside the station and 7 hours 55 minutes hours outside during the mission's only EVA. During the 28 May EVA, Jernigan (EV1) and Barry (EV2) transferred the US-built Orbital Transfer Device crane and elements of the Russian Strela crane from the cargo bay of Discovery to their locations on the exterior of the station. They also installed EVA foot restraints that could accommodate either American or Russian EVA footwear and three bags of tools and handrails for future assembly operations. An insulation cover was placed over a trunnion pin on Unity, they inspected one of two Early Communication Systems (E-Com) antennas on Unity, and finally photo-documented the exterior paint surfaces of both modules.

Upon entering the station, the crew were concerned over the quality of air circulation inside Zarya, but this was solved by changing the orientation of panel doors that were interrupting the flow of air around the station. Eighteen battery recharge controllers were replaced in Zarya and mufflers were installed over fans inside the FGB to reduce noise levels in the module. The crew also transferred over 1,618 kg of logistics across to the ISS, including clothing, sleeping bags, spare parts, medical equipment and 318 litres of water. They also installed the first of a series of strain gauges, which would be important as the station expanded to record the stress on docking interfaces, and cleaned filters and checked smoke detectors. Transferred in the opposite direction was 90 kg of equipment (198 items), which was moved back into Discovery for the return to Earth.

The day before undocking, the RCS on Discovery were pulsed 17 times to boost the station's orbit slightly, pending the arrival of the next Shuttle (which turned out to be a year later). Discovery was undocked from ISS on 4 June and after flying two circuits around the station for photo-documentation, the crew prepared for the return to Earth. One of their last tasks prior to landing was the release of a small reflective satellite, which would be a target for student observations around the world.

Milestones

212th manned space flight
124th US manned space flight
94th Shuttle mission
28th flight of Discovery
2nd Shuttle ISS mission
1st Discovery ISS mission
1st Shuttle mission to dock with ISS

STS-93

Int. Designation	1999-040A
Launched	23 July 1999
Launch Site	Pad 39B, Kennedy Space Center, Florida
Landed	27 July 1999
Landing Site	Runway 33, Shuttle Landing Facility, KSC, Florida
Launch Vehicle	OV-102 Columbia/ET-099/SRB BI-097/SSME #1 2012; #2 2031; #3 2019
Duration	4 days 02 hrs 49 min 37 sec
Call sign	Columbia
Objective	Deployment of Chandra X-Ray Observatory by IUS-27

Flight Crew

COLLINS, Eileen Marie, 42, USAF, commander, 3rd mission
Previous missions: STS-63 (1995); STS-84 (1997)
ASHBY, Jeffrey Shears, 45, USN, pilot
COLEMAN, Catherine Grace ("Cady"), 38, USAF, mission specialist 1, 2nd mission
Previous mission: STS-73 (1995)
HAWLEY, Steven Alan, 47, civilian, mission specialist 2, 5th mission
Previous missions: STS 41-D (1984); STS 61-C (1986); STS-31 (1990); STS-87 (1997)
TOGNINI, Michel Ange Charles, 49, French Air Force, mission specialist 3, 2nd mission
Previous mission: Soyuz TM15 (1992)

Flight Log

If the launch of STS-93 had occurred on time on 20 July, Eileen Collins, the first female commander of a US space mission, could have taken Columbia to orbit on the 30th anniversary of the Apollo 11 lunar landing (whose Command Module was also called Columbia). However, the launch was terminated at the $T-7$ second mark when more than double the permitted amount of hydrogen was detected in the aft engine compartment of the orbiter. System engineers in the firing room at KSC noted the indication and manually cut off the ground launch sequencers less than a second before SSME ignition. Post-abort evaluation determined that the reading was false. The next launch attempt, on 22 July, was scrubbed due to adverse weather conditions at KSC, but the launch attempt on 23 July was successful, the only delay being a communications problem with Columbia during the countdown which forced a seven minute slip in the launch time.

Eileen Collins, the first female Shuttle commander and first female commander of an American mission, looks over a checklist at the commander's station on the forward flight deck of Columbia during FD 1

Five seconds after leaving the pad, flight controllers noted a voltage drop in one of the electrical buses on the Columbia. As a result of the drop in voltage, one of two redundant main engine controllers on two of the three SSME (centre and right position) shut down. But the others performed nominally, supporting the climb to orbit. However, the orbit attained was 11.2 km lower than planned due to the premature cut-off of the SSME. This was later traced to a hydrogen leak in the #3 main engine nozzle, caused by the loss of an LO pin from the main injector during engine ignition. This had struck the hot wall of the nozzle and ruptured three LH coolant tubes.

Columbia's manoeuvring engines were used subsequently to raise the orbit to its proper altitude, allowing the deployment of the primary payload into its desired orbit. The Chandra X-Ray Observatory (formerly known as the Advanced X-Ray Astrophysical Facility, or AXAF) was successfully deployed using its two-stage IUS on FD 1. The IUS propelled the observatory into an operational orbit of approximately 10,000 × 140,000 km – at its farthest, almost one-third of the way to the Moon – in an orbital period of about 64 hours. This would permit the telescope to make 55 hours of uninterrupted observations each orbit. The primary mission of Chandra was scheduled to last five years through to 2004, although this was subsequently extended to ten years of operational activity until 2009.

During the remainder of the mission, secondary payloads and experiments were activated. These included the South-Western UV Imaging System (SWUIS) used to obtain UV imagery of Earth, the Moon, Mercury, Venus and Jupiter. The crew monitored several plant growth experiments and collected data from a biological cell culture experiment. They also evaluated the Treadmill Vibration Information System, which measured vibrations and the changes in microgravity levels caused by on-orbit exercise periods. This was important for gathering data to ensure that exercise periods on ISS did not disrupt delicate instruments and experiments. The crew also evaluated high-definition TV equipment for future use on both the Shuttle and ISS, which conformed to the latest industry standards for TV products. Tognini, who visited the Mir space station in 1992, spoke over the radio with his colleague and fellow countryman Jean-Pierre Haigneré, who was on the fifth of his six-month stay on the Russian Mir space station. Collins and Ashby also evaluated the Portable In-flight Landing Operations Trainer (PILOT), which utilised a laptop computer, simulation software and a joystick combination to provide refresher and skills training to the commander and pilot prior to performing the actual landing.

Milestones

213th manned space flight
125th US manned space flight
95th Shuttle mission
26th flight of Columbia
1st female Shuttle commander and 1st US female crew commander (Collins)
Shortest scheduled flight since 1990

STS-103

Int. Designation	1999-069A
Launched	19 December 1999
Launch Site	Pad 39B, Kennedy Space Center, Florida
Landed	27 December 1999
Landing Site	Runway 33, Shuttle Landing Facility, KSC, Florida
Launch Vehicle	OV-103 Discovery/ET-101/SRB BI-099/SSME #1 2053; #2 2043; #3 2049
Duration	7 days 23 hrs 10 min 47 sec
Call sign	Discovery
Objective	3rd Hubble Servicing Mission (HST-SM 3A)

Flight Crew

BROWN Jr., Curtis Lee, 43, USAF, commander, 6th mission
Previous missions: STS-47 (1992); STS-66 (1994); STS-77 (1996); STS-85 (1997); STS-95 (1998)
KELLY, Scott Joseph, 35, USN, pilot
SMITH, Steven Lee, 40, civilian, mission specialist 1, 3rd mission
Previous missions: STS-68 (1994); STS-82 (1997)
CLERVOY, Jean-François André, 41, civilian, ESA mission specialist 2, 3rd mission
Previous missions: STS-66 (1994); STS-84 (1997)
GRUNSFELD, John Mace, 41, civilian, mission specialist 3, 3rd mission
Previous missions: STS-67 (1995); STS-81 (1997)
FOALE, Colin Michael, 42, civilian, mission specialist 4, 5th mission
Previous missions: STS-45 (1992); STS-56 (1993); STS-63 (1995); STS-84/86 (1997)
NICOLLIER, Claude, 55, civilian, ESA mission specialist 5, 4th mission
Previous missions: STS-46 (1992); STS-61 (1993); STS-75 (1996)

Flight Log

This mission faced nine scrubs or delays due to mechanical issues or the weather, before finally reaching orbit. The mission had been scheduled for June 2000, but when the third of six gyroscopes on Hubble had failed, the service mission was divided into two separate Shuttle missions and the first was advanced. STS-103 was now due to fly in October 1999 and the second mission would follow in 2001. In mid-August, Shuttle management decided to inspect the wiring of the Shuttle fleet after the incidents during the STS-93 launch in July. As a result, STS-103 had a new launch date of 23 October, but the amount of work required to complete the repairs saw the launch put back further to 19 November. This gave NASA the option of launching either STS-103 or

Astronauts Mike Foale (left) and Claude Nicollier (in the RMS) install a Fine Guidance Sensor into a protective enclosure in the payload bay during the second EVA

the Space Radar Topography Mission (STS-99) first. On 13 November, Hubble was placed in safe mode when a fourth gyroscope failed. It was left pointing its arrays constantly at the Sun to generate electrical power, pending the service mission.

During the first weeks of December, seven new launch dates were set before the vehicle finally left the pad. The causes for the delays varied from the discovery of a 1.5-cm-long drill bit lodged in main engine #3 (the engine was replaced on the pad), to additional wiring damage in an umbilical between the Orbiter and ET, to the Thanksgiving holiday. There were also repairs to a dented LH main propulsion line, the inspection and verification of a number of welds in pressure lines, and problems with the weather. With the launch planned for 19 December, mission managers had decided to restrict the mission duration to eight days instead of the planned ten, to ensure that all flight and ground systems were secured for transition to year 2000. Shuttle computers are unable to operate over the change of year, and with the change to year 2000 expected to highlight additional glitches, NASA wanted to ensure it did not have a vehicle flying or linked to active ground systems at this time.

The first few days in the orbiter consisted of adjustments to the orbit and preparations for the work ahead. It took 30 orbits to reach the Hubble, which was captured by RMS on 21 December. Three EVAs were completed by the crew. Steve Smith (EV1) and John Grunsfeld (EV2) completed the first and third excursions, while Mike Foale (EV3) and ESA Astronaut Claude Nicollier (EV4) performed the mission's second EVA. During EVA 1 (22 Dec, 8 hours 16 minutes), the astronauts replaced the three Rate Sensor Units which each contained two of the gyroscopes, and installed six cell-phone-sized Voltage/Temperature Improvement Kits between the telescope's six ten-year-old batteries and its solar arrays. These would prevent overheating and overcharging of the batteries. EVA 2 (23 Dec, 8 hours 10 minutes) saw the astronauts install a new computer in the telescope which was 20 times faster than its older unit. They also installed a new fine-guidance sensor. The final EVA (24 Dec, 8 hours 8 minutes) included the installation of a transmitter to send scientific information from the telescope to the ground, replacing the one that had failed the year before. This was a delicate operation, as the transmitter was not part of the telescope and was not designed to be replaced by the astronauts. However, using specially designed tools, they achieved the task, demonstrating the value of utilising humans to effect repairs and servicing on units that otherwise could not be replaced or repaired. The final EVA also saw the installation of a solid state digital recorder, to replace one of the older mechanical reel-to-reel recorders.

Hubble was released back into orbit on Christmas Day 1999. This was only the third time that an American crew had been in space at Christmas aboard an American spacecraft. The first was the historic Apollo 8 mission around the Moon in 1968 and the second was during the third and final Skylab (SL-4) mission in 1973. In addition, astronauts John Blaha and Dave Wolf had spent Christmas and New Year aboard Mir with Russian colleagues in 1996 and 1997 respectively.

Milestones

214th manned space flight
126th US manned space flight
6th Shuttle mission
27th flight of Discovery
42nd US and 74th flight with EVA operations
3rd HST service mission
1st ESA astronaut to perform EVA from Shuttle (Nicollier)

STS-99

Int. Designation	2000-010A
Launched	11 February 2000
Launch Site	Pad 39A, Kennedy Space Center, Florida
Landed	22 February 2000
Landing Site	Runway 33, Shuttle Landing Facility, KSC, Florida
Launch Vehicle	OV-105 Endeavour/ET-92/SRB BI-100/SSME #1 2052; #2 2044; #3 2047
Duration	11 days 5 hrs 39 min 41 sec
Call sign	Endeavour
Objective	Acquisition of high-resolution topographical map of Earth's land masses (between 60°N and 56°S) by radar

Flight Crew

KREGEL, Kevin Richard, 43, civilian, commander, 4th mission
Previous missions: STS-70 (1995); STS-78 (1996); STS-87 (1997)
GORIE, Dominic Lee Pudwell, 42, USN, pilot, 2nd mission
Previous mission: STS-91 (1998)
THIELE, Gerard Paul Julius, 46, civilian, ESA mission specialist 1
KAVANDI, Janet Lynn, 40, civilian, mission specialist 2, 2nd mission
Previous mission: STS-91 (1998)
VOSS, Janice Elaine, 43, civilian, mission specialist 3, payload commander, 5th mission
Previous missions: STS-57 (1993); STS-63 (1995); STS-83 (1997); STS-94 (1997)
MOHRI, Mamoru Mark, 52, civilian, Japanese, mission specialist 4, 2nd mission
Previous mission: STS-47 (1992)

Flight Log

The Shuttle Radar Topography Mission (SRTM) used modified versions of the radar instruments that had flown on the two SRL Shuttle missions in 1994. In addition to providing the topographical radar images of Earth, the mission also tested new technologies for the deployment of large, ridged structures in space, and recorded measurements of their distortion to an extremely high precision. Space-borne imaging radar from the Shuttle had previously been flown on STS-2 (SIR-A) in 1981 and STS 41-G (SIR-B) in 1984, as well as a German experiment on STS-9 (Spacelab 1) in 1983 and the two SRL Shuttle missions (SIR-C) in 1994.

The launch date of STS-99 was originally set for 16 September 1999, but was postponed until October due to the Shuttle fleet wiring concerns and the subsequent remedial action. With so much work to do on the wiring issue, it was decided to launch

Part of the Shuttle Radar Topography Mission hardware is photographed through Endeavour's aft flight deck windows about half-way through the scheduled 11-day SRTM flight. The mast, only partially visible in the centre, is actually 61 metres in length

STS-99 no earlier than 19 November, and for a while, either the radar mission or the Hubble service mission (STS-103) could have flown first. In October, it was decided to fly the Hubble mission before STS-99. The SRTM launch therefore slipped to 13 January 2000 and then, after a review, to 31 January. That attempt was scrubbed at the $T - 9$ minute mark due to adverse weather. Then the launch moved from 9 to 11 February in order to work on some minor technical issues, three of which had to be addressed during the planned $T - 9$ minute hold on the day of launch delaying lift-off by about 14 minutes.

Once in orbit, the crew configured the vehicle for its orbital science mission. This included extending the SRTM to its full mast length of 61 metres from the payload bay over Endeavour's left wing. After checking out the orbiter and payload, the mapping began some 12 hours into the mission. The crew worked in two 12-hour shifts, with Kregel, Thiele and Kavandi as Red Shift and Gorie, Voss and Mohri as Blue Shift, and the mission was flown using an attitude hold period for radar mapping and flying the orbiter in a tail-first configuration. One pair of radar antennas were in the payload bay, with the other pair at the end of the boom, providing stereo images of the ground the vehicle flew over in C-band and X-band wavelengths, recording data in two wavelengths from two locations simultaneously. This would also provide 3D maps after the mission that were thirty times more accurate than any previous attempts.

Tests were also made on gas jets located at the end of the boom to absorb the firing of the orbiter's thrusters. Alignment of the radar sensors was vital for accurate data, but it was essential to reduce the strain on the mast when the vehicle was moved. By firing a brief RCS burn, the mast deflected slightly backwards and then rebounded forward. Once returned to vertical, a stronger RCS thrust was applied, arresting the mast's motion but increasing the orbital speed of the vehicle. It was noted on FD 2 that orbiter propellant usage had been higher than expected due to the failure of a cold gas thruster system on the end of the mast to offset the gravity gradient torque. This meant that more propellant was being used to maintain the attitude of the vehicle for data sweeps. Measurements were taken to reduce fuel expenditure and it was determined that enough propellant could be saved to complete the full mission as planned.

At the end of data gathering on FD 10, a total of 222 hours and 23 minutes of mapping had been achieved, covering 99.98 per cent of the planned mapping area once and 94.6 per cent of it twice. There remained only 207,000 km^2 (80,000 miles^2) in scattered areas uncovered, but most of this was in North America, which had previously been well-mapped. Over 123.2 million km^2 (47.6 million miles^2) had been mapped, with enough data on 32 high-density tapes aboard Endeavour to fill 20,000 CDs, or the entire book content of the US Library of Congress. It was estimated that it would take over two years to fully process the data.

Also aboard Endeavour was a student experiment called EarthKam, which took 2,715 digital photos during the mission through an overhead flight deck window. Students from 84 participating middle schools around the world could select photo targets and receive images via the Internet, supporting their class work in Earth science, geography, maths and space sciences. The landing was achieved on the second opportunity at KSC, with the first attempt having been waived off due to high crosswinds at the SLF.

Milestones

215th manned space flight
127th US manned space flight
97th Shuttle mission
14th flight of Endeavour
5th Shuttle mission featuring imaging radar

SOYUZ TM30

Int. Designation	2000-018A
Launched	4 April 2000
Launch Site	Pad 1, Site 5, Baikonur Cosmodrome, Kazakhstan
Landed	16 June 2000
Landing Site	40 km from Arkalyk
Launch Vehicle	R7 (11A511U); spacecraft serial number (7K-M) 204
Duration	72 days 19 hrs 42 min 16 sec
Call sign	Yenisey (Yenisey)
Objective	Mir EO-28 programme; evaluation of Mir complex for potential further operational (commercial) use

Flight Crew

ZALETIN, Sergei Viktorovich, 37, Russian Air Force, commander
KALERI, Aleksandr Yuriyevich, civilian, flight engineer, 3rd mission
Previous missions: Soyuz 14 (1992); Soyuz TM24 (1996)

Flight Log

After months of uncertainty, the funding for a new expedition to the Mir complex, the 29th, came from a new private corporation based in Amsterdam in the Netherlands. MirCorp was 60 per cent owned by RKK Energiya and shared with the venture capital group Gold and Appel and other entrepreneurs. The crew assigned to return to Mir were authorised by the Russian government to complete their mission, which included two new Progress cargo craft and a programme that could have extended into August 2000, a 70- to 90-day mission. MirCorp had also indicated that they had received authorisation from Energiya to lease Mir for space tourism, in-orbit advertising, industrial production and science, turning the station into a commercial space facility. NASA, quite understandably, was not happy about this sequence of events.

Soyuz TM30 launched from Baikonur on 4 April and docked with the station two days later. After a check of onboard systems and hatch integrity, the two cosmonauts opened the hatches and floated into the station, occupying it for the first time in eight months. For most of the first two weeks, the cosmonauts were occupied with housekeeping and bringing the station back into operation. One of their primary objectives was to establish whether the aging station could support the plans that MirCorp was promoting. An EVA was completed on 12 May and lasted for 4 hours 52 minutes. This was to be the final excursion from the Mir complex. The crew performed a test of a sealing compound on the hull of the station, an operation that had been delayed from previous missions. They also inspected and detailed the condition of the station, dismantled experiments and retrieved sample cassettes.

The final Mir resident crew of Zaletin (left) and Kaleri was launched to the Station aboard Soyuz TM30

During their stay aboard Mir, the EO-28 crew was able to perform research in biomedicine, space sciences, materials sciences, Earth sciences and technology. While discussions went on about what mission would follow, the cosmonauts began to pack up their experiment results and prepare the station for a period of inactivity, as for only the fourth time in its history, it would be left without a crew. MirCorp had planned a period of inactivity after this flight to secure new funds and to plan future operations based on the reports and feedback from this recent mission. In the early hours of 16 June, the two cosmonauts undocked from Mir and prepared for their return journey, leaving Mir alone in orbit. The crew reported that the station was in good shape and that there seemed to be no reason why it could not be revisited again. Of course, the subject of raising the funds to support these revisits was not mentioned.

Delays in securing the funding resulted in launch schedule slippage. The first space tourist was announced as American Dennis Tito, who scheduled to visit the station in 2000. This slipped to 2001 and funding from MirCorp was not forthcoming. Energiya now looked to de-orbit the station in the spring of 2001 and a new Progress was launched to direct the station to a fiery re-entry. On 15 February 2001, the Mir base block celebrated its fifteenth year in space. To ensure it fell on unpopulated areas, the Progress engines were fired for 20 minutes on 23 March and again twice more on succeeding orbits. The station entered the atmosphere and burned up, with some of the larger items of debris falling into the Pacific some 3,000 km east of New Zealand. The final communiqué from the Russians was that "Mir ceased to exit." It was time to look towards ISS, but in the ten years between the arrival of the Soyuz TM8 crew on 7 September 1989 and the departure of the Soyuz TM29 crew on 27 August 1999, Mir had been occupied continuously for 3,640 days 22 hours 52 minutes, a record the ISS will not break until October 2010.

Milestones

216th manned space flight
89th Russian manned space flight
82nd manned Soyuz mission
29th manned Soyuz TM mission
28th and final Mir resident crew
35th Russian and 75th flight with EVA operations
1st mission funded under MirCorp

STS-101

Int. Designation	2000-027A
Launched	19 May 2000
Launch Site	Pad 39A, Kennedy Space Center, Florida
Landed	29 May 2000
Landing Site	Runway 15, Shuttle Landing Facility, KSC, Florida
Launch Vehicle	OV-104 Atlantis/ET-102/SRB BI-102/SSME #1 2043; #2 2054; #3 2049
Duration	9 days 21 hrs 10 min 10 sec
Call sign	Atlantis
Objective	ISS Assembly flight 2A.2a; second ISS logistics mission

Flight Crew

HALSELL Jr., James Donald, 43, USAF, commander, 5th mission
Previous missions: STS-65 (1994); STS-74 (1995); STS-83 (1997); STS-94 (1997)
HOROWITZ, Scott Jay, 43, USAF, pilot, 3rd mission
Previous missions: STS-75 (1996); STS-82 (1997)
WEBER, Mary Ellen, 37, civilian, mission specialist 1, 2nd mission
Previous mission: STS-70 (1995)
WILLIAMS, Jeffery Nels, 42, USA, mission specialist 2
VOSS, James Shelton, 51, US Army, mission specialist 3, 3rd mission
Previous missions: STS-44 (1991); STS-53 (1992); STS-69 (1995)
HELMS, Susan Jane, 42, USAF, mission specialist 4, 4th mission
Previous missions: STS-54 (1993); STS-64 (1994); STS-78 (1996)
USACHEV, Yuri Vladimirovich, 42, civilian, Russian, mission specialist 5, 3rd mission
Previous missions: Soyuz TM18 (1994); Soyuz TM23 (1996)

Flight Log

The delays in preparing the Russian Service Module meant that it would not be launched until July 2000. This gave rise to maintenance concerns with Unity/Zarya, which by now had been in orbit for over 12 months. Specifically, there was concern about the shelf life and reliability of Zarya's batteries, which had not been designed to provide power for so long without the addition of the Service Module's power systems. A crew had been assigned to STS-101 (2A.2) and the mission was due to follow the launch and docking of the Service Module prior to arrival of the first resident crew. But with the delay, it was decided to split the STS-101 mission and original crew between two flights. The revised STS-101 mission (designated 2A.2a) would fly first in April 2000, and then after the docking of the Russian Service Module

Jeffery Williams hangs on to one of the newly installed handrails on the ISS PMA-2 during the 21 May EVA with James Voss

the new mission, STS-106 (2A.2b), would complete the preparations for receiving the first resident crew.

The original crew for STA-101 was Halsell, Horowitz, Weber and Williams, plus Ed Lu and Russian cosmonauts Yuri Malenchenko and Boris Morukov. The two cosmonauts had trained extensively on SM systems and activation, including unloading the first Progress re-supply craft planned to dock to the SM shortly after it became part of the ISS. Lu and Malenchenko had also trained for an EVA together, so it seemed sensible to utilise this training to reassign them to the STS-106 mission after the launch of the SM and Progress. It was also decided that in order to further prepare the station by completing get-ahead tasks for the first resident crew, and to give the second resident crew experience aboard the station, the latter would fly to ISS aboard STS-101 as part of the Shuttle crew. This would also be of benefit in smoothing the hand-over operations between the first and second crew the following year. Therefore, Usachev, Voss and Helms found themselves visiting ISS a year earlier than planned.

The launch of STS-101 did not occur in April, thanks to three launch delays caused by high winds at KSC and the overseas emergency landing sites. The mission was designated a "home improvement house call" and after its eventual launch on 19 May, STS-101 docked with the station the following day. On 21 May, Williams (EV1) and Voss (EV2) performed the mission's only EVA, of 6 hours 44 minutes, during which they secured the US-built crane and installed the final Strela units on PMA-1. After replacing a faulty communications antenna, they also installed handrails and a camera cable.

STS-101 remained docked to ISS for over 138 hours, with the hatches open for a total of 80 hours. During this time, the crew installed four new batteries and associated electronics in Zarya, as well as 10 new smoke detectors, four cooling fans, additional computer cables and a new communications memory unit and power distribution box for the US-built communications system. They also transferred over almost 1,500 kg of logistics into the Unity and Zarya modules, including clothing, tools, can openers, sewing kits, trash bags, a treadmill and exercise bicycle ergometer, an IMAX film camera and four 45-litre water containers for use by the resident crews. A further 590 kg of logistics was transferred the other way, back to Atlantis for return to Earth.

Prior to undocking, the orbit of ISS was boosted by a further 45 km by firing the onboard engines of Atlantis in three sessions. The new orbit was approximately 383 × 370 km. Once again, a fly-around of the station was performed after the orbiter undocked in order to photo-document the condition of the station.

Milestones

217th manned space flight
128th US manned space flight
Shuttle mission
21st flight of Atlantis
3rd Shuttle ISS mission
1st Atlantis ISS mission

STS-106

Int. Designation	2000-053A
Launched	8 September 2000
Launch Site	Pad 39B, Kennedy Space Center, Florida
Landed	20 September 2000
Landing Site	Runway 15, Shuttle Landing Facility, KSC, Florida,
Launch Vehicle	OV-104 Atlantis/ET-103/SRB BI-102/SSME #1 2052; #2 2044; #3 2047
Duration	11 days 19 hrs 12 min 15 sec
Call sign	Atlantis
Objective	ISS assembly flight 2A.2b

Flight Crew

WILCUTT, Terence Wade, 50, USMC, commander, 4th mission
Previous missions: STS-68 (1994); STS-79 (1996); STS-89 (1998)
ALTMAN, Scott Douglas, 41, USN, pilot, 2nd mission
Previous mission: STS-90 (1998)
LU, Edward Tsang, 37, civilian, mission specialist 1, 2nd mission
Previous mission: STS-84 (1997)
MASTRACCHIO, Richard Alan, 40, civilian, mission specialist 2
BURBANK, Daniel Christopher, 39, USCG, mission specialist 3
MALENCHENKO, Yuri Ivanovich, 38, Russian Air Force, mission specialist 4, 2nd mission
Previous mission: Soyuz TM19 (1994)
MORUKOV, Boris Vladimirovich, 49, civilian, Russian, mission specialist 5

Flight Log

The Russian Service Module Zvezda docked to the aft port of Zarya on 26 July 2000, and was followed by Progress M1-3 at the aft port of Zvezda on 8 August. Zvezda was critical to the early occupation of ISS because it provided flight control and orbit maintenance functions. Zvezda also included the crew quarters for the early resident crews, and EVA facilities prior to the arrival of the airlock modules. With the arrival of Zvezda, a resident crew could remain on the station without the Shuttle being docked to it.

STS-106 docked to Unity on 10 September and remained there for 189 hours, during which the hatches were opened for over 129 hours. On the mission's only EVA on 10 September (6 hours 14 minutes), Lu (EV1) and Malenchenko (EV2) connected nine power and data communication cables between Zvezda and Zarya as well as installing the station's compass, a 1.82-metre magnetometer which showed the station in respect to the Earth. They also ventured farther than any tethered crew member had

The ISS configuration as of September 2000, photographed by the departing STS-106 during a fly-around manoeuvre. From left, US Unity Node, Zarya Control Module, Zvezda Service Module, Progress M1-3

during a Shuttle EVA, over 30.4 metres above the cargo bay along the side of Zvezda and Zarya.

Work inside the station focused on reconfiguring Zvezda for operational use by removing launch bolts and restraints and installing voltage and current stabilisers inside the module. To save weight at launch, only five of the eight batteries had been installed, and the STS-106 crew installed the other three. They also installed components of the Elektron system designed to separate water into oxygen and hydrogen. The three tons of logistics transfers and numerous maintenance tasks took up most of the crew's time during the docked phase. Managers monitoring onboard consumables were able to approve an extra day of docked operations to help ease the burden. The items transferred included six water containers, all of the food stores for the first resident crew, office supplies, onboard environmental supplies, a vacuum cleaner and a computer with monitor.

Atlantis undocked, after a further re-boost to the station's orbit, on FD 11. This was followed by a fly-around of the station before commencing the preparations for the flight home. This undocking and fly-around manoeuvre, like those during the undocking of the Shuttle from Mir, is normally performed by the Shuttle pilot, giving them experience in flying the orbiter in preparation for a future rendezvous and docking mission as commander.

Milestones

218th manned space flight
129th US manned space flight
99th Shuttle mission
22nd flight of Atlantis
43rd US and 76th flight with EVA operations
3rd Shuttle ISS mission
2nd Atlantis ISS mission

STS-92

Int. Designation	2000-062A
Launched	11 October 2000
Launch Site	Pad 39A, Kennedy Space Center, Florida
Landed	24 October 2000
Landing Site	Runway 22, Edwards AFB, California
Launch Vehicle	OV-103 Discovery/ET-104/SRB BI-104/SSME #1 2045; #2 2053; #3 2048
Duration	12 days 21 hrs 43 min 47 sec
Call sign	Discovery
Objective	ISS assembly mission 3A; Zenith Truss (Z1) PMA-3

Flight Crew

DUFFY, Brian, 47, USAF, commander, 4th mission
Previous missions: STS-45 (1992); STS-57 (1993); STS-72 (1996)
MELROY, Pamela Ann, 39, USAF, pilot
CHIAO, Leroy, 40, civilian, mission specialist 1, 3rd mission
Previous mission: STS-65 (1994); STS-72 (1996)
McARTHUR Jr., William Surles, 49, US Army, mission specialist 2, 3rd mission
Previous missions: STS-58 (1993); STS-74 (1995)
WISOFF, Peter Jeffrey Karl, 42, civilian, mission specialist 3, 4th mission
Previous missions: STS-57 (1993); STS-68 (1994); STS-81 (1997)
LOPEZ-ALEGRIA, Michael Eladio, 42, USN, mission specialist 4, 2nd mission
Previous mission: STS-73 (1995)
WAKATA, Koichi, 37, civilian, Japanese, mission specialist 5, 2nd mission
Previous mission: STS-72 (1996)

Flight Log

The original launch date for this mission (5 October) was rescheduled to 9 October when film reviews of the STS-106 launch revealed that the right-hand ET-to-orbiter attach bolt had failed to retract correctly. While this problem was being resolved, an orbiter LO pogo accumulator re-circulation valve located in the MPS failed to respond correctly and required replacement. The second launch attempt was postponed due to high winds at the pad area preventing the safe fuelling of the ET. The following day, a ground support equipment pin and tether, used on access platforms, was observed on the ET-to-orbiter LO feed line. Because of the risk of potential damage during launch, a further 24-hour delay was called.

Discovery docked with ISS on 13 October and remained there for the next 165 hours. However, such was the EVA demand on this crew that only 27 hours

As Discovery separates from ISS, a crew member records this view with the new additions visible. At the top, most of the Z1 Truss is visible, while in the centre is the PMA-2 on the Unity Node, and beneath is the newly installed PMA-3 also on Unity. The solar arrays are on the Russian segment

was spent with internal hatches into ISS open. With the docking achieved, the crew used the RMS to lift the Zenith (Z1) Truss from the payload bay and onto the uppermost (zenith) docking port of Unity. Once completed, the crew confirmed the integrity of the seals and then opened the internal upper hatch to secure grounding connections between the Truss and the station. With that task completed, the EVA programme could begin.

The EVAs were completed by two pairs of astronauts. Chiao (EV1) and McArthur (EV2) performed the first and third excursions (15 Oct for 6 hours 28 minutes, and 17 Oct for 6 hours 48 minutes), while Wisoff (EV3) and Lopez-Alegria (EV4) completed the second and fourth (16 Oct for 7 hours 7 minutes, and 18 Oct for 6 hours 56 minutes). Both teams supported each other's EVAs from inside the orbiter. During the EVAs, the crews connected electrical umbilicals for power to heaters and electrical conduits in the Z1 Truss, relocated two communication antennas and installed a tool box for use during future on-orbit construction activities. On the second EVA, the PMA-3 was installed on Unity and the Z1 Truss was prepared for the future attachment of solar arrays, beginning with the flight of STS-97. The astronauts also installed two DC-to-DC-converters on top of the Z1 Truss which converted electricity generated by the solar arrays to the correct voltage. They tested a manual berthing mechanism, deployed a tray that would provide power for the US Laboratory Module (scheduled for delivery on STS-98) and remove a grapple feature from Z1. They also performed further tests of the SAFER units.

Following the completion of the EVAs, the crew began work inside the station, continuing the transfer of supplies and logistics for the first resident crew, who were scheduled to be the next docking mission at the station. The STS-92 crew also successfully tested the four control moment gyros used to orientate the station as it orbits the Earth. Microbial samples were taken from surfaces inside the station to check for contamination and they cleaned surfaces and storage containers with fungicidal wipes to inhibit microbial growth.

The original landing attempts on 22 October were waived off due to excessive crosswinds at the SLF. The winds remained high for the aborted 23 October landing attempt at the Cape and rain within the 50 km limit at Edwards meant the crew had to spend another day in space. With excessive winds still preventing any landing at the Cape, the rain at Edwards held off to allow the Shuttle to land there instead.

Milestones

219th manned space flight
130th US manned space flight
100th Shuttle mission
28th flight of Discovery
44th US and 77th flight with EVA operations
5th Shuttle ISS mission
2nd Discovery ISS mission

SOYUZ TM31

Int. Designation	2000-017A
Launched	31 October 2000
Launch Site	Pad 1, Site 5, Baikonur Cosmodrome, Kazakhstan
Landed	21 March 2001 (via STS-102)
Landing Site	Runway 15, Shuttle Landing Facility, KSC, Florida
Launch Vehicle	R7 (11A511U); spacecraft serial number (7K-M) 205
Duration	140 days 23 hrs 38 min 55 sec
Call sign	Alpha (ISS)/Uran (Uranus) (Soyuz TM31)
Objective	1st ISS resident crew (3R); 1st ISS Soyuz launch (1S); worked with STS-97 and STS-98 crews

Flight Crew

GIDZENKO, Yuri Pavlovich, 38, Russian Air Force, Soyuz commander, 2nd mission
Previous mission: Soyuz TM22 (1995)
KRIKALEV, Sergei Konstaninovich, 42, civilian, flight engineer, 5th mission
Previous missions: Soyuz TM7 (1988); Soyuz TM12 (1991); STS-60 (1994); STS-88 (1998)
SHEPHERD, William McMichael, 51, Captain USN, NASA ISS commander, 4th mission
Previous missions: STS-27 (1988); STS-41 (1990); STS-52 (1993)

Flight Log

This was the pioneering mission that began the permanent occupation of the International Space Station, the first of a rotational system of crews planned to work aboard ISS in shifts of three to four months initially, increasing to six months. The inclusion of the Russians in the ISS project brought experience and hardware to the programme, enabling the station to be launched and manned earlier than by using the original elements envisaged under the Space Station Freedom programme. By using the Zarya and Zvezda modules and support from Progress re-supply vehicles, a crew could now remain on board what was essentially Mir-2 before its expansion with US and other international elements. The other advantage was the decision to use the Soyuz transport craft as a crew ferry and on-orbit rescue vehicle.

The best place to train for space flight is in space itself, and by that definition, the most logical place to make sure space equipment works as designed is in orbit. Therefore, the objectives of the first crew were to make the small ISS habitable and to discover what worked and what did not. Once the safety and security of their residency was confirmed, the first crew arrived aboard Soyuz TM31 on 2 November 2000 after a two-day flight from Baikonur. Bill Shepherd, the first ISS commander,

The first ISS resident crew enjoying the wonders of microgravity and fresh fruit during their busy residency

was only the second American astronaut to ride into space on an R7 and though the Soyuz remained docked to ISS for the majority of their stay, the crew would eventually come home aboard the Shuttle. The Soyuz would only be used for an emergency return, if necessary.

Most of their work involved setting up the station and evaluating procedures and systems before the real science work commenced with the addition of the US Destiny lab and several logistics flights in 2001. This crew hosted two Shuttle visiting missions before receiving the crew that would bring them home. The first, STS-97, began the assembly of the station's solar array systems while the second, STS-98, brought the US Destiny lab. Though engineering and systems activation was a priority of this residency, science was not totally neglected. A joint US/Russian science programme included Earth observations, technology and protein crystal growth experiments, and radiation studies. Due to the extended duration of this mission, medical studies were also included. There were also educational experiments and a joint Russian/German materials science experiment programme. A full complement of science would have to await the delivery of dedicated research facilities and modules and more power from the solar arrays.

The day before the crew docked, Progress M1-3 had been undocked automatically from the rear port of Zvezda, allowing the Soyuz to take its place. On

18 November, Progress M1-4 docked with the station. This was later undocked and re-docked using the TORU system, a test of the cosmonaut-controlled docking system that had given so much trouble during Mir in 1997. A third Progress was sent to the crew (Progress M44) towards the end of their residency. In between, the crew temporarily vacated the station to move Soyuz TM31 from Zvezda's rear port (where the Progress-compatible refuelling system was located) to the vacant nadir port of Zarya, an operation that took about 30 minutes and freed up the rear port for the Progress M44 re-supply flight.

During the residency, Shepherd kept a log of events, which was later posted (in a censored form) on the NASA website and provided a fascinating insight into life aboard the station. As the crew worked with both US and Russian ground stations, they used GMT as time on board, with the official language as English. Weekends were observed where possible and rest days and Earth "holidays" (Christmas and Easter) were also observed.

Limited communications with ground stations in Russia, locating equipment and the noise levels were early problems identified, problems which, according to the Russians, were normal in the development of a new station. Prior to the installation of the solar arrays during STS-97 in December 2000, the Unity module was off limits to allow power that would heat the module to be diverted to keep the four CMG gyroscopes heated. After the installation of the P6 structure, access to Unity was allowed, offering more room to place equipment and the chance to finally retrieve the logistics stowed by earlier crews. With the arrival of Destiny, more volume was available, but with only a month of the first residency remaining, the real work in the new module and the expanded science programme would have to await the next resident crew. They duly arrived aboard STS-102 in March 2001. After years of training and delays, the ISS-1 crew had successfully configured the station to a position where a new crew could take over and science could begin in earnest. Though they were reluctant to hand over the helm with so much still to do, it was time to leave those details to a fresh crew and come home. Mission successful.

Milestones

220th manned space flight
90th Russian manned space flight
83rd manned Soyuz mission
30th manned Soyuz TM mission
1st Soyuz ferry ISS mission
1st ISS resident crew

STS-97

Int. Designation	2000-078A
Launched	30 November 2000
Launch Site	Pad 39B, Kennedy Space Center, Florida
Landed	11 December 2000
Landing Site	Runway 15, Shuttle Landing Facility, KSC, Florida
Launch Vehicle	OV-105 Endeavour/ET-105/SRB BI-103/SSME #1 2043; #2 2054; #3 2049
Duration	10 days 19 hrs 58 min 20 sec
Call sign	Endeavour
Objective	ISS Assembly flight 4A; P6 solar arrays

Flight Crew

JETT Jr., Brent Ward, 42, USN, commander, 3rd mission
Previous missions: STS-72 (1996); STS-81 (1997)
BLOOMFIELD, Michael John, 41, USAF, pilot, 2nd mission
Previous mission: STS-86 (1997)
TANNER, Joseph Richard, 50, civilian, mission specialist 1, 3rd mission
Previous missions: STS-66 (1994); STS-82 (1997)
GARNEAU, Marc Joseph Jean-Pierre, 51, civilian, Canadian mission specialist 2, 3rd mission
Previous missions: STS 41-G (1984); STS-77 (1996)
NORIEGA, Carlos Ismael, 41, USMC, mission specialist 3, 2nd mission
Previous mission: STS-84 (1997)

Flight Log

The first set of US-provided solar arrays were delivered to the ISS on this mission. Docking with ISS occurred during FD 3 at the newly installed (STS-92) PMA-3 located on the nadir port of Unity. This new port would allow the temporary relocation of the forward PMA-2 to the Z1 Truss during the next Shuttle mission (STS-98) until after the attachment of the Destiny laboratory. Endeavour remained docked to the station for 167 hours, but only had the internal hatches open for a total of 24 hours, leaving little time to visit with their neighbours. The ISS-1 crew essentially followed their own timetable and programme during docked operations, supporting the activities of the STS-97 crew when required.

The P6 array was lifted from the payload bay of Endeavour by the RMS and parked in an "overnight" position to warm its components in the Sun. Inside the Shuttle, the crew opened the hatch into Unity to leave supplies and computer hardware for later relocation inside the station. Checks of EVA equipment and support hardware were also completed and verified during the day. The three EVAs (with rest

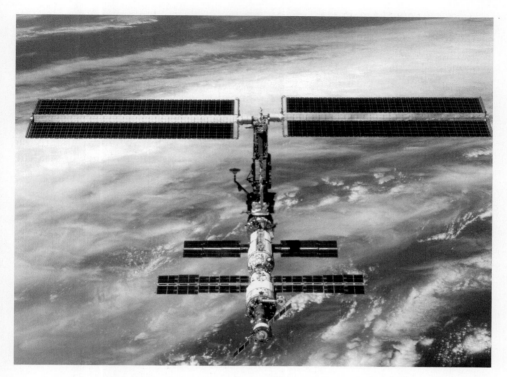

A view from the departing Shuttle back towards the ISS. The station now displays the deployed solar arrays delivered and installed by the crew of STS-97

days in between) would cover the installation and connection of solar arrays, preparation of the docking port for the attachment of Destiny on the next mission, installation of the Floating Potential Probe (FPP) to measure the electrical potential surrounding the station, and the installation of a camera cable outside the Unity.

During the first EVA (3 Dec for 7 hours 23 minutes), the EVA crew of Tanner (EV1) and Noriega (EV2) mated the P6 array to the station's Z1 Truss. Despite some delay in unfurling the arrays (a problem solved by recycling the deployment and retraction process), the starboard and port sides of the array were successfully unfurled over two days to their full length of 111.9 m long × 34 m wide. These were two of the eight arrays that would power ISS. Photovoltaic radiators to dissipate surplus heat from onboard electronics were also deployed. The second EVA (5 Dec for 6 hours 37 minutes) saw the two astronauts reconfigure electrical connections from the P6 arrays to the US segment, and prepare PMA-2 for its relocation during STS-98. After Destiny had been docked to the station, PMA-2 would be relocated to the rear of the laboratory to allow further Shuttle dockings at that location. For STS-97, the EVA programme continued (7 Dec for 5 hours 10 minutes) by moving an S-band antenna and releasing radiator launch restraints. They also increased the tension on the solar arrays during their third EVA. Following an Earth-based tradition conducted when a building reaches its final height, the EVA crew attached an image of an evergreen tree on a transfer bag to the FPP, a symbolic "topping out" of the ISS. Several get-ahead tasks were also completed before the men came back inside.

During the day inside ISS, the two crews completed a welcome ceremony and briefings, followed by structural tests of the array. A series of logistics transfer operations were completed, as was the removal of refuse back into Endeavour to make room on board the station. The undocking of Endeavour from ISS occurred without incident, and after the fly-around manoeuvre, the Shuttle departed, leaving the ISS-1 crew to continue their programme, enjoy the first festive holiday aboard the station and mark the change of year. The new year would bring a change of resident crews and further expansion of the station.

Milestones

221st manned space flight
131st US manned space flight
101st Shuttle flight
15th flight of Endeavour
45th US and 78th flight with EVA operations
6th Shuttle ISS mission
2nd Endeavour ISS mission

8

The Fifth Decade: 2001–2006

STS-98

Int. Designation	2001-006A
Launched	7 February 2001
Launch Site	Pad 39A, Kennedy Space Center, Florida
Landed	20 February 2001
Landing Site	Runway 22, Edwards AFB, California
Launch Vehicle	OV-104 Atlantis/ET-106/SRB BI-105/SSME #1 2052; #2 2044; #3 2047
Duration	12 days 21 hrs 21 min 0 sec
Call sign	Atlantis
Objective	ISS assembly flight 5A; delivery of US laboratory Destiny

Flight Crew

COCKRELL, Kenneth Dale, 50, civilian, commander, 4th mission
Previous missions: STS-56 (1993); STS-69 (1995); STS-80 (1996)
POLANSKY, Mark Lewis, 44, civilian, pilot
CURBEAM Jr., Robert Lee, 38, USN, mission specialist 1, 2nd mission
Previous mission: STS-85 (1997)
IVINS, Marsha Sue, 49, civilian, mission specialist 2, 5th mission
Previous missions: STS-32 (1990); STS-46 (1992); STS-62 (1994); STS-81 (1997)
JONES, Thomas David, 46, civilian, mission specialist 3, 4th mission
Previous missions: STS-59 (1994); STS-68 (1994); STS-80 (1996)

Flight Log

STS-98 had been rolled to the launch pad on 3 January 2001, but after concerns were raised over the integrity of SRB cables, it was taken back to the VAB 19 January (the original day of launch). A series of 36 "wiggle tests" were completed on the cables before they were cleared for the new launch date of 7 February. The vehicle was back on the pad by 26 January. After docking with ISS on FD 3 at the PMA-3 location on Unity, the hatches were opened for the initial transfer of logistics to the station. During four hours of joint operations with the ISS-1 crew, the Shuttle crew transferred three 45.4 litre water bags, a spare Zvezda computer, cables for powering up Destiny, personal gifts for the crew from their families, fresh food and new movies. Atlantis remained docked with ISS for 165 hours, with the hatches opened for a total of 63 hours during that time.

During FD 4, the US laboratory Destiny was successfully relocated to the forward port of Unity by RMS. This was done by first using the robotic arm to remove the PMA-2 from the front port and relocate it to the Z1 Truss. The RMS then returned to the payload bay to lift the 16-ton laboratory out to mate it with the front port of the Unity Node. An automatic bolt system permanently secured the lab to the

In the grasp of the Shuttle RMS, the US laboratory Destiny is moved from the payload bay of Atlantis. This photo was taken during the first EVA by Tom Jones, who monitored the laboratory's relocation to Node 1 along with fellow EVA astronaut Bob Curbeam

Node. The next day, ISS-1 commander Bill Shepherd and STS-98 commander Ken Cockrell entered Destiny for the first time, activating systems and filming onboard scenes with the IMAX camera. They activated air systems, fire extinguishers, alarm systems and computers. The PMA-2 was relocated from the Z1 Truss to the front of Destiny two days later.

The three EVAs by Jones (EV1) and Curbeam (EV2) (10 Feb for 7 hours 34 minutes, 12 Feb for 6 hours 50 minutes, and 14 Feb for 5 hours 25 minutes) supported the move of Destiny, with the two astronauts connecting electrical, data and coolant lines. They also provided visual clues for the relocation of hardware, as Ivins operated the RMS from the aft flight deck of the Shuttle. The astronauts placed covers over the launch restraints pins that had held Destiny in the payload bay and attached a vent to part of the lab's vent systems, as well as installing handrails and sockets on the exterior to support future EVAs. They also inspected the PMA-2 and Destiny connections and solar array connections. In an evaluation of a possible EVA rescue situation, Jones and Curbeam also tested how difficult it would be for a spacewalker to carry an immobile crew member back to the Shuttle airlock. During the first EVA, a small amount of frozen ammonia crystals leaked as Curbeam attached the coolant line. It soon dissipated and posed no problems, but decontamination actions were followed by both the EVA crew (Curbeam remained in the Sun for 90 minutes while Jones brushed off his colleague's suit) and the Shuttle crew (Cockrell, Polansky and Ivins wore oxygen masks for 20 minutes after re-pressurisation of the airlock, which had been partially pressurised to flush out any contaminants).

After the third EVA, the crew opened the internal hatches again to complete the transfer of logistics. In all 1,360 kg of supplies were transferred to the station and around 386 kg of trash (used batteries, packaging material, empty food containers and other items) were brought back to Atlantis. The orbit of the station was also re-boosted by the engines of the Atlantis once again prior to undocking. The addition of Destiny increased the internal volume of the habitable modules by $10.75 \, m^3$, the combined habitable modules now totalling over $36.9 \, m^3$, more volume than any other manned spacecraft in history.

Launched inside the laboratory were five systems racks that provided electrical power, cooling water, air revitalisation and temperature–humidity control. There were locations for up to 18 additional science racks, with six due for launch on the next mission. In addition to science facilities, the new module provide ISS with ECLSS, a thermal control system, a guidance and control system, navigation and electrical power systems, and support for EVA, robotics communications and tracking. Destiny's arrival greatly expanded the scope of operations at the station.

Milestones

222nd manned space flight
132nd US manned space flight
102nd Shuttle mission
23rd flight of Atlantis
7th Shuttle ISS mission
3rd Atlantis ISS mission
46th US and 79th flight with EVA operations

STS-102

Int. Designation	2001-010A
Launched	8 March 2001
Launch Site	Pad 39B, Kennedy Space Center, Florida
Landed	21 March 2001
Landing Site	Runway 15, Shuttle Landing Facility, KSC, Florida
Launch Vehicle	OV-103 Discovery/ET-107/SRB BI-106/SSME #1 2048; #2 2053; #3 2045
Duration	12 days 19 hrs 51 min 57 sec
Call sign	Discovery
Objective	ISS Assembly mission 5A.1; delivery of ISS-2 resident crew and return of ISS-1 crew; MPLM-1 Leonardo logistics flight

Flight Crew

WETHERBEE, James Donald, 48, USN, commander, 5th mission
Previous missions: STS-32 (1990); STS-52 (1992); STS-63 (1995); STS-86 (1997)
KELLY, James McNeil, 36, USAF, pilot
THOMAS, Andrew Sydney Withiel, 49, civilian, mission specialist 1, 3rd mission
Previous mission: STS-77 (1996); STS 89/91 (1996)
RICHARDS, Paul William, 36, mission specialist 2

ISS-2 crew up only:

VOSS, James Shelton, 51, US Army, mission specialist 3, ISS-2 flight engineer 1, 5th mission
Previous missions: STS-44 (1991); STS-53 (1992); STS-69 (1995); STS-101 (2000)
HELMS, Susan Jane, 42, USAF, mission specialist 4, ISS-2 flight engineer 2, 5th mission
Previous missions: STS-54 (1993); STS-64 (1994); STS 78 (1996); STS-101 (2000)
USACHEV, Yuri Vladimirovich, 42, civilian, Russian mission specialist 5, ISS-2 and Soyuz commander, 4th mission
Previous missions: Soyuz TM18 (1994); Soyuz TM23 (1996); STS-101 (2000)

ISS-1 crew down only:

KRIKALEV, Sergei Konstaninovich, 42, civilian, Russisn ISS-1 flight engineer, mission specialist 3, 5th mission
Previous missions: Soyuz TM7 (1988); Soyuz TM12 (1991); STS-60 (1994); STS-88/ISS (1998)

The ten astronauts making up the STS-102, ISS Expedition 1 and 2 crews assemble inside the Destiny laboratory for a group portrait to mark the first exchange of resident crews on the station. In foreground from left Gidzenko, Krikalev and Shepherd (ISS-1 crew); Helms Usachev and Voss (ISS-2 crew). At rear are the STS-102 crew from left Jim Kelly, Paul Richards, Jim Wetherbee and Andy Thomas. Note the ship's bell above the head of Thomas, rung to signify crew arrivals and departures

SHEPHERD, William McMichael, 51, USN, ISS-1 commander, mission specialist 4, 4th mission
Previous missions: STS-27 (1988); STS-41 (1990); STS-52 (1993)
GIDZENKO, Yuri Pavolich, 38, Russian Air Force, ISS-1 Soyuz commander, mission specialist 5, 2nd mission
Previous mission: Soyuz TM22 (1995)

Flight Log

This mission delivered the second resident crew to the station, along with almost 5 tons of supplies carried aboard the first Multi-Purpose Logistics Module (MPLM). Almost 1 ton of unwanted material was returned to Earth in the MPLM at the end of the mission. Docking with the ISS at the PMA-2 location on 10 March was followed a couple of hours later by the opening of hatches, with all ten crew members greeting each other in the spacious Destiny laboratory.

Two EVAs were completed during the mission. The first by Voss (EV1) and Helms (EV2) (11 Mar for 8 hours 56 minutes) was before they began their residency aboard the station and set a record for the longest EVA in Shuttle history. Their task was to prepare PMA-3 to be moved from Unity to make room for MPLM Leonardo, moving an antenna from the Common Berthing Mechanism (CBM) to allow the PMA-3 to be stowed there while Leonardo was being unloaded. A Lab Cradle Assembly was also relocated from the payload bay of Discovery to the side of Destiny where it would form the base of the Space Station RMS (SSRMS) to be delivered on the following Shuttle mission (STS-100). The second EVA by Thomas (EV3) and Richards (EV4) (13 Mar for 6 hours 21 minutes) included installation of an External Stowage Platform (ESP) for spare parts and the attachment of a spare ammonia coolant pump to the platform. Heater, power and control cables were also connected in preparation for the station robotic arm delivery.

During the almost 214 hours docked to ISS (a record), the hatches were open for a total of 142 hours. Leonardo was moved from the payload bay of Discovery to the CBM on 11 Mar. It was relocated back in the payload bay on 18 March after the mission had been extended a day to ensure the module was correctly emptied and properly repacked for entry and landing. The official transfer of crew members between ISS-1 and 2 was staggered over several days to allow necessary briefings to be completed without interrupting the significant logistics transfer between the vehicles. Usachev replaced Gidzenko on 10 March, Voss swapped with Krikalev on 11 March and Helms with Shepherd on 14 March, with Shepherd passing formal command of ISS to Usachev on 19 March. Each resident crew member relocated their personal Soyuz couch liners in the process of exchange. The returning ISS-1 crew took their places in recumbent seats on the mid-deck of Discovery (first evaluated during Shuttle–Mir missions) for the return to Earth. The successful operations allowed a smooth hand-over between the crews with as little interruption to station activities as possible. This was the first time a complete ISS resident crew had been exchanged on the Shuttle (something that had occurred on Mir only once, in 1995).

Milestones

223rd manned space flight
133rd US manned space flight
103rd Shuttle mission
29th flight of Discovery
8th Shuttle ISS mission
3rd Discovery ISS mission
47th US and 80th flight with EVA operations
1st Shuttle resident crew exchange mission
1st flight of Multi-Purpose Logistics Module
1st flight of MPLM-1 Leonardo

ISS EO-2

Int. Designation	N/A
Launched	8 March 2001 (aboard STS-102)
Launch Site	Pad 39B, Kennedy Space Center, Florida
Landed	22 August 2001 (aboard STS-105)
Landing Site	Shuttle Landing Facility, KSC, Florida
Launch Vehicle	STS-102
Duration	167 days 6 hrs 40 min 49 sec
Call sign	Flagman (Flagship)
Objective	ISS-2 expedition crew programme

Flight Crew

USACHEV, Yuri Vladimirovich, 42, civilian, ISS-2 and Soyuz commander, 4th mission
Previous missions: Soyuz TM18 (1994); Soyuz TM23 (1996); STS-101 (2000)
VOSS, James Shelton, 51, US Army, ISS-2 flight engineer 1, 5th mission
Previous missions: STS-44 (1991); STS-53 (1992); STS-69 (1995); STS-101 (2000)
HELMS, Susan Jane, 42, USAF, ISS-2 flight engineer 2, 5th mission
Previous missions: STS-54 (1993); STS-64 (1994); STS-78 (1996); STS-101 (2000)

Flight Log

On 10 March, the second ISS crew arrived at the station and formed part of a ten-person crew aboard ISS until the departure of Discovery on 19 March. Following the departure of STS-102, the ISS-2 crew was fully occupied with the commencement of science work and activation of the Destiny laboratory, as well as the introduction of robotics on the station with the delivery of the SSRMS (Canadarm2) during the STS-100 mission.

The science programme for this flight encompassed 38 investigations – 18 American and 20 Russian experiments – in plant biology, biology, assessment of the ISS environment, radiation studies, Earth observations, biotechnology, protein crystal growth, medical, materials sciences, technology and education. In addition to the science programme, the crew would host the crews of STS-100, 104 and 105, as well as the first visiting mission by Soyuz TM32 to exchange their Soyuz ferry. The latter included the first space tourist (space flight participant), American businessman Dennis Tito. To support the science work, the ISS Payload Operations Center located at NASA's Marshall Space Flight Center in Huntsville, Alabama, began round-the-clock operations with this residency. Additional science racks for Destiny were delivered via the MPLM carried by STS-100 and STS-105.

April was a busy month for the ISS-2 crew, with the undocking of Progress M44, the relocation of Soyuz TM31 from the Zarya nadir port to the Zvezda aft port, the

The second ISS resident crew pose for the camera. L to r Helms, Usachev and Voss

flight of STS-100 delivering Canadarm2, and the first Soyuz Taxi mission, Soyuz TM32. In May, they received the Progress M1-6 cargo craft and the following month (8 Jun), Voss and Usachev completed a 19-minute IVA in the Zvezda docking node moving a storage site in the docking node to the nadir docking port position. This had originally been in the front docking port of Zvezda, which was now permanently docked to the rear of Zarya, and would be more useful in the nadir location of the service module instead. During the IVA, Helms remained in the US segment in case she needed to evacuate to Soyuz TM32, which was docked to the Zarya nadir port. She would have awaited her two colleagues there, as they would have been able to enter the Soyuz via the OM EVA hatch in an emergency (a procedure only evaluated during the Soyuz 5/4 docking and EVA transfer mission in 1969). In the event, this procedure was not called upon.

In July, STS-104 delivered the Joint Airlock (Quest). During the docked operations, Helms and Voss used the new SSRMS to remove the airlock from the bay of the Shuttle and position it on Unity, where it was permanently attached by the STS-104 EVA crew members. The following month, the ISS-2 crew were relieved by the ISS-3 crew, who arrived on the STS-105 mission that would take the second expedition crew home after 167 days in space.

Milestones

2nd ISS resident crew
1st ISS EO crew to be launched by Shuttle

STS-100

Int. Designation	2001-016A
Launched	19 April 2001
Launch Site	Pad 39A, Kennedy Space Center, Florida
Landed	1 May 2001
Landing Site	Runway 22, Edwards AFB, California
Launch Vehicle	OV-105 Endeavour/ET-108/SRB BI-107/SSME #1 2054; #2 2043; #3 1049
Duration	11 days 21 hrs 31 min 14 sec
Call sign	Endeavour
Objective	ISS assembly mission 6A; delivery of Space Station Remote Manipulator System (SSRMS) Canadarm2; MPLM-2 logistics mission

Flight Crew

ROMINGER, Kent Vernon, 44, USN, commander, 5th mission
Previous missions: STS-73 (1995); STS-80 (1996); STS-85 (1997); STS-96 (1999)
ASHBY, Jeffrey Shears, 46, USN, pilot, 2nd mission
Previous mission: STS-93 (1999)
HADFIELD, Chris Austin, 41, Canadian Air Force, mission specialist 1, 2nd mission
Previous mission: STS-74 (1995)
PHILLIPS, John Lynch, 50, civilian, mission specialist 2
PARAZYNSKI, Scott Edward, 39, civilian, mission specialist 3, 4th mission
Previous missions: STS-66 (1994); STS-86 (1997); STS-95 (1998)
GUIDONI, Umberto, 46, civilian, ESA mission specialist 4, 2nd mission
Previous mission: STS-75 (1996)
LONCHAKOV, Yuri Valentinovich, 36, Russian Air Force, mission specialist 5

Flight Log

Endeavour docked to the ISS on 21 April with an international crew comprising astronauts from the US, Canada, Italy and Russia. In the payload bay was the second Italian-made MPLM (Raffaello) with 3.4 tons of cargo, including two more science racks for Destiny. Also aboard the orbiter was the Space Station Remote Manipulator System (SSRMS), Canadarm2. The crew would not open the hatches to greet each other this time, in order to preserve respective air pressure levels in the two spacecraft as the Endeavour crew prepared for their EVAs. Endeavour would remain docked to the station for 195 hours.

During the first EVA (22 Apr for 7 hours 10 minutes), the RMS lifted a Spacelab pallet to a cradle on the Destiny lab, where Hadfield (EV1) and Parazynski (EV2)

Chris Hadfield, the first Canadian to perform an EVA, is seen near the SSRMS Canadarm2, the new robotics tool for the ISS

removed the UHF antenna and installed it on the lab structure. They then unfolded the Canadarm2 and attached one end to Destiny while the arm was still secured on the pallet. Computer cables were then attached to the arm to give it computer communications with the lab. The next day, ISS-2 astronauts Helms and Voss commanded the new arm to "walk off" the pallet and grab an electrical grapple fixture on the lab, which would supply data, power and telemetry to the arm. As the crew relocated the station arm, the Shuttle's RMS was used to relocate the Raffaello logistics module to the station for later transfer of its 2,700 kg of cargo. This included two science racks and three US commercial payloads. The second EVA (24 Apr for 7 hours 40 minutes) saw the two astronauts connect power and Data Grapple Fixture circuits on Destiny for the station arm's operations. Old hardware was removed from the exterior of the station and spares were relocated from the Shuttle to a storage rack on the US lab. The astronauts also rewired power and data connections for the SSRMS. Despite some early problems with a back-up system (resolved by disconnecting and reconnecting the cables at the base of the arm), the power paths in both primary and redundant modes were completed.

A problem with station computers meant that some communications had to be routed through the Space Shuttle and while the problem was investigated, some work on the SSRMS was curtailed. Suspected software errors were thought to be the cause of the communication problems. The STS-100 docked mission to ISS was extended a day to support station operations while the problem was resolved over the next couple of days. On FD 9, the MPLM was returned by RMS to the payload bay of Endeavour. Inside was 726 kg of material for the return to Earth. The next day, the computer problems had been resolved, and the SSRMS was commanded to take the 1,361 kg pallet from its cradle on Destiny and "hand it over" to the Shuttle RMS, which replaced it in the payload bay. The exchange of the pallet from arm to arm was the first ever robotic-to-robotic transfer in space.

On the same day, the crew were informed of the launch of Soyuz TM32 carrying the first space tourist, Dennis Tito. The Shuttle crew undocked from ISS on 29 April and 14 hours later, with STS-100 still in orbit, the Soyuz TM crew docked to ISS. Aboard Endeavour, the crew received a weather waive-off from the Cape for 24 hours and the following day, with no improvement in the weather at the Cape, the first landing opportunity at Edwards AFB in California was taken up instead.

Milestones

224th manned space flight
134th US manned space flight
104th Shuttle mission
16th flight of Endeavour
9th Shuttle ISS mission
3rd Endeavour ISS mission
48th US and 81st flight with EVA operations
2nd MPLM flight
1st flight of MPLM-2 Raffaello
1st Canadian EVA (Hadfield)
1st robotic-to-robotic transfer in space

SOYUZ TM32

Int. Designation	2001-017A
Launched	21 April 2001
Launch Site	Pad 1, Site 5, Baikonur Cosmodrome, Kazakhstan
Landed	30 April 2001 (in TM31)
Landing Site	83 km northeast of Arkalyk
Launch Vehicle	R7 (11A511U); spacecraft serial number (7K-M) 206
Duration	7 days 22 hrs 4 min 8 sec
Call sign	Kristall (Crystal)
Objective	ISS flight 2S/EP1 visiting mission 1; 1st ISS Soyuz ferry exchange mission; 1st space "tourist" mission (Tito)

Flight Crew

MUSABAYEV, Talgat Amangeldyevich, 50, Russian Air Force, commander, 3rd mission
Previous missions: Soyuz TM19 (1994); Soyuz TM27 (1998)
BATURIN, Yuri Mikhailovich, 51, Russian political aide, flight engineer, 2nd mission
Previous mission: Soyuz TM28 (1998)
TITO, Dennis, 60, civilian, US space flight participant

Flight Log

Soyuz TM32 was scheduled to be the mission that exchanged the Soyuz (TM31) that had carried the first resident crew to the station six months before. The mission was also known as Taxi-1. However, just to confuse things, the Russians decided to rename these missions (and the Progress flights) by order of flight sequence to the new station. So Soyuz TM31 became Soyuz 1 (the original of which actually flew in 1967) and the first Progress (M1-3) became Progress 1 (which originally flew in 1978). Soyuz visiting missions had been conducted for some time with their own space stations, with Russian cosmonaut crews exchanging the Soyuz craft every six months at the end of the vehicle's "shelf-life". Beginning with ISS-2, main crews for the new station would be launched and landed on the Shuttle, though all crews would be trained on Soyuz entry and landing procedures, at least for the foreseeable future as the station was being constructed. The crew would include one experienced commander, one rookie flight engineer and, since Soyuz carried a third seat, the opportunity to fly a second Russian or international cosmonaut to ISS. Russia also quickly realised that this was a seat worth selling. The idea of flying fare-paying passengers on short visiting missions became a possibility near the end of the Mir programme, but that station was de-orbited before a "space tourist", or space flight participant (SFP) had the chance to pay the US$20 million price tag for the privilege. With ISS, the opportunity for such flights was once again available.

The first "space tourist", Dennis Tito (left), is safely back on Earth after his $20 million flight into space. Musabayev (centre) and Baturin are shown with him, having exchanged the TM31 craft for the fresher TM32

NASA was not happy about less than fully trained crew members flying to the ISS. As the station was under construction, there was the concern about the tourist endangering the resident crew or affecting the smooth operation of the mission. However, as the Russians were flying the passenger, NASA had no veto over the decision. Instead, they did try to curtail such tourists' station access to the Russian elements only. The first to fly in this way was American businessman and millionaire Dennis Tito, who had been planning a trip to Mir and was offered a ride to ISS to compensate.

After much media coverage, the TM32 crew arrived at ISS at the nadir port of Zarya. Though TV views of the crew transfer to the station were received, NASA indicated that only Russian-supplied images would be available, not theirs. The short programme of the Kristall crew included Earth observations, a seed and biology experiment, a food digestive experiment, crystal experiments and "commemorative activities". On board the station, the Taxi crew were called cosmonaut researchers 1 (Musabayev), 2 (Baturin) and 3 (Tito). Tito also used video and still cameras to record activities on board the station. In a Russian press conference, Tito mentioned that he had indeed visited the US segment several times. Despite reports that he suffered from space sickness, Tito indicated that he "liked space." Both Musabayev and Usachev reported that Tito had not hindered the work of the main crew and had assisted in meal preparations. Voss and Helms were advised by NASA to keep their activities with Tito to formal participation. In light of the visit of Tito, NASA amended its procedures for approving visiting "tourists" to ISS. Future visitors would now have to participate in basic safety and awareness training at JSC, and those "tourists" who followed Tito would have to develop a stronger scientific programme to occupy them during their week on the ISS.

The crew came home in the TM31 spacecraft on 6 May. A faulty infrared vertical sensor reportedly caused them to overshoot their landing site by 56 km. The landing was a hard one, so hard that Musabayev had to check whether his $20 million passenger was still alive! The mission certainly generated further interest in "Soyuz seats for sale" and provided the possibility of short visiting missions to ISS to exchange the Soyuz and, while there, conduct small science programmes that would not interrupt the main work of the resident crew. This was the continuation of the Interkosmos programme that the Russians had developed over 20 years before.

Milestones

225th manned space flight
91st Russian manned space flight
84th manned Soyuz mission
31st manned Soyuz TM mission
2nd Soyuz ISS ferry mission
1st Soyuz visiting mission
1st space flight participant (Tito)

STS-104

Int. Designation	2001-028A
Launched	12 July 2001
Launch Site	Pad 39B, Kennedy Space Center, Florida
Landed	24 July 2001
Landing Site	Runway 15, Shuttle Landing Facility, KSC, Florida
Launch Vehicle	OV-104 Atlantis/ET-109/SRB BI-108/SSME #1 2056; #2 0251; #3 2047
Duration	12 days 18 hrs 36 min 39 sec
Call sign	Atlantis
Objective	ISS assembly flight 7A; delivery of Joint Airlock module Quest

Flight Crew

LINDSEY, Steven Wayne, 40, USAF, commander, 3rd mission
Previous missions: STS-87 (1997); STS-95 (1998)
HOBAUGH, Charles Owen, 39, USMC, pilot
GERNHARDT, Michael Landen, 44, civilian, mission specialist 1, 4th mission
Previous missions: STS-69 (1995); STS-83 (1997); STS-94 (1997)
REILLY II, James Francis, 47, civilian, mission specialist 2, 2nd mission
Previous mission: STS-89 (1998)
KAVANDI, Janet Lynn, 42, civilian, mission specialist 3, 3rd mission
Previous missions: STS-91 (1998); STS-99 (2000)

Flight Log

This mission completed the second phase of ISS assembly by providing an airlock facility that would allow EVAs to be conducted from the station without the need for a Shuttle to be docked to it. Atlantis docked with ISS on 13 July and would remain there for the next 196 hours. The mission's three EVAs supported the installation of the Joint Airlock module (called Quest) on the station.

The first EVA (14 Jul for 5 hours 59 minutes) saw Gernhardt (EV1) and Reilly (EV2) remove insulation covers from berthing mechanisms on the airlock and install bars to what would be the attachment points for the four high pressure gas tanks. Using Canadarm2, ISS-2 crew member Susan Helms then lifted the airlock out of the payload bay of Atlantis and installed it on the right side of Unity. The EVA crew then attached heating cables from the station to the airlock and positioned foot restraints to support future EVAs. Three days later (17 Jul for 6 hours 29 minutes), the EVA crew returned to Quest to install the three tank assemblies on the airlock with the help of both the station and Shuttle robotic arm systems. The final EVA four days later (21 Jul for 4 hours 2 minutes) was the first out of the Quest airlock. During the

The Quest airlock is in the process of being installed onto the starboard side of Unity Node 1 of ISS. The airlock, delivered by STS-104, is being moved by ISS-2 FE Susan Helms using the SSRMS Canadarm2 from a control panel located in the US Destiny laboratory

excursion, the EVA crew installed further tank assemblies on the outside of the station, again with the help of both RMS systems. In the days between each EVA, nitrogen and oxygen lines were tested and valves installed inside the facility to connect Quest to the ISS ECLSS. A computer was also installed to manage the airlock systems. Air bubbles in a coolant line caused a water spill and its clean-up postponed other activities for a day. A leaky air circulation valve was replaced and the hatch was relocated from its initial location between the airlock and the Unity Node hatch to become an inner EVA hatch between the equipment lock (storage and servicing of EVA equipment) and the crew lock (access to open space). A dry run using the Quest systems was performed prior to formal inauguration on 21 July, just hours after the 32nd anniversary of the first lunar EVA on Apollo 11 on 20 July 1969.

Undocking from ISS occurred in the early hours of 22 July, and after a 24-hour waive-off due to weather conditions in Florida, Atlantis achieved a night time landing back at the Cape the following day. Between July 2000 and the arrival of Quest, over 69,853 kg of hardware had been added to the station. This included the Zvezda module, the Z1 Truss (STS-92), PMA-3 (STS-92), the P6 assembly and solar arrays (STS-97), the US Destiny lab (STS-98), Canadarm2 (STS-100) and now Quest (STS-104). This would allow the station to conduct independent operations. Though future assembly missions and re-supply missions would still be required, the crews on board the station could now conduct expanded programmes of science and station-based EVAs, as well as routine maintenance and housekeeping duties.

Milestones

226th manned space flight
135th US manned space flight
105th Shuttle mission
24th flight of Atlantis
49th US and 82nd flight with EVA operations
10th Shuttle ISS mission
4th Atlantis ISS mission
1st EVA from Quest airlock

STS-105

Int. Designation	2001-035A
Launched	10 August 2001
Launch Site	Pad 39A, Kennedy Space Center, Florida
Landed	22 August 2001
Landing Site	Runway 15, Shuttle Landing Facility, KSC, Florida
Launch Vehicle	OV-103 Discovery/ET-100/SRB BI-109/SSME #1 2052; #2 2044; #3 2045
Duration	11 days 21 hrs 13 min 52 sec
Call sign	Discovery
Objective	ISS assembly flight 7A.1; MPLM-1 logistics mission; delivery of ISS-3 crew; return of ISS-2 crew

Flight Crew

HOROWITZ, Scott Jay, 44, USAF, commander, 4th mission
Previous missions: STS-75 (1996); STS-82 (1997); STS-101 (2000)
STURCKOW, Frederick Wilford, 41, USMC, pilot, 2nd mission
Previous mission: STS-88 (1998)
FORRESTER, Patrick Graham, 44, USAF, mission specialist 1
BARRY, Daniel Thomas, 47, civilian, mission specialist 2, 3rd mission
Previous missions: STS-72 (1996); STS-96 (1999)

ISS-3 crew up only:
CULBERTSON Jr., Frank Lee, 52, civilian, mission specialist 3, ISS-3 commander, 3rd mission
Previous missions: STS-38 (1990); STS-51 (1993)
TYURIN, Mikhail Vladislavovich, 41, civilian, Russian mission specialist 4, ISS-3 flight engineer
DEZHUROV, Vladimir Nikolayevich, 39, Russian Air Force, mission specialist 5, ISS-3 Soyuz commander, 2nd mission
Previous mission: Soyuz TM21 (1995)

ISS-2 crew down only:
VOSS, James Shelton, 51, US Army, ISS-2 flight engineer 1, mission specialist 3, 5th mission
Previous missions: STS-44 (1991); STS-53 (1992); STS-69 (1995); STS-101 (2000)
HELMS, Susan Jane, 42, USAF, ISS-2 flight engineer 2, mission specialist 4, 5th mission
Previous missions: STS-54 (1993); STS-64 (1994); STS-78 (1996); STS-101 (2000)
USACHEV, Yuri Vladimirovich, 42, civilian, Russian ISS-2 and Soyuz commander, mission specialist 5, 4th mission
Previous missions: Soyuz TM18 (1994); Soyuz TM23 (1996); STS-101 (2000)

Inside Destiny, the crews of STS-105 and Expeditions 2 and 3 pose for a traditional in-flight joint crew portrait. On the left is the outgoing ISS-2 crew, in the centre is the STS-105 crew and to the right the incoming ISS-3 crew. From bottom right going clockwise: Sturckow and Forrester (both STS-105), Usachev, Voss and Helms (ISS-2 crew), Horowitz and Barry (both STS-105), and Tyurin, Dezhurov and Culbertson (ISS-3 crew)

Flight Log

Lightning and thick cloud, together with the risk of showers led to the original launch attempt on 9 August being scrubbed. The threat of bad weather the next day meant the launch window was opened five minutes earlier and the mission launched without incident. Docking with ISS took place on FD 3 and Discovery would remain linked to the station for 188 hours. The Leonardo MPLM was moved across to ISS on FD 4, where it would be unloaded over several days. The cargo included 3,000 kg of equipment, supplies and material. There were 12 racks of experiments and equipment in the module, six of which were Re-supply Stowage Racks that carried equipment, clothing, food and supplies. There were also four Storage Re-supply Platforms for logistics supplies and hardware, and two Express Racks that included smaller payloads for delivery to the station. Old hardware and used equipment was moved back to Leonardo for the trip back to Earth. On this flight, the ISS-2 crew's belongings were part of the return cargo. In total, some 1,360 kg of material was brought back to Earth.

Two EVAs (by Barry – EV1 and Forrester – EV2) were completed (16 Aug for 6 hours 16 minutes and 18 August for 5 hours 29 minutes), during which the Shuttle EVA crew installed the Early Ammonia Servicer (EAS), which included spare ammonia for use in the station's coolant system if required. During the second EVA, the crew prepared for the delivery of the S0 Truss (planned for 2002) by installing heater cables and handrails on both sides of the Destiny lab. During FD 5, Discovery took over control of ISS while Zvezda received upgraded software from Russian flight control. The command of the station reverted to Zvezda when the new software had been loaded and checked.

Official hand-over between ISS-2 and ISS-3 crew members took place on FD 6 (17 Aug), which included a series of briefings and exchange of Soyuz seat liners in Soyuz TM32. Discovery undocked from ISS on 20 August and after the usual fly-around and separation manoeuvre, the Shuttle crew released a small science satellite, called Simplesat, by means of spring ejection from a GAS canister in the payload bay. The orbiter landed on the second of two Florida opportunities, with the first having been waived off due to bad weather.

Milestones

227th manned space flight
136th US manned space flight
106th Shuttle mission
30th flight of Discovery
50th US and 83rd flight with EVA operations
11th STS ISS mission
4th Discovery ISS mission
3rd MPLM flight
2nd MPLM 01 Leonardo flight
2nd Shuttle ISS resident crew exchange mission

ISS EO-3

Int. Designation	N/A (launched on STS-105)
Launched	10 August 2001
Launch Site	Pad 39A, Kennedy Space Center, Florida
Landed	17 December 2001 (aboard STS-108)
Landing Site	Shuttle Landing Facility, KSC, Florida
Launch Vehicle	STS-105
Duration	128 days 20 hrs 44 min 56 sec
Call sign	Uragan (Hurricane)
Objective	ISS-3 expedition crew programme

Flight Crew

CULBERTSON Jr., Frank Lee, 52, civilian, US ISS-3 commander, 3rd mission
Previous missions: STS-38 (1990); STS-51 (1993)
DEZHUROV, Vladimir Nikolayevich, 39, Russian Air Force, ISS-3 Soyuz commander, 2nd mission
Previous mission: Soyuz TM21 (1995)
TYURIN, Mikhail Vladislavovich, 41, civilian, Russian ISS-3 flight engineer

Flight Log

The third resident crew, under the command of American astronaut Frank Culbertson, arrived at the station aboard STS-105 on 12 August for a four-month stay. During their residency, having assumed official command of the station on 17 August, the crew received two Progress re-supply vehicles, the Russian EVA and docking compartment and the second Soyuz Taxi crew. Four EVAs were also completed, but they did not receive any visiting Shuttle missions until STS-108 arrived to take them home, bringing the ISS-4 crew up to replace them.

The science programme for the mission involved 15 US experiments, 22 Russian, 3 Japanese, 4 French, a Canadian and an ESA experiment, some of which would be performed during the visiting mission. The research fields included biology, the environment of the ISS, radiation, Earth observations, geophysics, protein crystal growth, tissue growth, medical, materials science, technology and education.

Progress M45 arrived on 23 August and remained docked until 22 November. It would be replaced by Progress M1-7, which initially failed to hard-dock with the station and required an unplanned EVA to remove an obstruction to allow the docking operation to be completed. The other unmanned docking with the station during this residency was with Progress M-SO1, which delivered the Russian Docking Compartment (called Pirs, or "Pier"). A few days after docking, the instrumentation and propulsion sections of the Progress were discarded, leaving the forward module section of the vehicle (Pirs) permanently attached to Zvezda's nadir port (which had

The formal picture of the ISS-3 crew posing for the traditional pre-launch crew photo in Shuttle launch and entry suits proudly displaying the residency logo. L to r: Tyurin, Culbertson, Dezhurov.

been prepared during the IVA of ISS-2 on 8 June). This new module would be used by crews performing EVAs from the Russian segment of the station.

The first EVA from Pirs was conducted by Dezhurov and Tyurin (8 Oct for 4 hours 58 minutes) and involved the installation of external equipment to bring the facility up to operational status. The cosmonauts also commenced construction of the Strela boom (GStM-1) designed for further movements across the station's Russian segments, and installed some antennas. During the second EVA (15 Oct for 5 hours 51 minutes), the cosmonauts deployed several small experiment and materials sample cassettes on the outside of the station. These were long-duration exposure facilities that would be retrieved and replaced at regular intervals on future EVAs. The Soyuz TM32 was relocated to the rear Pirs docking port on 18 October in preparation for the arrival of the second visiting crew. The two Russian cosmonauts and French ESA astronaut Claudie Haigneré (formerly André-Deshays) arrived on 23 October for a week of science activities. They would depart in the older Soyuz TM32 spacecraft leaving a new vehicle for the ISS-3 crew to use in the event of an emergency. The third EVA (12 Nov for 5 hours 5 minutes) involved Frank Culbertson, who accompanied Dezhurov outside to continue the installation of handrails and cables on the outside of Pirs and Zvezda. They also tested the extension of the Stella cargo crane, as well as inspecting and photographing a solar battery which had failed to deploy from Zvezda after entering orbit and had remained stuck. The final EVA (3 Dec for 2 hours 46 minutes) was completed by Dezhurov and Tyurin and was unplanned. They inspected the aft port of Zvezda where Progress M1-7 had failed to hard-dock and reported the presence of a foreign object, a sealing ring left behind by the previous Progress (M45) when it undocked. Ground command signalled the Progress M1-7 to extend its docking probe, allowing the cosmonauts to use a cutting tool to remove the tangled rubber ring and free up the system to allow the Progress to successfully hard-dock with the station. A few days later, STS-108 arrived, and the third expedition drew to a close.

Frank Culbertson had been a former US manager of the Shuttle-Mir programme and had used this experience to evaluate the working conditions on the new station for future US resident crew planning, experiment and logistics activities. One of the more difficult periods on board the station was in the aftermath of the 11 September 2001 attacks on the US. Informed of the day's events during a medical check-up, the station later flew over New York and the crew saw the huge plume of smoke and dust from the attack and subsequent collapse of the World Trade Center buildings. Regular updates were sent to the crew following the attacks.

Milestones

3rd ISS resident crew
2nd ISS EO crew to be launched by Shuttle
1st EVAs by ISS resident crew

SOYUZ TM33

Int. Designation	2001-048A
Launched	21 October 2001
Launch Site	Pad 1, Site 5, Baikonur Cosmodrome, Kazakhstan
Landed	31 October 2001 (in Soyuz TM32)
Landing Site	180 km from the town of Dzhezkazgan
Launch Vehicle	R7 (11A511U); spacecraft serial number (7K-M) 207
Duration	9 days 20 hrs 0 min 25 sec
Call sign	Derbent (Derbent)
Objective	ISS mission 3S; second Soyuz Taxi (3S) mission; Soyuz ferry exchange; French *Andromède* visiting mission.

Flight Crew

AFANASYEV, Viktor Mikhailovich, 52, Russian Air Force, commander, 4th mission
Previous missions: Soyuz TM11 (1990); Soyuz TM18 (1994); Soyuz TM29 (1999)
HAIGNERÉ, Claudie, 44, civilian, flight engineer, 2nd mission
Previous mission: Soyuz TM24 (1996)
KOZEEV, Konstantin Mirovich, 33, civilian, flight engineer

Flight Log

The announcement of French cosmonaut Claudie Haigneré (formerly André-Deshays) to the crew of the second Soyuz Taxi mission had been made in December 2002. With the back-up cosmonaut crew of the first Taxi mission recycled to this flight, this would be a far less controversial mission than that of Dennis Tito, as Haigneré had previously completed a visit to the Mir station in 1996. For her new mission, she would be conducting the ESA "*Andromède*" programme of experiments, sponsored by the French Space Agency CNES. Soyuz TM33 docked to ISS on 23 October.

During the week aboard the station, the Derbent cosmonauts assisted their French colleague with her science programme and brought the small cargo of supplies and equipment from the Soyuz to the station. They also exchanged their personal seat liners with those in the TM32 spacecraft, which they would use to return to Earth at the end of their mission. They tested the systems and controls of the returning spacecraft and took air samples from inside the Russian segment of the station for analysis back on Earth, as well as participating in a number of small experiments and research tasks.

Haigneré's science programme included two experiments devoted to the observation of Earth and the study of the ionosphere. There were three experiments in life sciences – in the fields of neuroscience, physiology and developmental biology. There

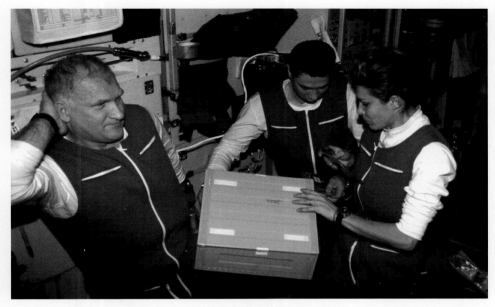

The Soyuz TM33 crew members examine a storage container in Zvezda during the week aboard the station. From left are Victor Afanasyev, Konstantin Kozeev and Claudie Haigneré

were two physics experiments prepared jointly by ESA and the German Space Agency, and two technology experiments and operational experiments designed to flight-test equipment and develop procedures, initiated by the European Astronaut Corps (EAC).

French high schools also devised educational experiments for the mission, a continuation of similar educational experiments flown during the 1999 French mission to Mir. These educational experiments included a wide range of studies to support science teaching in schools.

Throughout the mission, a French project team located at the Moscow TsUP Control Centre and another at the Toulouse Space Centre provided daily mission support to the flight crew, with two audio links per day and permanent monitoring between ISS and ground control. Haigneré received briefing and debriefing sheets via email from the new CADMOS (Centre for the Development of Microgravity Applications and Space Operations) based at Toulouse. Two CNES specialists monitored the docking of Soyuz and activities at TsUP, while at JSC in Houston, French ESA astronaut Michel Tognini provided interface between the support team in France and the one in Moscow, updating them with American activities and monitoring of the mission.

The Derbent crew returned in the Soyuz TM32 spacecraft on 31 October 2001 after a ten-day mission, leaving the newer Soyuz TM33 spacecraft for the ISS-4 crew and further highlighting the benefit of including small programmes of experiments on such Taxi missions in the future.

Milestones

228th manned space flight
92nd Russian manned space flight
85th manned Soyuz mission
32nd manned Soyuz TM mission
3rd ISS Soyuz mission (3S)
2nd Taxi flight
2nd visiting mission
1st European (French) woman to board ISS

STS-108

Int. Designation	2001-054A
Launched	5 December 2001
Launch Site	Pad 39B, Kennedy Space Center, Florida
Landed	17 December 2001
Landing Site	Runway 15, Shuttle Landing Facility, KSC, Florida
Launch Vehicle	OV-105 Endeavour/ET-111/SRB BI-110/SSME #1 2049; #2 2043; #3 2050
Duration	11 days 19 hrs 36 min 45 sec
Call sign	Endeavour
Objective	ISS assembly flight UF-1; MPLM-2 logistics flight; ISS resident crew exchange mission

Flight Crew

GORIE, Dominic Lee, 44, USN, commander, 3rd mission
Previous missions: STS-91 (1998); STS-99 (2000)
KELLY, Mark Edward, 37, USN, pilot
GODWIN, Linda Maxine, 49, civilian, mission specialist 1, 4th mission
Previous missions: STS-37 (1991); STS-59 (1994); STS-76 (1996)
TANI, Daniel Michio, 40, civilian, mission specialist 2

ISS-4 crew up only:

ONUFRIYENKO, Yuri Ivanovich, 40, Russian Air Force, mission specialist 4, ISS-4 and Soyuz TM commander, 2nd mission
Previous mission: Soyuz TM23 (1996)
BURSCH, Daniel Wheeler, 44, USN, mission specialist 5, ISS-4 flight engineer 1, 4th mission
Previous missions: STS-51 (1993); STS-68 (1994); STS-77 (1996)
WALZ, Carl Erwin, 46, USAF, mission specialist 5, ISS-4 flight engineer 2, 4th mission
Previous missions: STS-51 (1993); STS-65 (1994); STS-79 (1996)

ISS-3 crew down only:

CULBERTSON Jr., Frank Lee, 52, civilian, ISS-3 commander, mission specialist 3, 3rd mission
Previous missions: STS-38 (1990); STS-51 (1993)
TYURIN, Mikhail Vladislavovich, 41, civilian, Russian ISS-3 flight engineer, mission specialist 4
DEZHUROV, Vladimir Nikolayevich, 39, Russian Air Force, ISS-3 Soyuz commander, mission specialist 5, 2nd mission
Previous mission: Soyuz TM21 (1995)

Another change of shift on ISS and the traditional group photo in Destiny. At rear, left to right STS-108 crew Godwin, Kelly, Gorie and Tani. In front, l to r ISS-4 crew Walz, Onufriyenko and Bursch, and ISS-3 crew Culbertson, Tyurin and Dezhurov

Flight Log

Originally scheduled for launch on 3 December, the launch was postponed for 24 hours on 29 November in order the allow the ISS-3 crew to complete an extra unplanned EVA to clear the obstruction preventing Progress M1-7 from hard-docking with the station. The 4 December launch was postponed at the $T - 5$ minute point due to unfavourable weather in the KSC area, which remained throughout the duration of the launch window. After the successful launch, Endeavour docked to ISS during FD 3 (7 December) and remained linked to the station for the next 189 hours. There was one EVA, which was conducted by Godwin and Tani from the Shuttle airlock (instead of Quest) on FD 6. During the EVA (10 Dec for 4 hours 12 minutes), the astronauts installed insulation on the solar array rotation mechanism and retrieved antenna covers that had been stowed in a storage location on the outside of the station for return to Earth, and which may be returned for reuse on the station at a future date. They also performed a number of get-ahead tasks for the extensive EVAs planned for the coming year.

The flight was extended to 12 days in order to complete all the assigned maintenance and logistics transfer tasks assigned it. During several days of logistics transfer, the combined crew moved over 2,700 kg from the mid-deck of Endeavour and the Raffaello logistics module on to the station. This included over 385 kg of food, 453 kg of clothing, 136 kg of experiments and associated equipment, 362 kg of EVA hardware, and 272 kg of medical equipment. Over 900 kg of trash, unwanted gear and equipment was placed in the module for return to Earth, and in addition to the exchange of ISS resident crew personal items, the mid-deck of Endeavour was used for the return of several experiment results and samples from the research conducted during the ISS-3 residency. There were also several experiments conducted in the mid-deck of Endeavour during the mission, some of which would be transferred to the station while the others would return on the Shuttle.

While in orbit, the combined crew of astronauts and cosmonauts took time out to pay tribute to the victims of the 11 September attacks in the United States and the rescuers and investigation teams still working on the aftermath of the tragic day.

The official hand-over between ISS resident crews occurred on 13 December amid a week of briefings and exchange activities. One of the three Shuttle Inertial Measurement Units (IMU-2), the orbiter's primary navigation units, experienced a problem on 12 December and was taken off line. Only two of the units were working at the time to save electricity, so IMU-3 was brought back on line to support operations. The failed unit worked after this exchange but remained off line for the rest of the flight without impact upon the mission. Prior to return to Earth after undocking from the station, the crew deployed a small satellite (Starshine 2) from a GAS canister located in the payload bay. It was estimated that over 30,000 students from 650 schools in 26 countries would track the satellite during its eight months orbiting the Earth.

Milestones

229th manned space flight
137th US manned space flight
107th Shuttle mission
17th flight of Endeavour
51st US and 84th flight with EVA operations
12th Shuttle ISS mission
4th Endeavour ISS mission
4th MPLM flight
2nd MPLM-2 Raffaello flight
1st utilisation flight

ISS EO-4

Int. Designation	N/A (launched on STS-108)
Launched	5 December 2001
Launch Site	Pad 39B, Kennedy Space Center, Florida
Landed	19 June 2002 (aboard STS-111)
Landing Site	Edwards AFB, California
Launch Vehicle	STS-108
Duration	195 days 19 hrs 38 min 12 sec
Call sign	Skif (Scythian)
Objective	ISS-4 expedition programme

Flight Crew

ONUFRIYENKO, Yuri Ivanovich, 40, Russian Air Force, ISS-4 and Soyuz TM commander, 2nd mission
Previous mission: Soyuz TM23 (1996)
BURSCH, Daniel Wheeler, 44, USN, ISS-4 flight engineer 1, 4th mission
Previous missions: STS-51 (1993); STS-68 (1994); STS-77 (1996)
WALZ, Carl Erwin, 46, USAF, ISS-4 flight engineer 2, 4th mission
Previous missions: STS-51 (1993); STS-65 (1994); STS-79 (1996)

Flight Log

The fourth expedition featured an experiment programme of 26 US and 28 Russian experiments in the fields of bioastronautics research, physical sciences, space product development, fundamental space biology, Earth observations, education and technology. Many of these experiments were carried over from previous expeditions.

During their stay aboard the station, the crew performed three EVAs. The first two (14 Jan for 6 hours 3 minutes and 25 Jan for 5 hours 59 minutes) were conducted from the Pirs facility on the Russian segment. The final one (20 Feb for 5 hours 47 minutes) was conducted out of the Quest airlock on the American segment. During the first EVA, Onufriyenko and Walz relocated the cargo boom for the Strela crane from PMA-1 to the outside of Pirs. An amateur radio antenna was also installed on Zvezda. EVA-2 saw Onufriyenko and Bursch install six deflector shields on the Service Module thrusters, install a further antenna and retrieve various science packages from Zvezda. The final EVA was conducted by the two American members of the ISS-4 crew on the 40th anniversary of the first flight of John Glenn, the first American to orbit the Earth. Walz and Bursch performed the EVA from the Quest airlock, the first without a docked Shuttle in support. The objective was to prepare the area for installation of the S0 Truss during STS-110 in April 2002, relocating the tools to be used and inspecting the exterior of the station.

The ISS-4 inside the US Destiny lab. L to r Walz, Onufriyenko, Bursch

During the mission, the crew would oversee the upgrading of software as well as receiving the Progress M1-8 re-supply craft. Their visitors arrived on the STS-110 mission that delivered the S0 Truss assembly and on the Soyuz TM33 Taxi mission. Prior to the arrival of Soyuz TM34, the crew had to relocate TM33 from the nadir port of Zarya to redock with the Pirs docking port.

The crew experienced several power problems and a number of failures of the Elektron oxygen generator. To conserve oxygen reserves in the tanks of Quest, the crew resorted to burning solid fuel oxygen generators in the Zvezda module while they attempted repairs to Elektron. This took several days during May. They also experienced difficulties with the operation of the Canadarm2, including the failure of the wrist roll joint, which would require replacement.

They were relieved on the station in June by the ISS-5, who arrived on STS-111, the mission that would take the ISS-4 crew back to Earth. Both the Americans in the ISS-4 crew surpassed Shannon Lucid's previous US space flight endurance record (188 days), eventually clocking 196 days. In their careers, both men had flown four times and by the end of their ISS mission, Walz had logged 231 days in space while Bursch had accumulated 227 days. Onufriyenko had also completed two long missions and brought his experience to 389 days by the end of this one.

Milestones

4th ISS resident crew
3rd ISS EO crew to be launched by Shuttle
1st EVA from Quest without a Shuttle docked to the station

STS-109

Int. Designation	2002-010A
Launched	1 March 2002
Launch Site	Pad 39A, Kennedy Space Center, Florida
Landed	12 March 2002
Landing Site	Runway 33, Shuttle Landing Facility, KSC, Florida
Launch Vehicle	OV-102 Columbia/ET-112/SRB BI-111/SSME #1 2056; #2 2053; #3 2047
Duration	10 days 22 hrs 11 min 9 sec
Call sign	Columbia
Objective	4th Hubble Service Mission (HST SM 3B)

Flight Crew

ALTMAN, Scott Douglas, 42, USN, commander, 3rd mission
Previous missions: STS-90 (1998); STS-106 (2000)
CAREY, Duane Gene, 44, USAF, pilot
GRUNSFELD, John Mace, 43, civilian, mission specialist 1, payload commander, 4th mission
Previous missions: STS-67 (1995); STS-81 (1997); STS-103 (1999)
CURRIE, Nancy Jane, 43, US Army, mission specialist 2, 4th mission
Previous missions: STS-57 (1993); STS-70 (1995); STS-88 (1998)
LINNEHAN, Richard Michael, 44, civilian, mission specialist 3, 3rd mission
Previous missions: STS-78 (1996); STS-90 (1998)
NEWMAN, James Hansen, 45, civilian, mission specialist 4, 4th mission
Previous missions: STS-51 (1993); STS-69 (1995); STS-88 (1998)
MASSIMINO, Michael James, 39, civilian, mission specialist 5

Flight Log

The scheduled launch on 28 February was postponed 24 hours before tanking operations commenced when adverse weather conditions threatened launch criteria. Waiting 24 hours also gave the launch team the option of back-to-back launch opportunities, but they did not need them as launch occurred without delay on 1 March. Following the launch, controllers noted a degradation of the flow rate in one of two freon coolant loops which help dissipate heat from the orbiter. After a management review, the mission was given a "go" for its full duration. The problem had no impact on the crew's activities and the vehicle de-orbited nominally.

Hubble was grappled and secured in the payload bay by the RMS on 2 March (FD 2). A series of five EVAs were completed by the crew, working in pairs. Grunsfeld (EV1) and Linnehan (EV2) completed EVAs 1, 3 and 5, while Newman (EV3) and Massimino (EV4) completed EVAs 2 and 4. When not performing an EVA, the resting

John Grunsfeld (right) and Richard Linnehan signal the close of the fifth and final EVA at Hubble. One more service mission is planned for the telescope in 2008

team also acted as IV crew for those who were outside, and serviced, cleaned and prepared their own equipment ready for their next excursion. Each EVA was supported by Nancy Currie operating the RMS, with Altman and Carey photo-documenting the activities.

During the first EVA (4 Mar for 7 hours 1 minute), the astronauts removed the older starboard solar array from the telescope (attached during STS-61 in December 1993) and installed a new third-generation array. The old (retracted) array was then stowed in Columbia's payload bay for return to Earth for analysis of its condition after nine years in space. During EVA 2 (5 Mar for 7 hours 16 minutes), the new port array was installed, together with a new Reaction Wheel Assembly after the removal of the older array. The astronauts also installed thermal blankets on Bay 6, door stop extensions on Bay 5 and foot restraints to assist with the next EVA. EVA 2 also included a test of bolts located on the aft shroud doors. The lower two bolts were found to need replacing, which they accomplished successfully. EVA 3 (6 Mar for 6 hours 48 minutes) was delayed by a fault in Grunsfeld's suit, but after changing the HUT, they continued with the EVA programme. This included replacing the Power Control Unit (PCU) with a new unit capable of handling 20 per cent of power output generated from the new arrays. The extracted PCU was the original launched on the telescope in 1990, and this operation required the telescope to be powered down. This was the first time since its launch that Hubble had been turned off. The astronauts removed all 36 connectors to the old PCU and stowed it in the payload bay before attaching the new unit within 90 minutes. One hour later, the new unit passed its tests and Hubble came back to life. EVA 4 (7 Mar for 7 hours 18 minutes) completed the first science instrument upgrade of the mission by removing the last original instrument on the telescope, the Faint Object Camera, and installing the Advanced Camera for Surveys. They also installed the first element of an environmental cooling system, called the Electronics Support Module (ESM). The rest of the system would be installed the following day. The final EVA (8 Mar for 7 hours 32 minutes) saw the installation of the Near Infrared Camera and Multi-Object Spectrometer (NICMOS) in the aft shroud and the connection of cables to the ESM. They also installed the Cooling System Radiator on the outside of Hubble and fed radiator wires through the bottom of the telescope to connections on NICMOS.

Hubble was released by the RMS on 9 March (FD 9) and the next day was a rest day for the astronauts. During the day, they took the opportunity to speak with the ISS-4 crew (Yuri Onufriyenko, Carl Walz and Dan Bursch). FD 11 saw a full systems check before landing at the first opportunity at the Cape on FD 12, rounding out a highly successful mission. At this time, there was a further Hubble service mission on the manifest (HST SM #4) in 2004 or 2005, with a close-out mission in 2010. The options of either bringing the telescope back to Earth for eventual display in a museum or leaving it in orbit, boosted to a higher apogee to reduce atmospheric drag, were still being considered when Columbia was lost in February 2003. It looked as though Hubble was likely be abandoned when its systems eventually failed, but there was also growing support both inside and outside of NASA to devote one Shuttle mission to revisit the telescope before the Shuttle fleet is retired in 2010. In

October 2006, a return to Hubble was authorised for 2008 due to public and scientific demand for keeping the telescope working for as long as possible.

Milestones

230th manned space flight
138th US manned space flight
108th Shuttle mission
27th flight of Columbia
52nd US and 85th flight with EVA operations
4th Hubble service mission (3B)
EVA duration record for single Shuttle mission (35 hrs 55 min)

STS-110

Int. Designation	2002-018A
Launched	8 April 2002
Launch Site	Pad 39B, Kennedy Space Center, Florida
Landed	20 April 2002
Landing Site	Runway 33, Shuttle Landing Facility, KSC, Florida
Launch Vehicle	OV-104 Atlantis/ET-114/SRB BI-112/SSME #1 2048; #2 2051; #3 2045
Duration	10 days 19 hrs 42 min 44 sec
Call sign	Atlantis
Objective	ISS assembly flight 8A; delivery of S0 Truss and Mobile Transporter

Flight Crew

BLOOMFIELD, Michael John, 43, USAF, commander, 3rd mission
Previous missions: STS-86 (1997); STS-97 (2000)
FRICK, Stephen Nathaniel, 37, USN, pilot
WALHEIM, Rex Joseph, 39, USAF, mission specialist 1
OCHOA, Ellen Lauri, 43, civilian, mission specialist 2, 4th mission
Previous missions: STS-56 (1993); STS-66 (1994); STS-96 (1999)
MORIN, Lee Miller Emile, 49, USN, mission specialist 3
ROSS, Jerry Lynn, 54, civilian, mission specialist 4, 7th mission
Previous missions: STS 61-B (1985); STS-27 (1988); STS-37 (1991); STS-55 (1993); STS-74 (1995); STS-88 (1998)
SMITH, Steven Lee, 43, civilian, mission specialist 5, 4th mission
Previous missions: STS-68 (1994); STS-82 (1997); STS-103 (1999)

Flight Log

The 4 April launch was terminated an hour into the tanking due to a leak in a LH vent line on the Mobile Launcher Platform at the pad. Following repairs to the line, the launch was rescheduled to 8 April but was delayed on the day due to drop-outs in a back-up launcher processing system. After reloading the data, the launch was achieved with just 11 seconds remaining in the launch window.

Docking with ISS occurred on FD 3 (10 Apr) and over the next 170 hours, the Shuttle crew completed 4 EVAs to install the S0 Truss. During and in between the EVAs, the astronauts transferred supplies, equipment and experiments to the station, and brought back trash and unwanted hardware. They also transferred 45 kg of oxygen and 22 kg of nitrogen to the storage tanks in Quest to re-pressurise the airlock following EVA operations. A total of 664 kg of water was transferred to the ISS, along with an experimental plant growth chamber which replaced a crystal growth experi-

Steve Smith works inside the S0 Truss, newly installed on ISS. Rex Walheim (out of frame) worked in tandem with Smith during the mission's third EVA

ment that would be returned to Earth. The crew also transferred a new freezer for future crystal sample storage.

During FD 4, the S0 Truss was lifted out of the payload bay of Atlantis by Ellen Ochoa using the station's RMS, assisted by ISS-4 crew member Dan Bursch. It was located onto a clamp at the top of the Destiny lab, where it would serve as a platform on which other trusses would be attached and additional solar arrays mounted. The truss also included navigation devices, computers, coolant and power systems for additional laboratories and facilities to be added to the station later.

The EVAs were completed by two pairs of STS-110 astronauts. The first and third were performed by Smith (EV1) and Walheim (EV2), while the second and fourth were conducted by Ross (EV3) and Morin (EV4) (dubbed the "Silver" Team, as they were both grandfathers). EVA 1 (11 Apr for 7 hours 48 minutes) focused mainly on electrical and structural connections of the truss to the station after it had been moved from Atlantis's payload bay. The astronauts attached four mounting struts, deployed avionics trays and connected cables from Destiny to the new addition to the station. EVA 2 (13 Apr for 7 hours 30 minutes) saw the astronauts bolt the final two struts to the lab. Launch support panels and clamps were removed and a back-up device with an umbilical reel for the Mobile Transporter railcar was also installed. EVA 3 (14 Apr for 6 hours 27 minutes) was used to reconfigure electrical connections from the US lab to the truss for powering the Canadarm2. Clamps were also released on the Mobile Transporter cart during this EVA. EVA 4 (16 Mar for 6 hours 37 minutes) saw the installation of a 4.267-metre beam called the Airlock Spur from the S0 Truss to Quest to provide a quick pathway for future EVA astronauts. Floodlights, work platforms and electrical connections were also installed and connected in this final excursion of the mission.

Initial tests of the Mobile Transporter (railcar) were successfully completed on FD 8. ISS-4 crew member Walz commanded the transporter, using a laptop computer to move it to a work site about 5.2 metres down a rail that spanned the entire length of the 13.4 m truss. Then it was moved to a second site and back to the first. The unmanned cart moved about 22 m in total at a rate of about 3.5 cm per second. Automatic latching did not occur due to the railcar lifting slightly, but manual latching was successfully achieved. This unit would be extended over the coming missions and would be used to ease the translation of astronauts and equipment down the length of the completed truss in future years.

Milestones

231st manned space flight
139th US manned space flight
109th Shuttle mission
25th flight of Atlantis
53rd US and 86th flight with EVA operations
13th ISS Shuttle mission
5th Atlantis ISS mission
1st person to make 7 space flights (Ross)
US career EVA record of 58 hrs 18 minutes on 9 EVAs over 4 missions (Ross)

SOYUZ TM34

Int. Designation	2002-020A
Launched	25 April 2002
Launch Site	Pad 1, Site 5, Baikonur Cosmodrome, Kazakhstan
Landed	5 May 2002 (in Soyuz TM33)
Landing Site	26 km southeast of Arkalyk
Launch Vehicle	R7 (11A511U); spacecraft serial number (7K-M) 208
Duration	9 days 21 hrs 25 min 18 sec
Call sign	Uran (Uranus)
Objective	ISS mission 4S; Soyuz ferry exchange; Soyuz visiting mission 3; Italian *Marco Polo* research mission; South African SFP science mission

Flight Crew

GIDZENKO, Yuri Pavolich, 40, Russian Air Force, commander, 3rd mission
Previous missions: Soyuz TM22 (1995); ISS-1 (2000)
VITTORI, Roberto, 37, Italian Air Force, flight engineer
SHUTTLEWORTH, Mark, 28, civilian, South African space flight participant

Flight Log

This mission successfully exchanged the older Soyuz TM33 spacecraft for a "fresh" return capsule at ISS. In addition, Italian ESA astronaut Vittori completed a science programme for the Italian Space Agency and the second private fare-paying space flight participant (Shuttleworth) also became the first citizen from South Africa to fly into space. Soyuz TM34 docked with ISS at the Zarya nadir docking port on 27 April.

In addition to the exchange of personal effects and flight hardware required to bring home the older Soyuz and leave the newer vehicle for the ISS-4 resident crew, the two Russian commanders worked on medical experiments and a joint Russian/German/French plasma crystal experiment during the week of joint activities. Vittori's activities under the *Marco Polo* science programme included 23 sessions with four biomedical experiments. These included the relationship of the health of the individual to possible reductions in working capacity, an in-orbit test of the functional capability of a new integrated garment, and a medical experiment on the effects of space radiation on the functional state of the central nervous system and into the working capacity of the test subject. There was also a study of the vegetative regulation of arterial pressure and heart rate.

Mark Shuttleworth was determined to make his visit to the space station more scientifically rewarding and valuable than that of the previous space flight participant, Dennis Tito. He actively developed a programme of life science experiments utilising Russian equipment already aboard the station, as well as bringing four South African

The Soyuz Taxi-3 crew onboard the ISS. L to r Soyuz Commander Yuri Gidzenko, ESA astronaut Roberto Vittori and South African SFP Mark Shuttleworth

university-developed experiments with him on Soyuz. One of these would focus on stem cell research. Shuttleworth also took saliva samples from himself and his crew mates as part of the embryo and stem cell development experiment.

At the end of the week's visit, the crew packed their experiment results and cargo into the DM of Soyuz TM33. Soyuz has a limited cargo return capability and could only return 50 kg worth of cargo, of which 15 kg was allocated to the return of Italian experiment results and data. The landing of TM33 occurred without incident on 5 May, ending another highly successful visiting ferry exchange mission. It was hoped that these missions would become a regular occurrence twice a year at the station, but not with Soyuz TM. A new variant of Soyuz was waiting in the wings, and this version would have the capacity to carry taller crew members, incorporate upgrades to onboard systems and hardware and be capable of a longer orbital service life of up to a year. The flight of TM34 was therefore the last of a series that had first flown in space in May 1986.

Milestones

232nd manned space flight
93rd Russian manned space flight
86th manned Soyuz mission
33rd manned Soyuz TM mission
4th ISS Soyuz mission (3S)
3rd Soyuz ISS taxi flight
3rd ISS visiting mission
Final Soyuz TM mission
1st South African citizen in space (Shuttleworth)

STS-111

Int. Designation	2002-028A
Launched	5 June 2002
Launch Site	Pad 39A, Kennedy Space Center, Florida, USA
Landed	19 June 2002
Landing Site	Runway 22, Edwards AFB, California
Launch Vehicle	OV-105 Endeavour/ET-113/SRB BI-113/SSME #1 2050; #2 2044; #3 2054
Duration	13 days 20 hrs 35 min 56 sec
Call sign	Endeavour
Objective	ISS assembly mission UF2; MPLM logistics flight; ISS resident crew exchange

Flight Crew

COCKRELL, Kenneth Dale, 52, civilian, commander, 5th mission
Previous missions: STS-56 (1993); STS-69 (1995); STS-80 (1996); STS-98 (2001)
LOCKHART, Paul Scott, 46, USAF, pilot
CHANG-DIAZ, Franklin Ramon de Los Angeles, 52, civilian, mission specialist 1, 7th mission
Previous missions: STS 61-C (1986); STS-34 (1989); STS-46 (1992); STS-60 (1994); STS-75 (1996); STS-91 (1998)
PERRIN, Philippe, 39, French Air Force, mission specialist 2

ISS-5 crew up only:
KORZUN, Valery Nikolayevich, 49, Russian Air Force, mission specialist 3, ISS-5 and Soyuz commander, 2nd mission
Previous mission: Soyuz TM24 (1996)
WHITSON, Peggy Annette, 42, civilian, mission specialist 4, ISS-5 science officer
TRESCHEV, Sergei Vladimiriovich, 43, civilian, Russian mission specialist 5, ISS-5 flight engineer

ISS-4 crew down only:
ONUFRIYENKO, Yuri Ivanovich, 40, Russian Air Force, ISS-4 and Soyuz commander, mission specialist 3, 2nd mission
Previous mission: Soyuz TM23 (1996)
BURSCH, Daniel Wheeler, 44, USN, ISS-4 flight engineer 1, mission specialist 4, 4th mission
Previous missions: STS-51 (1993); STS-68 (1994); STS-77 (1996)
WALZ, Carl Erwin, 46, USAF, ISS-4 flight engineer 2, mission specialist 5, 4th mission
Previous missions: STS-51 (1993); STS-65 (1994); STS-79 (1996)

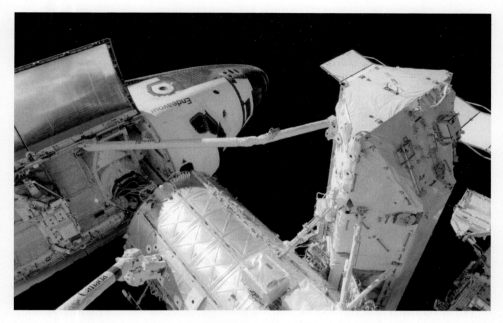

Endeavour is shown docked to the ISS at the PMA-2 on the forward end of the Destiny laboratory. A portion of the Canadarm2 is visible in the lower left corner and the Endeavour RMS is in full view stretched out with the S0 truss at its end

Flight Log

The original 29 May launch date for this mission was scrubbed due to adverse weather conditions and the rescheduled 31 May launch was also cancelled, this time due to the discovery on 30 May of pressure differences in the left OMS pod on the Endeavour. The component would be replaced on the pad and the launch date changed to 4 June, but the unique nature of this problem meant additional work to build a test fixture, so the launch had to be delayed 24 hours to 5 June. The docking with ISS occurred on FD 3 (7 June) and the vehicles remained docked together for 189 hours. Later on FD 3, the official change-over of ISS resident crew occurred, with the ISS-4 crew formally ending their 182-day residence aboard the station and becoming part of the STS-111 crew. The ISS-5 crew, now with their seat liners in the Soyuz TM34 DM, were officially the resident crew aboard the station.

Three EVAs were completed by Chang-Diaz (EV1) and French astronaut Phillip Perrin (EV2). During their first EVA (9 Jun for 7 hours 14 minutes), they installed power and data cables and a grapple fixture to the P6 Truss, which would be used to relocate it to its final position. They also retrieved six micrometeoroid shields from Endeavour's payload bay for temporary stowage on PMA-1. These would be installed on Zvezda by a later crew. They also inspected and photographed the failed CMG on the Z1 Truss and removed thermal blankets from the Mobile Base System before positioning it above the Mobile Transporter to thermally condition it prior to installation on the next EVA. During the second EVA (11 Jun for 5 hours), the two astronauts connected primary and back-up video and data cables between the MT railcar and MBS, and deployed an auxiliary grapple fixture on the MBS. This was called the Payload Orbital Replacement Unit Accommodation (POA) and is designed to grapple future payloads and hold on to them as they are moved across the station's truss atop the MBS. The attachment of four bolts completed the installation of the MBS and the crew also relocated a TV camera for better views of station assembly and maintenance operations. During the final EVA (13 Jun for 7 hours 17 minutes), the astronauts replaced the faulty Canadarm2 wrist roll joint with a new unit that had been brought up to the station with them in the Shuttle's payload bay. The faulty joint was stowed in the payload bay for return to Earth. This repair restored the station's RMS system to operational status.

MLPM Leonardo was moved from the payload bay of Endeavour to the side of the Unity module on 8 June and remained there for unloading and loading until it was returned to the payload bay on 14 June. In the transfer of logistics, cargo, hardware and supplies to the station, the crew relocated 3,652 kg from Leonardo and a further 453 kg from lockers on Endeavour's mid-deck. For the return to Earth, the MPLM was filled with 2,117 kg of equipment, waste and items no longer needed and a further 453 kg of returned material was located in the mid-deck lockers of the Shuttle. Among the items moved over to the station was a new science rack to house microgravity experiments, and a glove box that would permit the station crew to begin a series of experiments that required isolation conditions.

The Shuttle landed at Edwards AFB after three days of trying to land at the Cape. Low clouds, rain and thunderstorms cancelled KSC landing attempts on 17, 18 and 19 June, forcing the decision to land at Edwards AFB and giving the Shuttle and ISS-4 crews an additional two days in space.

Milestones

233rd manned space flight
140th US manned space flight
110th Shuttle mission
18th flight of Endeavour
54th US and 87th flight with EVA operations
14th Shuttle ISS mission
5th Endeavour ISS mission
5th MPLM flight
3rd MPLM-1 (Leonardo) flight
1st French Shuttle crew member EVA
1st French EVA from ISS (via Quest)

ISS EO-5

Int. Designation	N/A (launched on STS-111)
Launched	5 June 2002
Launch Site	Pad 39A, Kennedy Space Center, Florida
Landed	7 December 2002 (aboard STS-113)
Landing Site	Shuttle Landing Facility, KSC, Florida
Launch Vehicle	STS-111
Duration	184 days 22 hrs 14 min 23 sec
Call sign	Freget (Frigate)
Objective	ISS-5 expedition programme

Flight Crew

KORZUN, Valery Nikolayevich, 49, Russian Air Force, ISS-5 and Soyuz commander, 2nd mission
Previous mission: Soyuz TM24 (1996)
WHITSON, Peggy Annette, 42, civilian, ISS-5 science officer
TRESCHEV, Sergei Vladimiriovich, 43, civilian, Russian ISS-5 flight engineer

Flight Log

The fifth expedition to the ISS featured a science programme of 24 American and 29 Russian experiments. Whitson had the added privilege of performing an experiment during her mission on ISS for which she was principle investigator. The renal stone experiment was a research programme to study the possible formation of kidney stones during prolonged space flight. Whitson kept regular logs of her food intake and took a regular course of tablets of either potassium citrate or a placebo. By mid-July, the ESA glove box facility had been activated, but communication problems with the new unit meant that Whitson had to forego regular daily exercises for a couple of days while the problems were resolved.

During the residency, the crew received two Progress re-supply craft. In late June, Progress M1-8 was replaced by Progress M46, which delivered 2,580 kg of cargo for the crew and 825 kg of fuel. Three months later, Progress M1-9 replaced the M46 ferry and brought over 2,600 kg of cargo, including equipment for the ESA *Odessa* science programme in November. These regular re-supply flights were the lifeline of the station's main crew, supplementing the heavy lift capability of the Shuttle, and serving as an orbital refuse collection service once the new cargo had been unpacked.

August was mainly focused on EVAs. The first (16 Aug for 4 hours 25 minutes) saw Whitson and her commander start late due to a caution and warning signal that indicated a fault on their Orlan pressure suits. Recycling the pre-EVA operations to fix the problem meant that the EVA started 1 hour and 43 minutes late. The two crew members used the Strela boom to access the work area to place six (of an eventual 23)

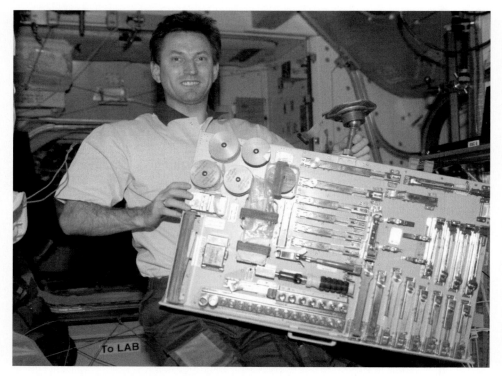

Cosmonaut Sergei Treshchev, ISS-5 flight engineer, holds a special pallet containing various tools used for orbital repairs and DIY aboard the station

micrometeoroid protection panels on the Zvezda module. Due to the late start, the installation of a Kromka detector, and the gathering of samples of thruster residue on the surface of Zvezda caused by other thrusters on the module, would be rescheduled for later EVAs. The second excursion (26 Aug for 5 hours 21 minutes) was also delayed 27 minutes, this time by a small leak from the pressure seals between Zvezda transfer compartments and where Pirs was docked to it. Recycling the hatch valves seemed to solve the problem. The cosmonauts set up TV cameras to record their activities, as well as an external Japanese experiment for specialists back in Japan. They also deployed the Kromka-2 deflector plate evaluator and retrieved an earlier plate to be returned to Earth for analysis, as well as deploying the final two ham radio antennas.

The ISS-5 crew received the STS-112 Shuttle crew in October (who delivered the S1 Truss), as well as the fourth visiting crew in the new spacecraft Soyuz TMA in November. After just over a week aboard the station, the visiting crew departed in the older TM34, marking the final re-entry and landing of that variant of the venerable Soyuz. Shortly after the departure of the visiting crew, STS-113 arrived with the replacement ISS-6 resident crew, returning home with the ISS-5 crew.

During their residency, the ISS-5 crew encountered and overcame a number of equipment problems, and conducted repairs and maintenance. Whitson wrote a series of journals about life and work on board ISS that were posted on the NASA web site and provided a fascinating insight into life aboard the station. On 16 September, NASA designated her the first NASA science officer, a designation that would be assigned to an American member of each crew from now on. She later wrote that the title was fine, apart from the number of emails she had received from friends all likening her to Mr. Spock, the science officer of the USS Enterprise in the original *Star Trek*.

Milestones

5th ISS resident crew
4th ISS EO crew to be launched by Shuttle
1st designated NASA science officer (Whitson)

STS-112

Int. Designation	2002-047A
Launched	7 October 2002
Launch Site	Pad 39B, Kennedy Space Center, Florida
Landed	18 October 2002
Landing Site	Runway 33, Shuttle Landing Facility, KSC, Florida
Launch Vehicle	OV-104 Atlantis/ET-115/SRB BI-115/SSME #1 2048; #2 2051; #3 2047
Duration	10 days 19 hrs 58 min 44 sec
Call sign	Atlantis
Objective	ISS assembly flight 9A; delivery and installation of S1 Truss and Crew Equipment Translation Aid (CETA) cart A

Flight Crew

ASHBY, Jeffrey Shears, 48, USN, commander, 3rd mission
Previous missions: STS-93 (1999); STS-100 (2001)
MELROY, Pamela Ann, 41, USAF, pilot, 2nd mission
Previous mission: STS-92 (2000)
WOLF, David Alan, civilian, mission specialist 1, 3rd mission
Previous missions: STS-58 (1993); STS-86/89 (1997)
MAGNUS, Sandra Hall, 37, civilian, mission specialist 2
SELLERS, Piers John, 47, civilian, mission specialist 3
YURCHIKIN, Fyodor Nikolayevich, 43, civilian, Russian mission specialist 4

Flight Log

Hurricane Lili, out in the Gulf of Mexico, had threatened mission control at JSC, and as the path of the storm could not be determined until late in its track, the decision was taken to power down the Houston centre. This meant that the original launch date for STS-112 of 2 October had to be rescheduled for 7 October. On launch, a back-up separation pyrotechnic system had to be used to release one of the SRBs from the launch platform when the primary charge failed to sever the hold-down bolts and release ground connections to the ET.

Docking of Atlantis to ISS was achieved on 9 October. The primary objective of this flight was the transfer and installation of the S1 Truss and the Crew Equipment Translation Aid (CETA) cart A. This was the first of two human-powered carts designed to traverse along the MBS rail, providing mobile work platforms for future EVA operations. The relocation of the truss was achieved the day after the docking, with MS Sandra Magnus and ISS SO Peggy Whitson using the Canadarm2 to relocate the truss at the starboard end of the S0 Truss by means of four remotely-controlled bolts.

The International Space Station as of October 2002. The departing Atlantis crew photographed the station following the undocking, and the newly added Starboard 1 (S1) Truss is visible in upper centre frame

During the first EVA (10 Oct for 7 hours 1 minute), Wolf (EV1) and Sellers (EV2) connected power, data and fluid lines and released launch bolts that allowed the S1 radiators to be orientated for optimum cooling. They also deployed a new S-band antenna near the end of the S1 Truss to increase voice communications capability with ground controllers. After releasing the launch restraints on the CETA-A, the two astronauts installed S1's nadir external camera. EVA 2 (12 Oct for 6 hours 4 minutes) featured further work with CETA, the installation of 22 Spool Positioning Devices (SPD) on the ammonia cooling line connections and a second exterior camera, this time on the Destiny lab, and the preparation and checking of equipment to support the attachment of the next truss section. EVA 3 (14 Oct for 6 hours 36 minutes) focused on connecting ammonia lines, removing a structural support clamp and installing SPDs on a pump assembly, as well as removing a bolt that prevented the activation of a cable cutter on the Mobile Transporter.

In between the EVAs, the STS-112 crew worked with the ISS-5 crew to transfer 816 kg of logistics and supplies to the ISS. Approximately the same mass was brought back by the Shuttle at the end of the mission. The crews also repaired the exercise treadmill vibration dampening system in Zvezda, adjusted protective circuits which measured electrical current on the S1 radiator assembly to a greater tolerance for its use in space, and removed and replaced a humidity separator in the Quest airlock which had been leaking. The crews moved new scientific experiments across to the station and relocated completed ones in the mid-deck of Atlantis. Seven water containers and a new protein crystal growth experiment were moved over to ISS, while liver cell samples were stowed carefully in the Shuttle. In addition, 123 kg of gaseous nitrogen was transferred in two batches (7 kg then 116 kg) from Atlantis to the station's storage tanks.

Milestones

234th manned space flight
141st US manned space flight
111th Shuttle mission
26th flight of Atlantis
55th US and 88th flight with EVA operations
15th Shuttle ISS mission
6th Atlantis ISS mission

SOYUZ TMA1

Int. Designation	2002-050A
Launched	30 October 2002
Launch Site	Pad 1, Site 5, Baikonur Cosmodrome, Kazakhstan
Landed	10 November 2002 (in TM34)
Landing Site	81 km north east of Arkalyk
Launch Vehicle	R7 (Soyuz FG); spacecraft serial number (7K design unknown) #211
Duration	10 days 20 hrs 53 min 09 sec
Call sign	Yenisei (Yenisey)
Objective	ISS mission 5S; Soyuz exchange mission; Soyuz visiting mission 4 (VC-4); Belgian *Odessa* mission

Flight Crew

ZALETIN, Sergei Viktorovich, 40, Russian Air Force, commander, 2nd mission
Previous mission: Soyuz TM30 (2000)
De WINNE, Frank, 41, Belgian Air Force, flight engineer 1
LONCHAKOV, Yuri Valentinovich, 37, Russian Air Force, flight engineer 2, 2nd mission
Previous mission: STS-100 (2001)

Flight Log

The crew for TMA1 seemed to be finalised in July 2002, with Mir veteran Sergei Zaletin and ESA Belgian astronaut Frank De Winne being joined by N Sync pop singer Lance Bass as the third space flight participant. This was the latest in a long line of suggested "millionaire" fare-paying cosmonauts for the flight. However, by 20 August, no payment from sponsors was forthcoming and Bass was removed from the crew.

To fill the seat and return the crew to a full complement of three, back-up commander Yuri Lonchakov was reassigned at short notice to fly the mission, the inaugural flight of the new Soyuz TMA1 spacecraft. TMA (Transport, Modification, Anthropometric) featured changes to allow taller and smaller crew members to fly in it, which meant that many of the American astronauts that had previously been unsuitable for Soyuz or ISS missions could now be considered for TMA training, a timely factor that became very fortunate in the next few months. Internal systems and provisions for comfort would allow crew members between 1.5 and 1.9 m tall, instead of the previous 1.64 and 1.82 m in the TM craft.

Soyuz TMA1 was the first new variant of Soyuz to fly without a prior unmanned flight, and it docked with ISS on 1 November. During the week aboard the station, the two Russian cosmonauts briefed the Russian ISS-5 crew members on the features of

The first TMA crew pose for a group photo with the ISS-5 resident crew. In foreground is ISS-5 commander Valeri Korzun, in middle row is TMA1 commander Sergei Zalyotin (left) and Belgian ESA astronaut Frank De Winne. In the back row l to r are ISS-5 FE Peggy Whitson, TMA1 FE Yuri Lonchakov and ISS-5 FE Sergei Treshev

the new spacecraft, assisted them with their work in the Russian segment, and participated in a small Russian science programme. They also assisted their Belgian colleague with his work. De Winne, the Belgian astronaut, conducted an ESA programme under the codename of *Odessa* that comprised 23 experiments. The programme featured research in the fields of biology, human physiology, physical sciences and education. He also talked with six students from universities in Scotland, Italy and The Netherlands who were at the ESA Taxi Flight Operations Coordination Centre (TOCC) at ESTEC in The Netherlands.

The crew landed in the TM34 spacecraft, the final descent of that variant of vehicle. According to Zaletin, the landing itself was "a little hard" as the vehicle hit the ground and tumbled a few times before coming to a halt. It was the first Russian night landing in ten years. The new TMA1 spacecraft, now docked to the ISS, would provide a return capability for the ISS-5 crew in the event of an emergency. This capability was passed over to the ISS-6 crew in December, though they had not expected to use it.

Milestones

235th manned space flight
94th Russian manned space flight
87th manned Soyuz mission
1st manned Soyuz TMA mission
5th ISS Soyuz mission (5S)
4th ISS Taxi flight
4th ISS visiting mission
1st manned flight of (R7) Soyuz FG launch vehicle

STS-113

Int. Designation	2002-052A
Launched	23 November 2002
Launch Site	Pad 39A, Kennedy Space Center, Florida
Landed	7 December 2002
Landing Site	Runway 33, Shuttle Landing Facility, KSC, Florida
Launch Vehicle	OV-105 Endeavour/ET-116/SRB BI-114/SSME #1 2050; #2 2044; #3 2045
Duration	13 days 18 hrs 48 min 38 sec
Call sign	Endeavour
Objective	ISS assembly mission 11A; delivery and installation of P1 Truss and the CETA-B cart; ISS resident crew exchange mission

Flight Crew

WETHERBEE, James Donald, 49, USN, commander, 6th mission
Previous missions: STS-32 (1990); STS-52 (1992); STS-63 (1995); STS-86 (1997); STS-102 (2001)
LOCKHART, Paul Scott, 46, USAF, pilot, 2nd mission
Previous mission: STS-111 (2002)
LOPEZ-ALEGRIA, Michael Eladio, 44, USN, mission specialist 1, 3rd mission
Previous missions: STS-73 (1995); STS-92 (2000)
HERRINGTON, John Bennett, 44, USN, mission specialist 2

ISS-6 crew up only:

BOWERSOX, Kenneth Duane, 45, USN, mission specialist 3, ISS-6 commander, 5th mission
Previous missions: STS-50 (1990); STS-61 (1993); STS-73 (1995); STS-82 (1997)
BUDARIN, Nikolai Mikhailovich, 49, civilian, Russian mission specialist 4, ISS-6 flight engineer and Soyuz commander, 3rd mission
Previous missions: Mir EO-19/STS-71 (1995); Soyuz TM27 (1998)
PETTIT, Donald Roy, 47, civilian, mission specialist 5, US ISS-6 science officer

ISS-5 crew down only:

TRESCHEV, Sergei Vladimiriovich, 43, civilian, Russian ISS-5 flight engineer, mission specialist 3
KORZUN, Valery Nikolayevich, 49, Russian Air Force, ISS-5 and Soyuz commander, mission specialist 4, 2nd mission
Previous mission: Soyuz TM24 (1996)
WHITSON, Peggy Annette, 42, civilian, ISS-5 science officer, mission specialist 5

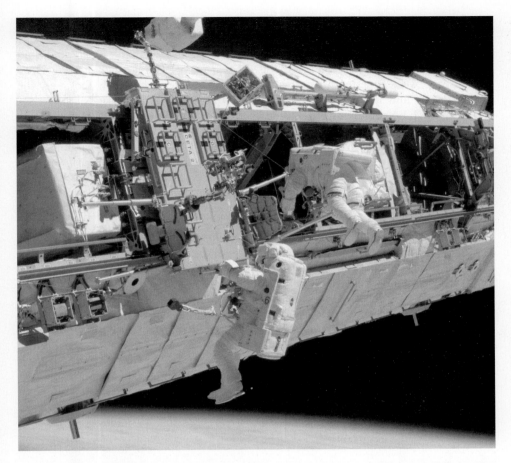

John Herrington (left) and Michael Lopez-Alegria work on the newly installed Port One (P1) Truss on ISS. Herrington is holding handrails attached to the CETA-B (2) cart

Flight Log

Higher than allowed oxygen levels detected in the orbiter's mid-body postponed the 11 November launch attempt, which was reset for 18 November. Technicians found a fatigued flexible hose to be the cause and this was replaced, but during the repair a platform impacted the RMS in the payload bay. The necessary inspections delayed the launch to 22 November. Bad weather at the TAL sites forced a further 24-hour delay in launch before the mission finally got off the ground. Docking with the station was achieved during FD 3 (25 November) and the following day, the P1 Truss was relocated to the port end of the S0 Truss and automatically bolted in place. Unbeknown at the time, this would be the last time a Shuttle docked to the station for over 30 months.

The three EVAs on this mission were conducted in support of the truss installation, as well as for a number of other tasks. EVA 1 (26 Nov for 6 hours 45 minutes) saw Lopez-Alegria (EV1) and Herrington (EV2) complete the electrical, power and fluid connections to the P1 Truss and install the SPD, ensuring that the quick disconnection mechanisms were functioning correctly. They also released launch locks on the CETA-B cart and installed the Node Wireless video system External Transceiver Assembly (WETA) antenna. This would give reception for the EVA helmet cameras without the presence of a Shuttle. The second EVA (on Thanksgiving Day, 28 Nov, for 6 hours 10 minutes) saw the crew continue the electrical and plumbing connections and the installation of a second WETA on P1. The CETA-B (or 2) cart was also installed on the S1 Truss and secured to the CETA-A (or 1) cart. The third EVA (30 Nov for 7 hours) saw the completion of the SPD installation and troubleshooting of the stalled MT. The astronauts also freed and deployed a UHF antenna that had become snagged during relocation work.

During the flight, the ISS-5 crew handed over the command of the station to the ISS-6 crew and together, both resident crews worked to repair faulty hardware and install new equipment in the station. STS-113 had delivered over 1,969 kg of hardware and supplies to ISS, including new science experiments. The Shuttle's engines were used in a series of burns to raise the orbit of the station. The formal hand-over of resident crews occurred shortly after the docking and the completion of safety briefings. Though no one knew it at the time, this would be the last Shuttle-based resident crew exchange for some time. Following undocking on 2 December, the Shuttle encountered four days of landing attempts being waived off due to bad weather, before finally making it on 7 December. This was the first time a mission had received three consecutive days of landing cancellations. This was also the last successfully completed Shuttle mission (launch to landing) for over two-and-a-half years.

Milestones

236th manned space flight
142nd US manned space flight
112th Shuttle mission
19th flight of Endeavour
56th US and 89th flight with EVA operations
16th Shuttle ISS mission
6th Endeavour ISS mission
5th Shuttle ISS resident crew exchange mission
1st native American (Chickasaw nation) to fly and walk in space (Herrington)

ISS EO-6

Int. Designation	N/A (launched on STS-113)
Launched	23 November 2002
Launch Site	Pad 39A, Kennedy Space Center, Florida
Landed	3 May 2003 (in TMA1)
Landing Site	South of Aral Sea near Turgai in Kazakhstan, 460 km short of its planned landing site
Launch Vehicle	STS-113
Duration	161 days 1 hr 14 min 38 sec
Call sign	Persey (Persey) – Soyuz TMA
Objective	ISS-6 expedition programme

Flight Crew

BOWERSOX, Kenneth Duane, 45, USN, ISS-6 commander, 5th mission
Previous missions: STS-50 (1990); STS-61 (1993); STS-73 (1995); STS-82 (1997)
BUDARIN, Nikolai Mikhailovich, 49, civilian, Russian ISS-6 flight engineer and Soyuz commander, 3rd mission
Previous missions: Mir EO-19/STS-71 (1995); Soyuz TM27 (1998)
PETTIT, Donald Roy, 47, civilian, US ISS-6 science officer

Flight Log

The sixth resident crew for ISS had arrived aboard STS-113 and had expected a Shuttle landing at the end of their mission. However, the loss of Columbia and her crew on 1 February 2003 resulted in the Shuttle fleet being grounded and the crew having to use the TMA1 spacecraft for the landing. They were replaced by the ISS-7 "caretaker crew", who arrived in TMA2. Shortly before launch on STS-113, the original NASA science officer, Don Thomas, was medically disqualified from the mission after reports stated that he had been subjected to sufficient radiation on his previous space flights that a long-duration flight could exceed his maximum allowable life time dose. Thomas was replaced by his back-up, Don Pettit who, due to the late change in the crew, had to wear the clothing already aboard the station and eat the menu already selected for Thomas. However, this was a minor issue to overcome as the crew settled down to a planned four-month tour. There were no planned visiting missions scheduled until the arrival of STS-114 to bring them home.

The science programme for this sixth residency included more than 20 new or existing investigations, for which the crew would devote over 240 hours to research time. The research fields included medicine, materials, plant science, commercial biotechnology and manufacturing. In addition, Don Pettit took time to include a series of science demonstrations that were supplemented by postings on the internet

Wearing Russian Sokol suits in preparation for the unexpected return to Earth on the TMA1 spacecraft, the ISS-6 crew are in the Russian Zvezda module on ISS. L to r ISS NASA science officer Don Pettit, ISS flight engineer and Soyuz commander Nikolai Budarin, and ISS commander Ken Bowersox

and dubbed "Saturday Morning Science" (similar to those activities conducted during the Skylab 3 mission in 1973).

There were two EVAs accomplished during the mission, though only one was originally planned. The first should have included Budarin, who would have been the first Russian to perform an EVA in an American suit from Quest. However, the EVA was delayed from 12 December to 15 January when US medical specialists ruled that Budarin did not meet the criteria for US EVAs (although he was not disqualified by Russian doctors from conducting any Russian-based EVA!). He was replaced on the EVA by Pettit. The 15 January EVA (6 hours 51 minutes) saw the two Americans release the remaining launch restraint from the P1 Truss radiator system that had been delivered by STS-113. They then witnessed its deployment, controlled from MCC Houston. Inspection and cleaning occupied the next part of their EVA, which closed with the retrieval of tools for a storage box on the Z1 Truss, and a health check on the ammonia reserves delivered in 2001 and stored on the P6 Truss. EVA 2 was added after the loss of the Columbia and conducted on 8 April (6 hours 26 minutes) to ensure that the next crew would only need to perform an EVA to deal with an emergency. The two astronauts reconfigured cables and continued external outfitting of the station pending the resumption of assembly flights. In addition to connecting electrical conduits, the crew inspected faulty heater cables, replaced a failed power control module on the MT and rerouted two of the four CMG cables on the Z1 Truss, as well as completing several smaller chores and get-ahead tasks.

Progress M1-9 was undocked from the station on 1 February, the same day that the crew were told of the loss of Columbia and the day before their new Progress resupply craft was launched. Progress M47 arrived on 4 February to deliver 2,568 kg of various cargos to restock the station. There was 870 kg of propellant, 70 kg of water, 50 kg of oxygen and 1,328 kg of dry cargo. Significantly more limited than the payload capability of the Shuttle, it would nonetheless be these unmanned re-supply missions that would be crucial to maintaining operations at ISS over the next two to three years. The engines of Progress M47 were also used to reboost the station's orbit slightly.

TMA2 docked with the station on 28 April, carrying a two-man crew to take over the residency. The ISS-6 crew departed on 3 May in TMA1, with a severely restricted return payload capability. A computer failure led to a highly ballistic descent and resulted in the spacecraft landing far off target. The crew experienced almost 8 G during the entry. It took over two hours for rescue teams to reach the crew, who had exited the Descent Module. They had been unable to communicate due to a broken antenna on the DM, so in future, all crews would be issued with mobile satellite phones. Pettit was much the worst for wear of the three crew members after the landing, being photographed sitting on the ground looking clearly distressed and taking longer to recover than his colleagues. The problem was traced to the yaw gyroscope experiencing a gimbal lock when its angular rate exceeded 54 degrees. The crew were praised for their professionalism, and changes would be incorporated from TMA3 to prevent the problem re-occurring. Of course, TMA2 was already in orbit.

Milestones

6th ISS resident crew
5th ISS EO crew to be launched by Shuttle
1st return to Earth in Soyuz (TMA) by NASA astronauts (Bowersox/Pettit)
1st NASA astronauts to return to Earth in a non-US spacecraft

STS-107

Int. Designation	2003-003A
Launched	16 January 2003
Launch Site	Pad 39A, Launch Complex 39, Kennedy Space Center, Florida
Landed	N/A; crew and vehicle lost during re-entry over east Texas
Landing Site	N/A
Launch Vehicle	OV-102 Columbia/ET-93/SRB BI-116/SSME #1 2055; #2 2053; #3 2049
Duration	15 days 22 hrs 20 min 22 sec (up to time of data loss from vehicle on 1 February 2003)
Call sign	Columbia
Objective	International scientific research mission carrying the new Research SpaceHab Double Module (RDM) configuration

Flight Crew

HUSBAND, Rick Douglas, 45, USAF, commander, 2nd mission
Previous mission: STS-96 (2000)
McCOOL, William Cameron, 41, USN, pilot
BROWN, David McDowell, 46, USN, mission specialist 1
CHAWLA, Kalpana, 41, civilian, mission specialist 2, 2nd mission
Previous mission: STS-87 (1997)
ANDERSON, Michael, 43, USAF, mission specialist 3, payload commander, 2nd mission
Previous mission: STS-89 (1998)
CLARK, Laurel Blair Salton, 41, USN, mission specialist 4
RAMON, Ilan, 48, Israeli Air Force, payload specialist 1

Flight Log

The STS-107 research mission had been delayed several times due to changes in the manifest. In June 1997, a research module flight was scheduled for the third quarter of 2000. This was a dedicated mission to train scientists to take full advantage of the ISS research capabilities and to reduce the gap between the last planned Shuttle Spacelab science mission (STS-90) and the start of in-depth science research aboard ISS (around 2001). In 1998, STS-107, a multi-discipline flight, was scheduled for launch in 2000 with a follow-on mission authorised in 2001. In October 2002, the second mission was cancelled and the funds reallocated to support STS-107. The original schedule proposed the mission's launch on 11 January 2001, but it would be two years and 13 delays before the flight finally lifted off on 16 January 2003. Though there were many delays, only a few were orbiter-related, specifically the removal of the Triana

Recovered from a roll of unprocessed film discovered by searchers from the debris, this image shows the crew striking a flying pose for the camera. From left bottom: Chawla, Husband, Clark and Ramon (the Red Team). From left top row: Brown, McCool and Anderson (Blue Team)

Earth observation satellite and its replacement with the Fast Reaction Experiments Enabling Science, Technology, Applications and Research (FREESTAR) payload and an extension of Columbia's Orbiter Maintenance Down Period (OMDP) by six months to correct problems with wiring. This saw the STS-109 Hubble service mission given higher priority. After Columbia had returned from STS-109 in March 2002, preparations began for the STS-107 mission, which was planned for 11 July. However, when cracks were found in a propellant duct liner aboard Atlantis, checks were made on the rest of the fleet and similar cracks were found on Columbia. The necessary repair work was accomplished during the summer, but this pushed the STS-107 mission into the new year.

STS-107 marked the first flight of the Research DM SpaceHab configuration. This was a modification to the basic single or double module used for research and logistics. The RDM, outfitted as a laboratory, could carry 61 Space Shuttle lockers (27.2 kg of experiments) and six double racks (635 kg of experiments). It had two viewing ports and facilities to carry experiments on its flat external top surface. The STS-107 RDM included nine commercial payloads, four ESA payloads linked to ISS risk mitigation experiments, and 18 NASA Office of Biological and Physical Research payloads. On top of the module were three technology experiments and in the payload bay were six further experiments, including the FREESTAR that carried space physics and space sciences experiments and 11 student experiments from schools across the US. This was an international payload, with experiments sponsored by NASA, the US commercial sector, ESA, the Canadian Space Agency, the German Space Agency, the USAF, and schools in Australia, China, Israel, Japan, Liechtenstein and the USA. The crew were kept very busy conducting all these experiments, operating double twelve-hour shift pattern, with Husband, Chawla, Clark and Ramon forming the Red Shift and McCool, Brown and Anderson the Blue Shift. The return from the experiments and research onboard the RDM and the comments from the astronauts themselves all indicated a highly successful flight, and there was a strong possibility that a re-flight may have been authorised, based on the quality and quantity of the experiments conducted on this flight.

When Columbia started the return home on 1 February, it appeared to have been another highly successful mission and there were high hopes for the future US space programme and continued research on ISS . However, 16 minutes from the planned landing at the Cape, all contact with the crew was lost. The vehicle had broken up in mid-air high above east Texas, claiming the lives of all seven astronauts. What followed was a seven-month enquiry, including a four-month search across Texas and neighbouring states to recover debris from the tragedy. Almost 85,000 items of orbiter debris were shipped to KSC for reconstruction to assist in determining what had caused the loss of the vehicle and her crew. Approximately 38 per cent of the dry weight of OV-102 was eventually recovered.

Post-flight investigations indicated that a piece of foam insulation had become detached from the ET at launch and had struck the orbiter's left wing as it ascended, creating either damage or a hole that seriously compromised the structural and thermal integrity of the vehicle at that point. Analysis of post-launch footage of the incident, as well as internal emails concerning the integrity of the vehicle, seemed to have underestimated the severity of the collision. STS-107 carried no RMS and there was no provision for EVA. Nor was the mission due to visit ISS, so there was no easy way of conducting an external inspection of the vehicle. When telemetry indicated to ground controllers that there was an off-nominal situation during the return to Earth, it was too late to do anything about it as the fiery plasma of re-entry breached the damaged wing and ripped Columbia apart. The tragedy, the second fatal Shuttle mission of the programme, came just days after the still-painful anniversaries of the loss of the Apollo 1 crew (27 January 1967) and the Challenger crew (28 January 1986). It also signalled the end of the Shuttle programme, with the fleet scheduled to be decommissioned in 2010. In the meantime, it would be a long and difficult job to maintain the ISS and a hard, painful two-and-a-half-year effort to restore the Shuttle to flight. Even then, the uncertainties remained.

Milestones

237th manned space flight
143rd US manned space flight
113th Shuttle mission
28th and last flight of Columbia
15th SpaceHab mission (9th double module)
1st flight of the Research DM (RDM) SpaceHab configuration
1st Israeli citizen in space (Ramon)
2nd Shuttle mission to end in fatalities

SOYUZ TMA2

Int. Designation	2003-016A
Launched	26 April 2003
Launch Site	Pad 1, Site 5, Baikonur Cosmodrome, Kazakhstan
Landed	27 October 2003 (with TMA3 FE Duque)
Landing Site	38 km from Arkalyk
Launch Vehicle	R7 (Soyuz FG); spacecraft serial number 212
Duration	184 days 22 hrs 46 min 09 sec
Call sign	Agat (Agate)
Objective	ISS-7 caretaker expedition programme

Flight Crew

MALENCHENKO, Yuri Ivanovich, 41, Russian Air Force, ISS-7 and Soyuz commander, 3rd mission
Previous missions: Soyuz TM19 (1994); STS-101 (2000)
LU, Edward Tsang, 39, civilian, US ISS-7 science officer, 3rd mission
Previous mission: STS-84 (1997); STS-101 (2000)

Flight Log

With the loss of Columbia in February 2003, it would be necessary to use the Soyuz TMA spacecraft to launch and return resident crews to the station for the time being until the Shuttle was declared operational again. Due to the limited supply capability for replenishing logistics on the station with the Shuttle fleet grounded, it was also determined that resident crews would now consist of only two persons, launched to the station every six months. The pairing would consist of one American and one Russian crew member, rotating the command position with each mission. The previously identified three-person ISS crews were reassigned as two-person teams, and the third seat was assigned to European astronauts flying short visiting missions during the exchange of crews for the time being. As the crews and launch manifest were changed and America began the investigation into the STS-107 accident, ESA announced that they had delayed the launch of their next astronaut to ISS for six months under agreement with Russia. Therefore, TMA2 would fly with only the two ISS-7 crew members aboard for six-month residency, with no planned EVAs and no other visitors.

The two crew members would be known as "caretaker" crews, able to maintain the systems of the complex, to prevent unmanned loss of control and to sustain consumables, but with limited capacity to conduct science programmes. The work programme of ISS-7 was designed not to overload the crew, as no two-person crew had resided on the station before. The ISS predecessors, Salyut and Mir, had been operated by two-person crews, but ISS was much larger and more complex. The crew

NASA ISS science officer Ed Lu (left) and station commander Yuri Malenchenko work the controls of the Canada2 RMS from inside Destiny lab

operated what science experiments were already aboard and continued the programme of Earth observations, with Lu officially designated the NASA science officer. There were also regular maintenance and housekeeping chores to be accomplished and the crew would receive the Progress M1-10 and M48 re-supply vehicles. Ed Lu continued the series of personal recollections of events on the station, which were posted on the web.

A demonstration of American EMU suiting was completed on 28 May, proving that the two men could suit up and remove suits without the assistance of a third person, should they be required to complete an emergency EVA. Some of the tasks were postponed due to problems with Lu's suit that required further investigation, but it was a useful training exercise. In late June, Lu communicated by radio with former ISS-5 science officer Peggy Whitson, who was commanding a diving expedition to the NEEMO undersea habitat. This is used to develop extreme environment exploration techniques, with diving crews including NASA astronauts (both with and without flight experience) and engineers or flight controllers, to compare space flight training and flight experiences to undersea exploration.

On 10 August, "space history" was made when Malenchenko was "married" via a TV a link to his fiancée who was in Houston, Texas. The marriage had been arranged for August prior to Malenchenko being reassigned to the flight as a result of crew reshuffling after Columbia was lost. It was too late to cancel the legalities, so the wedding was authorised, with Lu acting as best man. This history-making event would be the first and last such ceremony according to officials at TsUP, and subsequent cosmonaut contracts would be amended to include a clause that no wedding would be performed while they were in space.

The ISS-8 two-person resident crew arrived at the station in October, along with ESA astronaut Pedro Duque. He would return in TMA2 with the ISS-7 crew. Their landing occurred without incident.

Milestones

238th manned space flight
95th Russian manned space flight
88th manned Soyuz mission
2nd manned Soyuz TMA mission
6th ISS Soyuz mission (6S)
1st resident caretaker ISS crew (2 person)
1st resident crew with no planned EVAs since ISS-1
Lu celebrates 40th birthday in space (1 Jul)
1st space "wedding" (10 Aug)

SHENZHOU 5

Int. Designation	2003-045A
Launched	15 October 2003
Launch Site	Jiuquan Satellite Launch Complex, Gobi Desert region, northwest China
Landed	16 October 2003
Landing Site	Dorbod Xi, Siziwang grasslands of the Gobi Desert, Inner Mongolia
Launch Vehicle	CZ-2F Shenjian (Long March) (flight 5)
Duration	21 hours 26 minutes
Call sign	Unknown
Objective	Man-rating of the Shenzhou spacecraft for human flight; qualification and man-rating of the launch complex, launch spacecraft compatibility, orbital flight re-entry, flight control, and land-based recovery with a human passenger aboard

Flight Crew

YANG, Liwei, 38, Chinese PLA Air Force, command pilot

Flight Log

On 16 October 2003, after years of speculation and months of expectation, China became the third nation to achieve its own manned space flight capability with the launch of Yang Liwei aboard Shenzhou 5. The historic 21.5-hour flight was preceded by four unmanned test flights between November 1999 and January 2003. At the time, China had always stated that its first manned flight would occur before 2005. As the flight of Shenzhou 4 approached, it was indicated that this would be the final unmanned test flight and if successful, the first manned launch would occur before the end of 2003. Following the successful flight of Shenzhou 4, several reports revealed that the hardware was in preparation to support the first manned flight. From the team of yuhangyuans selected for the programme in 1998, a training team of three were selected, one of which would make the initial flight.

In late August 2003, the Shenzhou 5 spacecraft and its CZ-2F launch vehicle were delivered to the vehicle assembly building at the Jiuquan launch site. On 11 October, the stacked vehicle was moved to the launch pad, an operation that took 1 hour and 25 minutes. On 14 October, the training team of three yuhangyuan candidates were revealed to the Chinese press, but only after the prime pilot had been selected. The back-ups were Zhai Zhigang and Nie Haisheng, while the man destined to be the first Chinese national in space was Yang Liwei.

Yang Liwei, the first Chinese national in space

The final preparations for launch occurred in the early hours of 15 October, with Yang, suited up for the flight, travelling in a transfer bus to the launch site. Yang's entry into the spacecraft inside the launch shroud was similar to that of the Soyuz cosmonauts, through the Orbital Module and down into the Descent Module. Unlike the Russians however, this event was filmed. In Soyuz, there is simply no room in the crew access area for a camera system to be installed and this process has thus never been seen. For American missions, the launch has always been in full public view; during the Soviet era of launches, everything was kept secret. For the first Chinese manned launch, although the event was well publicised, the launch would not be covered live.

The launch occurred at 09:00 Beijing Time (BT − GMT + 8 hours). After two minutes, the launch escape tower ejected, no longer needed as the spacecraft could make an emergency return on its own. Sixteen seconds later, the strap-on boosters separated from the first-stage core. At 2 minutes 39 seconds into the flight, the first-stage core was shut down and separated from the vehicle as the second-stage engine and vernier engines took over, propelling the vehicle towards orbit. At 3 minutes 20 seconds, the two halves of the launch shroud used for aerodynamic purposes through the atmosphere were separated. The shut-down of the second stage occurred at 7 minutes 41 seconds Ground Elapsed Time (GET) and for the next 2 minutes 2 seconds, the vernier engines gave the Shenzhou the final nudge into orbit, before shutting down as the spacecraft separated. The launch had taken about 10 minutes. After 22 minutes in orbit, the two pairs of solar panels were deployed by ground command to generate electrical power. One pair was located on the rear Instrument Module, with the other pair on the forward Orbital Module. It had taken 12 years and one month after authorising the Project 921 programme for the first yuhangyuan to finally reach orbit.

Though an Orbital Module was present, it was announced for this flight that Yang would remain in the DM. His flight programme included the operation of a set of instruments and monitoring of space systems and functions. The spacecraft operated primarily on pre-set programmes with little input from the pilot. Tests of the communications system were combined with TV views from inside the spacecraft and outside the window. The flight duration was announced as only one day for this first manned test, but included three meals and two rest periods for the yuhangyuan. Attached to Yang's body were medical sensors which recorded his condition and transmitted the data to a medical team on Earth. The primary purpose of this mission was to man-rate the spacecraft and system, so extensive operations would wait for later missions. The next important stage was to get Yang home.

On 16 October, following his second rest period, preparations for re-entry and landing began during the fourteenth orbit of the spacecraft. At 05:04 a.m. BT, the command for retrofire to initiate the return to Earth programme was issued from a tracking ship located in the South Atlantic Ocean. About 332 minutes later the OM separated from the spacecraft and continued in orbital flight. Two minutes later, retro-rockets on the spacecraft fired, bringing the spacecraft out of orbit. The Instrument Module was separated about 21 minutes later, as the DM containing Yang plummeted to Earth. Following re-entry, the drogue parachute was deployed 11 minutes after the separation of the modules. The main parachute was deployed five minutes after the drogue and 2 minutes after the heat shield separated from the base of the vehicle. Seven minutes later, at 06:23 a.m. BT, the Descent Module of Shenzhou 5 landed safely after a flight of 21 hours 26 minutes.

Yang was taken back to Beijing for medical examination and mission debriefings. Now a national hero, from November he began a programme of public appearances as a major personality and the face of China's manned space programme. While Yang took the Chinese space programme to the public, behind the scenes work continued both for the next flight and also with the OM of Shenzhou 5. The OM, unlike that of Soyuz, was capable of independent manoeuvrable flight for some months, and Shenzhou 5's OM was packed with instruments and equipment that were tested after the yuhangyuan had returned to Earth, for about six months. These modules are expected to be of significance on future flights in the series, and are linked to the expected Chinese national space station programme.

Milestones

239th manned space flight
1st Chinese manned space flight
1st manned flight of CZ-2F launch vehicle
3rd nation to develop independent manned orbital flight
5th flight of Shenzhou spacecraft
1st manned flight of Shenzhou

SOYUZ TMA3

Int. Designation	2003-047A
Launched	18 October 2003
Launch Site	Pad 1, Site 5, Baikonur Cosmodrome, Kazakhstan
Landed	30 April 2004
Landing Site	60 km northeast of Arkalyk
Launch Vehicle	R7 (Soyuz FG); spacecraft serial number 213
Duration	194 days 18 hrs 23 min 43 sec
Call sign	Inguly (Inguly) – Soyuz TMA
Objective	ISS-8 caretaker expedition programme; ESA *Cervantes* Visiting Crew 5 programme

Flight Crew

FOALE, Colin Michael, 46, civilian, US ISS-8 commander and science officer, 6th mission
Previous missions: STS-45 (1992); STS-56 (1993); STS-63 (1995); STS 84/86 (1997); STS-103 (1999)
KALERI, Aleksandr Yuriyevich, 47, civilian, Russian ISS-8 Soyuz commander, 4th mission
Previous missions: Soyuz 14 (1992); Soyuz TM24 (1996); Soyuz TM30 (2000)
DUQUE, Pedro Francisco, 40, civilian, EAS Soyuz TM3 flight engineer 1, 2nd mission
Previous mission: STS-95 (1998)

Flight Log

Despite being termed a "caretaker crew", this residence would conduct a significant amount of science research with available hardware, routine maintenance and housekeeping, and an EVA. One of the reasons Foale did not continue the trend of regular email postings from ISS was that he was just too busy. It was planned to launch no fewer than three Progress re-supply vehicles during this residence to "stock up the station" in the absence of the Shuttle. However, funding difficulties (reminiscent of the latter days on Mir) meant that only one re-supply craft (M1-11) was actually launched. Careful management of onboard resources meant that the experienced crew, each with long-duration flights on Mir behind them, did not need to break into the new Progress supplies, although other equipment delivered was urgently needed. Significant maintenance was a priority with this crew, and with only two crewmembers they had a lot more work to do than would a crew of three.

Their flight to ISS was accompanied by Spanish ESA astronaut Pedro Duque. His *Cervantes* programme comprised 24 experiments requiring 40 hours of his 8 days aboard ISS. The experiments, conducted mainly in the Russian segment of the station,

ISS-8 crew Kaleri Foale and ESA astronaut Duque climb the steps to their Soyuz TMA3 spacecraft and a two-day flight to ISS. Duque would return with the ISS-7 crew a few days later but Kaleri and Foale were embarking on a six-month mission

consisted of 12 investigations in life sciences, three in physical sciences and Earth observations, five under education, two under technology and two ground-based experiments. Most were sponsored by the Spanish government, although some were re-flights from the October 2002 Belgian *Odessa* mission. At the end of his week on ISS, Duque returned to Earth in the TMA2 spacecraft with the ISS-7 crew on 28 October.

In November, Foale and Kaleri practised emergency ingress procedures, with Kaleri wearing an Orlan EVA suit to determine if a suited crewman could enter the Soyuz using internal hatches. Kaleri was guided (and pushed) by Foale. The test took longer than planned and was deemed unsuccessful, so a second test, this time with both astronauts wearing suits, was attempted and successfully accomplished on 19 February. The two crew members completed their only EVA on 26 February (3 hours 56 minutes) using Orlan suits and exiting via the Pirs airlock. With both men outside, the ISS complex was left unattended for the first time since November 2000. The pair installed a protective ring around the Pirs entry hatch which was meant to prevent snagging on the way out of or back into the airlock. The ring was removed after the EVA. The two men also installed European and Japanese scientific packages on the Zvezda before the EVA was terminated early due to problems with Kaleri's liquid cooling garment.

On 8 December, Mike Foale, who holds dual US and UK nationality, became the most experienced American astronaut, surpassing Carl Walz's career total of 230 days 13 hours 3 minutes and 38 seconds in four missions. Foale was now on his sixth mission but Kaleri, on his fourth mission, was still 238 days ahead of his colleague in career experience! The crew returned to Earth on Soyuz TMA3 with Dutch astronaut André Kuipers, who had arrived with the ISS-9 crew aboard Soyuz TMA4 at the end of April. Foale had by then logged over 374 days 11 hours in space, well ahead of all other American astronauts, but still far behind several cosmonauts, including Kaleri.

Milestones

240th manned space flight
96th Russian manned space flight
89th manned Soyuz mission
3rd manned Soyuz TMA mission
36th Russian and 90th flight with EVA operations
7th ISS Soyuz mission (6S)
2nd resident caretaker ISS crew (two-person)
Foale celebrates his 47th birthday in space (6 Jan)
Foale sets new career space flight record for an American astronaut at over 374 days

SOYUZ TMA4

Int. Designation	2004-013A
Launched	19 April 2004
Launch Site	Pad 1, Site 5, Baikonur Cosmodrome, Kazakhstan
Landed	24 October 2004 (with ISS-10 FE Shargin)
Landing Site	70 km north east of Arkalyk
Launch Vehicle	R7 (Soyuz FG); spacecraft serial number 214
Duration	187 days 21 hrs 16 min 9 sec
	10 days 20 hrs 52 min 46 sec (Kuipers)
Call sign	Altair (Altair) – Soyuz TMA
Objective	ISS-9 caretaker expedition programme; ESA (Dutch) Delta/Visiting Crew 6 programme

Flight Crew

PADALKA, Gennady Ivanovich, 40, Russian Air Force, Russian ISS-9 and Soyuz commander, 2nd mission
Previous mission: Soyuz TM28 (1998)
FINCKE, Edward Michael, 37, USAF, ISS-9 science officer
KUIPERS, André, 45, civilian, ESA Soyuz flight engineer

Flight Log

For the ninth main expedition to the ISS, the crew would conduct a programme of 24 US and 42 Russian experiments. Many of these were continuations of experiments delivered before the loss of Columbia, but there were four new investigations. The crew's flight plan envisaged 130 sessions, or over 200 hours, focusing on science on the station, in addition to the routine housekeeping and maintenance chores. During the first week aboard the station, the new resident crew worked with the outgoing ISS-8 crew and with Dutch astronaut André Kuipers, who had a package of ESA experiments under the Dutch Expedition for Life sciences, Technology and Atmospheric (DELTA) research programme. His research included five physiological experiments, five biology experiments, single investigations in microbiology, physical sciences, and Earth observations, three technology demonstrations and five educational projects. Kuipers completed his programme and returned to Earth with the ISS-8 crew on 30 April aboard TMA3.

Settling down to their own programme, the ISS-9 crew received no visiting crews but did receive the payloads delivered on Progress M1-11, M49 and M50. All the EVAs from this expedition were conducted from the Pirs module using Russian Orlan M suits. After working on suit repairs and servicing for over a month, their first EVA (24 Jun) was abandoned after 14 minutes because of a pressure drop in the main oxygen bottle of the Orlan M suit. Following successful repairs, the crew conducted

The TMA4/ISS-9 crew inside the ISS; l to r Padalka, Fincke and Kuipers

their first full EVA on 30 June (5 hours 40 minutes), during which a new circuit breaker was installed on the S0 Truss to power one of the four CMG. During the next excursion (3 Aug for 4 hours 30 minutes), the crew installed reflectors and communication units ready for the first ESA Automatic Transfer Vehicle (ATV, named "Jules Verne"), which is designed to carry seven tons of supplies to the station, boost the station's orbit, and remove waste materials for atmospheric burn-up. The first flight was planned for 2006 but was subsequently delayed to 2007 or 2008. The final ISS-9 excursion (3 Sep for 5 hours 21 minutes) saw the crew install three antennas to support the ATV docking on the rear port of the Zvezda module, and fit further handrails and tether guides for future EVAs.

On 18 June, Michael Fincke made space history by becoming the first US astronaut to become a father while in space. In Houston, his wife gave birth to their second daughter, with the astronaut listening to the delivery via his wife's cell phone and a relayed radio link through MCC-Houston. A video of the event was later sent up to the proud father.

The crew spent a significant amount of time in repairing onboard equipment, and Fincke also conducted troubleshooting diagnostics on the American EMU units after the previously reported cooling problems had been traced to water circulation pumps located inside the suits' integrated backpacks. Fincke removed the pump and videoed it for ground specialists to analyse the problem, but the pictures failed to reveal any obvious causes of the malfunction. Two new pumps were manifested for delivery on the next Progress mission (M50).

The crew also had ongoing problems with the Elektron oxygen generator that had been playing up for some time. Earlier in the ISS-9 residency, it briefly shut down twice, and by late August it would fail every three days or so and require a manual restart. These shutdowns were found to be centred on the liquid units (BZh in Russian) that held trapped gas inside micro-pumps, despite using purified water. The unit was put into a mode to increase oxygen production, which would raise the internal pressure of the station so that, should the unit need repairing, a lengthy shut-down would be possible without too much risk to the crew. These extensive repairs were conducted during September, with the crew installing an older unused unit, before eventually replacing the units to improve the performance of the Elektron system (although problems still remained as their residency drew to a close).

The crew completed their programme with several other maintenance tasks and returned to Earth with Russian test cosmonaut Yuri Shargin in October.

Milestones

241st manned space flight
97th Russian manned space flight
90th manned Soyuz mission
4th manned Soyuz TMA mission
37th Russian and 91st flight with EVA operations
8th ISS Soyuz mission (8S)
6th visiting mission (VC-6)
3rd resident caretaker ISS crew (2 person)

Before the next exchange of ISS resident crews occurred in October 2004, the first privately funded, non-government manned space flight took place – the sub-orbital flights of Spaceship One that won the Ansari X-Prize. These flights are covered in the Quest for Space chapter (Chapter 2).

SOYUZ TMA5

Int. Designation	2004-040A
Launched	14 October 2004
Launch Site	Pad 1, Site 5, Baikonur Cosmodrome, Kazakhstan
Landed	24 April 2005
Landing Site	85 km northeast of Arkalyk
Launch Vehicle	R7 (Soyuz FG); spacecraft serial number 215
Duration	192 days 19 hrs 0 min 59 sec; 9 days 21 hrs 29 min (Shargin)
Call sign	Tyan Shan (Tyan Shan – Soyuz TMA)
Objective	ISS expedition 10 programme; 7th visiting mission programme

Flight Crew

CHIAO, Leroy, 44, civilian, US ISS-10 commander and science officer, 4th mission
Previous missions: STS-65 (1994); STS-72 (1996); STS-92 (2000)
SHARIPOV, Salizhan Shakirovich, 40, Russian Air Force, ISS-10 flight engineer and Soyuz commander, 2nd mission
Previous mission: STS-89 (1998)
SHARGIN, Yuri Georgiyevich, 44, Russian Military Space Forces, Soyuz flight engineer

Flight Log

Russian businessman Sergei Polonsky was scheduled to become the third space flight participant on this flight, but he failed the medical and was replaced by cosmonaut Yuri Shargin. Shargin was a representative of the Russian Space Forces, and became the first member of that force to be selected for cosmonaut training in 1996. He had completed a full course of cosmonaut training, including simulations with the Mir complex. This was perhaps his only chance of a space flight, and he carried out his own programme of scientific studies, although it was stressed that none of these were military in nature as such investigations are banned on ISS. Earth observations and a daily biomedical programme were the main focus of his week on the station. Shargin returned to Earth on 24 October, along with the ISS-9 crew.

There had been a delay to the launch of TMA5, reportedly caused on 15 September by a small explosion in a separation bolt on the Soyuz docking ring. A leaking pressure membrane in a small tank of liquid hydrogen peroxide had also ruptured, and procuring the replacement from the TMA6 craft that was under construction at Energiya in Moscow delayed the planned 9 October launch to 14 October. Once on

ISS 10 crew: Sharipov (left) and Chiao (centre) with Yuri Shargin (right) review crew equipment during the final check of the Soyuz TMA5 spacecraft on 9 October 2004 at the Baikonur Cosmodrome, Kazakhstan

orbit, problems with an intermittent forward-firing thruster on TMA5 resulted in Sharipov having to conduct a manual docking with the station.

During the tenth expedition on ISS, the two-man crew would complete two Pirs-based EVAs, receive two Progress re-supply craft and relocate their TMA from the Pirs module (where they docked) to the Zarya module. There were also further problems with the Elektron unit, repairs to US EVA suits and Russian Orlan M suits, upgrades to the computer software, the breakdown of the station's toilet in the Zvezda living quarters, the replacement of a faulty heat exchanger in the Quest airlock module, and the relocation of the Canadarm2 to deal with. Fitted around all of this was the routine work with the experiments aboard the station and general house-keeping chores.

During the first EVA (26 Jan 2005 for 5 hours 28 minutes), Sharipov photographed the residue around vents on the outside of Zvezda, generated from by-products of the Elektron and carbon dioxide removal systems. The two men also completed the installation of a work platform and deployed a European commercial experiment and a Russian experiment. The second EVA (28 Mar for 4 hours 30 min-utes) included the installation of cables and antennas on Zvezda in support of future

ATV dockings and the deployment by hand of a small satellite to test new control techniques.

In late November, there were concerns that the food supplies onboard the station were less than expected. Three audits of onboard supplies indicated that the rations would run out by mid-January. It was intended that there should be a 45-day buffer of supplies to cope with any delay in the launch of the next re-supply vessel, but it was found that the records of inventory were not being kept as accurately as they should have been, probably due to the increased workload that the smaller crews were having to cover. Re-supply craft Progress M51 resolved the immediate problem by delivering 200 kg of food supplies, which would last until Progress M52 arrived in March.

On the ground, NASA was working toward the resumption of Shuttle flights, with the STS-114 mission planned for May 2005. Though no construction work would be conducted on the first two missions (designated Return-to-Flight missions 114 and 121), there would be an opportunity to deliver supplies, spares and other logistics, as well as removing the significant amount of unwanted gear that had built up since the departure of Endeavour (STS-113) in December 2002. During the early weeks of 2005, the crew began packing and stowing items that were to be returned in bundles on the floor of Zvezda and in a stowage rack in the Destiny lab.

Near the end of their tour of duty, the ISS-10 crew were joined by their replacements, the ISS-11 crew, who arrived on Soyuz TMA6 in April 2005. Also on board the new TMA was Italian astronaut Roberto Vittori, who brought some much-enjoyed Italian delicacies with him for the period of joint activities prior to returning to Earth with the ISS-10 crew aboard TMA5.

Milestones

242nd manned space flight
98th Russian manned space flight
91st manned Soyuz mission
5th manned Soyuz TMA mission
38th Russian and 92nd flight with EVA operations
9th ISS Soyuz mission (9S)
7th ISS visiting mission (VC-7)
4th resident caretaker ISS crew (2 person)

SOYUZ TMA6

Int. Designation	2005-013A
Launched	15 April 2005
Launch Site	Pad 1, Site 5, Baikonur Cosmodrome, Kazakhstan
Landed	11 October 2005
Landing Site	57 km northeast of Arkalyk
Launch Vehicle	R7 (Soyuz FG); spacecraft serial number 217
Duration	179 days 0 hrs 23 min; 9 days 21 hrs 21 min 2 sec (Vittori)
Call sign	Basalt (Basalt – Soyuz TMA)
Objective	ISS Expedition 11 programme; visiting crew 8 mission; ESA *Eneide* programme

Flight Crew

KRIKALEV, Sergei Konstaninovich, 46, civilian, Russian ISS-11 and Soyuz commander, 6th mission
Previous missions: Soyuz TM7 (1988); Soyuz TM12 (1991); STS-60 (1994); STS-88 (1998); ISS-1 (2000/01)
PHILLIPS, John Lynch, 54, civilian, US ISS-11 science officer, 2nd mission
Previous mission: STS-100 (2001)
VITTORI, Roberto, 40, Italian Air Force, Soyuz flight engineer, 2nd mission
Previous mission: Soyuz TM34 (2002)

Flight Log

The eleventh residency aboard ISS would be the first to receive a Shuttle mission since STS-113 in December 2002. The loss of Columbia and her crew had created the need to fly two-man resident crews, termed caretakers, because orbital operations were restricted by the availability of onboard supplies, and station expansion curtailed by the grounding of the Shuttle. This was the fifth such caretaker crew, but the docking of STS-114 in July signified that full station operations could soon be resumed.

Flying to the station with EO-11 was ESA astronaut Vittori, who completed 23 experiments in 91 sessions under the *Eneide* programme. His research consisted of studies in human physiology, biology, demonstrations of new technology and a programme of education demonstrations. Vittori would also participate in a number of ceremonial and public relations broadcasts, a feature of most crew exchanges and international missions in space. Vittori would return to Earth aboard TMA5 along with the two ISS-10 crew members.

Almost as soon as the new crew were left alone on the station, their time was taken up by problems with the Elektron system, which finally broke down in early May.

Change of shift on ISS. The ISS-10 and 11 crews give the thumbs-up to the continuation of manned station operations. L to r John Phillips and Sergey Krikalev (ISS-11), Leroy Chiao (ISS-10), Italian Roberto Vittori (ESA) and Salizhan Sharipov (ISS-10)

There were sufficient alternative oxygen supplies aboard the station to keep the crew supplied until the end of the year, with re-supply by the Shuttle and Progress, but it was still disconcerting as troubleshooting on the unit would be added to an already heavy work programme. Other maintenance included work on the treadmill vibration isolation system. The crew received their first re-supply craft (Progress M53) in June, which delivered 2,383 kg of cargo. This included 111 kg of oxygen and 420 kg of water, as well as 40 solid fuel oxygen generation cartridges and spare parts for the Elektron unit.

The crew also completed a range of science studies, including Earth observations, medical experiments, and microgravity research. In addition they tested new software that had been installed to support the Canadarm2 unit on the station and relocated their Soyuz from the Pirs module to the Zarya module on 19 July, freeing the docking compartment for their EVA the following month. The re-docking operation took approximately 30 minutes to complete. During these relocation manoeuvres, the station had to be prepared for autonomous flight, in case the Soyuz failed to re-dock and the crew were forced to return to Earth.

On 28 July, the STS-114 crew docked Discovery with the station, bringing much-welcomed supplies and visitors to the crew, and confidence in the resumption of Shuttle ISS assembly flights. In August, the Russian Vodzukh carbon dioxide removal system failed and while the Russian ground controllers investigated the problem and came up with a repair plan, the American unit in Destiny was activated to take over the operation. On 16 August Krikalev surpassed Sergei Avdeyev's career record of time logged in space (747 days 14 hours 14 minutes 11 seconds).

The two men performed their only EVA on 17 August (4 hours 57 minutes). For Phillips, it was the first excursion outside a spacecraft in space, but for veteran Krikalev, this was his eighth EVA. Their tasks included retrieving exterior sample cassettes, installing a new reserve TV camera for ATV dockings, and photographing exterior experiments and surfaces. Some of the activities planned for the EVA could not be accomplished due to lack of time and would be rescheduled for later crews. Following their EVA, the crew cleaned their equipment and resumed their scientific and maintenance work in preparation for their next visitors, the crew of Soyuz TMA7 and the third space flight participant. In their final days in orbit, the two men began packing up their research results, increased their exercise programme in preparation for the return to gravity and checked the systems of Soyuz TMA6. They would return to Earth with space flight participant Greg Olsen after he completed his week aboard the station.

During the descent of TMA6, there was a small pressure leak from the DM, which was noted before the undocking from ISS. An obstruction of some kind seemed to have prevented the air tight seal of the forward DM hatch into the OM. On module separation the leak continued to vent, although the crew were protected in their Sokol suits. The landing occurred successfully but Phillips had to be given smelling salts to prevent him drifting into unconsciousness. The astronaut later explained that he was far more uncomfortable in the DM than he had been in the Shuttle he had returned on in 2001. He was not sure if he became unconscious for a short time, but he knew his head was spinning. In images from the recovery operation, he certainly looked weak and pale.

Milestones

243rd manned space flight
99th Russian manned space flight
92nd manned Soyuz mission
6th manned Soyuz TMA mission
39th Russian and 93rd flight with EVA operations
10th ISS Soyuz mission (9S)
8th visiting mission (VC-7)
5th resident caretaker ISS crew (2 person)
Phillips is launched on his 54th birthday (15 Apr)
Krikalev celebrates his 47th birthday in space (27 Aug)
Krikalev sets a new cumulative record of 803 days 9 hrs 39 mins in space, on six missions
1st cosmonaut to fly six missions (Krikalev)
1st person to conduct a second residence aboard ISS (Krikalev)

STS-114

Int. Designation	2005-026A
Launched	26 July 2005
Launch Site	Pad 39B, Kennedy Space Center, Florida
Landed	9 August 2005
Landing Site	Runway 22, Edwards AFB, California
Launch Vehicle	OV-103 Discovery /ET-121/SRB BI-125/SSME #1 2057; #2 2054; #3 2056
Duration	13 days 21 hrs 32 min 48 sec
Call sign	Discovery
Objective	ISS mission LF-1; return-to-flight; MPLM logistics mission

Flight Crew

COLLINS, Eileen Marie, 48, USAF, commander, 4th mission
Previous missions: STS-63 (1995); STS-84 (1997); STS-93 (1999)
KELLY, James McNeal, 41, USAF, pilot, 2nd mission
Previous mission: STS-102 (2001)
NOGUCHI, Soichi, 40, civilian, Japanese mission specialist 1
ROBINSON, Stephen Kern, 49, civilian, mission specialist 2, 3rd mission
Previous missions: STS-85 (1997); STS-95 (1998)
THOMAS, Andrew Sydney Withiel, 53, civilian, mission specialist 3, 4th mission
Previous missions: STS-77 (1996); STS-89/91 (1996); STS-102 (2001)
LAWRENCE, Wendy Barrien, 46, USN, mission specialist 4, 4th mission
Previous missions: STS-67 (1995); STS-86 (1997); STS-91 (1998)
CAMARDA, Charles Joseph, 53, civilian, mission specialist 5

Flight Log

It was almost 30 months between the loss of Columbia and the launch of Discovery. During this time, apart from the investigation into the probable cause of the accident and the steps taken to reduce the risk of it happening again, the whole Shuttle programme had been evaluated and a new long-term goal established. After clearing the Shuttle for flight operations, it would be used to complete the construction of ISS by 2010. After that, the fleet would be retired and replaced by a Crew Exploration Vehicle (CEV). The CEV would be able to visit the station, but its primary role would be to return America to the Moon, hopefully by 2019 (the 50th anniversary of Apollo 11). American commitments to ISS would be met by 2016. Exactly what their involvement with the station would be after that is still to be decided.

The first launch attempt for STS-114 on 13 July was cancelled when an ET fuel sensor failed. Extensive troubleshooting delayed the mission, but the 26 July launch

Performing the first EVA beneath the belly of a Shuttle orbiter, astronaut Stephen Robinson, on the end of the ISS robotic arm Canadarm2 (out of frame), works to remove gap fillers protruding from the heat-shielding tiles of Discovery during the mission's third EVA

occurred on time and was the most extensively documented launch into space in history. A myriad of ground-based and high-altitude aircraft-borne cameras, ground radar systems, sensors and lasers on the ascending Shuttle and a TV system on the ET, all recorded spectacular shots of the ascent from the pad, SRB separation and Orbiter/ET separation. They also recorded the loss of some foam from the ET once more, which caused grave concern on the ground.

Discovery docked with ISS on 28 July. Prior to docking, Collins performed the first Rendezvous Pitch Manoeuvre some 183 metres from the station, at a rate of 0.75°/second, to allow the ISS-11 crew to photo-document the underside of the Discovery and its protective tiles. Analysis of the images revealed a little tile and foam damage, but the most serious problem appeared to be two protruding tile gap fillers, which might cause hot spots on entry. Several options were discussed and evaluated before the final solution was reached as the crew worked aboard the ISS. It was decided to allow the EVA crew to manually extract the fillers.

During the first EVA (30 Jul for 6 hours 50 minutes), Robinson (EV1) and Noguchi (EV2) worked with intentionally damaged tiles that had been brought up to space for the purpose of evaluating new repair procedures and equipment. They also installed a base and cabling for a stowage platform and rerouted power cables to CMG-2, one of the four gyroscopes that orientate the station. The second EVA (1 Aug for 7 hours 14 minutes) saw the removal of the failed CMG-1 and its replacement with a new unit, restoring ISS to four functioning units. The added third EVA (3rd Aug for 6 hours 1 minute) saw Robinson ride the RMS over the side of Discovery to remove the two gap fillers with his gloved hand, which was much easier than first thought. This was the first time that an astronaut had ventured underneath the Shuttle during an EVA. The final EVA also included installation of an external stowage platform on ISS and the deployment of a materials experiment package for long-term exposure to the harsh conditions of space.

During their 9 days of docked activities, the crew transferred significant logistics to the station. MPLM Raffaello was relocated to the side of Destiny on 29 July and unloaded over several days. It carried 1,710 kg of supplies and cargo, including the Human Research Facility 2 and the new CMG that was installed during EVA 2. Returned items included the 3.5 tons of material accumulated and stored since 2002, creating much welcomed volume inside the Zvezda when the waste had been relocated into Raffaello.

Discovery undocked from the station and performed a fly around, with both crews photographing each other's spacecraft before the Shuttle crew prepared for the much anticipated return from orbit. To add to the tension both on Earth and in orbit, the landing was delayed by two days after four opportunities to land at Florida were cancelled by bad weather. Much to everyone's relief, Discovery made a safe entry and landing at Edwards AFB. The mission had returned the fleet to space after the loss of Columbia, but there were still questions about the integrity of the foam fixtures. As a result, the next Shuttle mission would be delayed until the problems of foam coming off the ET were better understood.

Milestones

244th manned space flight
144th US manned space flight
114th Shuttle mission
31st flight of Discovery
57th US and 94th flight with EVA operations
17th Shuttle ISS mission
5th Discovery ISS mission
6th MPLM flight
3rd flight of MPLM-2 Raffaello

SOYUZ TMA7

Int. Designation	2005-039A
Launched	1 October 2005
Launch Site	Pad 1, Site 5, Baikonur Cosmodrome, Kazakhstan
Landed	9 April 2006
Landing Site	54 km northeast of Arkalyk
Launch Vehicle	R7 (Soyuz FG); spacecraft serial number 217
Duration	189 days 19 hrs 53 min 0 sec
	9 days 21 hrs 15 min 0 sec (Olsen)
Call sign	Rassvet (Sunrise, or Dawn – Soyuz TMA)
Objective	ISS-12 expedition programme; 9th visiting mission; 3rd SFP (Olsen) programme

Flight Crew

MCARTHUR Jr., William Surles, 54, US Army, ISS-12 commander and science officer, 4th mission
Previous missions: STS-58 (1993); STS-74 (1995); STS-92 (2000)
TOKAREV, Valery Ivanovich, 52, Russian Air Force, ISS-12 Soyuz commander and flight engineer, 2nd mission
Previous mission: STS-96 (1999)
OLSEN, Gregory Hammond, 60, civilian, US space flight participant

Flight Log

Greg Olsen had prepared for his mission with over 900 hours training, but had been medically disqualified from flying TMA6. His medical condition cleared up, however, and he was cleared to fly by Russian medical specialists. He was assigned to the next mission. During his week on ISS, Olsen did perform some experiments for ESA, but failed equipment and the lack of an export licence for an infrared camera prevented him completing his full programme. He did participate in the three live video and ham radio sessions with schools during the mission and spent his time Earth watching, exercising and photographing, mainly from the Russian segment. He was restricted in his activities, however, especially aboard US elements.

When the ISS-11 crew returned home, along with Olsen, the new resident crew got down to work with their own programme of experiments, maintenance and housekeeping, including two EVAs. The first of these saw the first use of US EVA suits and the Quest Airlock since April 2003. Tokarev became the first Russian to use US EVA equipment and facilities at the station. The two men installed a new camera at the end of the P1 Truss and removed a failed Rotary Joint Motor Controller (electronics box) and the Floating Potential Probe, which had been installed during STS-97 in 2000. They tossed this overboard, where it would orbit until re-entry about 100 days later.

ISS-7 crew wearing Sokol suits at launch-processing facility, Baikonur. L to r Olsen, Tokarev, McArthur

They also removed a failed circuit breaker in the Mobile Transporter and replaced it with a new unit. The second EVA was rescheduled from December to February to allow more time for preparation. This EVA was from Pirs, using Russian Orlan M suits, and featured the deployment of an Orlan SuitSat, which was an old suit outfitted with a transmitter and radio, beaming faint recorded voices of children to amateur radio operators for a few days. It was expected to orbit for several weeks before eventually re-entering the atmosphere. Other tasks performed during this EVA included relocation of the Strela grapple fixture to the PMA-3 on Unity, retrieval of exterior sample cassettes and external photography. A third EVA was cancelled as unnecessary.

During their stay on the station, the two men continued repairs on Elektron, the station toilet, and in the US laboratory Destiny. The crew also continued regular exercise and science work, along with unloading a Progress re-supply vehicle, performing tests with the Canadarm2, and a communications test in preparation for receiving ATV re-supply flights. McArthur's wife was the principle investigator for the educational payload operations and together they instigated a programme of science demonstrations for schools. Shuttle mission STS-114 had delivered much needed supplies and alleviated the clutter of unwanted gear, but extensive science operations would have to await the arrival of the next main crew in April and the third resident crew member (Thomas Reiter) on STS-121 in July 2006.

The crew had to relocate their Soyuz twice during the mission. The first relocation from the Pirs Module to the nadir port on Zarya took place on 18 November and lasted about 19 minutes. This freed up the Pirs docking compartment for use as the airlock for the EVA that was scheduled for 7 December but moved to February. The second relocation, on 20 March, lasted 22 minutes. The crew moved TMA7 from Zarya to the recently vacated rear port of Zvezda (Progress M54), to allow TMA8 to dock with Zarya so that it would not require relocation during the six-month mission. The ISS-12 crew came home with Brazilian astronaut Marcos Pontes in TMA7 on 8 April.

Milestones

245th manned space flight
100th Russian manned space flight
93rd manned Soyuz mission
7th manned Soyuz TMA mission
40th Russian and 95th flight with EVA operations
9th ISS Soyuz mission (9S)
9th visiting mission (VC-9)
6th resident caretaker ISS crew (2 person)
3rd space flight participant
Tokarev celebrates 53rd birthday (29 Oct) in space
1st Russian to perform EVA from Quest using EMU (Tokarev)

SHENZHOU 6

Int. Designation	2005-040A
Launched	12 October 2005
Launch Site	Jiuquan Satellite Launch Complex, Gobi Desert region, northwest China
Landed	16 October 2005
Landing Site	1 km from planned landing site near Dorbod Xi, Siziwang grasslands of the Gobi Desert, Inner Mongolia
Launch Vehicle	CZ-2F Shenjiun (Long March) flight 6
Duration	4 days 19 hrs 33 min
Call sign	Unknown
Objective	Extended manned test of Shenzhou spacecraft with crew of two

Flight Crew

FEI, Junlong, 40, Chinese PLA Air Force, commander
NIE, Haisheng, 40, Chinese PLA Air Force, operator

Flight Log

Five pairs of yuhangyuans trained for the second Chinese manned space flight. One month prior to the flight, three pairs were selected to continue training, one as prime crew and two as back-ups. The prime crew was announced shortly prior to launch, with their back-ups announced as Liu Boming and Jing Haipeng (Team 1), and Zhai Zhigang and Wu Jie (Team 2). Nie and Zhai had backed up Yang two years before.

The timing, duration and objectives of the flight were discussed in the western media for months, before an official release indicated an October flight of about five days. Improvements had been made to the launch vehicle, whose assembly was completed on 26 September. The spacecraft was joined to the launch vehicle on 4 October. There had been over 110 upgrades to Shenzhou 6 over its predecessor, in four specific areas: upgrading the spacecraft to support the flight of two crewmembers; upgrading the internal configuration of the crew compartment; improvements to the safety systems; and improvements to the system components. In all, over 40 new items of equipment and six pieces of software were incorporated into the Shenzhou 6. There had also been over 75 upgrades to the launch vehicle since Shenzhou 5.

The crew had arrived at the launch site 2 hours 45 minutes prior to launch and were sealed inside their spacecraft about 30 minutes later. The launch was nominal, with powered flight lasting 583 seconds. Twenty-one minutes into the mission, the Shenzhou propulsion system was used to adjust the initial 200 km orbit to one of 211 × 345 km. Shortly after launch, search teams were dispatched to recover the telemetry "black box" ejected from the launcher during the ascent. It was thought

The Shenzhou 6 crew of commander Fei Junlong (left) and pilot/operator Nie Haisheng

that it included telemetry that had not been downlinked to the ground. It was found 45 minutes after the launch.

During the flight, the two crew members conducted regular communication sessions with the ground, reporting on their personal condition and the status of the spacecraft, and speaking with their families. Several changes in orbital parameters were conducted during the five days in space, to test the operational capability of the Shenzhou orbital manoeuvring systems with a crew aboard. During the second orbit, the internal hatch into the OM was opened for the first time, and Fei was the first to enter. Three hours later, Fei swapped places with Nie. Very few details of the orbital activities of the crew were released, but the use of the OM and the extended flight allowed both men to take off their pressure suits for most of the orbital duration. Only one slept at any one time, however, with the second monitoring onboard systems.

The crew also shut and pressure-tested the internal hatch integrity, both for when the OM was separated and for the future when the OM would be used as an airlock to perform EVA. They also performed an experiment to test the reactions of the crew within the OM and DM, videotaping their somersaults and movement between the modules. The crew conducted a range of "scientific experiments", which included Earth observation and monitoring, and research on "biological and material sciences." Exact details were not forthcoming. The most important "experiment" in relation to future plans for EVA, docking, crew transfer and the creation of a small "space station", was the habitability studies conducted throughout the five days by both men.

As with the previous Shenzhou mission, the OM was separated and left in orbit prior to entry and landing. The "mission" of the OM was officially ended on 15 April 2006 after nearly 3,000 orbits of the Earth, giving mission controllers experience in the extended control of a spacecraft in Earth orbit.

Milestones

246th manned space flight
2nd Chinese manned space flight
1st Chinese two-person space flight
1st extended manned test flight of Shenzhou
Nie Haisheng celebrates his 41st birthday in space (13 Oct)

SOYUZ TMA8

Int. Designation	2006-009A
Launched	30 March 2006
Launch Site	Pad 1, Baikonur Cosmodrome, Kazakhstan
Landed	29 September 2006
Landing Site	90 km northeast of Arkalyk
Launch Vehicle	R-7 (Soyuz FG); Soyuz TMA 8 serial number 218
Duration	182 days 22 hrs 43 min 0 sec; 9 days 21 hrs 17 min 0 sec (Pontes)
Call sign	Karat (Carat – Soyuz TMA)
Objective	ISS Expedition 13 programme; Visiting Crew 10 programme – Brazilian "Cenrenario" mission

Flight Crew

VINOGRADOV, Pavel Vladimirovich, 52, civilian, Russian ISS-13 and Soyuz commander, 2nd mission
Previous mission: Soyuz TM26 (1997)
WILLIAMS, Jeffery Nels, 48, US Army, ISS-13 science officer, 2nd mission
Previous mission: STS-101 (2000)
PONTES, Marcos Caesar, 43, Brazilian Air Force, space flight participant

Flight Log

The appointment of the Brazilian astronaut to the crew came from a commercial agreement between the Brazilian space agency and the Russian space agency. The programme of scientific experiments under the *Centenario* label included eight small experiments being operated by Pontes in the Russian segment; one biomedical experiment, three biotechnology experiments, two engineering research experiments, and two educational experiments. The Brazilian also participated in a number of ceremonial and media activities as the first Brazilian in space. He returned to Earth on 9 April with the ISS-12 crew aboard TMA7.

The docking with ISS had occurred on 1 April. The hand over activities between the two main crews took a week before the ISS-12 crew and Pontes returned to Earth, leaving the new crew to continue the long ISS programme. As well as science work, ISS-13 conducted routine and unplanned maintenance, and exercised to maintain their condition during their six-month tour of duty. Earth resources and photography had long been an important programme from manned spacecraft and this flight was no exception. The crew photographed and observed the eruption of the Cleveland volcano on the Aleutian Islands in Alaska.

Their first EVA occurred on 2 June (6 hours 31 minutes) and included tasks on both the US and Russian segments. The EVA began from Pirs with the crew wearing

The first Brazilian astronaut, Marcos Pontes (centre), works aboard ISS during April 2006

Russian Orlan M suits. They installed a new valve nozzle on the side of Zvezda that would be used as a hydrogen exhaust from the Elektron oxygen generator. They also photographed the antenna to be used for ATV docking for analysis on Earth, to ensure they are correctly aligned when ATV operations begin. The crew also retrieved several exposure experiments and cassettes and removed a failed camera, replacing it with a new one on the MT.

In July, STS-121 visited ISS in the second of two return-to-flight missions. This time, the Shuttle delivered 3,356 kg of supplies to the station, as well as German ESA astronaut Thomas Reiter. He transferred to the main crew to work with ISS-13, returning the ISS crew to a complement of three for the first time since May 2003. Reiter would conduct the ESA Astrolab science programme while aboard the station. With the successful flight of STS-121, the expansion of the station would soon be resuming as the ISS-13 residence wound down. Reiter would continue with the ISS-14 crew for a few more weeks. On 3 August, Williams and Reiter completed an EVA (5 hours 54 minutes) that included the installation of hardware in preparation for future ISS assembly work, as well as deploying a number of instruments and experiments on the outside of the station.

In September 2006, the ISS-13 crew hosted the STS-115 crew for the first Shuttle assembly mission since STS-113 in November 2002. While docked to the station, the STS-115 crew added a further solar array truss and transferred logistics to support station operations.

On 19 September, the day after the STS-115 crew departed and the same day the Soyuz TMA9 crew were launched, the ISS-13 crew, after servicing the Elektron device in the service module, noted a small leak of KOH (potassium hydroxide, "caustic potash") electrolyte bubbles from the O_2 outlet nozzle. The crew immediately manually activated the fire alarm, which automatically shut down the ventilation system. Following the mission rules after such incidents, and as an extra precaution, all three men donned goggles, gloves, and surgical masks. The released caustic liquid (which was deemed to be Level 2 Toxicity – an "irritant") was immediately cleaned up with a cloth and no further leaks were noted. About 30 minutes later, the Vodzukh was activated, with a charcoal filter installed to scrub the air. Recorded air data remained well within acceptable values and protective gear was soon no longer required as onboard operations returned to the nominal schedule. At the time of the situation, ground controllers instigated a "spacecraft emergency" procedure to ensure that TDRS communication coverage would be at the highest priority. In the event this was not required and TDRS coverage was returned to normal.

The ISS-13 crew handed over to the ISS-14 crew the following week, completing several days of joint activities with them and space flight participant Anousheh Ansari. The ISS-13 crew landed in TMA8 with Ansari in the early hours of 29 September.

Milestones

247th manned space flight
101st Russian manned space flight
94th manned Soyuz mission
8th manned Soyuz TMA mission
41st Russian and 96th flight with EVA operations
10th ISS Soyuz mission (10S)
13th ISS resident crew (EO-13)
10th visiting mission (VC-10)
5th resident caretaker ISS crew (2 person – until July)
1st Brazilian citizen in space

STS-121

Int. Designation	2006-028A
Launched	4 July 2006
Launch Site	Pad 39B, Kennedy Space Center, Florida
Landed	17 July 2006
Landing Site	Runway 15, Shuttle Landing Facility, KSC, Florida
Launch Vehicle	OV-102 Discovery/ET-119/SRB BI-126/SSME #1 2045; #2 2051; #3 2056
Duration	12 days 18 hrs 37 min 54 sec
Call sign	Discovery
Objective	Second and final return to flight mission; utilisation and logistics flight (ULF1.1); MPLM logistics flight and partial ISS resident crew member delivery

Flight Crew

LINDSEY, Steven Wayne, 45, USAF, commander, 4th mission
Previous missions: STS-87 (1997); STS-98 (1998); STS-104 (2001)
KELLY, Mark Edward, 42, USN, pilot, 2nd mission
Previous mission: STS-108 (2001)
FOSSUM, Michael Edward, 48, civilian. mission specialist 1
NOWAK, Lisa Marie, 43, USN, mission specialist 2
WILSON, Stephanie Diana, 39, civilian, mission specialist 3
SELLERS, Piers John, 51, civilian, mission specialist 4, 2nd mission
Previous mission: STS-112 (2002)

ISS-13 crewmember up only:
REITER, Thomas, 48, German Air Force, ESA mission specialist 5, ISS-13 flight engineer 2, 2nd mission
Previous mission: Soyuz TM22 (1995)

Flight Log

Bad weather again delayed the launch of the Shuttle on 1 and 2 July, but after a one-day stand down, the launch resumed and STS-121 lifted off on America's 230th birthday, the first launch on Independence Day. Again the launch phase was dramatically caught on film and analysed for any serious debris impacts. Though some debris hits were noted, there were no serious impacts. Following the docking on 6 July, the crew relocated the Leonardo MPLM to the station, where 3,357 kg of logistics was transferred. This included a new heat exchanger, a new window and window seals for the Microgravity Science Glove Box, a US EMU suit, SAFER jet pack and personal items, and a new oxygen generator to be installed in Destiny. There was also a

After nine days of cooperative work, the crews of STS-121 and ISS-13 bid farewell to each other prior to the undocking of Discovery. In the foreground, Sellers says farewell to Reiter (ESA), who launched to ISS on STS-121 but remained on the station with the ISS-13 crew. At rear, Wilson says goodbye to ISS-13 cosmonaut Vinogradov

European Minus Eighty Degree Laboratory Freezer for storage of experiment samples prior to return to Earth. The MPLM was relocated back in the payload bay of Discovery on 14 July, by which time it had been filled with 2,086 kg of experiment samples, broken equipment and trash.

Three EVAs by Sellers (EV1) and Fossum (EV2) were completed during the docked phase of the flight. EVA 1 (8 Jul for 7 hours 31 minutes) included fitting a protection device for power, data and video cables on the S0 Truss. By routing cables through the Interface Umbilical Assembly, they could move the MT railcar to replace the trailing umbilical system with its power and data cable that had been inadvertently severed late in 2005. The astronauts then evaluated the RMS with the Orbiter Boom Sensor System, for potential use as a work platform during EVA to repair the damaged orbiter if required. EVA 2 (10 Jul for 6 hours 47 minutes) saw the crew restore the MT to full operation, while EVA 3 (12 Jul for 7 hours 11 minutes) focused on testing repairs on the Thermal Protection System Reinforced Carbon–Carbon panels, part of the evaluation of repair techniques that may be available to effect orbital repairs on future Shuttle missions. Photography of the samples (together with that of an area of Discovery's port wing) was conducted. These images would be returned to Earth for detailed analysis. The astronauts also relocated a fixed grapple bar on the integrated cargo carrier in the payload bay to a position on the ammonia tank inside the S1 Truss assembly, so that it could be moved by a future EVA crew at a later date.

Thomas Reiter had formerly exchanged to the ISS-13 crew shortly after the hatches between the vehicles had opened, moving his Soyuz seat liner into the DM of Soyuz TMA8. The more formal farewells occurred shortly prior to closure of the hatches and undocking of the Shuttle on 15 July. The two spacecraft had been docked for over 9 days, a day longer than planned as the mission was extended by a day to accommodate the third EVA. A final check of the surfaces of Discovery revealed nothing of concern and the safe re-entry and landing two days later gave an added boost to mission planners, eager to resume station construction with the next Shuttle mission.

Milestones

248th manned space flight
145th US manned space flight
115th Shuttle mission
32nd flight of Discovery
18th Shuttle ISS flight
58th US and 97th flight with EVA operations
6th Discovery ISS mission
7th MPLM flight
4th flight of MPLM-1 Leonardo
1st US launch on 4 July

STS-115

Int. Designation 2006-036A
Launched 9 September 2006
Launch Site Pad 39B, Kennedy Space Center, Florida
Landed 21 September 2006
Landing Site Runway 33, Shuttle Landing Facility, KSC, Florida
Launch Vehicle OV-104 Atlantis/ET-118/SRB BI-127/SSME: #1 2044, #2 2048; #3 2047
Duration 11 days 19 hrs 7 min 24 sec
Call sign Atlantis
Objective ISS assembly mission 12A; delivery and installation of P3/P4 Truss

Flight Crew

JETT, Brent, 47, USN, commander, 4th mission
Previous missions: STS-72 (1996); STS-81 (1997); STS-97 (2000)
FERGUSON, Chris, 44, USN, pilot
TANNER, Joe, 56, civilian, mission specialist 1, 4th mission
Previous missions: STS-66 (1994); STS-82 (1997); STS-97 (2000)
BURBANK, Dan, 45, USCG, mission specialist 2, 2nd mission
Previous mission: STS-106 (2000)
STEFANYSHYN-PIPER, Heidemarie, 43, USN, mission specialist 3
MACLEAN, Steve, 51, civilian, Canadian mission specialist 4

Flight Log

Set for a 29 August launch, the lift-off for the STS-115 mission to resume space station construction was postponed due to the proximity of tropical storm Ernesto. A decision was then made to roll back the STS-115 stack into the protection of the VAB for the duration of the storm as it passed KSC. This had a scheduling impact for the Russian launch of Soyuz TMA9 in September, but later the same day, NASA managers decided to reverse the decision and began moving the Shuttle back to the pad as weather predictions improved. On 6 September, a problem with Fuel Cell #1 in Atlantis was noted when a voltage spike in the coolant pump was recorded, threatening the planned 8 September launch. Analysis indicated that this was not a problem that would prevent the launch, but when a fuel cut-off sensor in the ET caused concern during the final minutes of the count, the mission was postponed 24 hours at the $T-9$ minute mark. After a nominal performance during tests, the launch was given the all clear to proceed, which it did without further incident.

The delay had resulted in a short postponement of the launch of Soyuz TMA9 to the station and the shortening of the STS-115 mission by a day.

Displaying a new set of wings, this photograph of ISS taken from the departing Shuttle reveals the newly-installed solar arrays delivered and installed by the crew of STS-115

The first day in orbit found the crew preparing equipment for the docking and EVA activities, as well as inspecting the thermal protection system on the orbiter. After analysis on the ground, no significant damage was found. Prior to docking on 11 September, the orbiter was flipped to allow the ISS-13 crew to observe and photo-document the TPS. Less than two hours after docking, the crew entered the station for the first time. At the end of the day, the first EVA crew of Tanner (EV1) and Stefanyshyn-Piper (EV2) "camped-out" for the night in the Quest airlock to purge their bloodstreams of nitrogen, which would shorten EVA preparations the next day. This pair completed the first and third EVAs of the mission, with Burbank (EV3) and Canadian Steve MacLean (EV4) completing the second EVA. All three EVAs (12 Sep for 6 hours 26 minutes; 13 Sep for 7 hours 11 minutes; and 15 Sep for 6 hours 42 minutes) were associated with the installation of the P3/P4 Truss and the deployment of the solar arrays and radiators.

No focused inspection of the Atlantis TPS was required after detailed analysis of the images from the crew, RMS and station inspections, so after the first two EVAs were completed, the crew rested for a couple of days and turned their attention to the transfer of logistics to and from the station. Such was the success of the first two EVAs, the crew managed to complete several get-ahead tasks along with their primary objectives.

On 17 September, Atlantis undocked from ISS after a visit lasting six days. Early the next morning, as the Shuttle began preparations for the return to Earth, Soyuz TMA9 was launched from Baikonur. With 12 space explorers in orbit at the same time on three different vehicles (six astronauts on board Atlantis, three on Soyuz and three on ISS), it was the most people in space at the same time since April 2001, when the ISS-2, STS-100 and Soyuz TM32 crews (totalling 13 crew members) were all aloft. The hatches were open for 5 days 21 hours and 57 minutes and during this time, the two crews transferred 362.88 kg of hardware and 473 kg of water into the station and returned 491 kg of unwanted hardware and trash. In addition, 90.72 kg of launch lock restraints and unnecessary hardware was placed in Progress M56 for disposal.

Another inspection of the TPS of Atlantis was completed the day after undocking and the following day, the Shuttle crew spoke with both the crew on ISS and the crew on the approaching Soyuz TMA9 craft in a three-way link up. On 19 September the mission was extended in order to re-check some of the TPS areas of Atlantis after small unidentified particles were found floating near the Shuttle. There were sufficient supplies to allow the mission to be extended until 22 September or for them to return to ISS for a possible rescue mission if anything untoward been found. However, analysis revealed no significant problems and Atlantis was cleared for landing on 21 September (the previous day had been ruled out due to weather concerns). In the event all went well, and Atlantis made a textbook landing at night at the SLF at the Cape.

During homecoming events in Houston on 21 September, Stefanyshyn-Piper collapsed twice and had to be assisted by officials and crew members. She was not taken to hospital and the effects were attributed to her adjustment to gravity after her first 12-day flight into space.

Milestones

249th manned space flight
146th US manned space flight
116th Shuttle mission
27th flight of Atlantis
19th Shuttle ISS mission
7th Atlantis ISS mission
59th US and 98th flight with EVA operations

SOYUZ TMA9

Int. Designation	2006-040A
Launched	19 September 2006
Launch Site	Pad 1 Site 5, Baikonur Cosmodrome, Kazakhstan
Landed	19 March 2007 (planned) EO-14
	29 September 2006 (Ansari with EO-13 crew)
Landing Site	Kazakhstan (planned)
Launch Vehicle	R7 (Soyuz FG); spacecraft serial number 219
Duration	182 days (planned) EO-14
	10 days 21 hrs 5 min 0 sec (Ansari)
Call sign	Vostok (East – Soyuz TMA)
Objective	ISS-14 research programme; Visiting Crew 11 programme

Flight Crew

LOPEZ-ALEGRIA, Michael Eladio, 48, USN, ISS-14 commander and science officer, 4th mission
Previous missions: STS-73 (1995); STS-92 (2000); STS-113 (2002)
TYURIN, Mikhail Vladislavovich, 46, civilian, Russian ISS-14 Soyuz commander, 2nd mission
Previous mission: ISS-3 (2001)
ANSARI, Anousheh, 40, civilian, US space flight participant

Flight Log

The original spaceflight participant on this mission was scheduled to be Japanese businessman Daisuke Enomato, but on 21 August he failed the pre-flight medical and was replaced by his back-up Anousheh Ansari. She is the Iranian-born naturalised American who co-founded Telecom Technologies Inc. in 1993, and the X-Prize sub-orbital space flight record attempts (won in 2005 by Spaceship One). During the flight to ISS, she reportedly suffered from Spaceflight Adaptation Syndrome, but seemed to recover successfully once on ISS to complete her research programme. This consisted of three TV broadcasts, amateur radio broadcasts, photo and video surveys of the Russian segment of ISS for education purposes, participation in two small ESA experiments and commemorative activities, including regular email contact to the Internet through her own website during the mission. She returned to Earth on 29 September with the ISS-13 crew in TMA8.

Soyuz TMA9 docked with ISS on 21 September to begin the six-month residency of the ISS-14 crew. During their stay, they will work with German ESA astronaut Thomas Reiter until he is replaced by NASA astronaut Sunita Williams during the STS-116 mission in December. Williams, like Reiter, will serve as ISS-14 flight

A million dollar ticket to space. Space flight participant Ansari is shown strapped to her Soyuz TMA seat in TMA9 shortly after entering orbit on the two-day flight to ISS with the Expedition Fourteen crew

engineer 2 and will continue to work with the ISS-15 crew until she in turn is replaced by another NASA astronaut during STS-118 in the spring of 2007.

The residency of ISS-14 will see a significant increase in onboard science activities with the return of a resident three-person crew. In addition, they will host Shuttle missions STS-116 and 117, receive two Progress re-supply missions and complete four EVAs. One of the spacewalks will be by Tyurin and Lopez-Alegria using the Pirs airlock and wearing Orlan suits. The other three will be by Lopez-Alegria and Williams out of the Quest airlock wearing US suits. The ISS-14 crew are expected to be relieved by the ISS-15 crew in March 2007, and will return to Earth with the next space flight participant (launched with the ISS-15 crew on TMA10) on 19 March 2007.

Milestones

250th manned space flight
102nd Russian manned space flight
95th manned Soyuz mission
9th manned Soyuz TMA mission
11th ISS Soyuz mission (11S)
11th visiting mission (VC-11)
4th space flight participant
1st female space flight participant

9

The Next Steps

With the successful flight of STS-114 in July 2005 and the second Return-to-Flight mission of STS-121 in July 2006, NASA revised the Shuttle manifest pending the retirement of the vehicle in 2010. There is also another servicing mission planned for the Hubble Space Telescope in 2008.

Table 9.1. ISS Assembly Manifest

Launch Date	Assembly Flight	Launch Vehicle	Element(s)
2006 Dec 14	12A	Discovery STS-116	P5 Truss SpaceHab single module Integrated Cargo Carrier (ICC)
2007 Feb 22	13A	Atlantis STS-117	S3/S4 Truss with Photovoltaic Radiator 3rd set of solar arrays and batteries
2007 May 1	ATV1	Ariane 5	European Automated Transfer Vehicle
2007 Jun 11	13A.1	Endeavour STS-118	SpaceHab single module S5 Truss External Stowage Platform 3 (ESP 3)
2007 Aug 9	10A	Atlantis STS-120	Node 2 Sidewall – Power and Data Grapple Fixture (PGDF)
2007 Oct	1E	Shuttle STS-122	Columbus European laboratory Multi-Purpose Experiment Support Structure – Non-Deployable (MPESS-ND)
2007 Dec	1J/A	Shuttle	Kibo Japanese Experiment Logistics Module – Pressurised Section (ELM-PS) Spacelab Pallet – Deployable 1 (SLP-D1) with Canadian Special Purpose Dextrous Manipulator, Dextre
2008 Feb	1J	Shuttle	Kibo Japanese Experiment Module – Pressurised Module (JEM-PM) Japanese Remote Manipulator System (JEM RMS)
2008 Jun	15A	Shuttle STS-119	S6 Truss Fourth set of solar arrays and batteries
2008 Aug	ULF2	Shuttle	Multi-Purpose Logistics Module (MPLM)
2008 Oct	2J/A	Shuttle	Kibo Japanese Experiment Module Exposed Facility (JEM EF) Kibo Japanese Experiment Logistics Module – Exposed Section (ELM-ES) Spacelab Pallet – Deployable 2 (SLP-D2)
Dec 2008	3R	Proton	Multipurpose Laboratory Module with European Robotic Arm (ERA)

Table 9.1 (*cont.*)

Launch Date	Assembly Flight	Launch Vehicle	Element(s)
2009 Jan	17A	Shuttle	Multi-Purpose Logistics Module (MPLM) Lightweight Multi-Purpose Experiment Support Structure Carrier (LMC) Three crew quarters, galley, second treadmill (TVIS2) Crew Health Care System (CHeCS 2)

Establish Six Person Crew Capability

Launch Date	Assembly Flight	Launch Vehicle	Element(s)
2009 Feb	HTV-1	H-IIA	Japanese H-II Transfer Vehicle
2009 April	ULF3	Shuttle	EXPRESS Logistics Carrier 1 (ELC 1) EXPRESS Logistics Carrier 2 (ELC 2)
2009 July	19A	Shuttle	Multi-Purpose Logistics Module (MPLM) Lightweight Multi-Purpose Experiment Support Structure Carrier (LMC)
2009 Oct	ULF4	Shuttle	EXPRESS Logistics Carrier 3 (ELC 3) EXPRESS Logistics Carrier 4 (ELC 4)
2010 Jan	20A	Shuttle	Node 3 with Cupola
2010 July	ULF5	Shuttle	EXPRESS Logistics Carrier 5 (ELC 5) EXPRESS Logistics Carrier 1 (ELC 1)

ISS Assembly Complete

Launch Date	Assembly Flight	Launch Vehicle	Element(s)
Under Review	9R	Proton	Research Module

Dates listed are subject to change. There will continue to be additional Progress and Soyuz flights for crew transport, logistics and re-supply.

THE IMMEDIATE FUTURE

The exchange of the 14th resident crew of ISS continues the occupation of the space station. A new crew is schedule to take over in March 2007 and Thomas Reiter, the German astronaut launched on STS-121 will continue to fly on the station with the EO-14 crew until STS-116 arrives with ISS-15 flight engineer Sunita Williams. Several European, Japanese and Canadian astronauts are expected to conduct short- and long-duration visits to the station in the coming years.

The Chinese have announced that their next mission will occur in 2008 and will feature an EVA. Rendezvous and docking, and the creation of a small (Salyut class) space station are their stated goals, along with eventual manned exploration of the Moon.

Several names have been identified as potential candidates for future space flight participants, although recent reports offering tourists the chance to perform a 90-minute EVA from ISS, or possibly fly on circumlunar flights are not likely to come to fruition for some considerable time.

Flight operations to ISS will continues, with roughly two Soyuz flights per year until around 2010 when the Shuttle retires. It is probable that Soyuz flights will continue and may well increase (given adequate funding) until the replacement Clipper spacecraft planned by the Russians is funded and tested. Until then, Soyuz will remain the ISS crew ferry vehicle and emergency crew return vehicle once Shuttle is retired. American participation on ISS is expected to change in 2016 with the completion of their primary research objectives in support of the *Vision for Space Exploration* plans that were announced in 2004.

Design of the new American Crew Exploration Vehicle is underway and the two launch vehicles – called Aries – have recently been revealed. These will support the manned American return to the Moon, hopefully by 2019 and the 50th anniversary celebrations of Apollo 11. These new landings will include extended surface exploration, leading to the creation of a scientific research station on the surface It is hoped that these studies will lead, eventually, to human expeditions to Mars.

The current (October 2006) manifest for flights between 2006 and 2011 can be found in Appendix C.

All of the missions and programmes, if they reach flight status, are for a future edition of this log. With the 50th anniversary of Yuri Gagarin's historic first flight into space less than five years away, the missions counting down towards that milestone have already begun. The story continues ...

SUMMARY

On a bright spring day in April 1961, a young Russian pilot climbed aboard a new type of vehicle – a manned spacecraft. He was about to attempt what no one had tried before. A former ballistic missile, adapted for carrying a man but not totally safe from error, was going to blast him on an eight-minute ride from Earth into space. For 108 minutes he would fly around his home planet, then endure, inside his protective spacecraft, the fiery heat of re-entry, before ejecting to descend by parachute to his native soil. In those 108 minutes, Yuri Gagarin moved from obscurity to one of the most famous names in human history. No matter how many people follow his trail from Earth, he will always be the first, the pioneer, the one who took mankind's first step out of the cradle. On any listing of most space experience in the 45 years since that flight, Gagarin's name will appear at the very bottom, but his achievement, his courage and his very persona will forever fly higher than any record book can show.

In the Cold War race for technical and national supremacy between America and the Soviet Union, their Arms Race spawned another race, to place the first person into space. Once that was done, their eyes turned towards our nearest solar neighbour, the Moon. This time, the Americans would win the race, but they would also come out losers. Though other missions under the Vostok, Voskhod, Mercury and Gemini programmes were planned it is probable that nothing more would have been achieved that could not have been achieved by later programmes, probably far more safely.

America's triumph with Apollo was short lived. In the spirit of determination and achievement that Kennedy's famous speech had engendered in the American psyche, great plans were laid for what would happen after the Moon landing goal had been achieved. The potential for extended duration missions in Earth orbit, orbital research and development flights, and reaching further targets was all lost in a wave of public apathy and political debate on the value of Apollo lunar programme once Apollo 11 had achieved Kennedy's goal. An expanded lunar exploration programme was abandoned, even with some of the hardware built and paid for. That hardware was placed in museums or left to rot, bygone icons of a forgotten era. America had other more pressing goals at home to think about that seemed to better justify, or at least consumed, the tax dollar.

For the Soviets, losing the Moon race was painful, but they turned their attention to a new target, the creation of a long term space station. Over the next thirty years, their programme and understanding of what it took to spend significant amounts of time in space grew, culminating with the Mir programme. Mir remained in orbit 15 years, and was permanently occupied for almost ten of them. Successive crews battled with shortages, failures and set backs, as well as huge success and hard-won achievement in stretching the human space experience from days and weeks, to months and years. If Apollo was the shining star of the first era of pioneering manned space exploration, then surely Mir was as bright a star in the second period as humans truly began to understand how to live and work in space.

Over in the United States America turned to the Space Shuttle. As with earlier programmes, this was envisaged as just one part pf a large space infrastructure. Grandiose plans included Earth and lunar orbital space bases, a lunar base, manned flights to Mars and even hotels and factories in orbit, all foreseen long before Shuttle ever flew. When it did, the reality of what it could actually do became readily apparent. And dreams remained dreams. The Shuttle could fly short research missions, capture, repair and redeploy space satellites, and fly mixed cargos into and out of space, but it could not do it as regularly or as cheaply as once thought. Shuttle could not reduce the cost per kilogram of reaching orbit, fly every two weeks, and launch everything America, and most of the world wanted to assign to it. And with no orbiting platform to deliver this cargo to, it became little more than an expensive and risky space truck. The loss of Challenger and her crew of seven was the final straw. Soon the commercial customers and military chiefs backed away from Shuttle as a new goal was set – a space station so large that it would need an international group of partners to build, support, and pay for it.

Space Station Freedom was another dream born from those visions of huge space cities in the 1950s and early 1960s. It was one thing aiming for a space station of this complexity, however, but quite another to build and pay for it. Costs, complications and problems grew to bursting point and by the early 1990s, the Space Shuttle, the space station, and even NASA itself, looked in dire straits. In Russia, years of papering over the cracks in both the space programme and the national economy finally caught up with them and the once-mighty Soviet Union and most of the communist world collapsed in an expensive and tragic mess.

Born from the turmoil was a new cooperative programme in space. Russia would join what was now the International Space Station programme. There was still a decade or so of hard work and sometimes fraught discussions, but one thing the ISS programme has shown, as anyone involved in it will underline, is that international teamwork and cooperation can achieve such a global and extensive goal. And the Shuttle could finally prove that it was capable of the task originally envisioned for it way back in those grandiose plans – supplying and constructing a space station. The loss of Columbia in 2003 has dealt a final blow to a Shuttle programme that has been flying for 25 years, although the infrastructure created by ISS will keep the programme going for a while longer.

By the 45th year of human space flight, the Shuttle was on the road to its second recovery, the crew complement of ISS was restored, tourists were paying a lot of money for the chance of making one short flight around the Earth, and a new player had entered the scene – China. The success of ISS is that it has been "international" and perhaps that is the way forward. Large national space programmes are relics of the past and cooperation across the globe in space may help with cooperation across the globe for more terrestrial goals. This book therefore records the trail from Gagarin to this 45th year; the successes and the failures, the milestones and the tragedies. We hope that it provides a handy reference of what has gone before as we stand on the edge of what could be about to happen.

As the 50th anniversaries of these first space flights approach between 2011 and 2021 – the first manned space flights, first EVAs, first docking, first lunar flights, first extended flights, and first space station – the future of human space flight seems to be forward-looking once again. Though the flight path may be unsteady, contingencies and back up plans have to be prepared, and mission objectives may change, Gagarin's trail is still bright and strong. And, as he said at the moment his rocket left Earth for the stars ... *"Poyekhali! ... Off we go!"*

Appendix A

The Log Book 1961–2006

This table lists all the manned space flights that were official astro-flights (X-15); X-Prize (Spaceship One); sub-orbital (Mercury-Redstone); launch pad aborts (Soyuz T10-1) and missions in progress (Soyuz 18-1; STS 51-L; STS-107). For space station residents and visiting missions, where crews have launched separately but returned on the same vehicle, individual durations are shown below the specific return mission.

Appendix A

Year	Mission	Crew at launch	Date (dd/mm/yy)	Duration (dd:hh:mm:ss)
1961	Vostok	Gagarin	12/04/61	00:01:48:00
	Mercury 3	Shepard	05/05/61 (sub-orbital)	00:00:15:28
	Mercury 4	Grissom	21/07/61 (sub-orbital)	00:00:15:37
	Vostok 2	Titov G.	06/08/61–07/08/61	01:01:18:00
1962	Mercury 6	Glenn	20/02/62	00:04:55:23
	Mercury 7	Carpenter	24/05/62	00:04:56:05
	X-15-3-62	White R.	17/07/62	00:00:10:00 (astro-flight)
	Vostok 3	Nikolayev	11/08/62–15/08/62	03:22:22:00
	Vostok 4	Popovich	12/08/62–15/08/62	02:22:57:00
	Mercury 8	Schirra	03/10/62	00:09:13:11
1963	X-15-3-77	Walker J.	17/01/63	00:00:10:00 (astro-flight)
	Mercury 9	Cooper	15/05/63–16/05/63	01:10:19:49
	Vostok 5	Bykovsky	14/06/63–19/06/63	04:23:06:00
	Vostok 6	Tereshkova	16/06/63–19/06/63	02:22:50:00
	X-15-3-87	Rushworth	27/06/63	00:00:10:00 (astro-flight)
	X-15-3-90	Walker J.	19/07/63	00:00:10:00 (astro-flight)
	X-15-3-91	Walker J.	22/08/63	00:00:10:00 (astro-flight)
1964	Voskhod	Komarov/Feoktistov/Yegorov	12/10/64–13/10/64	01:00:17:03
1965	Voskhod 2	Belyayev/Leonov	18/03/65–19/03/65	01:02:02:17
	Gemini 3	Grissom/Young	23/03/65	00:04:52:51
	Gemini 4	McDivitt/White	03/06/65–07/06/65	04:01:56:12
	X-15-3-138	Engle	29/06/65	00:00:10:00 (astro-flight)
	X-15-3-143	Engle	10/08/65	00:00:10:00 (astro-flight)
	Gemini 5	Cooper/Conrad	21/08/65–29/08/65	07:22:55:14
	X-15-3-150	McKay	29/09/65	00:00:10:00 (astro-flight)
	X-15-1-153	Engle	14/10/65	00:00:10:00 (astro-flight)
	Gemini 7	Borman/Lovell	04/12/65–16/12/65	13:18:35:01

	Gemini 6	Schirra/Stafford	15/12/65–16/12/65	01:01:51:54
1966	Gemini 8	Armstrong/Scott D.	16/03/66	00:10:41:26
	Gemini 9	Stafford/Cernan	03/06/66–06/06/66	03:00:20:50
	Gemini 10	Young/Collins M.	18/07/66–21/07/66	02:22:46:39
	Gemini 11	Conrad/Gordon	12/09/66–15/09/66	02:23:17:08
	X-15-3-174	Dana	01/11/66	00:00:10:00 (astro-flight)
	Gemini 12	Lovell/Aldrin	11/11/66–16/11/66	03:22:34:31
1967	Apollo 1	Grissom/White/Chaffee	27/01/67	(fatal pad fire prior to launch date)
	Soyuz 1	Komarov	23/04/67–24/04/67	01:02:47:52
	X-15-3-190	Knight	17/10/67	00:00:10:00 (astro-flight)
	X-15-3-191	Adams	15/11/67	00:00:10:00 (astro-flight – fatal)
1968	X-15-1-197	Knight	21/08/68	00:00:10:00 (astro-flight)
	Apollo 7	Schirra/Eisele/Cunningham	11/10/68–22/10/68	10:20:09:03
	Soyuz 3	Beregovoy	26/10/68–30/10/68	03:22:50:45
	Apollo 8	Borman/Lovell/Anders	21/12/68–28/12/68	06:03:00:42
1969	Soyuz 4	Shatalov	14/01/69–17/01/69	02:23:20:47
	Soyuz 5	Volynov/Yeliseyev/Khrunov	15/01/69–18/01/69	03:00:54:15
				(Yeliseyev/Khrunov 01:23:45:50)
	Apollo 9	McDivitt/Scott/Schweickart	03/03/69–13/03/69	10:01:00:54
	Apollo 10	Stafford/Young/Cernan	18/05/69–26/05/69	08:00:03:23
	Apollo 11	Armstrong/Collins M./Aldrin	16/07/69–24/07/69	08:03:18:35
	Soyuz 6	Shonin/Kubasov	11/10/69–16/10/69	04:22:42:47
	Soyuz 7	Filipchenko/Gorbatko/Volkov V.	12/10/69–17/10/69	04:22:40:23
	Soyuz 8	Shatalov/Yeliseyev	13/10/69–18/10/69	04:22:50:49
	Apollo 12	Conrad/Gordon/Bean	14/11/69–24/11/69	10:04:36:25

(continued)

784 Appendix A

Year	Mission	Crew at launch	Date (dd/mm/yy)	Duration (dd:hh:mm:ss)
1970	Apollo 13	Lovell/Swigert/Haise	11/04/70–17/04/70	05:22:54:41
	Soyuz 9	Nikolayev/Sevastyanov	01/06/70–19/06/70	17:16:58:55
1971	Apollo 14	Shepard/Roosa/Mitchell	31/01/71–09/02/71	09:00:01:57
	Soyuz 10	Shatalov/Yeliseyev/Rukavishnikov	23/04/71–25/04/71	01:23:45:54
	Soyuz 11	Dobrovolsky/Volkov V./Patsayev	06/06/71–30/06/71	23:18:21:43
	Apollo 15	Scott D./Worden/Irwin	26/07/71–07/08/71	12:07:11:53
1972	Apollo 16	Young/Mattingly/Duke	16/04/72–27/04/72	11:01:51:25
	Apollo 17	Cernan/Evans/Schmitt	06/12/72–19/12/72	12:13:51:59
1973	Skylab 2	Conrad/Kerwin/Weitz	25/05/73–22/06/73	28:00:49:49
	Skylab 3	Bean/Garriott/Lousma	28/07/73–25/09/73	59:11:09:04
	Soyuz 12	Lazarev/Makarov	27/09/73–29/09/73	01:23:15:32
	Skylab 4	Carr/Gibson E./Pogue	15/11/73–08/02/74	84:01:15:37
	Soyuz 13	Klimuk/Lebedev	18/12/73–26/12/73	07:20:55:35
1974	Soyuz 14	Popovich/Artyukhin	03/07/74–19/07/74	15:17:30:28
	Soyuz 15	Sarafanov/Demin	26/08/74–28/08/74	02:00:12:11
	Soyuz 16	Filipchenko/Rukavishnikov	02/12/74–08/12/74	05:22:23:35
1975	Soyuz 17	Gubarev/Grechko	11/01/75–09/02/75	29:13:19:45
	Soyuz 18-1	Lazarev/Makarov	05/04/75	00:00:21:27 (launch phase abort)
	Soyuz 18	Klimuk/Sevastyanov	24/05/75–26/06/75	62:23:20:08
	Soyuz 19	Leonov/Kubasov	15/07/75–21/07/75	05:22:30:51
	Apollo 18	Stafford/Brand/Slayton	15/07/75–24/07/75	09:01:28:24
1976	Soyuz 21	Volynov/Zholobov	06/07/76–24/08/76	49:06:23:32
	Soyuz 22	Bykovsky/Aksenov	15/09/76–23/09/76	07:21:52:17
	Soyuz 23	Zudov/Rozhdestvensky	14/10/76–16/10/76	02:00:06:35

1977	Soyuz 24	Gorbatko/Glazkov	07/02/77–25/02/77	17:17:25:58
	Soyuz 25	Kovalenok/Ryumin	09/10/77–11/10/77	02:00:44:45
	Soyuz 26	Romanenko/Grechko	10/12/77–16/03/78	96:10:00:07
1978	Soyuz 27	Dzhanibekov/Makarov	10/01/78–16/01/78	05:22:58:58
	Soyuz 28	Gubarev/Remek	02/03/78–10/03/78	07:22:16:00
	Soyuz 29	Kovalenok/Ivanchenkov	15/07/78–02/11/78	139:14:47:32
	Soyuz 30	Klimuk/Hermaszewski	27/06/78–05/07/78	07:22:02:59
	Soyuz 31	Bykovsky/Jaehn	26/08/78–03/09/78	07:20:49:04
1979	Soyuz 32	Lyakhov/Ryumin	25/02/79–19/08/79	175:00:35:37
	Soyuz 33	Rukavishnikov/Ivanov G.	10/04/79–12/04/79	01:23:01:06
1980	Soyuz 35	Popov/Ryumin	09/04/80–11/10/80	184:20:11:35
	Soyuz 36	Kubasov/Farkas	26/05/80–03/06/80	07:20:45:44
	Soyuz T2	Malyshev/Aksenov	05/06/80–09/06/80	03:22:19:30
	Soyuz 37	Gorbatko/Pham Tuan	23/07/80–31/07/80	07:20:42:00
	Soyuz 38	Romanenko/Tamayo-Mendez	18/09/80–26/09/80	07:20:43:24
	Soyuz T3	Kizim/Makarov/Strekalov	27/11/80–10/12/80	12:19:07:42
1981	Soyuz T4	Kovalenok/Savinykh	12/03/81–26/05/81	74:17:37:23
	Soyuz 39	Dzhanibekov/Gurragcha	22/03/81–30/03/81	07:20:42:03
	STS-1	Young/Crippen	12/04/81–14/04/81	02:06:20:53
	Soyuz 40	Popov/Prunariu	14/05/81–22/05/81	07:20:41:52
	STS-2	Engle/Truly	12/11/81–14/11/81	02:06:13:13
1982	STS-3	Lousma/Fullerton	22/03/82–30/03/82	08:00:04:45
	Soyuz T5	Berezovoy/Lebedev	13/05/82–10/12/82	211:09:04:32
	Soyuz T6	Dzhanibekov/Ivanchenkov/Chrétien	24/06/82–02/07/82	07:21:50:52
	STS-4	Mattingly/Hartsfield	27/06/82–02/07/82	07:01:09:31
	Soyuz T7	Popov/Serebrov/Savitskaya	19/08/82–27/08/82	07:21:52:24
	STS-5	Brand/Overmyer/Allen J./Lenoir	11/11/82–16/11/82	05:02:14:26

(continued)

Appendix A 785

786 Appendix A

Year	Mission	Crew at launch	Date (dd/mm/yy)	Duration (dd:hh:mm:ss)
1983	STS-6	Weitz/Bobko/Musgrave/Peterson	04/04/83–09/04/83	05:00:23:42
	Soyuz T8	Titov V./Strekalov/Serebrov	20/04/83–22/04/83	02:00:17:48
	STS-7	Crippen/Hauck/Fabian/Ride/Thagard	18/06/83–24/06/83	06:02:23:59
	Soyuz T9	Lyakhov/Alexandrov	27/06/83–23/11/83	149:10:46:01
	STS-8	Truly/Brandenstein/Bluford/Gardner/Thornton W.	30/08/83–05/09/83	06:01:08:43
	Soyuz T10-1	Titov V./Strekalov	26/09/83	(launch pad abort prior to lift off)
	STS-9	Young/Shaw/Garriott/Parker/Lichtenberg/Merbold	28/11/83–08/12/83	10:07:47:23
1984	STS 41-B	Brand/Gibson R./McNair/Stewart/McCandless	03/02/84–11/02/84	07:23:15:55
	Soyuz T10	Kizim/Solovyov V./Atkov	08/02/84–02/10/84	236:22:49:04
	Soyuz T11	Malyshev/Strekalov/Sharma	03/04/84–11/04/84	07:21:40:06
	STS 41-C	Crippen/Scobee/Hart/van Hoften/Nelson G.	06/04/84–13/04/84	06:23:40:06
	Soyuz T12	Dzhanibekov/Savitskaya/Volk	17/07/84–29/07/84	11:19:14:36
	STS 41-D	Hartsfield/Coats/Mullane/Hawley/Resnik/Walker C.	30/08/84–05/09/84	06:00:56:04
	STS 41-G	Crippen/McBride/Sullivan/Ride/Leestma/Scully-Power/Garneau	05/10/84–13/10/84	08:05:23:38
	STS 51-A	Hauck/Walker D./Allen J./Fisher A./Gardner D.	08/11/84–16/11/84	07:23:44:56
1985	STS 51-C	Mattingly/Shriver/Onizuka/Buchli/Payton	24/01/85–27/01/85	03:01:23:23
	STS 51-D	Bobko/Williams D./Griggs/Hoffman/Seddon/Garn/Walker C.	12/04/85–19/04/85	06:23:55:23
	STS 51-B	Overmyer/Gregory F./Lind/Thagard/Thornton W./Wang/van den Berg	29/04/85–06/05/85	07:00:08:46
	Soyuz T13	Dzhanibekov/Savinykh	06/06/85–26/09/85	112:03:12:06 (Savinykh 168:03:51:00)
	STS 51-G	Brandenstein/Creighton/Fabian/Nagel/Lucid/Baudry/Al-Saud	17/06/85–24/06/85	07:01:38:52
	STS 51-F	Fullerton/Bridges/Henize/Musgrave/England/Acton/Bartoe	29/07/85–06/08/85	07:22:45:26

	STS 51-I	Engle/Covey/van Hoften/Lounge/Fisher W.	27/08/85–03/09/85	07:02:17:42
	Soyuz T14	Vasyutin/Grechko/Volkov A.	17/09/85–21/11/85	64:21:52:08
				(Grechko 08:21:13:06)
	STS 51-J	Bobko/Grabe/Hilmers/Stewart/Pailes	03/10/85–07/10/85	04:01:44:38
	STS 61-A	Hartsfield/Nagel/Dunbar/Buchli/Buford/Furrer/Messerschmid/Ockels	30/10/85–06/11/85	07:00:44:53
	STS 61-B	Shaw/O'Connor/Ross/Cleave/Spring/Walker C./Neri-Vela	26/11/85–03/12/85	06:21:04:49
1986	STS 61-C	Gibson R./Bolden/Nelson G./Hawley/Chang-Diaz/Cenker/Nelson B.	12/01/86–18/01/86	06:02:03:51
	STS 51-L	Scobee/Smith M./Onizuka/Resnik/McNair/Jarvis/McAuliffe	28/01/86	00:00:01:13 (explosion during ascent)
	Soyuz T15	Kizim/Solovyov V.	13/03/86–16/07/86	125:00:00:56
1987	Soyuz TM2	Romanenko/Laveikin	06/02/87–29/12/87	326:11:37:57 (Laveikin 174:03:25:56)
	Soyuz TM3	Viktorenko/Alexandrov/Faris	22/07/87–30/07/87	07:23:04:55 (Alexandrov 160:07:16:58)
	Soyuz TM4	Titov V./Manarov/Levchenko	21/12/87–21/12/88	365:22:38:57 (Levchenko 07:21:58:12)
1988	Soyuz TM5	Solovyov A./Savinykh/Alexandrov	07/06/88–17/06/88	09:20:09:19
	Soyuz TM6	Lyakhov/Polyakov/Mohmand	29/08/88–07/07/88	08:20:26:27 (Polyakov 240:22:34:47)
	STS-26	Hauck/Covey/Lounge/Hilmers/Nelson G.	29/09/88–03/10/88	04:01:00:11
	Soyuz TM7	Volkov A./Krikalev/Chrétien	26/11/88–26/04/89	151:11:08:23 (Chrétien 24:18:07:25)
	STS-27	Gibson R./Gardner G./Mullane/Ross/Shepherd	02/12/88–06/12/88	04:09:05:35

Appendix A   787

(*continued*)

788 Appendix A

Year	Mission	Crew at launch	Date (dd/mm/yy)	Duration (dd:hh:mm:ss)
1989	STS-29	Coats/Blaha/Buchli/Springer/Bagian	13/03/89–18/03/89	04:23:38:50
	STS-30	Walker D./Grabe/Thagard/Cleave/Lee	04/05/89–08/05/89	04:00:56:27
	STS-28	Shaw/Richards R./Leestma/Adamson/Brown M.	08/08/89–13/08/89	05:01:00:09
	Soyuz TM8	Viktorenko/Serebrov	06/09/89–19/02/90	166:06:58:16
	STS-34	Williams D./McCulley/Lucid/Chang-Diaz/Baker E.	18/10/89–23/10/89	04:23:39:21
	STS-33	Gregory F./Blaha/Carter/Musgrave/Thornton K.	22/11/89–27/11/89	05:00:06:48
1990	STS-32	Brandenstein/Wetherbee/Dunbar/Ivins/Low	09/01/90–20/01/90	10:21:00:36
	Soyuz TM9	Solovyov A./Balandin	11/02/90–09/08/90	179:01:17:57
	STS-36	Creighton/Casper/Hilmers/Mullane/Thuot	28/02/90–04/03/90	04:10:18:22
	STS-31	Shriver/Bolden/McCandless/Hawley/Sullivan	24/04/90–29/04/90	05:01:16:06
	Soyuz TM10	Manakov/Strekalov	01/08/90–10/12/90	130:20:35:51
	STS-41	Ruchards/Cabana/Melnick/Shepherd/Akers	06/10/90–10/10/90	04:02:10:04
	STS-38	Covey/Culbertson/Springer/Meade/Gemar	15/11/90–20/11/90	04:21:54:31
	STS-35	Brand/Gardner G./Hoffman/Lounge/Parker/Durrance/Parise	02/12/90–10/12/90	08:23:05:08
	Soyuz TM11	Afanasyev/Manarov/Akiyama	02/12/90–26/05/91	175:01:51:42 (Akiyama 07:21:54:40)
1991	STS-37	Nagel/Cameron/Godwin/Ross/Apt	05/04/91–11/04/91	05:23:32:44
	STS-39	Coats/Hammond/Harbaugh/McMonagle/Bluford/Veach/Hieb	28/04/91–06/05/91	08:07:22:23
	Soyuz TM12	Artsebarsky/Krikalev/Sharman	18/05/91–10/10/91	144:15:21:50 (Sharman 07:21:14:20) (Krikalev 311:20:01:54)
	STS-40	O'Connor/Gutierrez/Bagian/Jernigan/Seddon/Gaffney/Hughes-Fulford	05/06/91–14/06/91	09:02:14:20
	STS-43	Blaha/Baker M./Lucid/Low/Adamson	02/08/91–11/08/91	08:21:21:25
	STS-48	Creighton/Reightler/Gemar/Buchli/Brown M.	12/09/91–18/09/91	05:08:27:38
	Soyuz TM13	Volkov A./Aubakirov/Viehbock	02/10/91–25/03/92	175:02:52:43 (Aubakirov/Viehbock 07:22:12:59)
	STS-44	Gregory F./Henricks/Voss J.S./Musgrave/Runco/		

Year	Mission	Crew	Dates	Duration
1992	STS-42	Grabe/Oswald/Thagard/Readdy/Hilmers/Bondar/Merbold	22/01/92–30/01/92	08:01:14:44
	Soyuz TM14	Viktorenko/Kaleri/Flade	17/03/92–10/08/92	145:14:10:32 (Flade 07:21:56:52)
	STS-45	Bolden/Duffy/Sullivan/Leestma/Foale/Frimout/Lichtenberg	24/03/92–02/04/92	08:22:09:28
	STS-49	Brandenstein/Chilton/Hieb/Melnick/Thuot/Thornton K./Akers	07/05/92–16/05/92	08:21:17:38
	STS-50	Richards R./Bowersox/Dunbar/Baker E./Meade/DeLucas/Trinh	25/06/92–09/07/92	13:19:30:04
	Soyuz TM15	Solovyov A./Avdeyev/Tognini	27/07/92–01/02/93	188:21:41:15 (Tognini 13:18:56:14)
	STS-46	Shriver/Allen A/Nicollier/Ivins/Hoffman/Chang-Diaz/Malerba	31/07/92–08/08/92	07:23:15:03
	STS-47	Gibson R./Brown C./Lee/Apt/Davis/Jemison/Mohri	12/09/92–20/09/92	07:22:30:23
	STS-52	Wetherbee/Baker M./Veach/Shepherd/Jernigan/MacLean	22/10/92–01/11/92	09:20:56:13
	STS-53	Walker D./Cabana/Bluford/Voss/Clifford	02/12/92–09/12/92	07:07:19:47
1993	STS-54	Casper/McMonagle/Harbaugh/Runco/Helms	13/01/93–19/01/93	05:23:38:19
	Soyuz TM16	Manakov/Poleshchuk	24/01/93–02/07/93	179:00:43:46
	STS-56	Cameron/Oswald/Foale/Cockrell/Ochoa	07/04/93–17/04/93	09:06:08:24
	STS-55	Nagel/Henricks/Ross/Precourt/Harris/Walter/Schlegel	26/04/93–06/05/93	09:23:39:59
	STS-57	Grabe/Duffy/Low/Sherlock/Wisoff/Voss J.E.	21/06/93–02/07/93	09:23:44:54
	Soyuz TM17	Tsibliyev/Serebrov/Haigneré J-P.	01/07/93–14/01/94	196:17:45:22 (Haigneré 20:16:08:52)
	STS-51	Culbertson/Readdy/Newman/Bursch/Walz	12/09/93–22/09/93	09:20:11:11
	STS-58	Blaha/Searfoss/Seddon/McArthur W./Lucid/Wolf/Fettman	18/10/93–01/11/93	14:00:12:32
	STS-61	Covey/Bowersox/Thornton K./Nicollier/Hoffman/Musgrave/Akers	02/12/93–12/12/93	10:19:58:37

(continued)

790 Appendix A

Year	Mission	Crew at launch	Date (dd/mm/yy)	Duration (dd:hh:mm:ss)
1994	Soyuz TM18	Afanasyev/Usachev/Polyakov	08/01/94–09/07/94	182:00:27:02 (Polyakov 437:17:58:31)
	STS-60	Bolden/Reightler/Davis/Sega/Chang-Diaz/Krikalev	03/02/94–11/02/94	08:07:09:22
	STS-62	Casper/Allen A./Thuot/Gemar/Ivins	04/03/94–18/03/94	13:23:16:41
	STS-59	Gutierrez/Chilton/Apt/Clifford/Godwin/Jones T.	09/04/94–20/04/94	11:05:49:30
	Soyuz TM19	Malenchenko/Musabayev	01/07/94–04/11/94	125:22:53:36
	STS-65	Cabana/Halsell/Hieb/Walz/Thomas D./Chiao/Mukai	08/07/94–23/07/94	14:17:55:00
	STS-64	Richards R./Hammond/Linenger/Helms/Meade/Lee	09/09/94–20/09/94	10:22:49:57
	STS-68	Baker M./Wilcutt/Smith S/Bursch/Wisoff/Jones T.	30/09/94–11/10/94	11:05:46:08
	Soyuz TM20	Viktorenko/Kondakova/Merbold	04/10/94–22/03/95	169:05:21:35 (Merbold 31:12:35:56)
	STS-66	McMonagle/Brown C./Ochoa/Tanner/Clervoy/Parazynski	03/11/94–14/11/94	10:22:34:02
1995	STS-63	Wetherbee/Collins E./Harris/Foale/Voss J.E./Titov V.	02/02/95–11/02/95	08:06:28:15
	STS-67	Oswald/Gregory W./Grunsfeld/Lawrence/Jernigan/Durrance/Parise	02/03/95–18/03/95	16:15:08:48
	Soyuz TM21	Dezhurov/Strekalov/Thagard	14/03/95–11/09/95	115:08:43:02
	STS-71	Gibson R./Precourt/Baker E./Harbaugh/Dunbar/Solovyov A./Budarin	27/06/95–07/07/95	09:19:22:17 (STS-71 crew only)
	Mir EO-19	Solovyov A./Budarin (launched on STS-71)	27/06/95–11/09/95	75:11:20:21
	STS-70	Henricks/Kregel/Thomas D./Currie/Weber	13/07/95–22/07/95	08:22:20:05
	Soyuz TM22	Gidzenko/Avdeyev/Reiter	03/09/95–29/02/96	179:01:41:46
	STS-69	Walker D./Cockrell/Voss J.S./Newman/Gernhardt	07/09/95–18/09/95	10:20:28:56
	STS-73	Bowersox/Rominger/Coleman/Lopez-Alegria/Thornton K./Leslie/Sacco	20/10/95–05/11/95	15:21:52:28
	STS-74	Cameron/Halsell/Hadfield/Ross/McArthur W.	12/11/95–20/11/95	08:04:30:44
1996	STS-72	Duffy/Jett/Chiao/Scott W./Wakata/Barry	11/01/96–20/01/96	08:22:01:47
	Soyuz TM23	Onufriyenko/Usachev	23/02/96–02/09/96	193:19:07:35

STS-75	Allen A./Horowitz/Hoffman/Cheli/Nicollier/ Chang-Diaz/Guidoni	22/02/96–09/03/96	15:17:40:21	
STS-76	Chilton/Searfoss/Sega/Clifford/Godwin/Lucid	22/03/96–31/03/96	09:05:15:53	(Lucid 188:04:00:11)
STS-77	Casper/Brown C./Thomas A./Bursch/Runco/Garneau	19/05/96–29/05/96	10:00:39:18	
STS-78	Henricks/Kregel/Linnehan/Helms/Brady/Favier/ Thirsk	20/06/96–07/07/96	16:21:47:45	
Soyuz TM24	Korzun/Kaleri/André-Deshays	17/08/96–02/03/97	196:17:26:13 (André-Deshays 15:18:23:37)	
STS-79	Readdy/Wilcutt/Apt/Akers/Walz/Blaha	16/09/96–26/09/96	10:03:18:26	(Blaha 128:05:27:55)
STS-80	Cockrell/Rominger/Jernigan/Jones T./Musgrave	19/11/96–07/12/96	17:15:53:18	
1997 STS-81	Baker M/Jett/Wisoff/Grunsfeld/Ivins/Linenger	12/01/97–22/01/97	10:04:55:21 (Linenger 132:04:00:21)	
Soyuz TM25	Tsibliyev/Lazutkin/Ewald	10/02/97–14/08/97	184:22:07:41	(Ewald 19:16:34:46)
STS-82	Bowersox/Horowitz/Tanner/Hawley/Harbaugh/Lee/ Smith S.	11/02/97–21/02/97	09:23:37:09	
STS-83	Halsell/Still/Voss J.E./Gernhardt/Thomas D./Crouch/ Linteris	04/04/97–08/04/97	03:23:12:39	
STS-84	Precourt/Collins E./Clervoy/Nicollier/Lu/ Kondakova/Foale	15/05/97–24/05/97	09:05:19:56	(Foale 144:13:47:21)
STS-94	Halsell/Still/Voss J.E./Gernhardt/Thomas D./Crouch/ Linteris	01/07/97–17/07/97	15:16:34:04	
Soyuz TM26	Solovyov A./Vinogradov	05/08/97–19/02/98	197:17:34:36	
STS-85	Brown C./Rominger/Davis/Curbeam/Robinson/ Tryggvason	07/08/97–19/08/97	11:20:26:59	
STS-86	Wetherbee/Bloomfield/Titov V./Parazynski/ Chrétien/Lawrence/Wolf	25/09/97–06/10/97	10:19:20:50	(Wolf 127:20:00:50)
STS-87	Kregel/Lindsey/Chawla/Scott W./Doi/Kadenyuk	19/11/97–05/12/97	15:16:34:04	

(continued)

Appendix A   791

792 Appendix A

Year	Mission	Crew at launch	Date (dd/mm/yy)	Duration (dd:hh:mm:ss)
1998	STS-89	Wilcutt/Edwards/Reilly/Anderson/Dunbar/ Sharipov/Thomas A.	22/01/98–31/01/98	08:19:46:54 (Thomas 140:15:12:06)
	Soyuz TM27	Musabayev/Budarin/Eyharts	29/01/98–25/08/98	207:12:51:02 (Eyharts 20:16:36:48)
	STS-90	Searfoss/Altman/Linnehan/Hire/Williams D./ Buckley/Pawelczyk	17/04/98–25/08/98	15:21:49:59
	STS-91	Precourt/Gorie/Kavandi/Lawrence/Chang-Diaz/ Ryumin	02/06/98–12/06/98	09:19:53:54
	Soyuz TM28	Padalka/Avdeyev/Baturin	13/08/98–08/02/99	198:16:31:20 (Avdeyev 379:14:51:10) (Baturin 11:19:41:33)
	STS-95	Brown C./Lindsey/Robinson/Parazynski/Duque/ Mukai/Glenn	29/10/98–07/11/98	08:21:43:56
	STS-88	Cabana/Sturckow/Ross/Currie/Newman/Krikalev	04/12/98–15/12/98	11:19:17:57
1999	Soyuz TM29	Afanasyev/Haigneré J-P./Bella	20/02/99–28/08/99	188:20:16:19 (Bella 07:21:56:29)
	STS-96	Rominger/Husband/Jernigan/Ochoa/Barry/Payette/ Tokarev	27/05/99–06/06/99	09:19:13:57
	STS-93	Collins E./Ashby/Hawley/Coleman/Tognini	23/07/99–27/07/99	04:02:49:37
	STS-103	Brown C./Kelly S./Grunsfeld/Smith S./Foale/ Nicollier/Clervoy	19/15/99–27/12/99	07:23:10:47
2000	STS-99	Kregel/Gorie/Thiele/Kavandi/Voss J.E./Mohri	11/02/00–22/02/00	11:05:39:41
	Soyuz TM30	Zaletin/Kaleri	04/04/00–16/06/00	72:19:42:16
	STS-101	Halsell/Horowitz/Weber/Williams J./Voss J.S./ Helms/Usachev	19/05/00–29/05/00	09:21:10:10
	STS-106	Wilcutt/Altman/Lu/Mastracchio/Burbank/ Malenchenko/Morukov	08/09/00–19/09/00	11:19:12:15
	STS-92	Duffy/Melroy/Lopez-Alegria/Wisoff/McArthur W./ Chiao/Wakata	11/10/00–24/10/00	12:21:43:47

	Soyuz TM31	Shepherd/Gidzenko/Krikalev (ISS-1)	31/10/00–21/03/01	140:23:38:55
	STS-97	Jett/Bloomfield/Tanner/Garneau/Noriega	30/11/00–11/12/00	10:19:58:20
2001	STS-98	Cockrell/Polansky/Curbeam/Jones T./Ivins	07/02/01–20/02/01	12:21:21:00
	STS-102	Wetherbee/Kelly J.M./Richards P./Thomas A.	08/03/01–21/03/01	12:19:51:57
	ISS-2	Usachev/Voss J.S./Helms	08/03/01–22/08/01	167:06:40:49
	STS-100	Rominger/Ashby/Hadfield/Parazynski/Guidoni/Phillips/Lonchakov	19/04/01–01/05/01	11:21:31:14
	Soyuz TM32	Musabayev/Baturin/Tito	28/04/01–06/05/01	07:22:04:08
	STS-104	Lindsey/Hobaugh/Gernhardt/Reilly/Kavandi	12/07/01–23/07/01	12:18:36:39
	STS-105	Horowitz/Sturckow/Barry/Forrester	10/08/01–22/08/01	11:21:13:52
	ISS-3	Culbertson/Dezhurov/Tyurin	10/08/01–22/08/01	128:20:44:56
	Soyuz TM33	Afanasyev/Haigneré C/Kozeev	21/10/01–17/12/01	09:20:00:25
	STS-108	Gorie/Kelly M./Godwin/Tani	05/12/01–17/12/01	11:19:36:45
	ISS-4	Onufriyenko/Bursch/Walz	05/12/01–19/06/02	195:19:38:12
2002	STS-109	Altman/Carey/Currie/Grunsfeld/Linnehan/Newman	01/03/02–12/03/02	10:22:11:09
	STS-110	Bloomfield/Frick/Walheim/Ochoa/Smith S/Morin/Ross	08/04/02–19/04/02	10:19:43:48
	Soyuz TM34	Gidzenko/Vittori/Shuttleworth	25/04/02–05/05/02	09:21:25:18
	STS-111	Cockrell/Lockhart/Chang-Diaz/Perrin	05/06/02–19/06/02	13:20:35:56
	ISS-5	Korzun/Whitson/Treschev	05/06/02–07/12/02	184:22:14:23
	STS-112	Ashby/Melroy/Wolf/Magnus/Sellers/Yurchikin	07/10/02–18/10/02	10:19:58:44
	Soyuz TMA1	Zaletin/De Winne/Lonchakov	30/10/02–10/11/02	10:20:53:09
	STS-113	Wetherbee/Lockhart/Lopez-Alegria/Herrington	23/11/02–07/12/02	13:18:48:38
	ISS-6	Bowersox/Budarin/Pettit	23/11/02–03/05/03	161:01:14:38

(*continued*)

794 Appendix A

Year	Mission	Crew at launch	Date (dd/mm/yy)	Duration (dd:hh:mm:ss)
2003	STS-107	Husband/McCool/Brown D./Chawla/Anderson Clark/Ramon	16/01/03–01/02/03	15:22:20:22
	Soyuz TMA2	Malenchenko/Lu (ISS-7)	26/04/03–27/10/03	184:22:46:09
	Shenzhou 5	Yang	15/10/03	00:21:26:00
	Soyuz TMA3	Foale/Kaleri (ISS-8)/Duque	18/01/03–30/04/04	194:18:23:43 (Duque 09:21:01:58)
2004	Soyuz TMA4	Padalka/Finke (ISS-9)/Kuipers	19/04/04–24/10/04	187:21:16:09 (Kuipers 10:20:52:46)
	SpaceShip1-60	Melvill	21/06/04	00:00:24:00 (X-Prize flight)
	Spaceship1-65	Melvill	29/09/04	00:00:24:00 (X-Prize flight)
	Spaceship1-66	Binnie	04/10/04	00:00:24:00 (X-Prize flight)
	Soyuz TMA5	Sharipov/Chiao (ISS-10)/Shargin	14/10/04–24/04/05	192:19:00:59 (Shargin 09:21:29:00)
2005	Soyuz TMA6	Krikalev/Phillips (ISS-11)/Vittori	14/04/05–11/10/05	179:00:23:00 (Vittori 09:21:21:02)
	STS-114	Collins E./Kelly J.M./Noguchi/Robinson/Thomas A./Lawrence/Camarda	26/07/05–09/08/05	13:21:32:48
	Soyuz TMA7	McArthur W./Tokarev (ISS-12)/Olsen	01/10/05–09/04/06	189:19:53:00 (Olsen 09:21:15:00)
	Shenzhou 6	Fei/Nie	12/10/05–16/10/05	04:19:33:00
2006	Soyuz TMA8	Vinogradov/Williams J (ISS-13)/Pontes	30/03/06–29/09/06	182:22:43:00 (Pontes 09:21:17:00)
	STS-121	Lindsey/Kelly M/Fossum/Nowak/Wilson/Sellers/Reiter	04/07/06–17/07/06	12:18:37:54 (STS 121 crew only) Reiter in space
	STS-115	Jett/Ferguson/Tanner/Burbank/Stefanyshyn-Piper/MacLean	09/09/06–21/09/06	11:19:07:24
	Soyuz TMA9	Lopez-Alegria/Tyurin (ISS-14)/Ansari	19/09/06–In Space	(182 days planned) (Ansari 10:21:05:00)

Appendix B

Cumulative Space Flight and EVA Experience

Table B.1. Duration log April 1961–September 2006.

Order of most spaceflight experience, up to 29 September 2006 and the end of ISS Expedition 13.

Name	Country	Flights	Time in Space (hrs:min)
Sergei K. Krikalev	USSR/Russia	6	19,258:57
Sergei V. Avdeyev	Russia	3	17,942:22
Valery V. Polyakov	USSR/Russia	2	16,312:34
Anatoly Y. Solovyov	USSR/Russia	5	15,624:13
Alexandr Y. Kaleri	Russia	4	14,637:53
Viktor M. Afanasyev	USSR/Russia	4	13,339:35
Yuri V. Usachev	Russia	4	13,232:28
Musa K. Manarov	USSR	2	12,985:32
Alexandr A. Viktorenko	USSR/Russia	4	11,741:46
Nikolai V. Budarin	Russia	3	10,585:26
Yuri V. Romanenko	USSR	3	10,238:21
Gennady I. Padalka	Russia	2	9,377:29
Alexandr A. Volkov	USSR	3	9,373:52
Yuri I. Onufriyenko	Russia	2	9,311:48
Vladimir G. Titov[1]	USSR/Russia	4	9,288:47
Vasily V. Tsibliyev	Russia	2	9,187:47
Valery G. Korzun	Russia	2	9,158:50
Pavel V. Vinogradov	Russia	2	9,136:18
Alexandr A. Serebrov	USSR/Russia	4	9,011:53
Leonid D. Kizim	USSR	3	8,993:59

(*continued*)

796   Appendix B

Table B.1 (*cont.*)

Name	Country	Flights	Time in Space (hrs:min)
C. Michael Foale	USA	6	8,970:08
Valery V. Ryumin	USSR/Russia	4	8,921:28
Vladimir A. Solovyov	USSR	2	8,686:51
Talgat A. Musabayev	Russia	3	8,021:49
Vladimir A. Lyakhov	USSR	3	7,998:49
Yuri P. Gidzenko	Russia	3	7,918:46
Yuri I. Malenchenko	Russia	3	7,744:52
Gennady M. Manakov	USSR/Russia	2	7,437:20
Alexandr P. Alexandrov	USSR	2	7,433:03
Gennady M. Strekalov[1]	USSR/Russia	5	6,622:26
Viktor P. Savinykh	USSR	3	6,066:39
Vladimir N. Dezhurov	Russia	2	6,029:28
Oleg Y. Atkov	USSR	1	5,686:50
Carl E. Walz	USA	4	5,533:03
Leroy Chiao	USA	4	5,505:41
Daniel W. Bursch	USA	4	5,446:14
William S. McArthur	USA	4	5,431:06
Shannon W. Lucid	USA	5	5,362:34
Valentin V. Lebedev	USSR	2	5,262:00
Vladimir V. Kovalenok	USSR	3	5,194:12
Kenneth D. Bowersox	USA	5	5,077:13
Anatoly N. Berezovoi	USSR	1	5,073:05
Susan J. Helms	USA	5	5,044:47
Jean-Pierre Haigneré	France	2	5,028:25
Edward Tsang Lu	USA	3	4,956:48
James S. Voss	USA	5	4,854:32
Salizhan S. Sharipov	Russia	2	4,839:47
Leonid I. Popov	USSR	3	4,814:55
Valery I. Tokarev	Russia	2	4,791:07
Jeffrey N. Williams	USA	2	4,627:53
John L. Phillips	USA	2	4,581:54
E. Michael Finke	USA	1	4,509:16
Alexandr I. Lazutkin	Russia	1	4,441:42
Peggy A. Whitson	USA	1	4,438:14
Sergei V. Treshchev	Russia	1	4,438:14
Yelena v. Kondakova	Russia	2	4,299:12
Thomas Reiter[2]	Germany	1	4,297:42
Alexandr N. Balandin	USSR	1	4,297:18
Alexandr F. Poleshchuk	Russia	1	4,296:44
Andrew S. W. Thomas	USA	4	4,182:03
Alexandr I. Laveikin	USSR	1	4,179:26
John E. Blaha	USA	5	3,874:49

Table B.1 (*cont.*)

Name	Country	Flights	Time in Space (hrs:min)
Donald R. Pettit	USA	1	3,865:15
William M. Shepherd	USA	4	3,823:51
David A. Wolf	USA	3	3,672:33
Alexandr S. Ivanchenkov	USSR	2	3,540:39
Vladimir A. Dzhanibekov	USSR	5	3,495:59
Anatoly P. Artsebarsky	USSR	1	3,471:22
Norman E. Thagard	USA	5	3,541:28
Frank L. Culbertson, Jr.	USA	3	3,446:50
Jerry M. Linenger	USA	2	3,435:50
Georgi M. Grechko	USSR	3	3,236:33
Mikhail V. Tyurin[2]	Russia	1	3,092:45
Gerald P. Carr	USA	1	2,017:16
Edward G. Gibson	USA	1	2,017:16
William R. Pogue	USA	1	2,017:16
Sergei V. Zaletin	Russia	2	2,008:35
Vital I. Sevastyanov	USSR	2	1,936:19
Pyotr I. Klimuk	USSR	3	1,890:19
Owen K. Garriott	USA	2	1,674:56
Alan L. Bean	USA	2	1,671:45
Jack R. Lousma	USA	2	1,619:14
Franklin R. Chang-Diaz	USA	7	1,600:24
James D. Wetherbee	USA	6	1,594:27
Kent V. Rominger	USA	5	1,586:58
Vladimir V. Vasyutin	USSR	1	1,557:52
Kenneth D. Cockrell	USA	5	1,548:30
Tamara E. Jernigan	USA	5	1,500:29
Jerry L. Ross	USA	7	1,392:55
Curtis L. Brown, Jr.	USA	6	1,358:54
Marsha S. Ivins	USA	5	1,342:26
F. Story Musgrave	USA	6	1,280:50
Thomas D. Jones	USA	4	1,271:53
Kevin R. Kregel	USA	4	1,265:22
James D. Halsell, Jr.	USA	5	1,260:35
Boris V. Volynov	USSR	2	1,255:20
Wendy B. Lawrence	USA	4	1,227:57
Jeffrey A. Hoffman	USA	5	1,210:51
Bonnie J. Dunbar	USA	5	1,209:24
Ulf D. Merbold	Germany	3	1,207:37
Steven W. Lindsey	USA	4	1,203:32
Vitaly M. Zholobov	USSR	1	1,182:24

(*continued*)

Table B.1 (*cont.*)

Name	Country	Flights	Time in Space (hrs:min)
Janice E. Voss	USA	5	1,181:51
Charles Conrad, Jr.	USA	4	1,179:39
Peter J. K. Wisoff	USA	4	1,064:51
Richard M. Linnehan	USA	3	1,049:49
Joseph R. Tanner	USA	4	1,045:16
Jean-Loup J.M. Chrétien	France	3	1,043:19
Donald A. Thomas	USA	4	1,042:14
James H. Newman	USA	4	1,042:10
Michael L. Gernhardt	USA	4	1,041:05
Scott J. Horowitz	USA	4	1,139:42
Claude Nicollier	Switzerland	4	1,029:02
Thomas T. Henricks	USA	4	1,026:40
Michael E. Lopez-Alegria[2]	USA	3	1,022:25
Scott L. Parazynski	USA	4	1,021:09
John M. Grunsfeld	USA	4	1,108:08
Terrence W. Wilcutt	USA	4	1,009:03
Brent W. Jett Jr.	USA	4	1,003:42
Nancy J. (Sherlock) Currie	USA	4	999:34
Brian Duffy	USA	4	978:39
Kathryn C. Thornton	USA	4	973:27
Michael A. Baker	USA	4	965:40
Stephen L. Smith	USA	4	960:17
Richard A. Searfoss	USA	3	947:19
Charles J. Precourt	USA	4	945:46
Scott D. Altman	USA	3	927:13
Linda M. Godwin	USA	4	918:12
Robert D. Cabana	USA	4	910:43
Andrew M. Allen	USA	3	904:05
Alexei A. Gubarev	USSR	2	899:37
Eileen M. Collins	USA	4	885:40
Robert L. Gibson	USA	5	868:18
Jay Apt	USA	4	847:10
John W. Young	USA	6	835:42
Stephen S. Oswald	USA	3	814:33
Richard N. Richards	USA	4	814:30
Thomas D. Akers	USA	4	813:45
Janet L. Kavandi	USA	3	812:11
Stephen K. Robinson	USA	3	807:42
John H. Casper	USA	4	805:32
Gregory J. Harbaugh	USA	4	797:40
Paul J. Weitz	USA	2	793:14
Mark C. Lee	USA	4	790:55

Table B.1 (*cont.*)

Name	Country	Flights	Time in Space (hrs:min)
Dominic L. Gorie	USA	3	789:11
Daniel C. Brandenstein	USA	4	789:07
Michael J. Bloomfield	USA	3	779:02
Steven A. Hawley	USA	5	767:44
Richard J. Hieb	USA	3	766:37
Kalpana Chawla[3]	USA	2	758:54
Vance D. Brand	USA	4	746:04
Daniel T. Barry	USA	3	735:29
Viktor V. Gorbatko	USSR	3	732:46
Margaret R. Seddon	USA	3	730:23
David M. Walker	USA	4	724:32
Steven R. Nagel	USA	4	721:36
James A. Lovell, Jr.	USA	4	715:05
G. David Low	USA	3	714:08
Carl J. Meade	USA	3	713:14
Kevin P. Chilton	USA	3	704:20
J.J. Marc Garneau	Canada	3	698:01
Vladislav N. Volkov	USSR	2	689:03
Ellen S. Baker	USA	3	687:31
Charles F. Bolden	USA	4	680:30
Joseph P. Kerwin	USA	1	672:50
William F. Readdy	USA	3	672:43
Guion S. Bluford Jr.	USA	4	688:35
Jean-François A. Clervoy	France	3	688:35
Michael R.U. Clifford	USA	3	666:21
Jeffrey S. Ashby	USA	3	664:20
Paul S. Lockhart	USA	2	663:25
Umberto Guidoni	Italy	2	663:12
Pierre J. Thuot	USA	3	654:45
N. Jan Davis	USA	3	650:16
Richard O. Covey	USA	4	644:11
Ronald J. Grabe	USA	4	627:42
Richard D. Husband[3]	USA	2	617:37
James M. Kelly	USA	1	614:24
Samuel T. Durrance	USA	2	614:15
Ronald A. Parise	USA	2	614:15
Michael P. Anderson[3]	USA	2	595:07
Winston E. Scott	USA	2	591:35
Mark E. Kelly	USA	2	590:15
Donald R. McMonagle	USA	3	585:15

(*continued*)

Table B.1 (*cont.*)

Name	Country	Flights	Time in Space (hrs:min)
Charles D. Gemar	USA	3	581:39
Daniel C. Burbank	USA	2	576:19
Georgi T. Dobrovolsky	USSR	1	570:22
Viktor I. Patsayev	USSR	1	570:22
Robert L. Curbeam, Jr.	USA	2	569:48
Pamela A. Melroy	USA	2	569:43
Frederick W. Sturckow	USA	2	568:32
Chiaki Mukai	Japan	2	567:39
Piers J. Sellers	USA	2	566:37
Eugene A. Cernan	USA	3	566:16
Robert L. Crippen	USA	4	565:48
Kenneth L. Cameron	USA	3	562:13
David R. Scott	USA	3	546:54
Yuri V. Lonchakov	Russia	2	546:24
Brewster H. Shaw, Jr.	USA	3	533:53
Kathryn D. Sullivan	USA	3	532:49
David C. Leestma	USA	3	532:33
Mario Runco, Jr.	USA	3	530:49
Koichi Wakata	Japan	2	524:25
Steven G. MacLean	Canada	2	521:33
Andrian G. Nikolayev	USSR	2	519:24
James F. Reilly II	USA	2	519:24
Claudi (Deshays) Haigneré	France	2	514:23
Thomas K. Mattingly II	USA	3	508:34
Thomas P. Stafford	USA	4	507:44
Catherine G. Coleman	USA	2	500:42
Valery F. Bykovsky	USSR	3	497:49
Oleg G. Makarov[4]	USSR	4	497:43
Leopold Eyharts	France	1	496:37
Carlos I. Noriega	USA	2	494:48
David C. Hilmers	USA	4	494:17
James F. Buchli	USA	4	490:25
Ellen L. Ochoa	USA	4	489:40
Sidney M. Gutierrez	USA	2	488:01
Henry W. Hartsfield, Jr.	USA	3	482:51
John M. Lounge	USA	3	482:23
Charles D. Walker	USA	3	477:56
Frank F. Borman II	USA	2	477:36
Roberto Vittori	Italy	2	474:46
Roger K. Crouch	USA	2	473:58
Gregory T. Linteris	USA	2	473:58
Susan L. Still	USA	2	473:58

Table B.1 (*cont.*)

Name	Country	Flights	Time in Space (hrs:min)
Yuri M. Baturin	Russia	2	473:23
Svetlana Y. Savitskaya	USSR	2	473:06
Reinhold Ewald	Germany	1	472:35
Michael L. Coats	USA	3	463:59
L. Blaine Hammond Jr.	USA	2	463:13
Robert A.R. Parker	USA	2	462:52
Byron K. Lichtenberg	USA	2	461:56
Mamoru M. Mohri	Japan	2	460:10
Frederick D. Gregory	USA	3	455:08
Mary Ellen Weber	USA	2	451:30
Michel A.C. Tognini	France	2	450:47
Pedro F. Duque	Spain	2	450:44
Valery N. Kubasov	USSR	3	449:59
Pavel R. Popovich	USSR	2	448:29
Bernard A. Harris, Jr.	USA	2	438:08
Charles L. Veach	USA	2	436:19
Frederick H. Hauck	USA	3	434:09
Yuri N. Glazkov	USSR	1	425:23
Ronald M. Sega	USA	2	420:35
George D. Nelson	USA	3	407:54
Charles E. Brady, Jr.	USA	1	405:48
Jean-Jacques Favier	France	1	405:48
Robert Brent Thirsk	Canada	1	405:48
John O. Creighton	USA	3	404:25
Chris A. Hadfield	Canada	2	402:02
William G. Gregory	USA	1	399:10
Karol J. Bobko	USA	3	386:04
Loren J. Shriver	USA	3	386:00
Bryan D. O'Connor	USA	2	383:19
Charles G. Fullerton	USA	2	382:51
William C. McCool[3]	USA	1	382:20
David M. Brown[3]	USA	1	382:20
Laurel B.S. Clark[3]	USA	1	382:20
Ilan Ramon[3]	Israel	1	382:20
Albert Sacco, Jr.	USA	1	381:52
Jay C. Buckey	USA	1	381:50
Kathryn P. Hire	USA	1	381:50
James A. Pawelczyk	USA	1	381:50
Dafydd R. Williams	Canada	1	381:50
Maurizio Cheli	Italy	1	377:41

(*continued*)

Table B.1 (*cont.*)

Name	Country	Flights	Time in Space (hrs:min)
Yuri P. Artyukhin	USSR	1	377:30
Takao Doi	Japan	1	376:34
Leonid K. Kadenyuk	Ukraine	1	376:34
Richard M. Mullane	USA	3	356:21
Martin J. Fettman	USA	1	344:13
Sally K. Ride	USA	2	343:48
James A. McDivitt	USA	2	338:57
James D.A. Van Hoften	USA	2	337:58
James P. Bagian	USA	2	337:54
Dale A. Gardner	USA	2	336:54
James C. Adamson	USA	2	334:22
Soichi Noguchi	Japan	1	333:32
Charles J. Camarda	USA	1	333:32
Philippe Perrin	France	1	332:36
Lawrence J. DeLucas	USA	1	331:30
Eugene H. Trinh	USA	1	331:30
John B. Herrington	USA	1	330:49
Kenneth S. Reightler, Jr.	USA	2	327:47
Guy S. Gardner, Jr.	USA	2	320:11
John M. Fabian	USA	2	316:03
Richard F. Gordon, Jr.	USA	2	315:53
Joseph P. Allen IV	USA	2	313:59
William E. Thornton	USA	2	313:18
Bruce McCandless II	USA	2	312:32
Bruce E. Melnick	USA	2	311:28
Mark L. Polansky	USA	1	309:21
Paul W. Richards	USA	1	307:52
Michael E. Fossum	USA	1	306:38
Lisa M. Nowak	USA	1	306:38
Stephanie D. Wilson	USA	1	306:38
Charles O. Hobaugh	USA	1	306:37
Ronald E. Evans, Jr.	USA	1	301:52
Harrison H. Schmitt	USA	1	301:52
Walter M. Schirra, Jr.	USA	3	295:13
James B. Irwin	USA	1	295:12
Alfred M. Worden, Jr.	USA	1	295:12
Robert F. Overmyer	USA	2	290:23
Buzz Aldrin	USA	2	289:54
Robert L. Stewart	USA	2	289:01
Donald E. Williams	USA	2	288:34
Patrick G. Forrester	USA	1	285:14
Vladimir V. Aksyonov	USSR	2	284:15

Table B.1 (*cont.*)

Name	Country	Flights	Time in Space (hrs:min)
Yuri V. Malyshev	USSR	2	284:02
Daniel M. Tani	USA	1	283:37
Igor P. Volk	USSR	1	283:14
Richard A. Mastracchio	USA	1	283:12
Boris V. Morukov	Russia	1	283:12
Christopher J. Ferguson	USA	1	283:07
Heidemarie Stefanyshyn-Piper	USA	1	283:07
Gerhard P.J. Thiele	Germany	1	269:40
Michael Collins	USA	2	266:06
Charles M. Duke, Jr.	USA	1	265:51
Duane G. Carey	USA	1	262:11
Michael J. Massimino	USA	1	262:11
Mary L. Cleave	USA	2	262:02
Anousheh Ansari	USA	1	261:05
Anatoly V. Filipchenko	USSR	2	261:05
Frank De Winne	Belgium	1	260:53
Andre Kuipers	Netherlands	1	260:53
Bjarni V. Tryggvason	Canada	1	260:27
R. Walter Cunningham	USA	1	260:09
Donn F. Eisele	USA	1	260:09
Sandra H. Magnus	USA	1	259:59
Fyodor N. Yurchikhin	Russia	1	259:59
Stephen N. Frick	USA	1	259:43
Rex J. Walheim	USA	1	259:43
Lee M.E. Morin	USA	1	259:43
Mark N. Brown	USA	2	249:28
Russell L. Schweickart	USA	1	241:01
Hans W. Schlegel	Germany	1	239:40
Ulrich Walter	Germany	1	239:40
Vladimir A. Shatalov	USSR	3	237:59
Robert C. Springer	USA	2	237:33
Yuri G. Shargin	Russia	1	237:29
Mark Shuttleworth	South Africa	1	237:25
Marcos C. Pontes	Brazil	1	237:17
Gregory H. Olsen	USA	1	237:15
Nikolai N. Rukavishnikov	USSR	3	237:11
Alexandr P. Alexandrov	Bulgaria	1	236:10
Konstantin M. Kozeev	Russia	1	236:00
Julie Payette	Canada	1	235:14
L. Gordon Cooper, Jr.	USA	2	225:15

(*continued*)

Table B.1 (*cont.*)

Name	Country	Flights	Time in Space (hrs:min)
Joe H. Engle[5]	USA	2	225:01
John H. Glenn, Jr.	USA	2	218:38
Millie E. Hughes-Fulford	USA	1	218:15
F. Andrew Gaffney	USA	1	218:15
Donald K. Slayton	USA	1	217:28
Alan B. Shepard, Jr.[6]	USA	2	216:17
Edgar D. Mitchell	USA	1	216:02
Stuart A. Roosa	USA	1	216:02
Alexei S. Yeliseyev	USSR	3	214:25
Dirk D.D.D. Frimout	Belgium	1	214:09
Joe F. Edwards, Jr.	USA	1	212:47
Abdul Ahad Mohmand	Afghanistan	1	212:27
Neil A. Armstrong	USA	2	206:00
Richard H. Truly	USA	2	199:22
Jon A. McBride	USA	1	197:24
Paul D. Scully-Powers	USA	1	197:24
Roberta K. Bondar	Canada	1	193:14
Ronald E. McNair[7]	USA	2	192:29
Anna L. Fisher	USA	1	191:45
Franco Malerba	Italy	1	191:11
Scott J. Kelly	USA	1	191:11
Mohammed Ahmed Faris	Syria	1	191:05
Loren J. Acton	USA	1	190:46
John-David F. Bartoe	USA	1	190:46
Roy D. Bridges	USA	1	109:46
Anthony W. England	USA	1	190:46
Karl G. Henize	USA	1	190:46
Mae C. Jemison	USA	1	190:30
Vladimir Remek	Czechoslovakia	1	190:17
Toktar O. Aubakirov	USSR	1	190:13
Franz Viehbock	Austria	1	190:13
Miroslaw Hermaszewski	Poland	1	190:04
Dennis Tito	USA	1	190:04
Anatoly S. Levchenko	USSR	1	189:58
Klaus-Dietrich Flade	Germany	1	189:57
Ivan Bella	Slovakia	1	189:57
Toyohiro Akiyama	Japan	1	189:55
Rakesh Sharma	India	1	189:41
Helen P. Sharman	Britain	1	189:14
Sigmund Jähn	Germany	1	188:49
Bertalan Farkas	Hungary	1	188:46
Arnoldo Tamayo Mendez	Cuba	1	188:43

Table B.1 (*cont.*)

Name	Country	Flights	Time in Space (hrs:min)
Pham Tuan	Vietnam	1	188:42
Dumitru D. Prunariu	Romania	1	188:41
Jugderdemidin Gurragcha	Mongolia	1	186:43
William F. Fisher	USA	1	170:18
Sultan bin Salman al-Saud	Saudi Arabia	1	169:39
Patrick P.R. Baudry	France	1	169:39
F. Richard Scobee[7]	USA	2	158:53
Wubbo J. Ockels	Netherlands	1	168:44
Rheinhard A. Furrer	Germany	1	168:44
Ernst W. Messerschmid	Germany	1	168:44
Alexei A. Leonov	USSR	2	168:33
Don L. Lind	USA	1	168:09
Lodewijk van den Berg	USA	1	168:09
Taylor G. Wang	USA	1	168:09
S. David Griggs	USA	1	167:55
Edwin J. Garn	USA	1	167:55
Terry J. Hart	USA	1	167:40
Thomas J. Hennen	USA	1	166:52
Rudolfo Neri Vela	Mexico	1	165:04
Sherwood C. Spring	USA	1	165:04
William A. Anders	USA	1	147:01
Judith A. Resnik[7]	USA	2	146:10
Robert J. Cenker	USA	1	143:04
C. William Nelson Jr.	USA	1	143:04
Fred W. Haise, Jr.	USA	1	142:55
John L. Swigert, Jr.	USA	1	142:55
William B. Lenoir	USA	1	122:14
Michael J. McCulley	USA	1	120:39
Donald H. Peterson	USA	1	120:24
Manley L. Carter, Jr.	USA	1	120:07
Georgi S. Shonin	USSR	1	118:42
Fei Junlong	China	1	115:33
Nie Haishengi	China	1	115:33
Edward H. White II	USA	1	97:56
William A. Pailes	USA	1	97:45
Georgi T. Beregovoi	USSR	1	94:51
Ellison S. Onizuka[7]	USA	2	79:46
Gary E. Payton	USA	1	78:33
Valentina V. Tereshkova	USSR	1	70:50
Vladimir M. Komarov	USSR	1	50:54

(*continued*)

Appendix B

Table B.1 (*cont.*)

Name	Country	Flights	Time in Space (hrs:min)
Lev S. Dyomin	USSR	1	48:12
Gennady V. Sarafanov	USSR	1	48:12
Valery I. Rozhdestvensky	USSR	1	48:06
Vyacheslav D. Zudov	USSR	1	48:06
Yevgeny V. Khrunov	USSR	1	47:49
Vasily G. Lazarev[4]	USSR	2	47:36
Georgi I. Ivanov	Bulgaria	1	47:01
Pavel I. Belyayev	USSR	1	26:02
Gherman S. Titov	USSR	1	25:18
Konstantin P. Feoktistov	USSR	1	24:17
Boris B. Yegorov	USSR	1	24:17
Yang Liwei	China	1	21:26
Virgil I. Grissom[8]	USA	2	5:09
M. Scott Carpenter	USA	1	4:56
Yuri A. Gagarin	USSR	1	1:49
Mike Melvill[9]	USA	2	0:48
Joseph A. Walker[5]	USA	3	0:30
Brian Binnie[10]	USA	1	0:24
William J. Knight[11]	USA	2	0:20
Robert M. White[12]	USA	1	0:10
Robert A. Rushworth[12]	USA	1	0:10
John B. McKay[12]	USA	1	0:10
William H. Dana[12]	USA	1	0:10
Michael J. Adams[12]	USA	1	0:10
Gregory B. Jarvis[7]	USA	1	0:01
S. Christa McAuliffe[7]	USA	1	0:01

Notes

[1] Does not include September 1983 launch pad abort
[2] In Space aboard ISS-14
[3] Includes STS-107 duration up to loss of signal
[4] Includes 5 Apr Anomaly – Soyuz 18-1 launch abort
[5] Includes X-15 flights, three times 10 minutes
[6] Includes Mercury 3 sub-orbital flight
[7] Inclusive of STS 51-L time through loss of signal
[8] Includes Mercury 4 sub-orbital flight
[9] Includes Spaceship One flight two times 24 minutes
[10] Includes Spaceship One flight
[11] Includes X-15 flight, two times 10 minutes
[12] Includes X-15 flight, 10 minutes

All flights into space from 12 April 1961 (over 50 miles or 80m) or accidents of intended orbital missions in progress are included. Launch pad aborts prior to launch are not included.

All data correct up to 29 September 2006.

Appendix B

Listings include

1961	Mercury 3 and Mercury 4 completed NASA sub-orbital flights
1962–1968	Thirteen X-15 missions that exceeded 50 mile (80 km) altitude
1975	Soyuz 18-1 launch abort – Mission in progress
1986	STS 51-L Challenger. Mission in progress
2003	STS 107 Columbia – Mission in progress
2004	3 Spaceship One test and X-Prize private commercial flights that exceeded 100-km altitude

Listings do not include

1959–1968	X-15 flights that did not exceed 50 miles (80 km) altitude.
1967	Apollo 1 pad fire – pre-mission training simulation
1983	Soyuz T10-1 pad abort – mission aborted prior to lift off.
1984–2006	Five Shuttle pad aborts prior to SRB ignition

Appendix B

Table B.2. EVA duration log April 1961–September 2006.

Order of most spaceflight experience, up to 29 September 2006 and the end of ISS Expedition 13. Includes all IVAs.

Name	Country	EVAs	Duration (hrs:min)
Anatoly Y. Solovyov	USSR/Russia	16	79:51
Sergei V. Avdeyev	Russia	13	59:52
Jerry L. Ross	USA	9	58:18
Joseph R. Tanner	USA	7	56:09
Viktor M. Afanasyev	USSR/Russia	9	50:05
Stephen L. Smith	USA	7	49:49
Nikolai V. Budarin	Russia	9	44:54
Yuri I. Onufriyenko	Russia	8	42:43
Talgat A. Musabayev	Russia	8	41:29
Sergei K. Krikalev	USSR/Russia	8	41:18
Piers J. Sellers	USA	6	41:10
John M. Grunsfeld	USA	5	37:45
Vladimir N. Dezhurov	Russia	9	37:23
Leroy Chiao	USA	6	36:17
James H. Newman	USA	5	35:56
Musa K. Manarov	USSR/Russia	7	34:34
Michael E. Lopez-Alegria[1]	USA	5	33:58
Pavel V. Vinogradov	Russia	8	32:50
Anatoly P. Artsebarsky	USSR	6	32:09
Alexandr A. Serebrov	USSR/Russia	10	31:52
Yuri V. Usachev	Russia	7	30:50
Thomas D. Akers	USA	4	29:40
Leonid D. Kizim	USSR	7	28:51
Vladimir A. Solovyov	USSR	7	28:51
F. Story Musgrave	USA	4	26:19
Mark C. Lee	USA	4	26:01
Jeffrey A. Hoffman	USA	4	25:02
William S. McArthur Jr.	USA	4	24:21
Eugene A. Cernan	USA	4	24:13
Daniel T. Barry	USA	4	23:49
David A. Wolf	USA	4	23:33
Alexandr Y. Kaleri	Russia	5	23:24
Michael L. Gernhardt	USA	4	23:16
Harrison H. Schmitt	USA	4	23:10
James S. Voss	USA	4	22:45
C. Michael Foale	USA	4	22:45
Gennady M. Strekalov	USSR/Russia	6	22:31
Valeri G. Korzun	Russia	4	22:19
Gennady I. Padalka	Russia	6	22:09
Charles M. Duke Jr.	USA	4	21:38
Michael E. Fossum	USA	3	21:29

Table B.2 (*cont.*)

Name	Country	EVAs	Duration (hrs:min)
Richard M. Linnehan	USA	3	21:21
Kathryn C. Thornton	USA	3	21:11
James D.A. Van Hoften	USA	4	20:45
John W. Young	USA	3	20:14
David R. Scott	USA	5	20:14
Stephen K. Robinson	USA	3	20:05
Soichi Noguchi	Japan	3	20:05
John B. Herrington	USA	3	19:55
Peter J.K. Wisoff	USA	3	19:53
Thomas D. Jones	USA	3	19:49
Robert L. Curbeam Jr.	USA	3	19:49
Alexandr S. Viktorenko	USSR/Russia	6	19:42
Winston E. Scott	USA	3	19:36
Franklin R.L.A Chang-Diaz	USA	3	19:31
Philippe Perrin	France	3	19:31
Carlos I. Noriega	USA	3	19:20
James B. Irwin	USA	4	19:14
Vasily V.V. Tsibliyev	Russia	6	19:10
Jeffrey N. Williams	USA	3	19:09
Carl E. Walz	USA	3	18:55
Vladimir G. Titov	USSR/Russia	4	18:47
Gregory J. Harbaugh	USA	3	18:29
Richard J. Hieb	USA	3	17:42
Pierre J. Thuot	USA	3	17:42
Yuri I. Malenchenko	Russia	3	17:21
James F. Reilly II	USA	3	16:30
Gerald P. Carr	USA	3	15:51
Edward G. Gibson	USA	3	15:20
E. Michael Fincke	USA	4	14:54
Scott E. Parazynski	USA	2	14:50
Chris A. Hadfield	Canada	2	14:50
Michael J. Massimino	USA	2	14:34
Thomas Reiter[2]	Germany	3	14:16
Rex J. Walheim	USA	2	14:15
Lee M.E. Morin	USA	2	14:07
Gennady M. Manakov	USSR/Russia	3	13:46
Owen K. Garriott	USA	3	13:44
Mikhail V. Tyurin[1]	Russia	3	13:35
William R. Pogue	USA	2	13:31
Heidemarie Stefanyshyn-Piper	USA	2	13:08
Takao Doi	Japan	2	12:42
Bruce McCandless II	USA	2	12:12
Robert L. Stewart	USA	2	12:12

(*continued*)

Table B.2 (*cont.*)

Name	Country	EVAs	Duration (hrs:min)
Sherwood C. Spring	USA	2	12:00
Daniel W. Bursch	USA	2	11:46
Patrick G. Forrester	USA	2	11:45
Joseph P. Allen IV	USA	2	11:42
Dale A. Gardner	USA	2	11:42
William F. Fisher	USA	2	11:34
Charles Conrad Jr.	USA	4	11:33
Valeri I. Tokarev	Russia	2	11:05
Jack R. Lousma	USA	2	11:01
Jerome Apt	USA	2	10:49
Alexandr N. Baladin	USSR	2	10:47
Alan L. Bean	USA	3	10:30
Yuri V. Romanenko	USSR	4	10:16
Linda M. Godwin	USA	2	10:14
Alexandr A. Volkov	USSR	2	10:09
Alexandr F. Poleshchuk	Russia	2	9:58
Salizhan S. Sharipov	Russia	2	9:58
Kenneth D. Bowersox	USA	2	9:46
Don R. Pettit	USA	2	9:46
Alan B. Shepard Jr.	USA	2	9:22
Edgar D. Mitchell	USA	2	9:22
George D. Nelson	USA	2	9:13
Susan J. Helms	USA	1	8:56
Alexandr I. Laveikin	USSR	3	8:48
Vladimir A. Dzhanibekov	USSR	2	8:35
Claude Nicollier	Switzerland	1	8:10
Buzz Aldrin	USA	4	8:09
Daniel Burbank	USA	1	7:11
Steven MacLean	Canada	1	7:11
Vladimir A. Lyakhov	USSR	3	7:08
Carl J. Meade	USA	1	6:51
Andrew S.W. Thomas	USA	1	6:21
Paul W. Richards	USA	1	6:21
Jean-Pierre Haigneré	France	1	6:19
Edward Tsang Lu	USA	1	6:14
Michael R.U. Clifford	USA	1	6:02
Jean-Loup J.M. Chrétien	France	1	5:57
Tamara E. Jernigan	USA	1	5:55
G. David Low	USA	1	5:50
Alexandr P. Alexandrov	USSR	2	5:45
Sergei V. Treshev	Russia	1	5:21
Frank L. Culbertson Jr.	USA	1	5:05
Svetlana Y. Savitskaya	USSR	1	5:00
Viktor P. Savinykh	USSR	1	5:00

Table B.2 (*cont.*)

Name	Country	EVAs	Duration (hrs:min)
Jerry M. Linenger	USA	1	4:57
John L. Phillips	USA	1	4:57
Sergei V. Zaletin	Russia	1	4:52
Bernard A. Harris Jr.	USA	1	4:39
Peggy A. Whitson	USA	1	4:25
Donald H. Peterson	USA	1	4:17
Daniel M. Tani	USA	1	4:12
Yuri P. Gidzenko	Russia	2	3:35
Kathryn D. Sullivan	USA	1	3:27
David C. Leestma	USA	1	3:27
Joseph P. Kerwin	USA	1	3:25
S. David Griggs	USA	1	3:00
Richard F. Gordon Jr.	USA	2	2:41
Anatoly N. Berezovoi	USSR	1	2:33
Valentin V. Lebedev	USSR	1	2:33
Neil A. Armstrong	USA	1	2:31
Paul J. Weitz	USA	2	2:21
Vladimir V. Kovalyonok	USSR	1	2:05
Alexandr S. Ivanchenkov	USSR	1	2:05
Michael Collins	USA	2	1:29
Georgi M. Grechko	USSR	1	1:28
Thomas K. Mattingly II	USA	1	1:24
Valery V. Ryumin	USSR	1	1:23
Russell L. Schweickart	USA	1	1:07
Ronald E. Evans Jr.	USA	1	1:06
Alfred M. Worden Jr.	USA	1	0:39
Yevgeny V. Khrunov	USSR	1	0:37
Alexei S. Yeliseyev	USSR	1	0:37
Edward H. White II	USA	1	0:21
Alexei A. Leonov	USSR	1	0:12

Notes

[1] In space aboard ISS-14
[2] In space as part of ISS-13/ISS-14 crew. Total includes 5 hr 54 min EVA during ISS-13.

Appendix C

Future Flight Manifest 2006–2011
(as at 1 October 2006)

814 Appendix C

Date	Mission	Flight	Country	Crew	Objective
2006					
Dec	STS-116 (117) Discovery (33)	12A.1 ISS-20	USA	Polansky (Cdr), Oefelein (Plt) Curbeam (MS), Higginbotham (MS), Patrick (MS), Fuglesang (MS-ESA), Williams S. (ISS FE up only)	3rd port truss segment (ITS P5); SpaceHab SM; Integrated Cargo Carrier
2007					
Feb	STS-117 (118) Atlantis (28)	13A ISS-21	USA	Sturckow (Cdr), Archambault (Plt), Reilly II (MS), Swanson (MS), Forrester (MS), Olivas (MS)	2nd starboard truss segment (ITS S3/S4); Photovoltaic Radiator (PVR); 3rd set of solar arrays and batteries
Mar	Soyuz TMA10	ISS-15	Russia	Kotov (TMA Cdr), Yurchikhin (ISS Cdr), plus SFP?	3rd EO crewmember Williams/Anderson/Tani
Jun	STS-118 (119)	13A.1	USA	Kelly S. (Cdr), Hobaugh (Plt), Williams D. (MS-CSA), Morgan (MS), Mastracchio (MS), Caldwell (MS)	SpaceHab single cargo module; 3rd starboard truss assembly (ITS S5); External Stowage Platform 3 (ESP3)
	Endeavour	ISS-22		Anderson (ISS FE up only), Williams S. (ISS FE down only)	
Aug	STS-120 (120) Atlantis (29)	10A ISS-23	USA	Melroy (Cdr), Zamka (Plt), Parazynski (MS), Wheelock (MS), Foreman (MS), Nespoli (MS-ESA) Tani (ISS FE up only), Anderson (ISS FE down only)	Node 2, Sidewall – Power and Data Grapple Fixture (PDGF)
Oct	Soyuz TMA11	ISS-16	Russia	Malenchenko (TMA Cdr), Whitson (ISS Cdr), Malaysian RC	3rd EO crewmember Tani/Eyharts/Thirsk/Wakata
Oct	STS-122 (121)	1E	USA	Frick (Cdr), Poindexter (Plt), Walheim (MS),	Columbus European Laboratory Module;

	Discovery (34)	ISS-24		Love (MS), Melvin (MS), Schlegel (MS-ESA), Eyharts (ESA ISS FE up only), Tani (ISS FE down only)	Multi-Purpose Experiment Support Structure – Non Deployable (MPESS-ND)
Dec	STS-123 (122) Endeavour (21)	1J/A ISS-25	USA	NASA crew to be named, Doi (MS – Jaxa), Thirsk (CSA ISS FE up only), Eyharts (ESA ISS FE down only)	Kibo Japanese Experiment Logistics Module – Pressurised Section (ELM-PS); Spacelab Pallet – Deployable 1 (SLP-D1)
2008					
Feb	STS-124 (123) Atlantis (30)	1J ISS-26	USA	No crew assigned Wakata (ISS FE Up only); Thirsk (CSA ISS FE down only)	Kibo Japanese Experiment Module Pressurised Module (JEM-PM); Japanese Remote Manipulator System (JEM RMS)
Mar	STS-TMA12	ISS-17	Russia	Volkov S. (TMA/ISS Cdr), Sharipov, and a third	3rd EO crewmember Wakata?
Apr	STS-125 (124) Discovery (35)	HST SM-04	USA	Altman (Cdr), Johnson G. (Plt), Grunsfeld (MS), Massimino (MS), Feustel (MS), McArthur K. (MS), Good (MS)	Hubble Space Telescope Service Mission 4
Jun	STS-119 (125) Endeavour (22)	15A	USA	Gernhardt (MS), other crew members to be named	4th starboard truss segment (ITS S6); 4th set of solar arrays and batteries
Aug	STS-126 (126) Atlantis (31)	ULF-2	USA	No crew assigned	Multi Purpose Logistics Module (MPLM) Donatello
Sep	Soyuz TMA13	ISS-18	Russia	Kaleri (TMA Cdr), NASA ISS Cdr plus SFP?	3rd EO crew member?
Sep	Shenzhou 7		China	3 person crew	1st Chinese EVA planned
Oct	STS-127? (127)	2J/A	USA	No crew assigned	Kibo Japanese Experiment Module Exposed Facility (JEM-EF); Japanese Experiment Logistics Module – Exposed Section (ELM-ES); Spacelab Pallet-Deployable 2 (SLP-D2)

(continued)

816   Appendix C

Date	Mission	Flight	Country	Crew	Objective
2009					
Jan	STS-128 ? (128)	17A	USA	No crew assigned	MPLM; Lightweight Multi-Purpose Experiment Support Structure Carrier (LMC); Three crew quarters, galley, second treadmill (TVIS2); Crew Health Care System 2 (CHeCS 2)
Establish six person crew capability on ISS					
Mar	Soyuz TMA13	ISS-19	Russia	Krikalev (TMA/ISS Cdr)?; Surayev (FE) plus ?	Additional EO crew members?
Apr	STS-129? (129)	ULF-3	USA	No crew assigned	EXPRESS Logistics Carrier 1 (ELC 1); EXPRESS Logistics Carrier 2 (ELC 2)
Jul	STS-130? (130)	19A	USA	No crew assigned	MPLM; Lightweight Multi-Purpose Experiment Support Structure Carrier (LMC)
Sep	Soyuz TMA14?	ISS-20	Russia	No crew assigned	
Sep ?	Shenzhou 8 & Shenzhou 9		China China	No crew assigned No crew assigned	Shenzhou 8 & 9 to perform first Chinese manned docking and creation of small short-stay space station
Oct	STS-131? (131)	ULF-4	USA	No crew assigned	EXPRESS Logistics Carrier 3 (ELC 3); EXPRESS Logistics Carrier 4 (ELC 4); two Shuttle-equivalent flights for contingency
2010					
Jan	STS-132? (132)	20A	USA	No crew assigned	Node 3 with Cupola
Mar	Soyuz TMA15?	ISS-21	Russia	No crew assigned	

Jul STS-133? (132) ULF-5 USA No crew assigned EXPRESS Logistics Carrier 5 (ELC 5); EXPRESS Logistics Carrier 6 (ELC 6); two Shuttle-equivalent flights for contingency

ISS Assembly complete – Shuttle fleet retired

Sep Soyuz TMA16 ? ISS-22 Russia No crew assigned

2011

Mar Soyuz TMA17 ? ISS-23 Russia No crew assigned

Apr *50th anniversary of Yuri Gagarin's flight aboard Vostok – 1st manned space flight.*

Appendix C 817

Appendix C

The following information was compiled with the help of Collect Space 7 Oct 2006, Robert Pearlman

Soyuz TMA-crewing 2007–2008

TMA10 ISS-15:	April 2007–September 2007
Commander	Oleg Kotov
FE1	Fyodor Yurchikhin
FE2a	Suni Williams (up on STS-116) until June 2007
FE2b	Clay Anderson (up on STS-118) until September 2007
FE2c	Dan Tani (up on STS-120) until October 2007
TMA11 ISS-16:	September 2007–March 2008
Commander	Yuri Malenchenko
FE1	Peggy Whitson
FE2a	Dan Tani (up on STS-120) until October 2007
FE2b	Leopold Eyharts (up on STS-122) until December 2007
FE2c	Bob Thirsk (up on STS-123) until March 2008
FE2d	Koichi Wakata (up on STS-124) until April 2008
TMA12 ISS-17:	March 2008–September 2008
Commander	Sergei Volkov
FE-1	Peggy Whitson (stays on ISS for 9 months returns on STS-119)
FE-2	Shalizhan Sharipov (launched on TMA-12)
FE-2b	Sandy Magnus (up on STS-119) until September 2008
FE-2c	Greg Chamitoff (up on STS-126) until November 2008.

Appendix D

A Selected Timeline

1961

Apr	Yuri Gagarin becomes the first person fly into space and completes one orbit
May	Alan Shepard becomes the first American in space on a sub-orbital flight
Aug	Gherman Titov is launched on the first 24-hour mission, of 17 orbits

1962

Feb	John Glenn becomes the first American to orbit the Earth, with 3 orbits
Jul	First X-15 flight to exceed 50 miles (Robert White)
Aug	Andrian Nikolayev sets new endurance record (3 days 22 hours)

1963

Jun	Valeri Bykovsky sets new endurance record (4 days 23 hours)
	Valentina Tereshkova becomes first woman in space (2 days 22 hours)
Aug	Highest X-15 flight (66.75 miles) – Pilot Joseph Walker

1964

Oct	First multi-person space crew (3) – Voskhod 1; First civilians in space

1965

Mar	Alexei Leonov becomes first person to walk in space
Mar	First US multi-person crew (2) on Gemini 3
Jun	Ed White becomes first American to walk in space

Aug	Gemini 5 sets new endurance record (7 days 22 hours)
	Cooper becomes first person to orbit Earth a second time
Dec	Gemini 7 set new endurance record (13 days 18 hours)
	First space rendezvous – Gemini 6 with Gemini 7

1966

Mar	First space docking – Gemini 8 with Agena target
Sep	Gemini 11 attains highest altitude of Earth orbital manned flight (850 miles)

1967

Jan 27	Three Apollo 1 astronauts killed in pad fire
Apr	Soyuz 1 pilot Vladimir Komarov killed during landing phase
Oct	X-15 fastest flight (4520 mph – Mach 6.7) (Pete Knight)
Nov	X-15 pilot Michael Adams is killed in crash of #3 aircraft after attaining 50.4 miles

1968

Aug	Thirteenth and final X-15 "astro-flight"
Oct	First three-man Apollo flight (Apollo 7)
	Schirra becomes first person to make three orbital spaceflights
Dec	Apollo 8 becomes first lunar orbital mission

1969

Jan	Soyuz 5/4 first manned docking and crew transfer (by EVA)
Mar	Manned test of LM in Earth orbit (Apollo 9)
May	Manned test of LM in lunar orbit (Apollo 10)
Jul	First manned lunar landing – Apollo 11
Oct	First triple manned spacecraft mission (Soyuz 6, 7, 8)
Nov	Second manned lunar landing Apollo 12

1970

Apr	Apollo 13 aborted lunar landing mission
	Lovell becomes first to fly in space four times
Jun	Soyuz 9 cosmonauts set new endurance record (17 days 16 hrs)

1971

Feb	Third manned lunar landing (Apollo 14)
Apr	Launch of world's first Space Station – Salyut (de-orbits Oct 1971)

Appendix D

Jun	First space station (Salyut) crew. Killed during entry phase (Soyuz 11)
Jul	Fourth manned lunar landing (Apollo 15)

1972

Apr	Fifth manned lunar landing (Apollo 16)
Dec	Sixth and final (Apollo) manned lunar landing (Apollo 17)

1973

Apr	Salyut 2 (Almaz) fails in orbit (de-orbits in 26 days)
May	Launch of unmanned Skylab (re-enters Jul 1979)
	First Skylab crew sets new endurance record of 28 days
Jul	Second Skylab crew increases endurance record to 59 day 11 hrs
Nov	3rd and final Skylab crew increases endurance record to 84 days 1 hr

1974

Jun	Launch of Salyut (Almaz) 3 (de-orbits Jan 1975)
Jul	First successful Soviet space station mission (Soyuz 14)
Dec	Launch of Salyut 4 (de-orbits Feb 1977)

1975

Apr	Soyuz 18 crew survive launch abort
Jul	Soyuz 19 and Apollo dock in space – first international mission

1977

Sep	Salyut 6 launched (de-orbits Jul 1982)
Dec	First Salyut 6 resident crew set new endurance record of 96 days 10 hrs

1978

Jan	First Soyuz exchange mission (Soyuz 27 for Soyuz 26)
Mar	First Soviet Interkosmos mission (Czechoslovakian)
	First non-Soviet, non-American person in space (Remek)
Jun	Second Salyut 6 crew sets new endurance record of 139 days 14 hrs

1979

Feb	Third Salyut 6 resident crew increases endurance record to 175 days

1980

Apr	Fourth Salyut 6 resident crew increases endurance record to 184 days 20 hrs
Jun	First manned flight of Soyuz T variant

1981

Apr	First Shuttle launch (Columbia STS-1) on 20th anniversary of Gagarin's flight
	John Young becomes first to make five space flights
Nov	First return to space by manned spacecraft (Columbia STS-2)

1982

Apr	Salyut 7 launched (de-orbits Feb 1991)
May	First Salyut 7 resident crew sets new endurance record of 211 days 9 hrs
Nov	First "operational" Shuttle mission, STS-5, is also the first four-person launch

1983

Apr	First flight of Challenger
Jun	Sally Ride becomes first US woman in space during STS-7, the first five-person launch
Sep	Soyuz T10-1 launch pad abort
Nov	First Spacelab mission – STS-9; first six-person launch
	John Young flies record sixth mission

1984

Feb	First use of MMU (STS 41-B) on untethered spacewalks
Feb	Third Salyut 7 resident crew sets new endurance record of 236 days 22 hrs
Jul	Svetlana Savitskaya becomes the first woman to walk in space (Soyuz T12/Salyut 7)
Aug	First flight of Discovery on STS 41-D
Oct	First seven-person launch (STS 41-G)
	Kathy Sullivan becomes first American woman to walk in space

1985

Jan	First classified DoD Shuttle mission (STS 51-C)
Jul	First Shuttle Abort-to-Orbit profile (STS 51-F)
Oct	First flight of Atlantis (STS 51-J)
Oct	First eight-person launch (STS 61-A)

1986

Jan	Challenger and its crew of seven lost 73 seconds after launch (STS 51-L)
Feb	Mir core module launched unmanned
Mar	First resident crew to Mir (Soyuz T15)

1987

Feb	Second Mir resident crew sets new endurance record of 326 days 11 hrs First manned Soyuz TM variant
Dec	First flight of over a year as third Mir resident crew sets endurance record of 365 days 22 hrs

1988

Sep	Shuttle Return-to-Flight mission (STS-26)

1990

Apr	Hubble Space Telescope deployment (STS-31)

1992

May	First flight of Endeavour (STS-49)

1993

Dec	First Hubble Service Mission (STS-61)

1994

Jan	Valery Polyakov sets new endurance record (437 days 17 hrs) for one mission (lands Mar 1995)
Feb	First Russian cosmonaut to fly on Shuttle (Krikalev STS-60)

1995

Feb	First Shuttle–Mir rendezvous STS-63/Mir Eileen Collins becomes first female Shuttle pilot
Mar	First American launched on Soyuz (Thagard – TM21)
Jul	First Shuttle docking with Mir (STS-71 – Thagard down)
Nov	Second Shuttle–Mir docking (STS-74)

1996

Mar	Third Shuttle–Mir docking (STS-76 – Lucid up)
Sep	Fourth Shuttle-Mir docking (STS-79 – Lucid down, Blaha up)
Nov	Longest Shuttle mission (17 days 15 hrs – STS-80)
	Musgrave becomes only astronaut to fly all five orbiters

1997

Jan	Fifth Shuttle–Mir docking (STS-81 – Blaha down, Linenger up)
Feb	Second Hubble service mission (STS-82)
May	Sixth Shuttle–Mir docking (STS-84 – Linenger down, Foale up)
Jun	Collision between unmanned Progress vessel and Mir space station damages Spektr module
Sep	Seventh Shuttle–Mir docking (STS-86 – Foale down, Wolf up)

1998

Jan	Eighth Shuttle–Mir docking (STS-89 – Wolf down, Thomas up)
Jun	Ninth and final Shuttle–Mir docking (STS-91 – Thomas down)
Oct	John Glenn returns to space aged 77, 36 years after his first space flight
Nov	First ISS element launched – Zarya FGB
Dec	First ISS Shuttle mission (STS-88)

1999

Jul	Eileen Collins becomes first female US mission commander (STS-93)
Aug	Mir vacated for first time in ten years
Dec	Third Hubble service mission (STS-103)

2000

Apr	Last (28th) Mir resident crew (72 days)
Oct	First ISS resident crew launched

2001

Mar	Mir space station de-orbits after 15 years service
Apr	Dennis Tito becomes first space flight participant, or "tourist"

2002

Mar	Fourth Hubble service mission (STS-109)
Apr	Jerry Ross becomes first person to fly seven missions in space
Oct	First manned flight of Soyuz TMA

Appendix D 825

2003

Feb   Columbia and crew of seven lost during entry phase of mission STS-107
Apr   ISS assumes two-person caretaker crews
Oct   First Chinese manned spaceflight (Shenzhou 5)
          Yang Liwei becomes first Chinese national in space

2004

Sep   Spaceship One flies to 337,500 ft (102.87 km)
Oct   Spaceship One flies to 367,442 ft (111.99 km) claiming $10 million X-Prize

2005

Jul   Shuttle Return-to-Flight mission 1 – STS-114
Oct   First Chinese two-man space flight – Shenzhou 6

2006

Jul   Second Shuttle Return-to-Flight mission – STS-121
Aug   ISS returns to three-person capability
          Resumption of ISS construction – STS-115

Bibliography

The authors have referred to their own extensive archives in the compilation of this book. In addition, the following publications and resources were of great help in assembling the data:

The Press Kits, News releases and mission information from NASA, ESA, CSA, RKK-Energiya, JAXA (NASDA), CNES, and Novosti have been invaluable resources for many years

Magazines:

Flight International 1961–2006
Aviation Week and Space Technology 1961–2006
BIS *Spaceflight* 1961–2006
Soviet Weekly/Soviet News 1961–1990
Orbiter, Astro Info Service 1984–1992
Zenit, Astro Info Service, 1985–1991
ESA Bulletin 1975–2006

British Interplanetary Society Books:

History of Mir 1986-2000; Mir: The Final Year Supplement, Editor Rex Hall 2000/ 2001
The ISS Imagination to Reality Volume 1 Ed Rex Hall 2002
The ISS Imagination to Reality Volume 2, Ed Rex Hall 2005

NASA Reports:

NASA Astronautics and Aeronautics, various volumes, 1961–1995

Mir Hardware Heritage, David S.F. Portree NASA RP-1357, March 1995.
Walking to Olympus: An EVA Chronology, David S.F. Portree and Robert C. Trevino, NASA Monograph in Aerospace history, #7 October 1997

NASA Histories:

1966	This New Ocean, a History of Project Mercury, SP-4201
1977	On the Shoulders of Titans: A history of Project Gemini, NASA SP-4203
1978	The Partnership: A history of Apollo–Soyuz Test Project, NASA SP-4209
1979	Chariots for Apollo: A history of manned lunar spacecraft, NASA SP-4205
1983	Living and working in space: A history of Skylab NASA SP 4208
1989	Where No Man Has Gone Before: a history of Apollo lunar exploration missions, NASA SP-4214
2000	Challenge to Apollo: the Soviet Union and the Space Race 1945–1974, Asif Siddiqi, NASA SP-2000-4408

Other Books:

1980	Handbook of Soviet Manned Space Flight, Nicholas L. Johnson, AAS Vol 48, Science and Technology Series
1981	The History of Manned Spaceflight, David Baker
1987	Heroes in Space: From Gagarin to Challenger, Peter Bond
1988	Space Shuttle Log: The First 25 Flights, Gene Gurney and Jeff Forte
1988	The Soviet Manned Space Programme, Phillip Clark
1989	The Illustrated Encyclopaedia of Space Technology, Chief Author Ken Gatland
1990	Almanac of Soviet Manned Space Flight, Dennis Newkirk
1992	At the Edge of Space: The X-15 Flight Program, Milton O. Thompson
1999	Who's Who in space: The ISS Edition, Michael Cassutt
2001	Space Shuttle, History and Development of the National STS Program, Dennis Jenkins

Springer–Praxis Space Science Series (which include extensive references and bibliographies for further reading)

1999	Exploring the Moon: The Apollo Expeditions, David M. Harland
2000	Disasters and Accidents in Manned Spaceflight, David J. Shayler
2000	The Challenges of Human Space Exploration, Marsha Freeman
2001	Russia in Space: The Failed Frontier, Brian Harvey
2001	The Rocket Men, Vostok & Voskhod, the First Soviet Manned Spaceflights, Rex Hall and David J. Shayler
2001	Skylab:; America's Space Station, David J. Shayler
2001	Gemini: Steps to the Moon, David J. Shayler
2001	Project Mercury: NASA's First Manned Space Programme, John Catchpole
2002	The Continuing Story of the International Space Station, Peter Bond

2002	Creating the International Space Station, David M. Harland and John E. Catchpole
2002	Apollo: Lost and Forgotten Missions, David J. Shayler
2003	Soyuz, a Universal Spacecraft, Rex Hall and David J. Shayler
2004	China's Space Programme: From Concept to Manned Spaceflight, Brian Harvey
2004	Walking in Space, David J. Shayler
2004	The Story of the Space Shuttle, David M Harland
2005	The Story of Space Station Mir, David M. Harland
2005	Women in Space: Following Valentina, David J Shayler and Ian Moule
2005	Space Shuttle Columbia: Her Missions and Crews, Ben Evans.
2005	Russia's Cosmonauts: Inside the Yuri Gagarin Training Center, Rex Hall, David J. Shayler and Bert Vis
2006	Apollo: The Definitive Source Book, Richard W. Orloff and David M. Harland
2006	NASA Scientist Astronauts, Colin Burgess and David J. Shayler

Printing: Mercedes-Druck, Berlin
Binding: Stein+Lehmann, Berlin